Liquid Crystals in Complex Geometries

Formed by polymer and porous networks

Liquid Crystals in Complex Geometries

Formed by polymer and porous networks

EDITED BY

GREGORY PHILIP CRAWFORD

AND

SLOBODAN ŽUMER

CRC Press
Taylor & Francis Group
Boca Raton London New York

CRC Press is an imprint of the
Taylor & Francis Group, an **informa** business
A TAYLOR & FRANCIS BOOK

CRC Press
Taylor & Francis Group
6000 Broken Sound Parkway NW, Suite 300
Boca Raton, FL 33487-2742

First issued in paperback 2020

ISBN-13: 978-0-367-57959-3 (pbk)
ISBN-13: 978-0-7484-0464-3 (hbk)

Visit the Taylor & Francis Web site at
http://www.taylorandfrancis.com

and the CRC Press Web site at
http://www.crcpress.com

British Library Cataloging in Publication Data

A catalogue record for this book is available from the British Library.

Library of Congress Cataloging Publication data are available

Front cover photograph shows a polarizing microscope photograph of a partially oriented polymer network suspended in an isotropic solvent. The network is assembled via photopolymerization of a diacrylate monomer in a nematic liquid crystal confined to a cylindrical capillary tube with diameter 300 μm. Although it only occupies 4% of the mixture, the network was successfully transferred in an isotropic solvent which replaced the liquid crystal matrix. The interference pattern shows that the details of the original escaped-radial director-field remained captured and undamaged. The photograph was taken by Dr Andre Scharkowski at the Liquid Crystal Institute at Kent State University, Kent, Ohio, USA. (See Chapter 4, *Polymer Network Assemblies in Nematic Liquid Crystals*.)

Rear cover photograph shows a reflective bistable cholesteric liquid crystal display designed and fabricated at the Liquid Crystal Institute at Kent State University, Kent, Ohio, USA. The active area is 4″ × 4″ with a resolution of 80 dots per inch. (Courtesy of Dr Jon Ruth.)

Cover design by Youngs Design In Production

Typeset in Times 9/11pt by Santype International Limited, Salisbury, Wiltshire

This book is dedicated to Professor J. W. Doane, with respect and kind regards.

Contents

Preface

Liquid crystals confined to spherical and cylindrical enclosures have intrigued scientists for nearly a century, yet the intricate mechanisms which govern their physical properties are only today becoming understood. An era of fascination with confined liquid crystals was revitalized in the mid-1980s with the discovery of polymer-dispersed liquid crystals (PDLCs). Liquid crystals confined to small droplets were no longer just curious systems from the basic science perspective but evolved into powerful materials for electro-optic applications. The best known application of these materials is the PDLC light shutter which transforms from a scattering, translucent appearance to an optically clear state upon application of an electric field. Tremendous efforts in explaining the role of surfaces on liquid crystal materials used in classical nematic and ferroelectric liquid crystal cells have contributed substantially to the current understanding of the physics of liquid crystals confined to well-defined curved geometries.

Interest in all aspects of these systems has developed at a pace which after 10 years still shows no signs of slackening. In the 1990s, a new period began with confinement of liquid crystals to more general random-type geometries formed by polymer and porous networks. The physical phenomena behind these confined systems are complicated and often mysterious. In addition to being fundamentally challenging, randomly confined systems have great potential in many new optics applications and are expected to play a dominant role in the future of flat-panel displays. We believe that this collection of contributed papers by internationally renowned researchers, exclusively covering soft condensed matter systems confined to complex curved geometries, will accelerate and stimulate the development of these exciting and powerful composite materials.

We are indebted to many friends and colleagues. We first wish to mention Professor Bill Doane to whom we dedicate this book. Beginning in 1985 Professor Doane aggressively and energetically pursued detailed experiments on PDLC systems at Kent State University as both a researcher and mentor of numerous students. Owing to his early efforts and his constant pursuit in understanding the physics and applications of confined liquid crystals, the field now has burgeoned into an important subdiscipline of liquid crystal science and technology. We wish to congratulate him on his 60th birthday and to extend to him our sincere appreciation for his role as a mentor (GPC) and colleague, and for his significant accomplishments as a physicist and technologist in the field of liquid crystals. We also wish to thank Elaine Landry and Renate Ondris-Crawford for their assistance and advice in preparing this treatise. Finally we are grateful to all of our esteemed colleagues who graciously devoted their time and expertise in preparing contributions for this book. If not for their fine contributions, such a collection of high-quality chapters in this rapidly expanding discipline would not be possible. We are confident that their dedication to this endeavour has been a rewarding experience.

G. P. Crawford
Palo Alto, CA, USA

S. Žumer
Ljubljana, Slovenia

Professor J. W. Doane

The dedication of this book, edited by two associates and contributed to by a group of international scientists, to Professor J. W. Doane is itself a clear statement of the respect of the scientific community for his work and also of the affection in which he is held by his fellow scientists. However, I am delighted to have been asked to write this foreword which provides me with a valued opportunity, both as a friend and a scientist who, like him, has worked for many years in the field of liquid crystals, to express my own high regard for him and his achievements and also to put the man in context for those readers who may be new to the field of liquid crystals.

J. William Doane was born in the Sand Hills of Western Nebraska on 26 April 1935, but his family later moved to Missouri. In high school, his interest in ham radio oriented him towards physics and it was in this subject that he graduated with a BS in 1956 from the University of Missouri. Two years' service as an artillery officer preceded his return to the graduate school to do his doctoral research. This was successfully completed in 1965 on a project which involved nuclear magnetic resonance (NMR) spectroscopy, but in the field of inorganic materials, not liquid crystals.

His introduction to the latter subject came when he joined the staff at Kent State University (KSU) in 1965 and met Professor Glenn Brown, already an established figure in the field of liquid crystal research who earlier that year had created the Liquid Crystals Institute (LCI). Also very significantly, he met there Professor James McGrath, who was already applying NMR techniques to liquid crystals. By 1974, he had become Professor in the Department of Physics at KSU, a department which Professor Alfred Saupe, another internationally recognised figure in the liquid crystal field, had joined in 1968. Professor Doane's association with the LCI and Professor Glenn H. Brown, its founder Director, was always close, and later in 1979, he combined with his role in physics the Associate Directorship of the Institute. In 1983, consequent upon the declining health of Glenn Brown, he assumed the position of Director of the LCI.

His immediate task as Director was to attract external funding to support the Institute, a job which he undertook with great vigour and success through making the LCI a centre of display activity. A wealth of research grants and contracts was secured over the years, ensuring the health and vigour of the Institute and culminating in the prestigious recognition of the Institute by the National Science Foundation as the NSF Center for Advanced Liquid Crystalline Optical Materials (ALCOM).

Those of us who have been involved in similar roles know well the degree of personal commitment and effort, not to mention scientific reputation, required to attain such achievements. The support had to be found without impairing his scientific impetus, and a glance through his extensive list of publications, review articles and books shows that while developing and obtaining security for the LCI, he was producing quality work in his laboratories. His earlier publications clearly reflect his background in NMR. Numerous papers on NMR studies of liquid crystals – many such as those on biaxiality of liquid crystals and studies of unusual phases in lyotropic systems providing very nice illustrations of the versatility of the NMR technique for studies of anisotropic systems – are interspersed with others reflecting his interest in phase transitions, diffusion in liquid crystals and the re-entrance phenomenon.

Although such studies were continued after 1986, that year marks a kind of watershed in his research direction, his first papers appearing then on light scattering effects from nematic microdroplets. From 1986, an increasing number of his publications and patents were concerned with microdroplet phenomena, and he must be regarded today as the international expert on polymer-dispersed liquid crystal (PDLC) systems, the devices which have emanated from them and continue to do so, and the whole subject of the anchoring and properties of liquid crystals confined in small cavities. NMR spectroscopy has of course continued to be a vital tool in much of this work.

At grass roots level, we of course refer to him as plain Bill Doane, and bearing in mind all that he has done, I think that an emerging title for Bill has to be Father of Polymer Dispersed Liquid Crystal Systems. However, not content with all these achievements Bill Doane has gone on most recently to apply his inventiveness and experience to studies of the fundamental science, development and applications of polymer-stabilized cholesterics and polymer-modified liquid crystals. Such systems are in fact the focus of much of this book, and emphasizing his vision, they are now used in reflective displays, which is where the LCD market will move in the future. Indeed, he is co-founder of the company Kent Display Systems (KDS) which is developing polymer-stabilized cholesteric texture (PSCT) reflective displays and serves on its Board of Directors.

His great contributions to science and technology have been recognized in many ways by, as we would say in England, the establishment. To mention but two awards, he received the Kent State University President's Medal for Outstanding Service in 1989 and the Arthur K. Doolittle Award of the Division of Polymer Materials of the American Chemical Society in 1990. A Fellow of the American Physical Society, he has held visiting research professorships in Ljubljana, Canberra and Sydney, and currently serves on the Board of the Society for Information Display and Imaging.

However, despite his extremely busy life, it is typical of the man that he has always been willing to give time freely for the benefit of the scientific community in which he works. As an example, he has since 1986 filled the role of Treasurer of the International Liquid Crystal Society in a thorough and responsible way.

In conclusion, and at this point addressing you directly Bill, it is a great pleasure to thank you personally and on behalf of science in general for the immense benefits you have brought to our understanding of liquid crystals, these Magic Systems of the old International Liquid Crystal Society song. You have combined the NMR skills of your early training with your own personal vision and awareness, and have opened our eyes to a range of new properties of liquid crystal systems and new possibilities for their commercial exploitation. On the occasion of your 60th birthday, we wish you well and express our hopes for a continuation of your scientific productivity in the years to come.

George W. Gray

List of Contributors

Fouad M. Aliev, Department of Physics and Materials Research Center, University of Puerto Rico, San Juan, PR 00931, USA

Mike P. Allen, H. H. Wills Physics Laboratory, University of Bristol, Royal Fort, Tyndall Avenue, Bristol BS8 1TL, UK

Jayanth R. Banavar, Department of Physics and Center for Materials Physics, 104 Davey Laboratory, The Pennsylvania State University, University Park, PA 16802, USA

Giovanni Barbero, Dipartimento di Fisica, Politecnico di Torino, Corso Duca degli Abruzzi 24, 10129 Torino, Italy

Tommaso Bellini, Dipartimento di Elettronica, Universita da Pavia, 27100 Pavia, Italy, and Condensed Matter Laboratory, Department of Physics, University of Colorado, Boulder, CO 80309, USA

Robert Blinc, Physics Department and J. Stefan Institute, University of Ljubljana, Jamova 39, 61000 Ljubljana, Slovenia

Philip J. Bos, Liquid Crystal Institute, Kent State University, Kent, OH 44242, USA

Dirk J. Broer, Philips Research Laboratories, Prof. Holstlaan 4, 5656 AA Eindhoven, The Netherlands

Liang-Chy Chien, Liquid Crystal Institute, Kent State University, Kent, OH 44242, USA

Marek Cieplak, Institute of Physics, Polish Academy of Sciences, 02-668 Warsaw, Poland

Noel A. Clark, Condensed Matter Labortory, Department of Physics, University of Colorado, Boulder, CO 80309, USA

Doug J. Cleaver, Division of Applied Physics, Sheffield Hallam University, Sheffield S1 1WB, UK, and Department of Chemistry, University of Southampton, Southampton SO17 1BJ, UK

Gregory P. Crawford, Xerox Palo Alto Research Center, 3333 Coyote Hill Road, Palo Alto, CA 94305, USA

George Durand, Laboratoire de Physique des Solides, Bât. 510, Centre Universitaire, 91405 Orsay Cedex, France

Rudolf Eidenschink, NEMATEL, Galileo-Galilei-Str. 10, 55129 Mainz, Germany

Daniele Finotello, Department of Physics and Liquid Crystal Institute, Kent State University, Kent, OH 44242, USA

Katalin Fodor-Csorba, Research Institute for Solid State Physics of the Hungarian Academy of Sciences, H-1525 Budapest, PP Box 49, Hungary

David Fredley, Liquid Crystal Institute, Kent State University, Kent, OH 44242, USA

Yeuk K. Fung, Liquid Crystal Institute, Kent State University, Kent, OH 44242, USA

Hiroshi Hasebe, Dainippon Ink & Chemicals, Inc., Ina-machi, Kitaadachi-gun, Saitama 362, Japan

Toru Hashimoto, Stanley Electric Co. Ltd, Yokohama, Kanagawa 225, Japan

Rifat A. M. Hikmet, Philips Research Laboratories, Prof. Holstlaan 4, 5656 AA Eindhoven, The Netherlands

Germano S. Iannacchione, Department of Physics and Liquid Crystal Institute, Kent State University, Kent, OH 44242, USA

Yasufumi Iimura, The Graduate School of Technology, University of Agriculture and Technology, Koganei, Tokyo 184, Japan

Yoshihisa Iwamoto, The Graduate School of Technology, University of Agriculture and Technology, Koganei, Tokyo 184, Japan

Antal Jákli, Max Planck Research Group, Liquid Crystal Systems, Mühlpforte 1, 06108 Halle/S, Germany

Kazuhisa Katoh, Stanley Electric Co. Ltd, Yokohama, Kanagawa 225, Japan

Heinz-S. Kitzerow, Iwan N. Stranski Institut, Technische Universität Berlin, Sekretariat ER 11, Strasse des 17 Juni 135, 10623 Berlin, Germany

Shusuke Kobayashi, The Graduate School of Technology, University of Agriculture and Technology, Koganei, Tokyo 184, Japan

Samo Kralj, Faculty of Mathematical Studies, University of Southampton, Southampton SO17 1BJ, UK, and Department of Physics, Faculty of Education, University of Maribor, Koroška 160, Maribor 62000, Slovenia

Marcus Kreutzer, Institute of Applied Physics, Technische Hochschule Darmstadt, Hochschulstr. 6, 64289 Darmstadt, Germany

Jianlin Li, Liquid Crystal Institute, Kent State University, Kent, OH 44242, USA

Amos Maritan, Dipartimento di Fisica, International School for Advanced Studies, Trieste, Italy

Igor Muševič, J. Stefan Institute, University of Ljubljana, Jamova 39, 61000 Ljubljana, Slovenia

Janez Pirš, J. Stefan Institute, University of Ljubljana, Jamova 39, 61000 Ljubljana, Slovenia

Sihai Qian, Department of Physics and Liquid Crystal Institute, Kent State University, Kent, OH 44242, USA

Jolly Rahman, Tektronix Inc., Beaverton, Oregon, USA

Aaron G. Rappaport, Condensed Matter Laboratory, Department of Physics, University of Colorado, Boulder, CO 80309, USA

Charles Rosenblatt, Department of Physics, Case Western Reserve University, Cleveland, OH 44106-7079, USA

Mika Škarabot, J. Stefan Institute, University of Ljubljana, Jamova 39, 61000 Ljubljana, Slovenia

T. J. Sluckin, Faculty of Mathematical Sciences, University of Southampton, Southampton SO17 1BJ, UK, and Isaac Newton Centre for the Mathematical Sciences, 20 Clarkson Road, Cambridge CB3 0EH, UK

Haruyoshi Takatsu, Dainippon Ink & Chemicals, Inc., Ina-machi, Kitaadachi-gun, Saitama 362, Japan

Britt N. Thomas, Condensed Matter Laboratory, Department of Physics, University of Colorado, Boulder, CO 80309, USA, and Exxon Research and Engineering, Annandale, NJ 08801, USA

Sanjay Tripathi, Reveo Corporation, 8 Skyline Drive, Hawthorne, NY 10532, USA

Anikó Vajda, Research Institute for Solid State Physics of the Hungarian Academy of Sciences, H-1525 Budapest, PP Box 49, Hungary

Marija Vilfan, J. Stefan Institute, University of Ljubljana, Jamova 39, 61000 Ljubljana, Slovenia

Nataša Vrbančič-Kopač, J. Stefan Institute, University of Ljubljana, Jamova 39, 61000 Ljubljana, Slovenia

John L. West, Liquid Crystal Institute, Kent State University, Kent, OH 44242, USA

Deng-Ke Yang, Liquid Crystal Institute, Kent State University, Kent, OH 44242, USA

Haiji Yuan, Kent Display Systems, 343 Portage Blvd., Kent, OH 44240, USA

Boštjan Žekš, J. Stefan Institute, University of Ljubljana, Jamova 39, 61000 Ljubljana, Slovenia

Slobodan Žumer, Physics Department, University of Ljubljana, Jadranska 19, 61000 Ljubljana, Slovenia

Historical Perspective of Liquid Crystals Confined to Curved Geometries

From freely suspended droplets to flat-panel displays

G. P. CRAWFORD and S. ŽUMER

Interest in liquid crystalline materials confined to curved geometries has expanded greatly in recent years because of their important role in new and emerging electro-optic technologies and their richness in physical phenomena (Doane, 1991). The discovery of the usefulness of these materials, conspicuous in electrically controllable light scattering windows (Doane, 1990) and reflective mode display technology (Yang *et al.*, 1994), has burgeoned into a mature reflective display technology and heralded an era of fascination with the confinement of organized fluids. Independent of the method used to constrain liquid crystals, phase separation (Doane *et al.*, 1986), encapsulation (Fergason, 1985; Drzaic, 1986), or permeation (Craighead *et al.*, 1982; Crawford *et al.*, 1991), these systems have one underlying common theme: a symmetry-breaking, non-planar confinement imposed by the surrounding matrix. In addition, confined liquid crystal systems differ from macroscopic bulk liquid crystals because of their large surface-to-volume ratio. Their composite nature profoundly affects the ordering of the liquid crystal molecules and their susceptibility to external fields, making them ideal for a host of new electro-optic applications (West, 1994) and intellectually challenging from the basic science standpoint (Kitzerow, 1994).

The first investigation of confined liquid crystals dates back to the early 1900s when interest in the macroscopic properties of bulk liquid crystals was still in its infancy. Lehmann (1904) successfully suspended supramicrometre nematic liquid crystal droplets in a viscous isotropic medium. By studying the birefringent textures with optical polarizing microscopy (see Figure 1.1(a)) he concluded that the specific director configuration within the spherical confining cavity depends on the liquid crystal material and the angle at which the liquid crystal molecules are anchored to the isotropic fluid interface.

Many years passed before scientists began studying liquid crystal droplets again. In 1969, Dubois-Violette and Parodi applied elastic theory, developed by Frank in 1958, to predict several stable nematic director-field configurations in spherical droplets. They demonstrated the stability of the bipolar and radial nematic director fields (see Figure 1.2). They also showed that in the presence of an applied electric field for materials with a positive dielectric anisotropy, the bipolar droplet reorients parallel to the electric field direction while the radial droplet experiences a configuration transition to the axial structure (see Figure 1.2). These theoretical efforts were stimulated by the experimental studies of Meyer (1969) who earlier observed such configurations.

In the early 1970s, efforts focused on the nematic director fields of liquid crystals constrained to a cylindrical environment. Cladis and Kléman (1972) applied elastic theory to predict stable

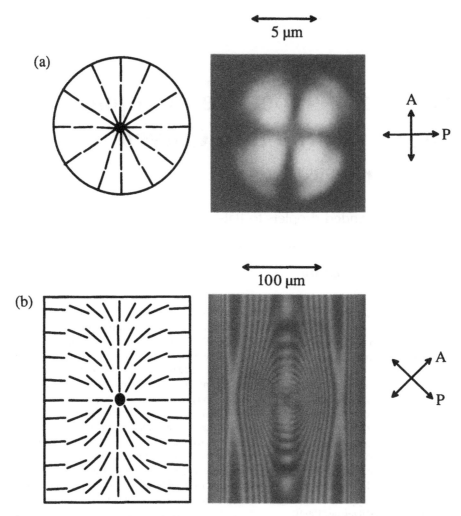

Figure 1.1 Two nematic director fields as viewed between crossed polarizers of an optical microscope that are stable in well-defined geometries when homeotropic anchoring conditions persist at the cavity wall. The radial droplet is 5 μm in diameter (a) and the capillary tube is 200 μm in diameter (b).

configurations in supramicrometre capillary tubes whose cylindrical walls were modified with a surface-aligning agent to induce perpendicular (homeotropic) boundary conditions. They predicted the existence of the escaped-radial director-field configuration presented in Figure 1.1(b). The nematic director is anchored perpendicular to the cavity and continuously becomes parallel to the cylinder axis; this configuration was found to be of lower energy than the planar – radial director field predicted earlier. Since it is energetically equivalent for the nematic director to escape in both directions, point defects often occur (see Figure 1.1(b)). Williams *et al.* (1972), Saupe (1973) and Meyer (1973) studied these point defects and characterized them as the radial and hyperbolic type.

These early milestones on confined liquid crystals led to many subsequent studies through the 1970s. Candau *et al.* (1973) were the first to study the spiral nature of chiral nematic droplets. Press and Arott (1974) applied elastic theory to nematic droplets to explain their

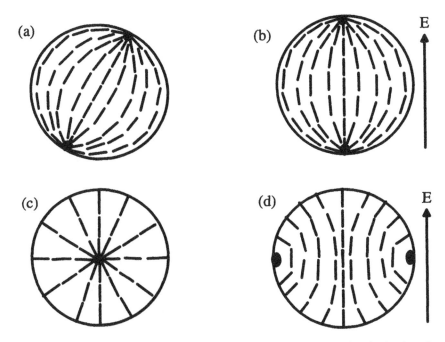

Figure 1.2 The various nematic director fields in spherical cavities: (a) the bipolar; (b) the aligned bipolar; (c) the radial; (d) the axial.

observation of the unusual twisted-star texture in the centre of a radial droplet when viewed between crossed polarizers. Cladis (1974) extended her earlier work by confining materials that exhibit a chiral-smectic A phase transition to capillary tubes. She observed a configuration transition from the escaped-radial configuration to the planar–radial configuration as the bend-to-splay elastic constant diverged. Melzer and Nabarro (1977) studied parallel anchoring conditions in capillary tubes but the task of obtaining uniform concentric boundary conditions proved to be a formidable obstacle. Cladis *et al.* (1979) proposed that the chiral nematic phase grows from a smectic A phase via a spiralling disclination when a material that exhibits a chiral nematic–smectic A transition is constrained to a capillary tube.

In the early 1980s, the liquid crystal droplet problem was re-examined from the basic standpoint. Volovik and Lavrentovich (1983) extensively studied the dynamics of creation, annihilation and transformation of topological defects (e.g. boojums, hedgehog and disclinations) in a closed system following predictions of topological theory. Kurik and Lavrentovich (1982) investigated the defect lines in chiral nematic droplets. Up to this point, confined liquid crystals were only curious systems from the basic science perspective.

In 1982, Craighead and coworkers disclosed a possible device which utilized confined liquid crystals as an electro-optic light valve: a microporous membrane permeated with a positive dielectric anisotropy liquid crystal. In the absence of an electric field, the permeated membrane is slightly scattering because of the mismatch between the indices of refraction of the membrane and the liquid crystal. Upon application of an applied voltage, the ordinary index of refraction of the liquid crystal is closely matched to that of the microporous matrix, and the permeated membrane becomes transparent for light propagating along the field direction. The general idea of index matching was conceived earlier by Hilsum (1976) for a different application. The device proposed by Craighead *et al.* (1982) was never commercialized because of inadequate contrast.

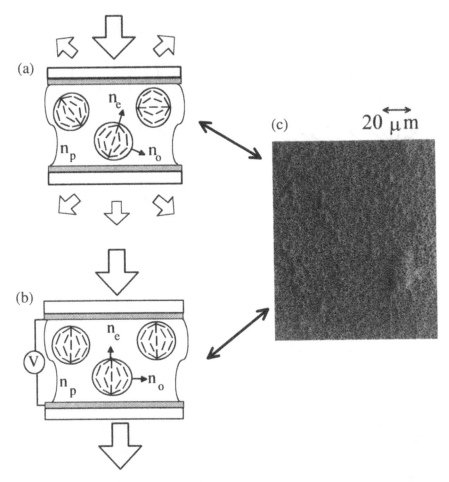

Figure 1.3 The operation of a polymer-dispersed liquid crystal (PDLC) display in the scattering state (a) and the transparent state (b). A scanning electron microscope (SEM) phoograph of a PDLC system is presented in (c) (courtesy of John West, Liquid Crystal Institute).

The usefulness of spherical nematic droplets was not appreciated until the mid-1980s when it was discovered that submicrometre and micrometre droplets could easily be dispersed in a rigid polymer binder (see Figure 1.3) by either phase separation (Doane *et al.*, 1986) or emulsification (Fergason, 1985; Drzaic, 1986). The dispersion materials formed by the phase separation technique became known as polymer-dispersed liquid crystals (PDLCs) and those formed by emulsification methods are known as nematic curvilinear aligned phase (NCAP) materials. We will use the acronym PDLC to describe both systems. By matching the ordinary refractive index of the liquid crystal, n_o, with that of the polymer binder, n_p, an electrically controllable light scattering medium transforms from a translucent, white appearance to a transparent appearance upon application of an applied voltage (see Figures 1.3(a) and (b)). The potential of PDLCs in high-contrast electro-optic light shutters revitalized the interest in liquid crystals confined to curved geometries and revealed many unanswered questions concerning the morphology of the polymer binder (see Figure 1.3 (c)), the effect of confining surfaces and the high surface-to-volume ratio (Erdmann *et al.*, 1990), and the configuration of the nematic director field within the spherical cavity (see Figures 1.4(a) and (b)) (Ondris-Crawford *et al.*, 1991).

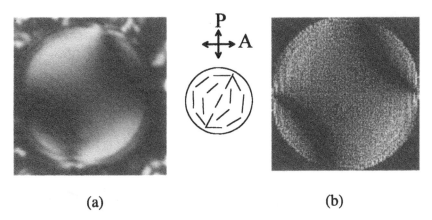

Figure 1.4 An optical polarizing microscopy texture of a 40 μm bipolar droplet (a) next to a simulated one (b). The orientation of the director field is schematically shown.

The pioneering discovery of PDLC materials for electro-optic applications suggested many applications and the interest in confined liquid crystals flourished. These dispersions allowed studies of liquid crystals in systems with a large surface-to-volume ratio where surface boundary effects become pronounced. It became apparent very early in the development of these materials that the details of the configuration within the supramicrometre and/or submicrometre droplets must be well understood. Optical polarizing microscopy is limited to droplet sizes greater than 5 μm; therefore, Golemme *et al.* (1988a) employed deuterium nuclear magnetic resonance (²H-NMR) to probe the details of the nematic director field in submicrometre-size cavities. They went on (Golemme *et al.*, 1988b) to study how the nematic–isotropic phase transition is modified under extreme confinement conditions, confirming the earlier theoretical predictions of Sheng (1976) who proposed in 1982 that the nematic-isotropic phase transition temperature becomes continuous at a critical enclosure size. Theoretical efforts on the phase transition in spherical droplets were later developed by Kralj *et al.* (1991). The particularly high surface-to-volume ratio of PDLC materials enabled the study of surface effects in confined liquid crystals by a wide variety of experimental techniques not conventionally used to study interfaces in the past. Vilfan *et al* (1988) used proton nuclear magnetic resonance to study the molecular dynamics of liquid crystals on the surfaces of microdroplets. Studies by Erdmann *et al.* (1990) unveiled the details of the radial-to-axial configuration transition in PDLC droplets and determined the molecular anchoring strength at the polymer cavity surface.

Concurrently, studies of a more applied nature were in progress to fabricate an improved liquid crystal light value. West (1988) and Vaz *et al.* (1987) initially characterized various liquid crystal-polymer dispersion systems for light valve applications. Montgomery and Vaz (1987) developed the methodology to determine the contrast ratio of PDLC systems and Lackner *et al.* (1988) published a comparison between the various methods to measure contrast in PDLC systems. Margerum *et al.* (1989) studied the effects of off-state alignment on electro-optic performance of PDLCs. The efforts of Wu *et al.* (1989) demonstrated that droplet shaping modifies the response time; they developed a simple phenomenological model to understand the effect of elliptical morphology on the dynamic switching process. The details of this approach were recently extended by Kelly and Palffy-Muhoray (1994).

The mechanisms of light scattering in PDLC materials were also being addressed by Žumer and Doane (1986) for submicrometre droplets and by Žumer (1988) for micrometre-sized droplets. In particular they investigated the effect of droplet structures and refractive index matching. Montgomery (1988) and Montgomery and Vaz (1989) also investigated scattering properties of PDLC films, and Žumer *et al.* (1989) proceeded to study the transmission properties. Whitehead *et al.* (1993) probed light scattering effects in a dispersion of micrometre

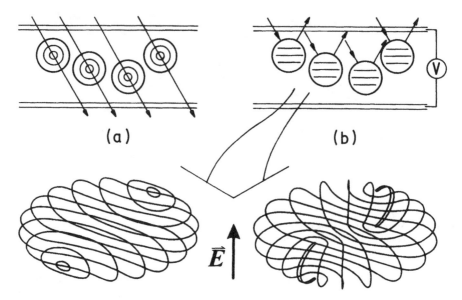

Figure 1.5 Conceptual illustration of the operation of a polymer-dispersed chiral liquid crystal (PDCLC) (a). Director fields of chiral nematic droplets with negative dielectric anisotropy in the external field where the applied field transforms chiral surfaces from spherical (a) to planar (b). Two possible structures are shown below.

aligned nematic droplets and Drzaic and Gonzales (1993) studied reflective index gradients and the light scattering performance in PDLC films. Kelly *et al.* (1992) investigated the effects of wavelength dependence of scattering in PDLC films, focusing on droplet-size effects.

The successful studies of nematic liquid crystal droplets led researchers to examine chiral nematic liquid crystals under confinement for reflective display applications. Crooker and Yang (1990) employed chiral nematic droplets for colour electro-optic displays. For materials with negative dielectric anisotropy and sufficiently short pitch, they showed that it is possible by applying electric fields to induce a texture showing bright selective reflection (see Figure 1.4). The director fields of chiral nematic droplets shown in Figure 1.5 (Bezič and Žumer, 1992; Bajc *et al.*, 1995) have been a topic of great interest because of their importance in colour reflective displays (Kato *et al.*, 1993a,b); they are discussed in detail in the chapter by Kitzerow (this volume).

Shortly after the first use of cholesteric liquid crystal compounds in dispersion materials Kitzerow *et al.* (1992) revealed that alignment of ferroelectric and antiferroelectric liquid crystal droplets can be controlled by a mechanical shearing technique during the photopolymerization process; Molsen and Kitzerow (1994) subsequently proceeded to demonstrate bistability in ferroelectric PDLCs. Pirš *et al.* (1992), Zyryanov *et al.* (1993), Lee *et al.* (1994) and Patel *et al.* (1994) also presented studies on ferroelectric PDLCs. There are two immediate advantages of ferroelectric PDLCs over their surface-stabilized ferroelectric liquid crystal counterpart: (i) thicker cell gaps can be employed, which simplifies manufacturing, and (ii) the polymer introduces added stability to protect against mechanical shock failure. Many of the ferroelectric PDLC studies are reviewed in the literature (Kitzerow, 1994) and are discussed here by Kitzerow (Chapter 8). The feasibility of ferroelectric PDLCs in fast switching applications is expected to lead to many exciting basic studies in the future.

Perhaps the most gratifying milestone of confined materials is their recent entry into the commercial market as privacy windows. Figure 1.6 shows the off-state (a) and the on-state (b) of a PDLC privacy window manufactured by Marvin Windows using films developed by the

Figure 1.6 A demonstration of a PDLC privacy window made with 3M™ Privacy Film. The laminated glass allows total privacy at the flip of a switch. The glass changes from a milky white appearance (a) to a transparent appearance (b). This application of PDLC materials is available from commercial glaziers and residential window manufacturers for applications ranging from building exteriors to conference rooms and bathrooms (courtesy of 3M, Viracon™ Privacy Glass).

3M Corporation. The large-scale shutters are suitable for indoor privacy windows, automobile windows, and even outdoor windows on double-pane glass. By incorporating a birefingent polymer binder, Doane *et al.* (1990) are working to eliminate the off-angle haze that becomes apparent at wide viewing angles; this approach has not yet been commercialized. Dyes have been incorporated into PDLC materials for colour and to improve contrast (Drzaic *et al.*, 1992; West and Ondris-Crawford, 1991).

In active matrix technology widely used for high-information-content displays, each pixel is connected to a non-linear electronic element so that the required level of voltage is maintained during the entire frame period to preserve the necessary level of contrast (Wu, 1994). There are several critical issues concerning PDLC materials for high-resolution displays using active matrix (AM) technology (Wu, 1994): (i) eliminating the hysteresis in the voltage response curve; (ii) lowering the driving voltage to be compatible with amorphous silicon (a-Si) and polysilicon (poly-Si) technologies; and (iii) increasing the resistivities for desired charge hold times. The PDLC materials developed by Merck (Coates *et al.*, 1993) can achieve these characteristics over a limited temperature range for the metal–insulator–metal (MIM) AM technology (Ginter *et al.*, 1993). Reamey *et al.* (1992) made substantial improvements in the emulsion-based materials. Jones *et al.* (1992) reported a direct-view AM-compatible emulsion system. Niiyama *et al.* (1993) developed materials to eliminate hysteresis and improve resistivity for poly-Si AM substrates. Recently, Vinouze *et al.* (1994) demonstrated an AM PDLC for reflective direct-view applications. It is an exciting time for PDLC materials as they are being developed for AM substrates for a wider scope of high-resolution applications.

Perhaps the most suitable application for PDLC materials is for projection light valves because off-angle haze and forward scattering are not an issue with a suitable design of the projection system. Since there is no need for polarizers on PDLC displays, the brightness of the projector is enhanced by at least a factor of two; in addition, brighter lamps can be employed since the light shutter operates on light scattering rather than an absorption principle. There have been several efforts to develop PDLC materials for projection light valves (Yaniv *et al.*, 1989; Kuniga *et al.*, 1990; Pirš *et al.*, 1990; Tomita and Jones, 1992; Ginter *et al.*, 1993; Tahizawa *et al.*, 1991; Kinugasa *et al.*, 1991; Tomita, 1993; Niiyama *et al.*, 1993; Coates, 1993).

A novel application of PDLC materials is in the general area of holography, optical computing, and networking. The underlying idea is to pattern structures in the droplet distributions on the order of the wavelength of visible light. One type of patterning scheme has been developed by Sutherland *et al.* (1993). A photocurable polymer causes the liquid crystal phase to separate into droplets contained in layers that orient perpendicular to the cell surface, thereby forming a switchable diffraction grating. Application of this technique is still in its infancy but appears to have an exciting future as more efforts are geared towards selectively phase separating PDLC materials.

One of the newest applications of PDLC materials is a colour reflective display that operates on the principle of Bragg reflection. By setting up an interference pattern within the sample along the substrate normal with visible light, Tanaka *et al.* (1994) managed to phase-separate liquid crystal droplets into Bragg planes (see Figure 1.7). In the off-state (see Figure 1.7(a)), the index of refraction of the polymer does not match the effective refractive index (average of the ordinary, n_o, and extraordinary, n_e, refractive index) of the liquid crystal droplets, thereby setting up a refractive index gradient. The optical interference of the multilayer structure reflects light at the Bragg wavelength and transmits the rest. The spacing of the Bragg planes is governed by the wavelength of the light used to perform the photopolymerization. In the on-state (see Figure 1.7(b)), the ordinary index of refraction of the liquid crystal, n_o, matches that of the polymer, n_p, and the refractive index gradient disappears; therefore the cell is transparent and all light is transmitted. There are many exciting possibilities of this new PDLC technology in direct-view and projection reflective display applications.

Throughout the development of PDLC materials for display applications, many other uses of these electrically controllable light scattering films became apparent. Palffy-Muhoray *et al.* (1990) investigated optical-field-induced switching in PDLC films. Thermally addressed PDLCs and their memory characteristics were reported (Yamaguchi and Sato, 1991; Yamaguchi *et al.*, 1992). McCarger *et al.* (1992) implemented a carefully engineered PDLC shutter into an infrared imaging system to replace the mechanical shutter. Mallicoat (1992) employed PDLC light valves in a sequential colour display system. The feasibility of dichroic dyes used in PDLCs to introduce colour and improved contrast was also investigated (Drzaic and Gonzales, 1992; West and Ondris-Crawford, 1992).

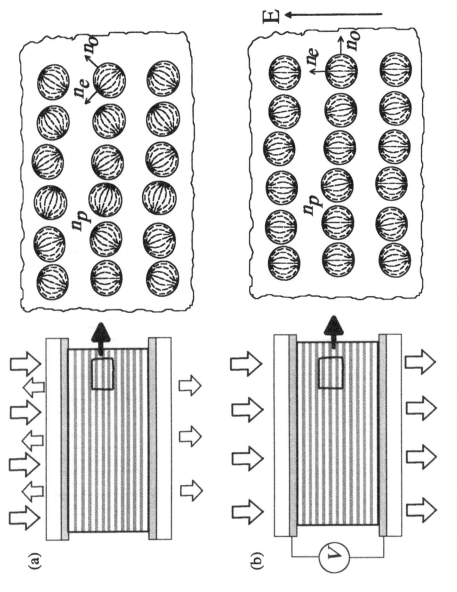

Figure 1.7 A conceptual illustration of the holographically formed PDLC in its reflecting state (a) and transparent state (b).

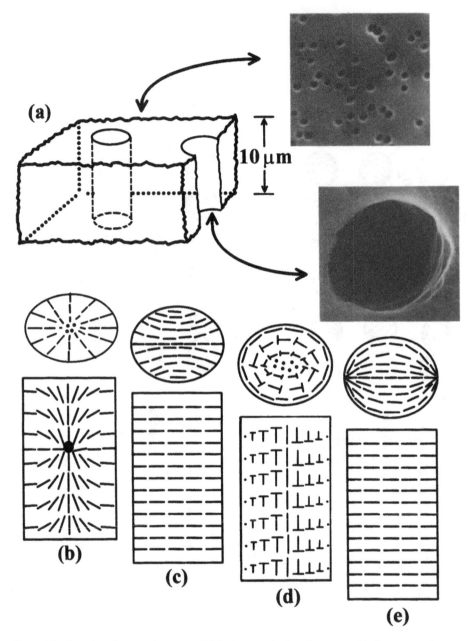

Figure 1.8 Scanning electron microscope (SEM) photograph of a microporous Nuclepore membrane. The various nematic director fields that have been confirmed in permeated Nuclepore membranes are the escaped radial with point defects (b), the planar–polar (c), the escaped–twisted (d) and the planar–bipolar (e).

Kato *et al*. (1993a,b) have fabricated a multipage display using a stack of PDLC films.

All of the basic and applied activity on spherical liquid crystal droplets led to a resurgence of interest in confined liquid crystals in general. Systems with simple cylindrical symmetry of

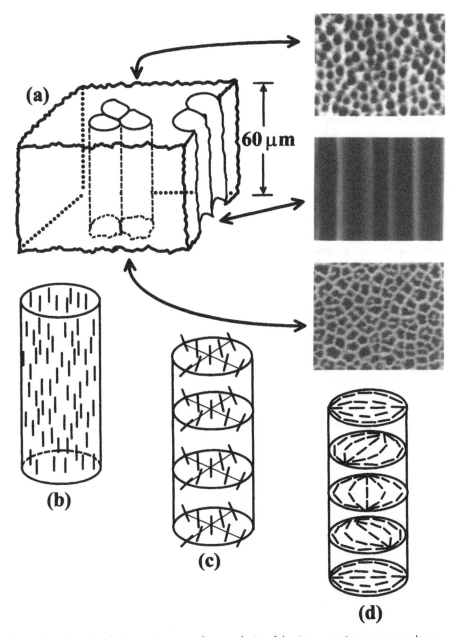

Figure 1.9 Scanning electron microscope photographs (a) of the Anopore microporous membrane. Some of the configurations that are possible are the parallel axial (b), the tilted axial (c) and the twisted–bipolar (d).

micrometre and submicrometre cavity sizes were employed to study surface elastic properties and surface-induced order. Nuclepore membranes (see Figure 1.8(a)) are polymer membranes with cylindrical channels penetrating through their 10 μm thickness (Chen *et al.*, 1980; Crawford *et al.*, 1992a). They are available in a wide variety of pore sizes ranging from 0.03 μm to 12.0 μm in diameter making them attractive media to permeate with liquid crystal materials.

For homeotropic boundary conditions, the escaped-radial with point defects (see Figure 1.8(b)) and the planar-polar director field (see Figure 1.8(c)) are found to exist depending on the cavity size and the surface treatment (Allender *et al.*, 1991; Crawford *et al.*, 1992b). From these studies the value of the saddle-splay surface elastic constant K_{24} and the polar molecular anchoring strength were determined. Studies on these homeotropic structures were extended by probing their director profile as the nematic–smectic A transition is approached from above (Ondris-Crawford *et al.*, 1992). By treating the inner cavity walls with a polyimide, which is well known to induce homogeneous anchoring conditions, the inner corrugations shown in Figure 1.8(a) are sufficient to introduce uniform concentric anchoring (i.e. the corrugations are analogous to the grooves induced by the rubbing process). The stable configurations found in polyimide treated cavities of Nuclepore membranes are the escaped–twisted (see Figure 1.8(d)) and the planar–bipolar (see Figure 1.8(e)) which allow the simultaneous determination of the saddle–splay surface elastic constant K_{24} and the polar and azimuthal molecular anchoring strength (Ondris-Crawford *et al.*, 1993). The supramicrometre capillary tube problem was also revisited in the 1990s in the spirit of the earlier studies by Cladis and Kléman (1972), Williams *et al.* (1972), Meyer (1973) and Saupe (1973). It was found that careful microscopic observations and texture simulations provided a measure of the bend-to-splay bulk elastic constant ratio (Scharkowski *et al.*, 1993) and K_{24} and the polar anchoring strength (Polak *et al.*, 1994). Chen and Liang (1991) investigated the configuration in closed capillary tubes by capping the ends of the tubes with an additional surface boundary.

The other microporous filter technology that is attractive for a confinement medium is the Anopore membrane (Furneaux *et al.*, 1989). Anopore membranes are thicker and have a larger porosity than their Nuclepore counterparts; the only disadvantage is that they are only commercially available with a pore diameter of 0.2 μm. Scanning electron microscopy photographs of the Anopore membrane are presented in Figure 1.9(a). The cross-section shows uniform cylindrical channels despite the 'honeycomb' nature of the bottom surface. Details on the morphology of Anopore membranes can be found in the literature (Crawford *et al.*, 1992a). The inner cavity walls are easily accessible to chemical treatments. These systems were useful for studying surface-induced order (Crawford *et al.*, 1991) and molecular anchoring and orientational wetting transitions (Crawford *et al.*, 1993). Iannacchione and Finotello (1992) used Anopore membranes to study the effect of confinement on the nematic–isotropic phase transition temperature with the heat capacity technique. Ondris-Crawford *et al.* (1994) studied chiral doped nematic compounds in Anopore membranes and found the parallel–axial structure for the pure nematic (Figure 1.9(b)), the radially twisted structure for low chirality (Fig 1.9(c)), and the twisted bipolar structure for higher chirality (Figure 1.9(d)). Schmiedel *et al.* (1994) proposed the existence of an axially twisted conic structure which can be stable for certain ratios of the elastic constants and surface anchoring strength (Ambrožič and Žumer, 1996).

Applications and basic science problems of PDLC materials and confined liquid crystals appear endless. We have only scratched the surface of the fine work on PDLCs and liquid crystals confined to well-defined geometries that has taken place in the past few years. We encourage readers to consult some of the review articles in the literature (Doane, 1990, 1991; Crawford and Doane, 1992, 1994; Bouteiller *et al.*, 1993; Kitzerow, 1994; West, 1994; Drzaic, 1995) for a more detailed overview. In the past few years confined liquid crystal studies have entered a new and exciting era by expanding the research focus to complex geometries that are more random in nature. The pioneering applied efforts of Hikmet (1991a,b) and Yang *et al.* (1992) have stimulated great interest in the structure of liquid crystals in complex geometries. Low concentrations of polymer are photopolymerized in a liquid crystal environment to modify the bulk properties of the materials. These polymer network assemblies are believed to form thin fibrils in the liquid crystal environment (Figure 1.10(a)). Their potential in reflecting displays (West *et al.*, 1993; Yang, *et al.*, 1994), electrically controllable light scattering displays (Hikmet 1991a,b; Yang *et al.*, 1992), ferroelectric displays (Pirš et al., 1995), TN and STN display (Bos *et al.* 1993; Fredley *et al.*, 1994; Hasebe *et al.*, 1994), active matrix displays (Pfeiffer *et al.*, 1994), projection displays (Fung *et al.*, 1993), bulk alignment (Jain and Kitzerow, 1994), and amorphous

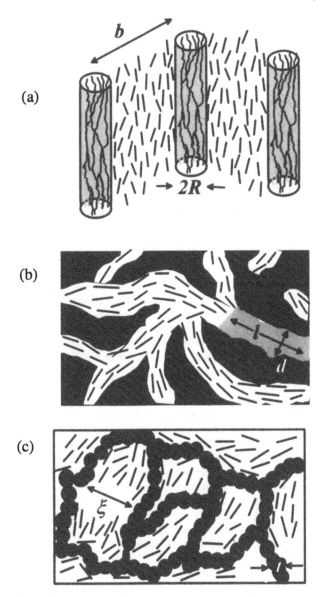

Figure 1.10 Liquid crystals under a complicated form of confinement. The polymer network assemblies are shown in (a) for a fibril morphology (pore segments) of radius R and lattice spacing b, the Vycor glass in (b) with mean diameter d and length l, and the silica aerogel in (c) with a compartment size of ξ.

alignment (Hashimoto *et al.*, 1994) is phenomenal. A new type of dispersion consisting of aggregates of inorganic spheres with bistable memory capability was reported by Kreuzer *et al.* (1993).

In addition, the potential of polymer network assemblies and aggregates of inorganic spheres in flat-panel displays and electro-optic applications has generated great interest in related systems with a random structure such as Vycor glass (see Figure 1.10(b)) and silica aerogel matrices (see Figure 1.10(c)). The knowledge acquired for well-defined systems (droplets and cylinders) has guided the focus on liquid crystals confined in complex geometries for both the

13

basic science and applications. Aliev and Breganov (1988) initiated many of the early studies of liquid crystals in Vycor-type micropores. The nematic–isotropic phase transition is found to be replaced by continual evolution of orientational order in Vycor and Vycor-type glass; two models have been proposed to understand this behaviour (Iannacchione *et al.*, 1993; Tripathi *et al.*, 1994). The structure and dynamics of liquid crystals confined to aerogel is also a fruitful area of basic research. Several studies have been performed to understand the dynamics (Wu *et al.*, 1992), the phase transition characteristics of the randomly confined liquid crystals (Bellini *et al.*, 1992), the structure of the confined liquid crystal (Clark *et al.*, 1993), the orientational order parameter (Zidanšek *et al.*, 1994), and theories based on mean field theory and Monte Carlo simulations (Maritan *et al.*, 1994).

The collection of papers in this book constitutes an effort to bring together the exciting research and development of liquid crystals confined to complex geometries that has taken place in only a few short years. Starting with Barbero and Durand in Chapter 2, surface effects in confined systems are briefly reviewed. The contributions by Hikmet (Chapter 3), Žumer and Crawford (Chapter 4), Yang and coworkers (Chapter 5), Jakli and coworkers (Chapter 6) and Vilfan and Vrbančič (Chapter 7) focus on experimental studies of polymer networks assembled in an anisotropic environment. A diverse set of models illustrates our current understanding of the polymer network morphology. Kitzerow (Chapter 8) covers different polymer-dispersed and polymer-stabilized chiral liquid crystals while Blinc and coworkers (Chapter 9) focus on confined and stabilized ferroelectric liquid crystals. Broer (Chapter 10) treats the chemistry and morphology of systems that are aligned in the liquid crystal state and then photo-polymerized. We include many application papers to show the importance of these materials and their future in flat-panel display technology. West (Chapter 11) and Yuan (Chapter 12) address critical issues of these materials in reflective flat-panel display technologies; in addition they show current applications of these materials and speculate on future trends. Bos and coworkers in Chapter 13 report on the use of very low concentration polymer networks to modify twisted-nematic performance and avoid the annoying stripe formation in certain super-twisted-nematic (STN) structures. Kobayashi and coworkers in Chapter 14 describe the use of the polymer networks to stabilize defect structures that form in amorphous aligned TN displays to enhance electro-optic performance. A new dispersion technology presented by Kreuzer and Eidenschink in Chapter 15 uses aggregates and agglomerates of nanometre particles in liquid crystals, and the surprising features and many potential applications are addressed. Further systems where the liquid crystal is permeated into the pores of a prefabricated complex matrix are also treated in detail. Finotello and coworkers (Chapter 16), Aliev (Chapter 17) and Tripathi and Rosenblatt (Chapter 18) present experimental studies on liquid crystals confined to porous glass where extreme confinement effects greatly modify the phase transition and dynamical behaviour of the liquid crystal. Bellini and coworkers (Chapters 19 and 20) focus on light scattering and x-ray measurements of liquid crystals confined to aerogel matrices, emphasizing the function of pore size and near phase transition temperatures. The many experimental results that have become available in the last two years on liquid crystals in complex systems have stimulated much theoretical interest. Sluckin and coworkers (Chapter 21) and Banavar and coworkers (Chapter 22) present a theoretical framework in which liquid crystals confined to complex geometries can be modelled.

References

ALIEV, F. M. and BREGANOV, M. N. (1988) Temperature hysteresis and dispersion of the dielectric constant of a nematic liquid crystal in micropores, *Sov. Phys. JETP* **47**, 117–120.

ALLENDER, D. W., CRAWFORD, G. P. and DOANE, J. W. (1991) Determination of the liquid crystal surface elastic constant K_{24} *Phys. Rev. Lett.* **151**, 1442–1445.

AMBROŽIČ, M. and ŽUMER, S. (1996) Stability of chiral nematic structures in cylindrical cavities, *Phys. Rev. E* (to appear).

BAJC, J., BEZIČ, J. and ŽUMER, S. (1995) Choral nematic droplets with tangential anchoring and negative dielectric anisotropy in an electric field, *Phys. Rev. E* **51**, 2176–2189.

BELLINI, T., CLARK, N. A., MUZNY, C. D., WU, L., GARLAND, C. W., SCHAEFER, D. W. and OLIVER, B. J. (1992) Phase behavior of the liquid crystal 8CB in a silica aerogel *Phys. Rev. Lett.* **69**, 788–791.

BEZIČ, J. and ŽUMER, S. (1992) Structures of the cholesteric liquid crystal droplets with parallel surface anchoring, *Liq. Cryst.* **11**, 593–619.

BOS, P. J., RAHMAN, J. A. and DOANE, J. W. (1993) A low threshold-voltage polymer network TN device, *Soc. Inf. Disp. Dig.* **XXIV**, 877–880.

BOUTEILLER, L., CHASTAING, E., LE BARNY, P., MASSOE, F. and ROBIN, P. (1994,) Liquid crystal polymer composites for display applications, *Rev. Tech. Thomson-CSF* **26**, 115–140.

CANDAU, S., LEROY, P. and DEBEAUVAIS, F. (1973) Magnetic field effects in nematic and cholestric droplets suspended in an isotropic liquid, *Mol. Cryst. Liq. Cryst.* **23**, 283–297.

CHEN, S.-H. and LIANG, B. J. (1991) Stability of a hedgehog nematic configuration in a small closed cylindrical cavity, *Appl. Phys. Lett.* **59**, 1173–1175.

CHEN, T., DIPIRRO, M.J., BHATTACHARYYA, B. and GASPARINI, F. M. (1980) Some properties of Nuclepore filters, *Rev. Sci. Instrum.* **51**, 846–849.

CLADIS, P. E. (1974) Study of the nematic and smectic A phases of a N-p-cyanobenzylidene-p-n-octyloxyaniline in tubes, *Phil. Mag.* **29**, 641–663.

CLADIS, P. E. and KLÉMAN, M. (1972) Non-singular disclinations of strength S = +1 in nematics, *J. Phys.* **40**, 591–598.

CLADIS, P. E., WHITE, A. E., and BRINKMAN, W. F. (1979) Non-singular defect structures near the smectic A transition, *J. Phys.* **40**, 325–335.

CLARK, N. A., BELLINI, T., MALZBENDER, R. M., THOMAS, B. M., RAPPAPORT, A. G., MUZNY, C. D., SCHAEFER, D. W. and HRUBESH, L. (1993) X-ray scattering study of smectic ordering in a silica aerogel, *Phys. Rev. Lett* **71**, 3505–3508.

COATES, D. (1993) Normal and reverse mode polymer dispersed liquid crystal devices, *Displays* **14**, 94–103.

COATES, D., GREENFIELD, S., GOULDING, M., BROWN, E. and NOLAN, P. (1993) Recent developments in materials for TFT/PDLC devices, *Proc. SPIE* **1911**, 2–14.

CRAIGHEAD, H. G., CHENG, J. and HACKWOOD, S, (1982) New display based on electrically induced index matching in an inhomogeneous medium, *Appl. Phys. Lett.* **40**, 22–24.

CRAWFORD, G. P. and DOANE, J. W. (1992) Polymer dispersed liquid crystals, *Condens. Matter News* **1**, 5–11.

CRAWFORD, G. P. and DOANE, J. W. (1994) Ordering and ordering transitions of confined liquid crystals, *Mod. Phys. Lett. B* **7**, 1758–1808.

CRAWFORD, G. P., STANNARIUS, R and DOANE, J. W. (1991) Surface induced orientational ordering in the isotropic phase of a liquid crystal material, *Phys. Rev. A* **44**, 2558–2569.

CRAWFORD, G. P., STEELE, L. M., ONDRIS-CRAWFORD, R. J., IANNACCHIONE, G. S., YEAGER, C. J., DOANE, J. W. and FINOTELLO, D. (1992a) Characterization of the cylindrical cavities of Anopore and Nuclepore membranes, *J. Chem. Phys.* **96**, 7788–7796.

CRAWFORD, G. P., ALLENDER, D. W. and DOANE, J. W. (1992b) Surface elastic and molecular anchoring properties of nematic liquid crystals, *Phys. Rev A* **45**, 8693–8708.

CRAWFORD, G. P., ONDRIS-CRAWFORD, R. J., ŽUMER, S. and DOANE, J. W. (1993) Anchoring and orientational wetting transitions of confined liquid crystals, *Phys Rev. Lett.* **70**, 1838–1841.

CROOKER, P. P. and YANG, D.-K. (1990) Polymer dispersed chiral liquid crystal color display, *Appl. Phys. Lett.* **57**, 2529–2531.

DOANE. J. W. (1990) Polymer dispersed liquid crystals, *Liquid Crystals: Applications and Uses*, ed. B. Bahadhur, Vol. 1, Ch. 14, Singapore: World Scientific.

DOANE. J. W. (1991) Polymer dispersed liquid crystals: Boojums at work, *MRS Bull.* **XVI**, 22–28.

DOANE, J. W., VAZ, N. S., WU, B. G. and ŽUMER, S. (1986) Field controlled light scattering from nematic microdroplets, *Appl Phys. Lett.* **48**, 269–271.

DOLINŠEK, J., JARH, O., VILFAN, M., ŽUMER, S., BLINC, R., DOANE, J. W. and CRAWFORD, G. P. (1991) Two-dimensional deuterium nuclear magnetic resonance of a polymeric-dispersed nematic liquid crystal, *J. Chem. Phys.* **95**, 2154–2161.

DRZAIC, P. S. (1986) Polymer dispersed nematic liquid crystal for large area displays and light valves, *J. Appl. Phys.* **60**, 2142–2148.

DRZAIC, P. S. (1995) Polymer dispersed liquid crystals: a look back, a look ahead, *Liq. Cryst. Today* **5**, 2–4.

DRZAIC, P. S. and GONZALES, A. M. (1992) Phenomenological scattering models for nematic droplet/polymer films: Refractive index and droplet correlation effects, *Mol. Cryst. Liq. Cryst.* **222**, 11–20.

DRZAIC, P. S., and GONZALES, A. M. (1993) Refractive index gradients and light scattering in polymer dispersed liquid crystal films, *Appl Phys. Lett.* **62**, 1332–1334.

DRZAIC, P. S., GONZALES, A. M., JONES, P. and MONTOYA, W. (1992) Dichroic-based displays from nematic dispersion, *Soc. Inf. Disp. Dig.* **XXIII**, 571–574.

DUBOIS-VIOLETTE, E. and PARADI, O. (1969) Emulsions nématiques effects de champ magnétiques et effects piézoélectriques, *J. Phys. Colloq.* C4-57–64.

ERDMANN, J. H., ŽUMER, S. and DOANE, J. W. (1990) Configuration transition in nematic liquid crystals confined to a small spherical cavity, *Phys Rev. Lett.* **64**, 1907–1910.

FERGASON, J. L. (1985) Polymer encapsulated nematic liquid crystals for display and light control applications, *Soc. Inf. Disp. Dig.* **XVI**, 68–69.

FRANK, F. C. (1958) Liquid crystals: On the theory of liquid crystals, *Discuss. Faraday Soc.* **25**, 19–28.

FREDLEY, D. S., QUINN, B. M. and BOS, P. J. (1994) Polymer stabilized SBE devices, *Proc Int. Display Research Conf.*, pp. 480–483.

FUNG, Y. K., YANG, D.-K., DOANE, J. W. and YANIV, Z. (1993) Projection display from polymer stabilized cholestric textures, *Proc. Eurodisplay 93*, pp. 157–160.

FURNEAUX, R. C., RIGBY, W. R. and DAVIDSON, A. P. (1989) The formation of controlled porosity membranes from anodically oxidized aluminum, *Nature* **30**, 147–149.

GINTER, E., LUDER, E., KALLFASS, T., HUTTELMAIER, S., DOBLER, M., HOCHBOLZER, D., COATES, D., TILLIN, M. and NOLAN, P. (1993) Optimized PDLCs for active-matrix addressed light valves in projection systems, *Proc. Eurodisplay 93*, pp. 105–108.

GOLEMME, A., ŽUMER, S., ALLENDER, D. W. and DOANE, J. W. (1988a) Continuous nematic-isotropic transition in submicrometer-size liquid crystal droplets, *Phys. Rev. Lett.* **61**, 1937–1940.

GOLEMME, A., ŽUMER, S., DOANE, J. W. and NEUBERT, M. (1988b) Deuterium NMR of polymer dispersed liquid crystals, *Phys. Rev. A* **37**, 559–569.

HASEBE, H., TAKATSU, H., IIMURA, Y. and KOBAYASHI, S. (1994) Effect of polymer network made of liquid crystal diacrylate on characteristics of liquid crystal display device, *Jpn. J. Appl. Phys.* **33**, 6245–6248.

HASHIMOTO, T., KATOH, K., HASEBE, H., TAKATSU, H., IWAMOTO, Y., IIMURA, Y. and KOBAYASHI, S. (1994) Polymer stabilized amorphous finished nematic liquid crystal display, *Proc. Int. Display Research Conf.*, pp. 484–486.

HIKMET, R. A. M. (1991a) Anisotropic gels and plasticized networks formed by liquid crystal molecules, *Liq. Cryst.* **9**, 405–416.

HIKMET, R. A. M. (1991b) 'Electrically induced light scattering from anisotropic gels, *J. Appl. Phys.* **68**, 4406–4012.

HILSUM, C. (1976) Electrooptic device, British Patent 1,442,360.

HOFFMANN, J. (1989) Inorganic membrane filter for analytic separations, *Am. Lab.* **21**, 70–73.

IANNACCHIONE, G. S. and FINOTELLO, D. (1992) Calorimetric studies of phone transistions in confined liquid crystals, *Phys. Rev. Lett.* **69**, 2094–2097.

IANNACCHIONE, G. S., CRAWFORD, G. P., ŽUMER, S., DOANE, J. W. and FINOTELLO, D. (1993) Randomly constrained orientational order in porous glass, *Phys. Rev. Lett.* **71**, 2595–2598.

JAIN, S. C. and KITZEROW, H.-S. (1994) Bulk-induced alignment of nematic liquid crystals by photopolymerization, *Appl. Phys. Lett.* **64**, 2946–2948.

JONES, P., MONTOYA, W., GARZA, G. and ENGLER, S. (1992) Performance of NCAP for high-resolution active-matrix reflective display, *Soc. Inf. Disp. Dig.* **XXIII**, 762–765.

KATO, K., TANAKA, K., TSURU, S. and SAKAI, S. (1993a) Color image formation using polymer dispersed cholesteric liquid crystal, *Jpn. J. Appl. Phys.* **32**, 4600–4604.

KATO, K., TANAKA, K., TSURU S. and SAKAI, S. (1993b) Multipage display using stacked polymer-dispersed liquid crystal films, *Jpn. J. Appl. Phys.* **32**, 4594–4599.

KELLY, J. R. and PALFFY-MUHORAY, P. (1994) The optical response of polymer dispersed liquid crystals, *Mol. Crys. Liq. Crys.* **243**, 11–29.

KELLY, J., WU, W. and PALFFY-MUHORAY, P. (1992) Wavelength dependence of scattering in PDLC films: droplet size effects, *Mol. Cryst. Liq. Cryst.* **223**, 251–261.

KINUGASA, N., YANO, Y., NAGAYAMA, H., YOSIDA, H. and WATANABE, H. (1991) NCAP liquid crystals for large-area projection screens, *Soc. Inf. Disp. Dig.* **XXII**, 598–601.

KITZEROW, H.-S. (1994) Polymer dispersed liquid crystals: From the nematic curvilinear aligned phase to ferroelectric films, *Liq. Cryst.* **16**, 1–31.

KITZEROW, H.-S., MOLSEN, H. and HEPPKE, G. (1992) Linear electrooptic effects in polymer dispersed ferroelectric liquid crystals, *Appl. Phys. Lett.* **60**, 3039–3095.

KRALJ, S., ŽUMER, Z. and ALLENDER, D. W. (1991) Nematic-isotropic phase transition in a liquid crystal droplet, *Phys. Rev. A* **43**, 2943–2952.

KREUZER, M., TSCHUDI, T., DE JEU, W. H. and EIDENSCHINK, R. (1993) A new bistable liquid crystal display with bistability and selective erasure using scattering in filled nematics, *Appl. Phys. Lett.* **62**, 1712–1714.

KUNIGA, M., YIRAI, Y., OOI, Y., NIIJAMA, S., ASAHAWA, T., MASUMO, K., KUMAI, H., YUKI, M. and GUNJIMA, T. (1990) A full-color projection TV using LC/polymer composite light valves, *Soc. Inf. Disp. Dig.* **XXI**, 227–230.

KURIK, M. V. and LAVRENTOVICH, O. D. (1982) Negative-positive monopole transitions in cholesteric liquid crystals, *Sov. Phys. JETP* **35**, 444–447 [Zh. Eksp. Teor. Fiz. **35**, 362].

LACKNER, A. M., MARGERUM, J. D., RAMOS, E., WU, S. T. and LIM, K. C. (1988) Contrast measurements for polymer dispersed liquid crystal displays, *Proc. SPIE* **958**, 73–79.

LEE, K. P., SUH, S.-W. and LEE, S. D. (1994) Fast linear electrooptic switching properties of polymer-dispersed ferroelectric liquid crystals, *Appl. Phys. Lett.* **64**, 718–720.

LEHMANN, O. (1904) Die Hauptstatze der Lehre von den Flussigen Kristallen, *Flussige Kristalle*, Leipzig.

MALLICOAT, S. (1992) Sequential color reflective display, *Soc. Inf. Disp. Dig.* **XXII**, 766–768.

MARGERUM, J. D., LACKNER, A. M., RAMOS, E., LIM, K. C. and SMITH, W.H. (1989) Effects of off-state alignment in polymer dispersed liquid crystals, *Liq. Cryst.* **5**, 1477–1487.

MARITAN, A., CIEPLAK, M., BELLINI, T. and BANAVAR, J. (1994) 'Nematic-isotropic transition in porous media, *Phys. Rev. Lett.* **72**, 4113–4116.

McCARGER, J. W., ONDRIS-CRAWFORD, R. J. and WEST, J. L. (1992) Polymer dispersed liquid crystal infrared light shutter, *J. Electron. Imaging* **1**, 22–28.

MELZER, D. and NABARRO, F. R. N. (1977) Optical studies of a nematic liquid crystal with circumferential surface orientation in a capillary, *Phil. Mag.* **35**, 901–906.

MEYER, R. B. (1969) Piezoelectric effects in liquid crystals, *Phys. Rev. Lett.* **22**, 918–921.

MEYER, R. B. (1973) On the existence of even indexed disclinations in nematic liquid crystals, *Phil. Mag.* **27**, 405–424.

MOLSEN, H., and KITZEROW, H.-S. (1994) Bistability in polymer dispersed ferroelectric liquid crystals, *J. Appl. Phys* **75**, 710–716.

MONTGOMERY, JR, G. P. (1988) Angle-dependent scattering of polarized light by polymer dispersed liquid crystal films, *J. Opt. Soc. Am. B* **5**, 774–784.

MONTGOMERY, JR, G. P. and VAZ, N. A. (1987) Contrast ratio of polymer dispersed liquid crystal films, *Appl. Opt.* **26**, 738–743.

MONTGOMERY, JR, G. P. and VAZ, N. (1989) Light scattering analysis of the temperature dependent transmitance of a polymer dispersed liquid crystal film in its isotropic phase, *Phys. Rev. A* **40**, 6580–6591.

NIYAMA, S., HIRAI, Y., KUNIGITA, M., KURMAI, H., WAKABAYASHI, T., IIDA, S. and GUNJIMI, T. (1993) Hysteresis and dynamic response effects on the image quality in a LCPC projection display, *Soc. Inf. Disp. Dig.* **XXIV**, 869–872.

ONDRIS-CRAWFORD, R. J., BOYKO, E. P., WAGNER, B. G., ERDMANN, J. H., ŽUMER, S. and DOANE, J. W. (1991) Microscope textures of polymer dispersed liquid crystals, *J. Appl. Phys.* **69**, 6380–6386.

ONDRIS-CRAWFORD, R. J., CRAWFORD, G. P., DOANE, J. W., ŽUMER, S., VILFAN, M. and VILFAN, I. (1992) Surface molecular anchoring in microconfined liquid crystals near the nematic-smectic A transition, *Phys. Rev. E* **48**, 1998–2005.

ONDRIS-CRAWFORD, R. J., CRAWFORD, G. P., ŽUMER, S. and DOANE, J. W. (1993) Curvature-induced configuration transition in confined liquid crystals, *Phys. Rev. Lett.* **70**, 194–197.

ONDRIS-CRAWFORD, R. J., AMBROŽIČ, M., DOANE, J. W. and ŽUMER, S. (1994) Pitch-induced transition of chiral nematic liquid crystals in submicrometer cylindrical cavities, *Phys. Rev. E* **50**, 4773–4779.

PALFFY-MUHORAY, P., LEE, M. A. and WEST, J. L. (1990) Optical field induced scattering in polymer dispersed liquid crystal films, *Mol. Cryst. Liq. Cryst.* **179**, 445–460.

PATEL, P., CHU, D., WEST, J. L. and KUMAR, S. (1994) Bistable switching in polymer dispersed ferroelectric smectic-C* displays, *Soc. Inf. Disp. Dig.* **XXV**, 845–847.

PFEIFFER, M., SUN, Y., YANG, D.-K., DOANE, J. W., SAUTTER, W., HOCHHOLZER, V., GINTER, E., LUEDER, E. and YANIV, Z. (1994) Design of PSCT materials for MIM addressing, *Soc. Inf. Disp. Dig.* **XXV**, 837–840.

PIRŠ, J., BLINC, R. MARIN, R., MUŠEVIČ, I., PIRŠ, S., ŽUMER, S. and DOANE, J. W. (1992) Volume stabilized ferroelectric liquid crystal display, *14th Int. Liquid Crystal Conf., Pisa, Italy, 1992, Book of Abstracts*, p. 278.

PIRŠ, J., BLINC, R., MARIN, B., MUŠEVIČ, I. and PIRŠ, S., (1995) Volume stabilized ferroelectric liquid crystal, *Liq. Cryst.* **264**, 2176–2185.

PIRŠ, J., OLENIK, M., MARIN, B., ŽUMER, S. and DOANE, J. W. (1991) Color modulator based on polymer dispersed liquid crystal light shutter, *J. Appl. Phys.* **68**, 3826–3831.

POLAK, R. D., CRAWFORD, G. P., KOSTIVAL, B. C., ŽUMER, S. and DOANE, J. W. (1994) Optical determination of the saddle-splay constant K_{24} in nematic liquid crystals, *Phys. Rev. E* **49**, R978–R981.

PRESS, M. J. and AROTT, A. S. (1974) Theory and experiment on configurations with cylindrical symmetry in liquid crystal droplets, *Phys. Rev. Lett.* **33**, 403–406.

REAMEY, R., MALLOY, K. and MONTOYA, W. (1992) Voltage-holding in nematic droplet-polymer films: Contributions of matrix polymer, liquid crystal, and morphology, *Soc. Inf. Disp. Dig.* **XXIII**, 796–772.

SAUPE, A. (1973) Disclinations and properties of the director fields in nematic and cholesteric liquid crystals, *Mol. Cryst. Liq. Cryst.* **21**, 211–238.

SCHARKOWSKI, A., CRAWFORD, G.P., ŽUMER, S. and DOANE, J. W. (1993) A method for the determination of the elastic constant ratio K_{33}/K_{11} in nematic liquid crystals, *J. Appl. Phys.* **73**, 7280–7287.

SCHMIEDEL, H., STANNARIUS, R., FELLER, G. and CRAMER, Ch.(1994) Experimental evidence of a conic helical liquid crystalline structure in cylindrical microcavities, *Liq. Cryst.* **17**, 323–332.

SHENG, P. (1976) Phase transition in surface-aligned nematic films, *Phys. Rev. Lett* **37**, 1059–1062.

SHENG, P.(1982) Boundary layer phase transition in nematic liquid crystals, *Phys. Rev. A* **26**, 1610–1617.

SUTHERLAND, R. L., NATARAJAN, L. V., TONDIGLIA, V. P. and BUNNING, T. J. (1993) Bragg gratings in an acrylate polymer consisting in periodic polymer dispersed liquid crystal planes, *Chem. Mater.* **5**, 1533–1538.

TAHIZAWA, K., KIKUCHI, H. and FUJIKAKE, H. (1991) NCAP liquid crystals for large area projection screens, *Soc. Inf. Disp. Dig.* **XXIV**, 598–601.

TANAKA, K., KATO, K., TSURU, S. and SAKAI, S. (1994) Holographically formed liquid crystal-polymer device for reflective color displays, *J. Soc. Inf. Disp.* **2**, 37–40.

TOMITA, A.(1993) Status of projection-type polymer-dispersed LCDs, *Soc. Inf. Disp. Dig.* **XXIV**, 865–868.

TOMITA, A. and JONES, P. (1992) Projection displays using nematic dispersions, *Soc. Inf. Disp. Dig.* **XXIII**, 579–582.

TRIPATHI, S., ROSENBLATT, C. and ALIEV, F. M. (1994) Orientational susceptibility in porous glass near a bulk nematic isotropic phase transition, *Phys. Rev. Lett.* **72**, 2725–2728.

VAZ, N. A., SMITH, G. W. and MONTGOMERY, G. P. (1987) A light control film composed of liquid crystal droplets dispersed in an epoxy matrix, *Mol. Cryst Liq. Cryst.* **146**, 17–34.

VILFAN, M., RUTAN, V., ŽUMER, S., LAHAJNAR, G., BLINC, R., DOANE, J.W. and GOLEMME, A.(1988) Proton spin lattice relaxation in nematic microdroplets, *J. Chem. Phys.* **89**, 597–604.

VINOUZE, B., BOZE, D., GUILBERT, M. and TRUBERT, C. (1994) Active matrix with polymer dispersed liquid crystal for reflective direct view liquid crystal displays, *Proc. Int. Display Research Conf.*, pp. 472–475.

VOLOVIK, G. E. and LAVRENTOVICH, O. D. (1983) Topological dynamics of defects: Boojums in nematic droplets, *Sov. Physics JETP* **85**, 1159–1166 [*Zh. Eksp. Teor. Fiz.* **85** 1997–2010].

WEST, J. L.(1988) Phase separation of liquid crystals in polymers, *Mol. Cryst. Liq. Cryst.* **157**, 427–441.

WEST, J. L.(1994) Liquid crystal dispersions, *Technological Applications of Dispersions*, ed. R. B. McKay, Ch. 10, New York: Marcel Dekker.

WEST, J. L. and ONDRIS-CRAWFORD, R. (1991) Characterization of polymer dispersed liquid crystal shutters by ultraviolet/visible and infrared absorption spectrum, *J. Appl. Phys.* **70**, 3785–3790.

WEST, J. L., AKINS, R. B., FRANCL, J. and DOANE, J. W.(1993) Cholesteric/polymer dispersed light shutters, *Appl. Phys. Lett.* **63**, 1471–1473.

WHITEHEAD, JR, J. B., ŽUMER, S. and DOANE, J. W. (1993) Light scattering from a dispersion of aligned nematic droplets, *J. Appl. Phys.* **73**, 1057–1065.

WILLIAMS, C. E., PIERANSKI, P. and CLADIS, P. E. (1972) Nonsingular S = +1 screw disclination lines in nematics, *Phys. Rev. Lett.* **29**, 90–92.

WU, B.-G., ERDMANN, J. H. and DOANE, J. W. (1989) Response times and voltages for PDLC light shutters, *Liq. Cryst.* **5**, 1453–1465.

WU, I-WEI (1994) High-definition displays and technology trends for TFT-LCDs, *J. Soc. Inf. Disp.* **2**, 1–14.

WU, X.-I. GOLDBURG, W. I. and LIU, M. X. (1992) Slow dynamics of isotropic-nematic phase transition in silica gels, *Phys. Rev. Lett.* **69**, 470–473.

YAMAGUCHI, R. and SATO, S. (1991) Memory effects of light transmission properties in polymer dispersed liquid crystal (PDLC) films, *Jpn. J. Appl. Phys.* **30**, L616–L618.

YAMAGUCHI, R., OOKAWARA, H. and SATO, S. (1992) Thermally addressed polymer dispersed liquid crystal displays, *Jpn. J. Appl. Phys.* **31**, L1093–L1095.

YANG, D.-K., CHIEN, L.-C. and DOANE, J. W. (1992) Cholesteric liquid crystal/polymer dispersions for haze-free light shutters, *Appl. Phys. Lett.* **60**, 3102–3104.

YANG, D.-K., WEST, J. L., CHIEN, L.-C. and DOANE, J. W. (1994) Control of reflectivity and bistability displays using cholesteric liquid crystal, *J. Appl. Phys.* **76**, 1331–1333.

YANIV, Z., DOANE, J. W., WEST, J. L. and TAMURA-LIS, W. (1989) Active matrix polymer dispersed liquid crystal display, *Proc. Japan Display 89*, pp. 572–575.

ZIDANŠEK, A., KRALJ, S., LAHAJNAR, G. and BLINC, R. (1994) Deuterium NMR study of liquid crystals confined in aerogel matrices, *Phys. Rev. E* **51**, 3332–3340.

ŽUMER, S. (1988) Light scattering from nematic droplets, *Phys. Rev. A* **37**, 4006–4015.

ŽUMER, S. and DOANE, J. W. (1986) Light scattering from a small nematic droplet, *Phys. Rev. A* **34**, 3373–3386.

ŽUMER, S., GOLEMME, A. and DOANE, J. W. (1989) Light extinction in a dispersion of small nematic droplets, *J. Opt. Soc. Am. A*, **6** 403–411.

ZYRYANOV, V. YA., SMORGON, S. L. and SHABANOV, V. F. (1992) Polymer-dispersed ferroelectric liquid crystals as display materials, *Soc. Inf. Disp. Dig.* **XXIII**, 776–777.

ZYRYANOV, V. YA., SMORGON, S. L. and SHABANOV, V. F. (1993) Electrooptics of polymer dispersed liquid crystals, *Ferroelectrics* **143**, 271–276.

2

Surface Anchoring of Nematic Liquid Crystals

G. BARBERO and G. DURAND

2.1 Introduction

Liquid crystals have been a fascinating subject of investigation since their discovery, at the end of last century. To start with, the interest was purely academic and seemed to fade away after the Second World War. In the 1960s, under pressure from engineers who demonstrated the potential of liquid crystals for displays, interest was rekindled, sustained by the curiosity of condensed matter scientists for intermediate ordered systems. For 20 years, most of the effort was devoted to the understanding of bulk properties; surface effects were oversimplified (the 'strong anchoring' hypothesis) and not fully analysed. In the 1980s, the development of new displays using mixed systems as polymer-dispersed liquid crystals (PDLCs) completely changed the scope: in these confined geometries, surface effects tend to dominate the bulk ones. In addition, the randomness of the systems suggested the use sometimes of a more stochastic description. In this chapter, we try to give a phenomenological unified description of the surface anchoring of nematic liquid crystals using a Landau-like model. To simplify, we restrict our analysis to planar interfaces between the nematic material and a solid substrate. We show that these boundaries can give anchorings of various strengths, not only through 'contact' forces, but with long-range electric effects as well. We introduce a section on stochastic anchorings, and we finally discuss the so-called K_{13} paradox which seems to remain a subject of discussion. Because of our unification, the presentation may appear a little formal, but the physical problems discussed are very practical. We think that, in the future, mastering weak and multistable anchorings could lead to the development of new, fast, bistable surface controlled displays, as recently demonstrated (Barberi et al., 1991, 1992, 1993).

2.2 Elastic Free-energy Density for Nematics in the Harmonic Approximation

Nematic liquid crystals (NLCs) formed by rod-like molecules are materials exhibiting spontaneous (ferroelectric) quadrupolar ordering. If α is a unit vector along each rod molecule, the polar order $\langle \alpha \rangle$ averaged on all molecules is zero since the nematic is not a dipolar ferroelectric. To define the mean orientation, we need to look at the mean quadrupolar order defined by

$$\frac{3}{2} \langle \alpha\alpha - \mathbf{I}/3 \rangle \tag{2.1}$$

where **I** is the identity tensor. We find that this molecular average is a tensor oriented along the 'director' **n** ($n^2 = 1$), which can be written as

$$\frac{3}{2}\langle \alpha\alpha - I/3 \rangle = \frac{3}{2} S\langle nn - I/3 \rangle \tag{2.2}$$

It is usual (but incorrect) to use **n** ($\equiv -n$) to define the local molecular orientation of the NLC. S is the scalar uniaxial order parameter (de Gennes and Prost, 1993). In what follows, we use the traceless tensor order parameter

$$Q_{ij} = (3/2)S(n_i n_j - (\tfrac{1}{3})\delta_{ij}) \tag{2.3}$$

having quadrupolar structure (de Gennes, 1969). If Q_{ij} is position independent the NLC state is called fundamental. For position-dependent Q_{ij} similarly, as in the classical theory of elasticity, it is possible to write an 'elastic' energy density (Alexe-Ionescu et al., 1993a). Let F be the free-energy density of the NLC phase. In order to separate the uniform part from the spatially dependent one, F may be considered a function of the kind

$$F = F(Q_{ij}, Q_{ij,k}) \tag{2.4}$$

where $Q_{ij,k} = \partial Q_{ij}/\partial x_k$ are the spatial derivatives of the tensor order parameter elements. In the case in which the Q_{ij} change slowly in space, the $Q_{ij,k}$ are small quantities. Consequently it is possible to rewrite (2.4) in the form

$$F(Q_{ij}, Q_{ij,k}) = F_u(Q_{ij}, 0) + F_e(Q_{ij,k}) \tag{2.5}$$

where $F_u(Q_{ij}, 0)$ is the free-energy density of the uniform NLC, whereas $F_e(Q_{ij,k})$ represents the elastic contribution to the total free-energy density. As is well known the uniform part of F depends only on the temperature. It can be written in Landau's approximation in order to study the nematic–isotropic phase transition (Priestley et al., 1976). By supposing that the NLC distortion is completely described in terms of $Q_{ij,k}$, the elastic contribution to F can be expanded in power series of the deformation tensor ($Q_{ij,k}$). In the harmonic approximation $F_e(Q_{ij,k})$ can be written as

$$F_e(Q_{ij,k}) = A_{ijk}Q_{ij,k} + \tfrac{1}{2}B_{ijklmn}Q_{ij,k}Q_{lm,n} + O(3) \tag{2.6}$$

where $O(3)$ represents terms which are, at least, of the third order in $Q_{ij,k}$. An expression of the kind (2.6) is used in the elastic theory of solid materials. In that case $Q_{ij,k}$ is substituted by the deformation tensor U_{ij} (Landau and Lifschitz, 1976). In the electrostatic theory, instead of $Q_{ij,k}$ the components of the electrostatic field E_i appear (Evangelista and Barbero, 1994). In thermal problems connected with thermal conductivity the role of the deformation tensor is played by the temperature gradient (Poincaré, 1888). In (2.6) the tensors **A** and **B** describe the elastic properties of the NLC. If $A \neq 0$, the ground state of the NLC is deformed. This kind of NLC is called cholesteric. The equivalent of tensor **A** in the elastic theory is connected with thermal dilation. In electrostatics this quantity is connected with the ferroelectric polar properties of the dielectric. The elastic tensors **A** and **B** have to be decomposed in terms of the elements of symmetry of the NLC under consideration. This means that in the bulk **A** and **B** have to be decomposed in terms of the tensor order parameter (**Q**), the identity tensor (δ) and antisymmetric tensor (ε). As stated before, in the bulk for the usual NLC, $A \equiv 0$. In the framework $S = \text{constant}$, $F_e(Q_{ij,k})$ reduces to the well-known Frank expression (Frank, 1958)

$$F_e = \tfrac{1}{2}[K_{11}(\text{div } n)^2 + K_{22}(n \text{ rot } n)^2 + K_{33}(n \times \text{rot } n)^2]$$
$$- K_{24} \text{div}(n \text{ div } n + n \times \text{rot } n) \tag{2.7}$$

We stress that in the case in which S is position independent the deformation tensor $Q_{ij,k}$ is

$$Q_{ij,k} = S(n_i n_{j,k} + n_j n_{i,k}) \tag{2.8}$$

i.e. $\nabla Q \propto \nabla n$. This means that if S is constant (2.6) is equivalent to

$$F_e(n_{i,j}) = \tfrac{1}{2}K_{ijlm}n_{i,j}n_{l,m} \tag{2.9}$$

which is quadratic in the spatial derivatives of the NLC director. In other words, linear terms in second-order derivatives of **n** are absent. Hence the term connected to K_{24} in (2.7) is quadratic in $n_{i,j}$. It is only apparently of second order in the spatial derivatives. Expression (2.9) holds only if $n_{i,j}$ are small quantities. This implies that **n** has to change over macroscopic regions. In the case where **n** changes abruptly over mesoscopic regions, (2.9) no longer represents a good approximation for $F(n_{i,j})$. In this situation it is necessary to extend (2.9). This can be done in two different ways. First, by considering in (2.9) anharmonic terms of the kind (Landau and Lifschitz, 1976)

$$\tfrac{1}{3}C_{ijklmn}n_{ij}n_{k,l}n_{m,n} \tag{2.10}$$

In this manner the basic equations of the problem are no longer linear. Second, by characterizing the NLC deformation not only in terms of $n_{i,j}$, but also in terms of $n_{i,jk}$, in this case terms of the kind

$$R_{ijk}n_{i,jk} + \tfrac{1}{2}M_{ijklmn}n_{i,jk}n_{l,mn} \tag{2.11}$$

also appear in (2.9). This procedure is standard in the elastic theory of solid materials (Nye, 1957; Alexe-Ionescu *et al.*, 1993a, Barbero and Durand, 1993; Alexe-Ionescu, 1993). We stress that in (2.11) the term quadratic in $n_{i,jk}$ plays the role of a saturation term. It is necessary in order to avoid divergences in the total free energy. Its use will be shown later when discussing the effect of the K_{13} splay bend curvature.

2.3 The Intrinsic and Extrinsic Parts of the Surface Free-energy Density

In the previous section we analysed the bulk free-energy density of the NLC separating it into uniform and elastic contributions. Of course real NLC samples are limited by a closed surface Σ. Near the surface the elastic tensors **A**, **B**, **C**, **R**, **M**, ... introduced in section 2.2 have to be decomposed also in terms of the elements of symmetry characterizing the substrate. In the harmonic approximation the spatial variation of Q_{ij} over mesoscopic lengths is supposed to be very small. This implies that the surplus of free-energy density due to the presence of the surface may be considered as a function of Q_{ij} only. Hence near the limiting surface there is a surplus of free-energy density which can be approximated as

$$\Delta F_s = \Delta F_s(Q_{ij}) \tag{2.12}$$

in the harmonic approximation. By supposing the substrate to be flat and isotropic (see Figure 2.1), the element of symmetry characterizing the surface is its geometrical normal **k**. In this framework from (2.12) one obtains

$$\Delta F_s = \beta_1 k_i Q_{ij}k_j + \beta_2 k_i Q_{ij}Q_{je}k_e + \beta_3 Q_{ij}Q_{ij} + \beta_4(k_i Q_{ij}k_2)^2 + \dots \tag{2.13}$$

to second order in the tensor order parameter (Goossen, 1985; Sluckin and Poniewierski, 1984; Barbero *et al.*, 1991). In (2.13) the phenomenological parameters β_i take into account the presence of the surface. They are temperature independent.

As is clear from the discussion reported above ΔF_s is partially due to the reduced symmetry of the NLC near the substrate and partially due to the direct interaction between the NLC and the substrate. The first part will be called the intrinsic part; the second is the extrinsic contribution. It is possible to separate ΔF_s in the two contributions by means of the following simple picture. Let ρ_n be the range of intermolecular forces yielding the NLC phase of the order of the size of a molecule. A molecule at a distance $z > \rho_n$ from the wall is submitted to a complete interaction with other NLC molecules (see Figure 2.2(a)). On the contrary if $z \sim \rho_n$ the interaction of the considered molecule with other NLC molecules is incomplete. Hence

23

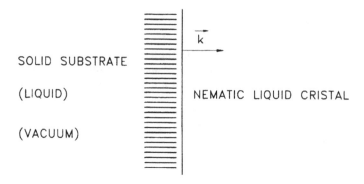

Figure 2.1 An NLC limited by a flat isotropic substrate, as a liquid. The element of symmetry characterizing the substrate is its geometrical normal **k**.

there exists a surface layer of molecular thickness ρ_n in which NLC molecules are submitted to an incomplete interaction (see Figure 2.2(b)). Furthermore a direct interaction between the NLC and the substrate may also exist, owing to van der Waals' interaction, or steric interaction. If ρ_S is the interaction range of these forces, the molecules in the surface layer $0 \leq z \leq \rho_S$ (see Figure 2.3) are submitted also to the direct interaction. ρ_S is of the order of ρ_n. Hence it is possible to define the surface energy in the following manner:

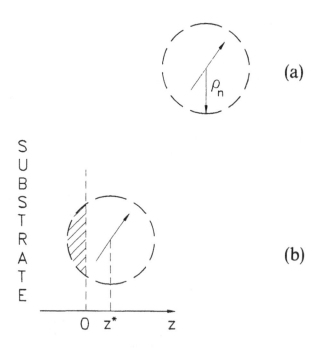

Figure 2.2 (a) NLC molecule in the bulk. ρ_n indicates the range of the intermolecular forces responsible for the NLC phase. ρ_n is of the same order of magnitude as the molecular dimension, since the contact forces are expected to be the most important. A molecule in the bulk has a complete interaction with the other NLC molecules (the interaction 'sphere' is complete). (b) NLC molecule near the limiting surface. If $z < \rho_n$ the interaction 'sphere' is incomplete. The shaded region does not contribute to the NLC–NLC interaction. The thickness of the surface layer is of the order of ρ_n, i.e. of molecular size.

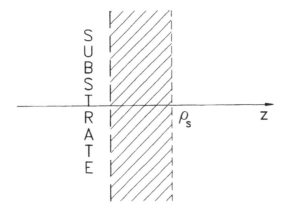

Figure 2.3 An NLC limited by a substrate. ρ_s is the range of the forces connected to the direct NLC–substrate interaction. Only the NLC molecules in the surface layer $0 \leqslant z \leqslant \rho_s$ are submitted to the direct interaction with the substrate. ρ_s is expected to be of the order of ρ_n.

$$f_S = \int_0^\rho \Delta F_S \, dz \tag{2.14}$$

where $\rho \sim \rho_n \sim \rho_s$. Since ρ is expected to be of the order of molecular length, in ΔF_S the tensor order parameter Q_{ij} may be considered to be nearly position independent. Consequently, by substituting (2.13) into (2.14) one has

$$F_S = \alpha_1 k_i Q_{ij} k_j + \alpha_2 k_i Q_{ij} Q_{je} k_e + \alpha_3 Q_{ij} Q_{ij} + \alpha_4 (k_i Q_{ij} k_i)^2, \tag{2.15}$$

where

$$\alpha_i = \int_0^\rho \beta_i \, dz \tag{2.16}$$

We stress that (2.15) is meaningful only if ρ is very small. This implies that in the direct interaction between the substrate and the NLC long-range interactions are excluded, as in the electrostatic one. These long-range contributions have to be analysed separately, as will be shown later when, for instance, the ionic contribution to the surface energy will be considered.

In the absence of bulk distortions, expression (2.15) shows that the NLC surface orientation depends on the temperature (Sluckin and Ponciewierski, 1984; Barbero *et al.*, 1991; Alexe-Ionescu *et al.*, 1992). This effect, known as the temperature surface transition, has been observed by different groups (Bouchiat and Langevin-Cruchon, 1971; Faetti and Fronzoni, 1978, Chiarelli *et al.*, 1983, 1984; di Lisi *et al.*, 1990; Flatischler *et al.*, 1992; Komitov *et al.*, 1992; Barbero *et al.*, 1993; Patel and Yokoyama, 1993). We stress that in the absence of bulk distortion (i.e. for $Q_{ij,k} \equiv 0$) from (2.15) we can deduce the surface easy tensor order parameter $\mathbf{Q}^e$ defined as the surface order parameter minimizing f_S. It is defined by

$$\frac{\partial f_S}{\partial Q_{ij}} = 0 \qquad \left(\frac{\partial^2 f_S}{\partial Q_{ij} \, \partial Q_{kl}} \right)_{Q_{ij} = Q_{ij}^e} > 0 \tag{2.17}$$

For an actual surface order parameter Q_{ij}^S not very different from Q_{ij}^e the surface energy f_S may be approximated by

$$f_S(Q_{ij}^S) = f_S(Q_{ij}^e) + \tfrac{1}{2} \tilde{W} (Q_{ij}^S - Q_{ij}^e)^2 \tag{2.18}$$

recently proposed by Nobili and Durand (1992). Expression (2.18) generalizes an old expression proposed by Rapini and Papoular long ago in 1969. According to them the presence of a surface is equivalent to a surface field. In the hypothesis of constant scalar order parameter, $f_S(n_i^S)$, where n_i^S is the surface NLC director, may be approximated by

$$f_S(n_i^S) = f_S(n_i^e) - \tfrac{1}{2}W(\mathbf{n}_S \cdot \mathbf{n}_e)^2 \tag{2.19}$$

where W is the surface anchoring strength (Barbero and Barberi, 1983). Expression (2.19) is similar to the one describing the magnetic free energy of an NLC in an external magnetic field (de Gennes and Prost, 1993).

As an example, imagine a solid plate imposing at its surface a uniaxial order S_0 along the direction $\mathbf{n}_0$ (for instance, in its plane). The generalized surface energy for an NLC of surface order S and surface orientation $\mathbf{n}$ would be written as

$$W_S = \tfrac{1}{2}W(\mathbf{Q} - \mathbf{Q}_0)^2$$

with $\mathbf{Q} = \tfrac{3}{2}S(\mathbf{nn} - \mathbf{I}/3)$ and $\mathbf{Q}_0 = \tfrac{3}{2}S_0(\mathbf{n}_0\mathbf{n}_0 - \mathbf{I}/3)$. W_S is then

$$W_S = \tfrac{9}{8}W[\tfrac{2}{3}S^2 + \tfrac{2}{3}S_0^2 - 2SS_0(\cos^2\theta - 1/3)]$$

where $\cos\theta = \mathbf{n}\cdot\mathbf{n}_0$. For $\mathbf{n} = \mathbf{n}_0$, W_S is just proportional to $(S - S_0)^2$, as it should be. For $\mathbf{n} \neq \mathbf{n}_0$, on the other hand, the optimum surface order parameter is no longer $S = S_0$ but is given by $dW_S/dS = 0$, i.e.

$$S = \tfrac{3}{2}S_0(\cos^2\theta - 1/3)$$

At the 'magic' angle $\theta = \cos^{-1}(1/3)$, $S = 0$ and the surface does not tend to induce any ordering. For a completely 'broken' angular anchoring $\mathbf{n}\perp\mathbf{n}_0$ (under the action of a strong field) $S = -S_0/2$. This is easy to understand. The nematic director must fulfil two apparently contradictory conditions: to be along $\mathbf{n}$ perpendicular to $\mathbf{n}_0$ but to place as many molecules along the surface 'field' $\mathbf{n}_0$. The general solution should be biaxial but in a uniaxial model, the way to do it is to choose a negative order parameter $-S_0/2$. The molecules are then distributed radially in a plane perpendicular to $\mathbf{n}_0$ (negative S), and half of them are on the average aligned along $\mathbf{n}_0$.

In general, putting back the optimum S into W_S, one obtains Rapini–Papoular (RP) terms in $\cos^2\theta$, plus a new term $\cos^4\theta$; the RP expression is just an approximation for small deviations $\theta < 1$. For large angular deviations, one should observe in W_S higher harmonic terms from θ–S coupling, as in the experiment carried out by Nobili and Durand (1992). One does observe a surface disordering [a change in S induced by the surface disorientation (change in θ)], leading to a new $\sin 4\theta$ term in the surface torque $dW_S/d\theta$. The effect is more visible with 'storing' anchoring, since the surface ordering results not from W_S alone, but from the action of the bulk ordering on the surface. For 'weak' anchoring, the RP approximation is good enough. 'Strong' and 'weak' anchoring are usually defined by comparing W with K/ξ, where ξ is the coherence length associated with the nematic–isotropic transition. $W \sim K/\xi$ is strong and $W \ll K/\xi$ is weak. K/W is the usual surface extrapolation length, which should be comparable with ξ for strong anchoring or be larger than ξ for weak anchoring.

2.4 Variational Problem to Find the Director Field

In the harmonic approximation the total free energy of an NLC sample of volume τ limited by a surface Σ is given, in the hypothesis $S = $ constant, by

$$G[\mathbf{n(r)}] = \iiint_\tau f(n_i, n_{i,j})\, d\tau + \oiint_\Sigma f_S(n_i)\, d\Sigma \tag{2.20}$$

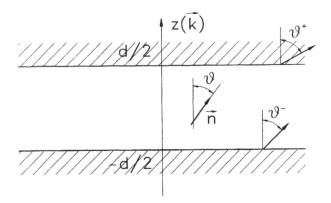

Figure 2.4 NLC sample limited by two flat parallel surfaces (slab). The NLC director is supposed to be everywhere parallel to the (x, z) plane. θ is the tilt angle between **n** and the **z** axis. ϑ^+ and ϑ^- are the surface tilt angles.

The first contribution in (2.20) is connected with the elastic deformation and with the free energy of the NLC in external fields; it is a bulk term. The second contribution is due to the presence of the limiting surface Σ. It represents the surface contribution to the total free energy. According to thermodynamics, the actual NLC orientation is the one minimizing (2.20). In the simple case where the sample is a slab of thickness d and **n** is everywhere parallel to the plane, we have $\mathbf{n} = \mathbf{n}(z)$, where z is the axis normal to the limiting walls (see Figure 2.4). By indicating the tilt angle as $\theta \equiv \cos^{-1}(\mathbf{n.k})$ and the surface of the limiting walls as A, the total free energy per unit surface is given by

$$g[\theta(z)] = \frac{G}{A} = \int_{-d/2}^{+d/2} F(\theta, \theta')\, dz + f_s^{(-)}(\theta^-) + f_s^{(+)}(\theta^+) \tag{2.21}$$

where $\theta' = d\theta/dz$ and $f_s^{\pm}(\theta^{\pm})$ represent the surface energies. Furthermore $\theta^{\pm} = \theta(\pm d/2)$ are the surface tilt angles. The first variation of $g[\theta(z)]$ is

$$dg[\theta(z)] = \int_{-d/2}^{d/2} \left(\frac{\partial F}{\partial \theta} - \frac{d}{dz}\frac{\partial F}{\partial \theta'} \right) \omega(z)\, dz + \left(\frac{-\partial F}{\partial \theta'} + \frac{\partial f_s^-}{\partial \theta^-} \right) \omega(-d/2)$$

$$+ \left(\frac{\partial F}{\partial \theta'} + \frac{\partial f_s^{(+)}}{\partial \theta^+} \right) \omega(d/2) \tag{2.22}$$

where $\omega(z)$ is an arbitrary function of C_1 class, i.e. continuous with its first-order derivative (Elsgolts, 1977). The requirement that $\delta g[\theta(z)] \equiv 0$, for all $\omega(z) \in C_1$, implies that (Elsgolts, 1977) $\theta(z)$ satisfies the Euler–Lagrange equation

$$\frac{\partial F}{\partial \theta} - \frac{d}{dz}\frac{\partial F}{\partial \theta'} = 0 \tag{2.23}$$

in the bulk ($-d/2 \leq z \leq d/2$) and the boundary conditions

$$-\frac{\partial F}{\partial \theta'} + \frac{\partial f_s^{(-)}}{\partial \theta^-} = 0 \quad \text{at } z = -d/2 \tag{2.24}$$

$$\frac{\partial F}{\partial \theta'} + \frac{\partial f_s^{(+)}}{\partial \theta^+} = 0 \quad \text{at } z = +d/2 \tag{2.25}$$

In the one constant approximation F is given by

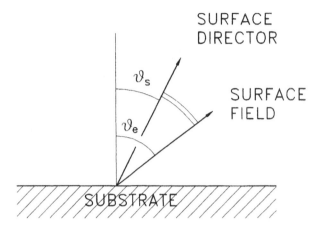

SURFACE
DIRECTOR

SURFACE
FIELD

ϑ_s

ϑ_e

SUBSTRATE

Figure 2.5 A solid substrate limiting an NLC. θ_e is the easy tilt angle. It is defined as the tilt angle minimizing the surface energy. θ_s is the actual tilt angle. θ_s depends, in general, on the bulk distortion.

$$F(\theta, \theta') = \tfrac{1}{2}K\theta'^2 + f(\theta) \tag{2.26}$$

where $f(\theta)$ takes into account the free energy of the NLC with the external field, and $\tfrac{1}{2}K\theta'^2$ is obtained from (2.7) in the framework $K_{11} = K_{22} = K_{33}$. If we suppose, furthermore, that $f_s^{(\pm)}$ may be approximated by the Rapini–Papoular expression (2.19), it follows that

$$f_s^{(\pm)} = \tfrac{1}{2}W^{(\pm)} \sin^2(\theta^\pm - \theta_e^\pm) \tag{2.27}$$

where $W^{(\pm)} > 0$ and $\theta_e^{(\pm)}$ are the surface easy tilt angles (see Figure 2.5). With these simplifying hypotheses the bulk equation (2.23) and the boundary conditions (2.24 and 2.25) become

$$K\theta'' - \frac{df}{d\theta} = 0 \quad -d/2 \leq z \leq d/2 \tag{2.23'}$$

which represents the bulk equilibrium of the mechanical torque, and

$$-K\theta' + \tfrac{1}{2}W^{(-)} \sin2(\theta^- - \theta_e^-) = 0 \quad z = -d/2 \tag{2.24'}$$

$$-K\theta' + \tfrac{1}{2}W^{(+)} \sin2(\theta^+ - \theta_e^+) = 0 \quad z = d/2 \tag{2.25'}$$

which state that the bulk elastic torque is equilibrated by the surface restoring torques.

Note that (2.23') is a second-order differential equation. Its general solution contains two integration constants to be determined by means of the two boundary conditions (2.24') and (2.25'). Hence the problem is well posed from the mechanical and mathematical points of view. We can then conclude that in the usual harmonic approximation of the elastic theory of NLC there are no fundamental problems in determining the NLC director field $\mathbf{n}(z)$ or the tilt angle $\theta(z)$.

2.5 Rapini–Papoular Approximation for the Surface Energy

As stated in section 2.3 one of the more popular expressions for f_s was proposed long ago by Rapini and Papoular (1969). According to these authors, for small deviations of the surface director from the easy axis, f_s may be approximated by (2.19) that from now on we write as

$$f_s^{RP} = -\tfrac{1}{2}W(\mathbf{n}_s \cdot \boldsymbol{\pi})^2 \tag{2.28}$$

indicating the easy axis by π and neglecting the isotropic part $f_S(\pi)$. As has been already stressed, (2.28) has been written in analogy with the magnetic interaction energy. This means that W, the anchoring energy strength, represents some kind of NLC anisotropy with respect to the 'surface field'. Expression (2.28) has been used by several authors to explain experimental results concerning thin NLC samples (Jérôme, 1991). However, deviations of f_S from f_S^{RP} have been observed in

(1) very thin hybrid nematic cells of very thin thickness (Barbero and Barberi, 1983; Elsgolts, 1977; Jérôme, 1991; Blinov *et al.*, 1991);

(2) homogeneous samples oriented by means of SiO evaporation (Yang *et al.*, 1982; Yokoyama and Van Sprang, 1985);

(3) homogeneous samples oriented by means of Langmuir–Blodgett films (Barbero and Petrov, 1994).

Furthermore other experimental investigations show that W in (2.28) seems to depend on the thickness of the sample (Yang *et al.*, 1982; Yokayama and Van Sprang, 1985, Barbero and Petrov, 1994; Blinov and Kabaenkov, 1987; Blinov and Sonin, 1984; Chuvyrov *et al.*, 1976, Churyrov and Lachirov, 1978; Chuvyrov, 1980; Kaznacheev and Sonin, 1983; Sonin and Kaznacheev, 1984).

The experimental observations reported above suggest that f_S^{RP} could be only a rough approximation of f_S. However, we will show that if the problems are correctly analyzed the situation is better. f_S^{RP} works relatively well for large deviations of $\mathbf{n}_S$ from π. This will be shown in the next sections.

2.6 Flexoelectric Contribution to the Surface Energy

Apparent deviations from f_S^{RP} have been observed in NLC samples in hybrid alignment. A cell of this kind (Figure 2.6) is characterized by opposite boundary conditions on the two limiting walls. In Figure 2.6 the lower surface has an easy axis which is homeotropic (π_H) whereas the upper one is planar (π_p). A cell of this kind has been studied in detail in the past because the strong distortion is not caused by an external field, but only with the thickness of the sample (Barbero and Barberi, 1983; Warenghem, 1984; Barbero and Simoni, 1984; Madhusudana and Durand, 1985; Hochbaum and Labes, 1982; Perez *et al.*, 1978; Proust and Ter-Minassian-Saraga, 1979; Proust and Perez, 1977; Dozov *et al.*, 1982, 1983; Barbero, 1984). In particular this cell has been used to study the dependence of f_S on the deviation of the surface director from the easy axis (Barbero *et al.*, 1984b).

The tilt angle profile in a hybrid cell may be easily obtained in the frame of the elastic theory of NLC described before. In this case the basic equations to solve are (2.23), (2.24) and (2.25). In the simple case in which the twist deformation is absent and $K_{11} \neq K_{33}$ the elastic free energy density is given by

$$F(\theta, \theta') = \tfrac{1}{2}K_{33}(1 - k \sin^2\theta)\theta'^2 \qquad (2.29)$$

where $k = 1 - (K_{11}/K_{33})$ is the elastic anisotropy. In the experimental arrangement (Barbero *et al.*, 1984a,b) the anchoring energy on the planar surface was weak, whereas the one on the homeotropic surface was strong. In this framework the total free energy per unit surface of the hybrid cell is

$$g[\theta(z)] = \int_0^d \tfrac{1}{2}K_{33}(1 - k \sin^2\theta)\theta'^2 \, dz + f_S(\theta_S) \qquad (2.30)$$

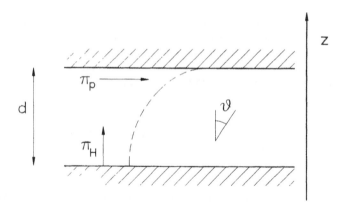

Figure 2.6 Hybrid aligned nematic cell. It is characterized by two opposite easy directions. The bulk distortion is due to the boundary conditions. In the case of weak anchoring energy the surface tilt angles depend on the thickness *d* of the sample.

instead of (2.21). In (2.30), $\theta_S = \theta(d)$ is the surface tilt angle on the planar surface, whereas $\theta(0) = 0$ due to the strong anchoring hypothesis (see Figure 2.6). The bulk equation (2.23) in this case is equivalent to

$$(1 - k \sin^2\theta)\theta'^2 = C^2 \quad 0 \le z \le d \tag{2.31}$$

where C is an integration constant. From (2.31) one deduces that

$$C = (1/d)I_k(0, \theta_S) \tag{2.32}$$

for the integration constant, and

$$I_k[0, \theta(z)]/I_k(0, \theta_S) = z/d \tag{2.33}$$

for the tilt angle $\theta(z)$. In (2.32) and (2.33) we have put

$$I_k(0, \theta) = \int_0^\theta \sqrt{1 - k \sin^2\mu} \, d\mu \tag{2.34}$$

The tilt angle on the planar wall, with weak anchoring energy, is determined by means of the boundary condition (2.25), which can be written as

$$\frac{K_{33}}{d}\sqrt{1 - k \sin^2\theta} \, I_k(0, \theta_S) + \frac{df_S(\theta_2)}{d\theta_S} = 0 \tag{2.35}$$

In the experiments of Barbero *et al.* (1984a,b) the tilt angle distribution is determined by using a wedge geometry, in order to have a continuous variation of the controlling field, i.e. of the thickness of the sample (see Figure 2.7). The tilt angle distribution is obtained by measuring the integrated path difference. Knowing $\theta(z)$, and hence θ_S, it is possible to obtain some information on the surface energy $f_S(\theta_S)$.

For a light beam polarized along the direction bisecting the **x** and **y** axes and normally incident on the cell, the path difference Δl between its **x** and **y** components, after it merges from the NLC sample, can be written as

$$\Delta l = \int_0^d \Delta n_{\text{eff}}(\theta) \, dz \tag{2.36}$$

where $\Delta n_{\text{eff}}(\theta) = n_{\text{eff}}(\theta) - n_o$ and $n_{\text{eff}}^{-2}(\theta) = \cos^2\theta/n_o^2 + \sin^2\theta/n_e^2$, n_o and n_e being the ordinary and extraordinary refractive indices, respectively. After some algebra it is possible to rewrite (2.36) in the form

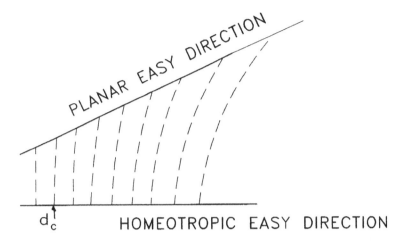

PLANAR EASY DIRECTION

HOMEOTROPIC EASY DIRECTION

Figure 2.7 Wedge geometry used in Barbero and Barberi (1983) and Barbero *et al.* (1984b) to study the dependence of the surface energy against the deviation of the director from the easy axis. The homeotropic alignment is strong and it is due to a silane treatment. For $d < d_c = K_{33}/W_P$ the sample is homogeneously homeotropic (W_P = anchoring energy on the planar wall). In this region, from the energy point of view, it is more convenient to break the anchoring on the planar wall than to have an elastic deformation (Barbero *et al.*, 1984b).

$$\Delta l = n_o \int_0^d \{[1 - R \sin^2\theta(z)]^{-1/2} - 1\} \, dz \tag{2.37}$$

with $R = 1 - (n_o/n_e)^2$. By using (2.32) and (2.33) we obtain finally that Δl is given by

$$\Delta l = n_o d\{[J_k(0, \theta_S)/I_k(0, \theta_S)] - 1\} \tag{2.38}$$

where

$$J_k(0, \theta_S) = \int_0^{\theta_S} [(1 - k \sin^2\mu)/(1 - R \sin^2\mu)]^{1/2} \, d\mu$$

Equation (2.38) gives Δl as a function of θ_S. A measurement of the optical path difference can then give information on θ_S and also on $f_S(\theta_S)$. Reported in Figure 2.8 are the experimental data of Δl and the theoretical prediction with a Rapini–Papoular expression for $f_S(\theta_S)$. The liquid crystal used is MBBA, with $k = 0.37$ (Oldano *et al.*, 1984) and $R = 0.239$ (Barbero *et al.*, 1977). As is evident the agreement is not good at all. However, in the previous analysis which coincides with the one published (Barbero and Barberi, 1983; Warenghem, 1984; Barbero *et al.*, 1984c; Madhusudana and Durand, 1985; Hochbaum *et al.*, 1982; Perez and Proust, 1978; Proust and Ter-Minassian-Saraga, 1979; Proust and Perez, 1977; Dozov *et al.*, 1982, 1983; Barbero, 1984) we have neglected the fact that NLCs are *flexoelectric*. This means that distorted NLCs exhibit an electric polarization proportional to the distortion ($n_{i,j}$) (Meyer, 1969). This phenomenon is the equivalent of the piezoelectricity in solid materials (Nye, 1957). The flexoelectric polarization is given by

$$P_i = d_{ijk} n_{j,k} \tag{2.39}$$

where d_{ijk} are the elements of the flexoelectric tensor. By taking into account the fact that NLCs are not polar ferroelectric, and hence **n** is equivalent to $-\mathbf{n}$, the flexoelectric tensor can be decomposed in the following manner:

$$d_{ijk} = \gamma_1 n_i n_j n_k + \gamma_2 n_i \delta_{jk} + \gamma_3 n_j \delta_{ik} + \gamma_4 n_k \delta_{ij} \tag{2.40}$$

31

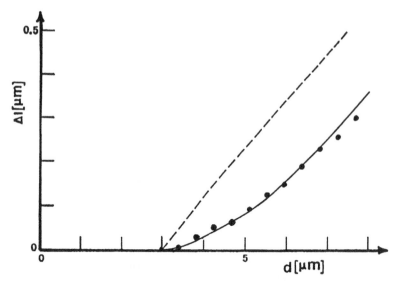

Figure 2.8 Experimental data of Barbero and Barberi (1983) and Barbero *et al.* (1984b) for the optical path diference Δ*l* against the thickness *d* of the sample for a hybrid cell. The dashed line is the theoretical curve according to the Rapini–Papoular approximation for the surface energy.

where γ_i are phenomenological parameters. Since $n_i n_i = 1$, $n_i n_{i,j} = 0$. It follows that by substituting (2.40) into (2.39) one obtains

$$P_i = \gamma_2 n_i n_{j,j} + \gamma_4 n_k n_{i,k} \qquad (2.41)$$

but $n_i n_{j,j} = n_i \text{ div } \mathbf{n}$ and $n_k n_{i,k} = -(\mathbf{n} \times \text{rot } \mathbf{n})_i$. Consequently (2.41) may be rewritten in covariant form as

$$\mathbf{P} = e_{11} \mathbf{n} \text{ div } \mathbf{n} - e_{33} \mathbf{n} \times \text{rot } \mathbf{n} \qquad (2.42)$$

where $e_{11} \equiv \gamma_2$ and $e_{33} \equiv \gamma_4$ are the flexoelectric coefficients (Madhusudana and Durand, 1985; Dozov *et al.*, 1982, 1983; Meyer, 1969).

Now we will show that the flexoelectric polarization changes the elastic anisotropy of the NLC. By taking into account the flexoelectric contribution to the total free energy, it is possible to show that the experimental data reported in Figure 2.8 can be fitted by supposing that $f_S = f_S^{RP}$ on the planar wall. With this aim let us evaluate the electric field connected with $\mathbf{P}$ given by (2.42). In this case

$$D_i = \varepsilon_{ij} E_j + 4\pi P_i \qquad (2.43)$$

where $\mathbf{D}$ is the dielectric displacement, $\mathbf{E}$ the intrinsic field, $\mathbf{P}$ the flexoelectric polarization and ε the dielectric tensor. The condition rot $\mathbf{E} = 0$ and the invariance of the problem for translations parallel to the $\mathbf{x}$ or $\mathbf{y}$ axes give $E_x = E_y = 0$. Then $D_y = 0$. By observing that $D_i = D_i(z)$, the condition div $\mathbf{D} = 0$, valid in the hypothesis that the NLC is a perfect insulator, gives

$$D_z = \text{constant} \qquad (2.44)$$

and this constant is zero because of the continuity of D_z. It follows from (2.43) that the bulk electric field in the sample is

$$E_z(z) = 2\pi \frac{e}{\varepsilon_{33}(\theta)} \sin(2\theta) \, \theta' \qquad (2.45)$$

where $\varepsilon_{33}(\theta) = \varepsilon_{\parallel} \cos^2\theta + \varepsilon_{\perp} \sin^2\theta$, $\varepsilon_{\parallel}$ and $\varepsilon_{\perp}$ being the dielectric coefficients parallel and perpendicular to **n**, $e = e_{11} + e_{33}$ (Deuling, 1974; Helfrich, 1974). The electric field (2.45) and the flexoelectric polarization (2.47) give an additional term in the bulk NLC free-energy density of the kind $-\frac{1}{2}P_z E_z$. From the previous calculations the total free energy per unit surface is given by

$$g[\theta(z)] = \int_0^d \frac{1}{2}K_{33}\chi(\theta)\theta'^2 \; dz + f_s(\theta_s) \tag{2.46}$$

where

$$\chi(\theta) = 1 - k \sin^2\theta + (k_f/4) \sin^2(2\theta)/[1 - (\varepsilon_a/\varepsilon_{\parallel})\sin^2\theta] \tag{2.47}$$

In (2.47) k is the above introduced elastic anisotropy, $\varepsilon_a = \varepsilon_{\parallel} - \varepsilon_{\perp}$ the dielectric anisotropy and

$$k_f = 4\pi \frac{e^2}{K_{33}\varepsilon_{\parallel}}$$

a new parameter describing the influence of the flexoelectric properties on the elastic behaviour of the NLC. By assuming $|e| \sim 7.5 \times 10^{-4}$ cgs units (Madhusudana and Durand, 1985; Marcerou and Prost, 1980; Dozov et al., 1984), $K_{33} \sim 0.7 \times 10^{-6}$ cgs units (Oldano et al., 1984) and $\varepsilon_{\parallel} \sim 4.7$ cgs units (Diguet et al., 1970) we obtain $k_f \sim 2$. It follows that the 'effective' elastic anisotropy is found to be of the order of $-k_f$, i.e. it is negative and very large. By considering (2.46) it is possible to re-analyse the experimental data of Figure 2.8, by supposing that $f_s = f_s^{RP} = \frac{1}{2}W \cos^2\theta_s$ on the planar wall (Barbero and Durand, 1986). Operating in the same manner used at the beginning of this section, one obtains that θ_s is given by

$$d/L = 2\tilde{I}_k(0, \theta_s)\sqrt{\chi(\theta_s)}/\sin(2\theta_s) \tag{2.48}$$

which is the equivalent of (2.35) when the flexoelectricity is taken into account and $f_s = f_s^{RP}$. In (2.48) $L = K_{33}/W$ is the so-called extrapolation length. Furthermore

$$\tilde{I}_k(0, \theta) = \int_0^\theta \sqrt{\chi(\theta)} \; d\theta \tag{2.49}$$

is the equivalent of $I_K(0, \theta)$ defined in (2.34). Now the optical path difference is found to be

$$\Delta l = n_0 d\{[\tilde{J}_k(0, \theta_s)/\tilde{I}_k(0, \theta_s)] - 1\} \tag{2.50}$$

where

$$\tilde{J}_k(0, \theta_s) = \int_0^{\theta_s} \left(\frac{\chi(\theta)}{1 - R \sin^2\theta}\right)^{1/2} d\theta$$

corresponds to $J_k(0, \theta)$ introduced before. In Figure 2.9 Δl is shown as a function of d for different values of k_f. The liquid crystal is MBBA and we assume $L = 3 \; \mu m$. We stress that Δl is very sensitive to k_f variations: $k_f = 0$ is the prediction of Rapini–Papoular without flexoelectricity presented above; $k_f = 2$ corresponds to $|e| \sim 6.5 \times 10^{-4}$ cgs units and $k_f = 4$ corresponds to $|e| \sim 9.5 \times 10^{-4}$ cgs units. On the contrary Δl is practically independent of k for k in the range $0 \leq k \leq 0.5$. Figure 2.10 shows the experimental data obtained on two samples having critical thicknesses (Barbero et al., 1984a,b; Warenghem, 1984) $L_1 = 1.8 \; \mu m$ and $L_2 = 3 \; \mu m$. The full curves are obtained from (2.50) with the material parameters of MBBA and $k_f = 3.5$, which implies $|e| \sim 8 \times 10^{-4}$ cgs units. This value is consistent with the one reported in Madhusudana and Durand (1985), Marcerou and Prost (1980) and Dozov et al. (1984).

In our analysis we have supposed that the NLC is a perfect insulator. This means that the Debye screening length λ_D is assumed to be very large with respect to the thickness of the sample. We now verify that λ_D is comparable with the extrapolation length. Therefore λ_D can be expressed by

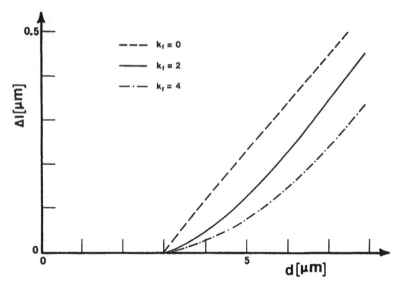

Figure 2.9 Optical path difference Δl against the thickness d of the sample in a hybrid cell with strong anchoring on the homeotropic wall for different values of $K_f = 4\pi e^2/K_{33}\varepsilon_{\parallel}$. The curve with $K_f = 0$ corresponds to the Rapini–Papoular prediction when the flexoelectricity is neglected.

$$\lambda_D^2 = D\tau$$

where D is the diffusion coefficient for ions in the NLC and τ the charge relaxation time. τ is observed as the maximum frequency at which one can induce electrohydrodynamical instabilities (Durand, 1976). Depending on the material conductivity, τ ranges from 10^{-1} to 10^{-3} s. D is of the order of 10^{-5} to 10^{-6} cm^2/s. With these values, one can estimate λ_D to

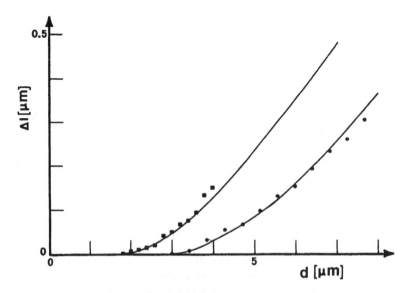

Figure 2.10 Best fit of the experimental data for Δl against d with the flexoelectric model.

range from 0.3 μm up to 10 μm. The anchoring extrapolation length L is known to range from 0.1 μm for strong anchoring up to a few micrometres for weak anchoring. One sees that λ_D can be comparable with (or larger than) L. Consequently for Δl near the critical thickness, the hypothesis of the perfect insulator works well.

To conclude this section we can state that apparent deviations from the f_S^{RP} observed in the hybrid cell can be explained by the fact that in the theoretical analysis the flexoelectricity had been neglected. We have shown that the main effect of the flexoelectric polarization on the elastic behaviour of the NLC is to introduce a very large (and negative) apparent elastic anisotropy. Taking into account the effective elastic anisotropy, it seems that the Rapini–Papoular expression works reasonably well.

2.7 Order Electric Contribution to the Surface Energy

Apparent deviations from the Rapini–Papoular model for the surface energy have been observed also in NLC samples oriented by means of SiO oblique evaporation (Yang and Rosenblatt, 1982; Yokoyama and Van Sprang, 1985). In these studies the thicknesses of the samples are usually very large with respect to λ_D and hence the flexoelectric effect is expected to play a small role. In fact for $d \gg \lambda_D$, the NLC can be considered as a conducting medium. In this case the flexoelectric charges are completely screened by the ions present in the NLC itself. However, in this case it is possible to show that by re-analysing in a correct way the experimental data of the above studies, the Rapini–Papoular expression works well. To do this we have just to remember that NLCs are ferroelectric quadrupolar materials. The density of the electrical quadrupolar tensor q_{ij} is proportional to the tensor order parameter of elements Q_{ij} (Barbero *et al.*, 1986, Durand, 1990, Osipov and Sluckin, 1992; Sluckin, 1995). Hence

$$q_{ij} = -\tilde{e}Q_{ij} \tag{2.51}$$

where $\tilde{e}$ is the quadrupolar density. $\tilde{e}$ is of the order of $-e$ introduced before, but is independent of temperature for $T \sim T_c$, the nematic–isotropic critical temperature, because the temperature effect is absorbed in Q_{ij}. By substituting expression (2.3) into (2.51) we obtain

$$q_{ij} = \tfrac{3}{2}\tilde{e}S(n_i n_j - \tfrac{1}{3}\delta_{ij}) \tag{2.52}$$

As is well known from electrostatic theory, a spatial variation of the electric quadrupolar tensor is a polarization given by

$$P_i = -q_{ij,j} \tag{2.53}$$

and, using (2.52),

$$P_i = \tfrac{3}{2}\tilde{e}(n_{i,j}n_j + n_i n_{j,j})S + \tfrac{3}{2}\tilde{e}(n_i n_j - \tfrac{1}{3}\delta_{ij})S_{,j} \tag{2.54}$$

It is possible to rewrite (2.54) in a covariant form

$$\mathbf{P} = \tfrac{3}{2}\tilde{e}\{S(\mathbf{n}\,\mathrm{div}\,\mathbf{n} - \mathbf{n} \times \mathrm{rot}\,\mathbf{n}) + [\mathbf{n}(\mathbf{n}\cdot\nabla S) - \tfrac{1}{3}\nabla S]\} \tag{2.55}$$

The first term on the RHS of (2.55) coincides with the flexoelectric polarization reported in (2.42) when $e_{11} = e_{33} = \tfrac{3}{2}\tilde{e}S$. The second term is new. We can conclude, from (2.55), that a distorted NLC exhibits an electric polarization which is partially due to

(1) spatial variation of $\mathbf{n}$, at $S = $ constant;
(2) spatial variation of S, at $\mathbf{n} = $ constant.

The contribution (1) is the usual flexoelectric polarization; the contribution (2) has been called 'order electric' polarization (Barbero *et al.*, 1986; Durand, 1990; Osipov and Sluckin, 1992; Sluckin, 1995).

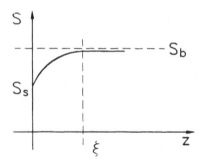

Figure 2.11 Spatial variation of the scalar order parameter with respect to z. For $z \to \infty$, $S \to S_b$, the bulk value of S. It depends only on the temperature of the system. For $z \to 0$, $S \to S_s$, it depends on the temperature, on the physical properties of the substrate and on the roughness of the surface. ξ is the NLC coherence length in the nematic phase.

Let us now consider the experimental situation used by Yang and Rosenblatt (1982) and Yokoyama and Van Sprang (1985). Owing to the SiO evaporation, the scalar order parameter on the surface is expected to be lower than the one in the bulk (Barbero and Durand, 1991a,b). It relaxes to its bulk value over the coherence length ξ; **n** practically does not change. This means that **P** given by (2.55) reduces to the order electric contribution

$$\mathbf{P}^{o} = \tfrac{3}{2}\tilde{e}[\mathbf{n}(\mathbf{n}\cdot\nabla S) - \tfrac{1}{3}\nabla S] \tag{2.56}$$

Since $S = S(z)$ (see Figure 2.11), only P_z is different from zero, and equal to

$$P_z^{o} = e(\cos^2\theta - \tfrac{1}{3})S' \tag{2.52}$$

where $e \sim \tfrac{3}{2}\tilde{e}$ and $S' = dS/dz$. By taking into account the fact that λ_D is usually very large with respect to ξ, the NLC may be considered as an insulating medium. Hence, in a similar way to the previous section, we have

$$D_z = \varepsilon_{zz}E_z + 4\pi P_z^{o} = 0$$

from which

$$E_z = -\frac{4\pi}{\varepsilon_{zz}}P_z^{o}$$

The dielectric free energy density connected to $\mathbf{P}^{o}$, localized in the surface layer of thickness ξ, is hence

$$-\tfrac{1}{2}P_z E_z = \frac{1}{2}\frac{4\pi}{\varepsilon_{zz}}e^2(\cos^2\theta - \tfrac{1}{3})^2 S'^2 \tag{2.58}$$

Since $\xi \ll d$, the thickness of the sample, it is possible to define an 'equivalent' surface energy as

$$\int_0^\infty -\frac{P_z E_z \, dz}{2} = \frac{4\pi}{2}\frac{e^2}{\varepsilon_{zz}}(\cos^2\theta - \tfrac{1}{3})^2 \frac{(S_b - S_s)^2}{\xi} \tag{2.59}$$

where S_b and S_s are the bulk and surface values of the scalar order parameter (see Figure 2.11). It follows that, even in the Rapini–Papoular approximation for the surface free energy, the total surface free energy is

$$\tfrac{1}{2}W \sin^2(\theta_S - \theta_e) + \frac{4\pi e^2}{\varepsilon_{33}}(\cos^2\theta - \tfrac{1}{3})^2 \frac{(S_b - S_s)^2}{\xi} \tag{2.60}$$

Equation (2.60) shows that the total surface energy contains a contribution in $\cos^4\theta$, beside the usual one in $\cos^2\theta$. As shown in Barbero *et al.* (1986) by means of (2.60) it is possible to find a good agreement between the experimental data of Yang and Rosenblatt (1982) and Yokoyama and Van Sprang (1985) and the theoretical predictions.

2.8 Stochastic Contribution to the Anchoring Energy

In sections 2.6 and 2.7 we have seen that possible deviations from the Rapini–Papoular expression for f_s can be caused by the flexoelectric or order electric effect. The above-mentioned calculations are based on the hypothesis that the surfaces are macroscopically homogeneous. Now we want to show that the random nature of the surface also introduces a fourth-order term as $(\mathbf{n_s}.\boldsymbol{\pi})^4$ in the effective surface energy, even if the flexoelectricity and the ferroelectric quadrupolar nature of the NLC are neglected.

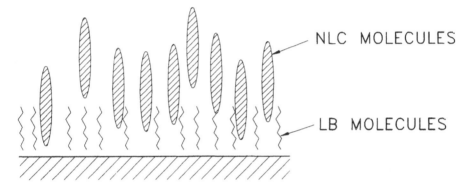

Figure 2.12 NLC oriented by means of an LB film. The surface NLC molecules are then oriented like the LB molecules. Then, they orient the bulk of the NLC by means of the anisotropic interaction which characterizes the nematic phase.

For the sake of simplicity, we consider a simple model in which the orienting film, like a Langmuir–Blodgett (LB) film, is monomolecular and formed by rod-like molecules (Figure 2.12). The mean orientation of the film is due to the interaction between the molecules forming the film itself and the substrate over which the film is deposited. Furthermore we suppose that the NLC orientation follows that of the orienting film. This is equivalent to supposing that the film is not compact. The molecules of the NLC may enter the holes (free places) present in the structure of the film (see Figure 2.12). In this manner the NLC molecules in contact with the film are oriented by steric interaction with the molecules of the film. Then this first layer of NLC molecules orients the bulk NLC by means of the anisotropic intermolecular interaction characterizing the NLC phase. This model of the LB film–NLC interface was originally proposed by Hiltrop and Stegemeyer (1978). It has been recently reconsidered by Komitov *et al.* (1994) and by Alexe-Ionescu *et al.* (1993b) to analyse temperature-induced surface transitions in NLC.

Let us consider a monomolecular film formed by rod-like molecules of length l. Let $\mathbf{m}$ be the direction of the molecular major axis. Owing to the interaction of the molecules and the substrate we have a film with two-dimensional order. One end of each rod is attached to the solid substrate in a quasi-regular pattern, whereas the other end is free. In this model, a molecule of the film may bend by an angle θ under the action of the direct interaction connected with the other molecules of the film (see Figure 2.13). In order to obtain the elastic energy for the monomolecular film, the starting hypothesis is the assumption of a two-body interaction of the kind

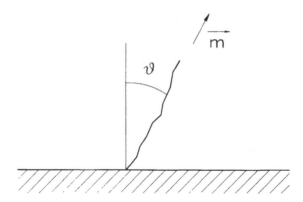

Figure 2.13 LB molecule in contact with the solid substrate. It interacts with other molecules of the film and with molecules of the substrate.

$$u(\mathbf{m}, \mathbf{m}', \mathbf{r})\, d\Sigma\, d\Sigma' \tag{2.61}$$

between the surface elements $d\Sigma\, d\Sigma'$, where $\mathbf{m}$ and $\mathbf{m}'$ are the direction of the molecular major axes at $\mathbf{R}$ and $\mathbf{R}'$, respectively, and $\mathbf{r} = \mathbf{R}' - \mathbf{R}$.

Following a standard procedure (Barbero *et al.*, 1991) from (2.61) the energy density of the film $f(\mathbf{R})$ at the point $\mathbf{R}$ is obtained by integrating $\frac{1}{2}u$ over the surface (mean field approximation). Simple calculations give (Alexe-Ionescu *et al.*, 1994a)

$$f(\mathbf{R}) = f_0(\mathbf{R}) + k^{LB}_{ijmn}m_{i,j}m_{m,n} \tag{2.62}$$

similar to (2.9). In (2.62) k^{LB}_{ijmn} are the elements of the surface elastic constants of the LB film, and $f_0(\mathbf{m})$ is the surface energy density of a uniformly oriented film ($\mathbf{m}$ position independent). Of course the film is not isolated, because it is in contact with the solid substrate. Hence we have to take into account also the direct interaction between the molecules of the film and the molecules of the substrate. In the hypothesis of the isotropic solid substrate, the surface free energy due to the direct interaction is of the kind $\psi(\mathbf{m}, \mathbf{k})$, where $\mathbf{k}$ is the geometrical normal to the substrate. This energy depends on the van der Waals interaction, dielectric interaction, and so on. It also depends on the physical properties of the solid substrate. In the ideal case of a homogeneous surface, ψ is position independent, but as is well known, real surfaces are never homogeneous. More precisely, on average they have approximately the same properties, but from point to point they change in a more or less stochastic manner. Hence we can write

$$\psi(\mathbf{m}\cdot\mathbf{k}, \mathbf{r}) = \psi_{av}(\mathbf{m}\cdot\mathbf{k}) + \delta\psi(\mathbf{m}\cdot\mathbf{k}, \mathbf{r}) \tag{2.63}$$

where $\delta\psi(\mathbf{m}\cdot\mathbf{k}, \mathbf{r})$ takes into account the stochastic part of the direct interaction between the film and the solid substrate (due, for instance, to free ions or local irregularities). The total surface energy (due to the intrinsic part f_0 and the direct interaction) is then

$$\gamma = f_0(\mathbf{m}\cdot\mathbf{k}) + \psi(\mathbf{m}\cdot\mathbf{k}, \mathbf{r}) = \gamma_0(\mathbf{m}\cdot\mathbf{k}) + \delta\psi(\mathbf{m}\cdot\mathbf{k}, \mathbf{r}) \tag{2.64}$$

where $\gamma_0(\mathbf{m}\cdot\mathbf{k}) = f_0(\mathbf{m}\cdot\mathbf{k}) + \psi_{av}(\mathbf{m}\cdot\mathbf{k})$ is the uniform part of the total surface energy. It represents the anisotropic part of the surface tension of the film–substrate interface. Let us imagine that the uniform film on the considered surface tends to be in homeotropic alignment (i.e. $\mathbf{m} \parallel \mathbf{k}$). In a first approximation we assume for $\gamma_0(\mathbf{m}\cdot\mathbf{k})$ the Rapini–Papoular expression

$$\gamma_0(\mathbf{m}\cdot\mathbf{k}) = -\tfrac{1}{2}W(\mathbf{m}\cdot\mathbf{k})^2$$

The total surface energy is then given by

$$\gamma = -\tfrac{1}{2}W(\mathbf{r})(\mathbf{m}\cdot\mathbf{k})^2 \tag{2.65}$$

where

$$W(\mathbf{r}) = W + \Delta W(\mathbf{r}) \tag{2.66}$$

in which $\Delta W(\mathbf{r})$ is the stochastic contributions to $W(\mathbf{r})$. Let us suppose now that $\mathbf{m}$ is everywhere parallel to the $(\mathbf{x}, \mathbf{z})$ plane. This hypothesis is very restrictive but, since we want only to analyze the renormalization of the surface energy introduced by the stochastic component of the direct film–substrate interaction, it can be reasonably accepted. By indicating with θ the angle made by $\mathbf{m}$ with $\mathbf{k}$, the total energy of the film is given by

$$\phi = \iint_{\Sigma} \left[\tfrac{1}{2}K_{\mathrm{LB}}(\nabla\theta)^2 + \tfrac{1}{2}W(\mathbf{r})\sin^2\theta\right]\,dz\,d\Sigma \tag{2.67}$$

where z is the surface of the film, $W(\mathbf{r})$ is given by (2.66) and $\nabla = \mathbf{i}(\partial/\partial x) + \mathbf{j}(\partial/\partial y)$. In (2.67) K_{LB} takes into account the elastic properties of the film (see eqn. (2.62)). The equilibrium distribution of $\mathbf{m}$ in the film minimizes ϕ given by (2.67), and thus we obtain

$$K_{\mathrm{LB}}\nabla^2\theta - W(\mathbf{r})\sin\theta\cos\theta = 0 \tag{2.68}$$

where $\nabla^2 = \partial^2/\partial x^2 + \partial^2/\partial y^2$. By extracting the fluctuating part of $W(\mathbf{r})$, as we have done in (2.66), and putting $\theta(\mathbf{r}) = \Theta + \delta\theta(\mathbf{r})$, where Θ is position independent, we can linearize (2.68). In this limit eqn (2.68), in operational form, is written as

$$\hat{L}\,\nabla^2\theta(\mathbf{r}) = \frac{\Delta W(\mathbf{r})}{2K_{\mathrm{LB}}}\sin 2\Theta \tag{2.69}$$

where $\hat{L} = \Delta - \alpha\cos 2\Theta$ and $\alpha = W/K_{\mathrm{LB}}$; it is expected that $\alpha^{-1} \sim l_0^2$ where l_0 is the diameter of the LB molecules. $\delta\theta(\mathbf{r})$ may be determined by means of the Green function. Simple calculations give

$$\delta\theta(\mathbf{r}) = \int G(\mathbf{r}, \mathbf{r}')\,\frac{\Delta W(\mathbf{r}')}{2K_{\mathrm{LB}}}\sin 2\Theta\,d\mathbf{r}'$$

where $G(\mathbf{r}, \mathbf{r}')$ is the Green function of (2.69). Let us suppose that

$$\langle h(\mathbf{r})\,h(\mathbf{r}')\rangle = D_\gamma\exp[-(|\mathbf{r} - \mathbf{r}'|)/R_\gamma]$$

where $h(\mathbf{r}) = \Delta W(\mathbf{r})/W$ is the dispersion, and R_γ is the correlation length of the random distribution $h(\mathbf{r})$. This kind of correlation function satisfies the fundamental property of stochastic systems:

$$\lim_{r\to\infty}\langle h(\mathbf{r})\,h(0)\rangle = 0$$

Taking into account the previous correlation function, the effective surface energy defined by

$$\phi_{\mathrm{eff}} = \frac{1}{\Sigma}\int_{\Sigma}\phi(\mathbf{r})\,d\Sigma$$

is found to be

$$\phi_{\mathrm{eff}} = \tfrac{1}{2}W\sin^2\Theta + (2\pi)^2 D_\gamma(\alpha R_\gamma^2)K_{\mathrm{LB}}I(\alpha R_\gamma^2)\sin^2 2\Theta$$

where

$$I(\alpha R_\gamma^2) = \frac{1}{1 - \alpha R_\gamma^2} + \frac{\ln(\alpha R_\gamma^2)}{2(1 - \alpha R_\gamma^2)^{3/2}} - \frac{\ln[1 + (1 - \alpha R_\gamma^2)^{1/2}]}{(1 - \alpha R_\gamma^2)^{3/2}}$$

39

The previous expression for ϕ_{eff} clearly shows that the inclusion of a stochastic spatial variation of the surface field, caused by the direct interaction between the film and the substrate, gives rise to a new functional form of the effective surface energy. More precisely, the usual anchoring strength, in the Rapini–Papoular sense, is renormalized by

$$W_{\text{eff}} = W + \Delta W$$

where

$$\Delta W = \gamma(2\pi)^2 D_\gamma(\alpha R_\gamma^2) I(\alpha R_\gamma^2) W$$

Furthermore, a new term appears, proportional to $\sin^4\theta$ and characterized by a coefficient equal to ΔW, but with opposite sign. Until now we have limited our analysis to the energy of the film. However, as we have stressed above, in the hypothesis where the film is not compact the bulk orientation of the NLC is due to the steric interaction between the first NLC layer and the film. Consequently, the anchoring energy of the NLC coincides with the anchoring energy of the film. Of course the total surface energy of the NLC may contain, besides the steric term discussed above, other contributions due, for instance, to dispersion interactions. However, the steric one is usually the most important at the interface between the NLC and the film. Hence, at least for this kind of interface, the theory presented above is expected to work well.

We can also estimate the order of magnitude of ΔW introduced above. The surface elastic constant of the film may be obtained by multiplying the bulk elastic constant of the film $\tilde{K}$ by the thickness of the film l. We assume $\tilde{K}$ to be of the same order of magnitude as the NLC elastic constant: $\tilde{K} \sim 10^{-6}$ dyn (de Gennes and Prost, 1993) and $l \sim 10$ Å. Consequently $K_{\text{LB}} \sim \tilde{K} l \sim 10^{-13}$ erg. The anchoring energy is of the order of $W \sim 10^{-2}$ erg/cm². Furthermore, it follows from the equation defining the correlation function that

$$D_\gamma = \left(\frac{W}{K_{\text{LB}}}\right)^2 \left\langle \frac{\Delta W(\mathbf{r})}{W} \frac{\Delta W(0)}{W} \right\rangle = \left(\frac{W}{K_{\text{LB}}}\right)^2 \sigma$$

By assuming $D_\gamma \sim 10^{-1}$ and $I \sim 1$ a simple calculation gives $\Delta W \sim 10^{-2}$ erg/cm², i.e. of the same order of W. This simple estimation shows that the influence of the stochastic part may be very important.

2.9 Ion Adsorption and Surface Anchoring Strength

Recent experiments have shown that sometimes the surface energy strength seems to depend on the thickness of the sample (Blinov and Kabaenkov, 1987; Sonin and Kaznacheev, 1984). We shall show now that this thickness dependence of the surface energy strength is an artefact and can be explained by the electrostatic contribution of the adsorbed ions to the total energy of the NLC sample (Barbero and Durand, 1990a, b; Alexe-Ionescu et al., 1993c). Let us consider an NLC in thermal equilibrium. In the bulk the volume density of the positive ions is statistically equal to that of negative ions: $|\rho_0^+| = |\rho_0^-|$. When an NLC of finite thickness limited by two solid substrates is considered, the local electrical equilibrium is perturbed. In fact when the solid substrate is in contact with the NLC, a selective ion adsorption takes place. For instance, positive ions are attracted by the solid surface, whereas negative ions are repelled. In this situation a double layer, in the sense of Debye and Hückel (Israelachvili, 1991), of thickness of the order of the Debye screening length λ_{D}, exists near the bounding surfaces (see Figure 2.14). Here σ is the electric charge density adsorbed on the surface. The electric field near the surface is approximately given by

$$E(z) = 4\pi \frac{\sigma}{\varepsilon} \exp(-z/\lambda_{\text{D}}) \tag{2.70}$$

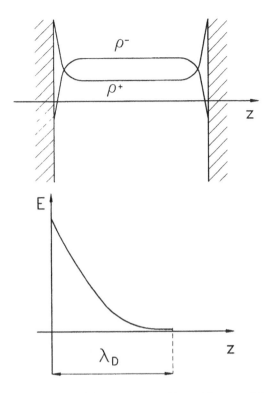

Figure 2.14 Equilibrium distributions of the ions in an NLC limited by two solid substrates. We suppose a selective adsorption of the positive ions (a). The surface density of adsorbed ions depends on the thickness of the sample. It gives rise to a surface electric field in two surface layers (b) whose thickness is of the order of the Debye screening length, λ_D.

in the continuous charge approximation. ε is the average relative dielectric constant of the NLC, and z is the distance of the considered point from the solid substrate. For symmetry the electric field is normal to the surface.

Since the NLC is an anisotropic material having $\varepsilon_\parallel \neq \varepsilon_\perp$ (where, as before, $\parallel$ and $\perp$ refer to **n**) the dielectric anisotropy $\varepsilon_a = \varepsilon_\parallel - \varepsilon_\perp$ is generally different from zero. The field given by (2.70) has therefore an orienting effect on the NLC (de Gennes and Prost, 1993). The related dielectric energy is

$$F_E = -\frac{1}{8\pi}\,\varepsilon_a(\mathbf{n}\cdot\mathbf{E})^2 = -\frac{1}{8\pi}\,\varepsilon_a E^2(z)\,\cos^2\theta \tag{2.71}$$

where θ is the tilt angle. The dielectric energy per unit area is obtained by integrating (2.72) from 0 to ∞. Taking into account (2.70) simple calculations give

$$f_E = \int_0^\infty F_E\,dz = -\pi\varepsilon_a(\sigma/\varepsilon)^2\,\lambda_D\,\cos^2\theta \tag{2.72}$$

with the hypothesis that θ is position independent. However, we want to stress that there is another term connected with a field of the kind given by (2.70). In fact, as is well known from electrostatic theory, a medium having a uniform electrical quadrupole density q_{ij} in a non-uniform field **E** has an electrical energy density given by

$$F_q = -q_{ij}E_{i,j} \tag{2.73}$$

where $E_{ij} = \partial E_i/\partial x_j$ is the spatial gradient of the electric field. As we have discussed above, for the NLC q_{ij} is given by (2.51). Consequently the energy term (2.73) for NLC materials can be written as

$$F_q = \tilde{e} Q_{ij} E_{i,j} \tag{2.74}$$

In the simplest case in which the dielectric field is of the kind $\mathbf{E} = E(z)\mathbf{k}$ where $E(z)$ is given by (2.70), F_q can be rewritten as

$$F_q = e(\cos^2\theta - \tfrac{1}{3})\frac{dE}{dz} \tag{2.75}$$

where $e = \frac{3}{2}\tilde{e}S$. The quadrupolar dielectric energy per unit surface is obtained by integrating (2.75) from 0 to ∞. By supposing, as before, the NLC orientation to be position independent, from (2.75) one obtains

$$f_q = \int_0^\infty F_q \, dz = -4\pi e \frac{\sigma}{\varepsilon}\cos^2\theta + \text{constant} \tag{2.76}$$

where the constant term is not important in our analysis, since it is independent of the NLC orientation. The total energy per unit surface, playing the role of 'effective' anchoring energy, is given by

$$f_{\text{eff}} = f_S + f_E + f_q \tag{2.77}$$

Let us now consider the case in which the easy axis is homeotropic. This means that in the Rapini–Papoular approximation $f_S^{RP} = -\tfrac{1}{2}W \cos^2\theta$ and (2.77) becomes

$$f_{\text{eff}} = -\tfrac{1}{2}W \cos^2\theta \tag{2.78}$$

where, from (2.72) and (2.76),

$$W_{\text{eff}} = W + 4\pi \frac{\sigma}{\varepsilon}\frac{\varepsilon_a}{2\varepsilon}(\lambda_D\sigma + 2e) \tag{2.79}$$

is the effective anchoring energy strength. It is important to stress that according to the sign of ε_a and e, W_{eff} may increase or decrease when the ion adsorption takes place. In the case where $\varepsilon_a < 0$ and the quadrupolar term is neglected, the ion adsorption gives rise to a destabilizing term, independently of the sign of the adsorbed charges, since this term is quadratic in σ. On the contrary, when the quadrupolar term is taken into account, the sign of σ is important, as discussed in Alexe-Ionescu et al. (1993c).

In Barbero et al. (1993b) by using a self-consistent method we have extended the Langmuir law valid for neutral particle adsorption (Kubo, 1967) to charged particles. In this manner we have shown that the adsorbed charge σ depends on the thickness of the NLC sample as (Barbero et al., 1993b)

$$\sigma = \Sigma \frac{d}{d + 2\lambda_D} \tag{2.80}$$

shown in Figure 2.15. In (2.80) σ depends on the conductivity of the NLC, on the adsorption energy and on the number of free sites $\Sigma/|q|$ on the surface ($|q| = $ electron charge). By substituting (2.80) in (2.79) we obtain the thickness dependence of the effective anchoring energy strength. It is important to stress that the two terms of electric origin appearing in (2.77), i.e. f_E and f_q, are usually of the same order of magnitude. In the case of 5CB analysed in (2.34) one has $\varepsilon_a \sim 13$, $\varepsilon \sim 12$ (Madhusudana and Pratibha, 1982), $e \sim 3 \times 10^{-3}$ cgs units (Maheswara-Murthy et al., 1993), $\sigma < 2$–3×10^2 esu/cm^2 (Maheswara-Murthy et al., 1993) and $\lambda_D \sim 10.6$ μm (Thurston et al., 1994). Consequently $f_q/f_E = 4e\varepsilon/(\varepsilon_a\sigma\lambda_D) > 1/2$, showing that the quadrupolar contribution to f_{eff} is of the same order as the ordinary dielectric

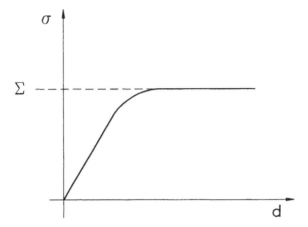

Figure 2.15 Surface density of adsorbed ions σ against the thickness of the sample d. For small d, σ is nearly proportional to d. On the contrary, for large d, σ tends to a saturation value, Σ, which depends on the adsorption energy and on the number of free sites on the surface.

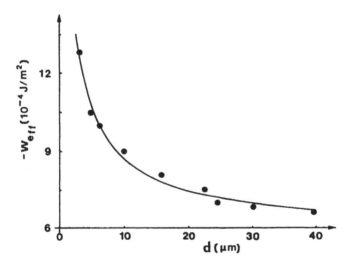

Figure 2.16 Experimental data by Blinov and Kabaenkov (1987) of W_{eff} against d concerning 5CB in the initial planar alignment and the theoretical curve obtained by means of the model based on the selective adsorption.

contribution. In Figure 2.16 the experimental data of Blinov and Kabaenkov (1987) are shown. The full curve is obtained by means of (2.80) and (2.79). The parameters of the best fit are $W = -2.2$ erg/cm^2, $\Sigma = 177$ esu/cm^2 and $e = 4.9 \times 10^{-3}$ cgs units. The agreement is fairly good. The example reported above shows that the thickness dependence of the anchoring energy strength may be interpreted by taking into account the selective ion adsorption.

2.10 Generalized Surface Energy for NLCs

In the previous sections we have shown that, in the harmonic approximation for the bulk free-energy density, the surface energy density may be considered as a local quantity. In other

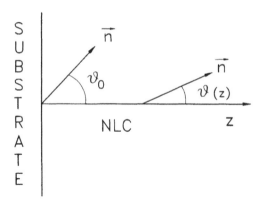

Figure 2.17   An NLC limited by a solid substrate placed at $z = 0$. The NLC occupies the half space $z > 0$.

words, if $F_e = F_e(Q_{ij,k})$, then $f_S = f_S(Q_{ij})$. In the framework $S = $ constant, if $F_e = F_e(n_{i,j})$, then $f_S = f_S(n_i)$. Some years ago Mada (1979) proposed, in the harmonic approximation for F_e, a surface free energy of the functional form $f_S = f_S(n_i, n_{i,j})$. This proposal was connected with the attempt to take into account the spatial variation of the elastic constants (Yokoyama *et al.*, 1987, Faetti, 1991; Alexe-Ionescu *et al.*, 1994b), but it is simple to show that the variational problem connected to the minimization of

$$G = \iiint_\tau F_e(n_{i,j}) \, d\tau + \oiint_\Sigma f_S(n_i, n_{i,j}) \, d\Sigma$$

is not well posed. To show this in detail let us consider an NLC sample occupying the half space $z > 0$. Let us suppose furthermore that **n** is parallel everywhere to the (**y, z**) plane (see Figure 2.17). In this case the total free energy per unit area of the NLC sample is

$$g[\theta(z)] = \frac{G}{A} = \int_0^\infty F(\theta, \theta') \, dz + f_S(\theta_0, \theta'_0) \tag{2.81}$$

where θ is the tilt angle and $\theta_0 = \theta(0)$, $\theta'_0 = d\theta/dz$. The first variation of (2.81), is given by

$$\delta g[\theta(z)] = \int_0^\infty \left(\frac{\partial F}{\partial \theta} - \frac{d}{dz} \frac{\partial F}{\partial \theta'} \right) \omega(z) \, dz$$

$$+ \left(-\frac{\partial F}{\partial \theta'} + \frac{\partial f_S}{\partial \theta_0} \right)_0 \omega(0) + \left(\frac{\partial f_S}{\partial \theta'_0} \right)_\infty \omega(0) + \lim_{z \to \infty} \left(\frac{\partial F}{\partial \theta'} \omega(z) \right) \tag{2.82}$$

If $\theta(z)$ extremizes $g[\theta(z)]$, then $\delta g[\theta(z)] \equiv 0$ for all $\omega(z)$ of the C_1 class. This implies that $\theta(z)$ is a solution of the Euler–Lagrange equation

$$\frac{\partial F}{\partial \theta} - \frac{d}{dz} \frac{\partial F}{\partial \theta'} = 0 \quad 0 \leq z < \infty \tag{2.83}$$

coinciding with (2.23), with the boundary conditions

$$-\frac{\partial F}{\partial \theta'} + \frac{\partial f_S}{\partial \theta_0} = 0 \tag{2.84}$$

$$\frac{\partial f_S}{\partial \theta'_0} = 0 \tag{2.85}$$

at $z = 0$, and

$$\frac{\partial F}{\partial \theta'} = 0 \qquad (2.86)$$

for $z \to \infty$. As discussed in section 2.4 eqn (2.83) is a second-order differential equation. Its general solution is of the kind $\theta(z) = \theta(z, C_1, C_2)$, i.e. it contains two integration constants. But now three boundary conditions have to be satisfied, represented by (2.84), (2.85) and (2.86). Consequently the problem is ill posed. The tilt angle $\theta(z)$ is, in general, discontinuous (Oldano and Barbero, 1985a, b; Barbero and Oldano, 1985). This discontinuity of $\theta(z)$ according to Hinov (1990) and Pergamenshchick (1993) is a mathematical artefact and hence physically not acceptable. To solve the problem connected with the minimization of functionals of the kind (2.81) they solved the differential equation (2.83), obtaining $\theta = \theta(z, C_1, C_2)$. After that they substituted this solution into (2.81). In this manner $g[\theta(z)]$ becomes an ordinary function at $g = g(C_1, C_2)$. The integration constants are then determined by minimizing g with respect to C_1 and C_2. As shown in Oldano and Barbero (1985a, b) and Barbero and Oldano (1985) the $\theta(z)$ determined in such a manner does not minimize (2.81). Moreover, it has been shown by Faetti (1994a, b) that the $\theta(z)$ determined by Hinov and Pergamenshchick, 1993) and does not describe an equilibrium situation, as the total torque is different from zero. In Barbero *et al.* (1989, 1993) and Barbero and Strigazzi (1989) it has been suggested to modify F entering into (2.81) in order to have a well-posed variational problem. According to the general recipe presented in section 2.2, if $f_S = f_S(\theta_0, \theta_0')$ this can be done considering $F = F(\theta, \theta', \theta'')$. A simple expression for the elastic energy density is

$$F = \tfrac{1}{2}K\theta'^2 + \tfrac{1}{2}K^*\theta''^2 \qquad (2.87)$$

where K^* is a new elastic constant. Expression (2.87) has been used by Barbero and coworkers to analyze the influence of surface anchoring energy depending on the gradient of the tilt angle on the NLC orientation. A simple analysis shows that $\theta(z)$ presents a sharp variation $\Delta\theta$ localized near the boundary surfaces over a length

$$l = \sqrt{K^*/K} \qquad (2.88)$$

The rapid angular variation $\Delta\theta$ is proportional to the actual surface tilt angle θ_S for small θ_S. Interestingly, $\Delta\theta$ is independent of l. In the limit of small molecular length l, $\Delta\theta$ could be considered as a surface discontinuity.

2.11 Intermolecular potential and elastic constants

We should stress that surface energies depending on n_i and $n_{i,j}$ may also be connected with bulk terms which can be integrated to the surface by means of Gauss's theorem. In fact, as shown by Nehring and Saupe, linear terms in second-order spatial derivatives may give a contribution to the bulk energy density of the same order as those connected to quadratic terms in first-order derivatives (Nehring and Saupe, 1971, 1972). According to Nehring and Saupe instead of (2.7) the elastic energy density for an NLC is given by

$$F = \tfrac{1}{2}[K_{11}(\text{div } \mathbf{n})^2 + K_{22}(\mathbf{n} \text{ rot } \mathbf{n})^2 + K_{33}(\mathbf{n} \times \text{rot } \mathbf{n})^2]$$
$$- K_{24} \text{ div}(\mathbf{n} \text{ div } \mathbf{n} + \mathbf{n} \times \text{rot } \mathbf{n}) + K_{13} \text{ div}(\mathbf{n} \text{ div } \mathbf{n}) \qquad (2.89)$$

in which the last term is linear in second-order spatial derivatives of $\mathbf{n}$. Consequently by substituting (2.89) into (2.20) one obtains that the effective surface energy is

$$f_{S_{\text{eff}}}(n_i, n_{i,j}) = f_S(n_i) + K_{13}(\mathbf{k} \cdot \mathbf{n}) \text{ div } \mathbf{n} \qquad (2.90)$$

i.e. it depends on n_i and $n_{i,j}$. According to our simple analysis if $K_{13} \neq 0$ a sharp variation of $\mathbf{n}$ is expected near the bounding surface (Oldano and Barbero, 1985a, b; Barbero and Oldano,

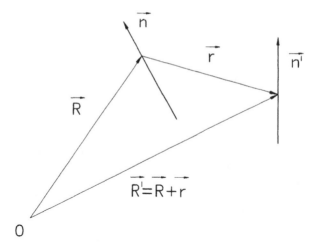

Figure 2.18 Element characterizing the interaction between two molecules of an NLC. **r** is the relative position of **n'** with respect to **n**. In the elastic limit $|\mathbf{n'} - \mathbf{n}|$ is supposed to be very small for $r \sim \rho_n$.

1985). To show this we have just to connect the elastic constants to the intermolecular interaction responsible of the nematic phase. Let us suppose that the intermolecular interaction is taken to be of the kind (see Figure 2.18)

$$V(\mathbf{n}, \mathbf{n'}, \mathbf{r}) = \sum_{a,b,c} J_{a,b,c}(r)(\mathbf{n} \cdot \mathbf{u})^a (\mathbf{n'} \cdot \mathbf{u})^b (\mathbf{n} \cdot \mathbf{n'})^c \tag{2.91}$$

where $a + c = 2p$ and $b + c = 2q$, p and q are integers and $\mathbf{u} = \mathbf{r}/r$ (Vertogen, 1983; Vertogen *et al.*, 1982). As usual, to obtain the elastic energy density it is necessary

(1)   to expand (2.91) as a power series of $\delta \mathbf{n} = \mathbf{n'} - \mathbf{n} = \mathbf{n}(\mathbf{R} + \mathbf{r}) - \mathbf{n}(\mathbf{R})$, to second order in **r**;
(2)   to expand $\delta \mathbf{n}$ to second order in **r**.

Routine calculations give (Vertogen, 1983; Barbero and Barberi, 1991), for the elastic constants entering into (2.89)

$$K_{11} = \frac{1}{2} \sum_{a,b,c} \frac{J_4(a, b, c)}{(a + b + 1)(a + b + 3)} \left(\frac{3ab}{a + b + 1} + c \right) \tag{2.92}$$

$$K_{22} = \frac{1}{2} \sum_{a,b,c} \frac{J_4(a, b, c)}{(a + b + 1)(a + b + 3)} \left(\frac{ab}{a + b - 1} + c \right) \tag{2.93}$$

$$K_{33} = \frac{1}{2} \sum_{a,b,c} \frac{J_4(a, b, c)}{(a + b + 3)} \left(\frac{ab}{a + b + 1} + c \right) \tag{2.94}$$

$$K_{24} = \frac{1}{4} \sum_{a,b,c} J_4(a, b, c) \left(\frac{3b + c}{(a + b + 1)(a + b + 3)} \right) \tag{2.95}$$

$$K_{13} = -\frac{1}{2} \sum_{a,b,c} J_4(a, b, c) \left(\frac{b}{(a + b - 1)(a + b + 3)} \right) \tag{2.96}$$

where

$$J_4(a, b, c) = 4\pi \int_0^\infty J_{a,b,c}(\mathbf{r})r^4 \, dr \tag{2.97}$$

From (2.96) one can deduce that if $b = 0$ then $K_{13} = 0$. This means that all the intermolecular interactions depending only on $\mathbf{n} \cdot \mathbf{n}'$ are characterized by $K_{13} \equiv 0$. This fact has an important consequence. In Barbero *et al.* (1989, 1993) it has been shown that the K_{13} elastic constant introduces a 'surface discontinuity' proportional to K_{13}/K. Let us consider a case in which $K_{13} \neq 0$, and hence V also depends on $(\mathbf{n} \cdot \mathbf{r})$ and $(\mathbf{n}' \cdot \mathbf{r})$. In this case the equilibrium configuration of a system comprising of only two molecules generally occurs for $\mathbf{n}'$ making a large angle with $\mathbf{n}$, if $\mathbf{r}$ and $\mathbf{n}'$ have been fixed. The parallel alignment appearing in the bulk of the NLC comes from averaging over all molecules. This obviously implies that a distortion must appear around a defect obtained by removal of a molecule. It would appear quite surprising if the lack of the interactions coming from such a molecule were to leave the equilibrium directions of all the surrounding molecules exactly the same. The distortion, weak or strong, occurs exactly because the remaining molecules will search for a new minimum energy and therefore cannot increase the free energy. A larger distortion is obtained by removing the outside molecules. This has been tested recently by a computer simulation with the use of the induced dipole–induced dipole intermolecular potential (Barbero *et al.*, 1995). There, the total free energy of an NLC has been estimated by directly considering the intermolecular interaction responsible for the nematic phase. The director profile obtained by minimizing this energy has been found to be practically coincident with that arising from the generalization proposed in Barbero *et al.* (1989) and Barbero and Strigazzi (1989). Hence it is possible to infer that in the surface boundary layer it is not possible to apply the usual first-order elastic theory, because of the 'discontinuity' localized over a semi-microscopic length connected to the splay–bend elastic constant.

The discussion reported above shows that a surface angular discontinuity may exist in the limit of a weak second-order elastic constant K^*. If one were able to observe such a rapid variation of orientation close to an obliquely orienting surface, one would be able to measure K_{13}. Since these sharp variations of angle are expected to occur on a mesoscopic length, one can reasonably include the K_{13} effect in the phenomenological expression of the surface energy. A model has in fact been made to interpret in this way temperature-induced surface transitions (Barbero and Durand, 1993).

2.12 Conclusion

The work presented in this chapter is a review of the efforts, mostly made in Orsay and Torrino, to understand the anchoring properties of nematic liquid crystals on solid amorphous substrates. To start with, the problem was to describe the angular dependence of the anchoring energy f_S which tends to keep the surface molecules along a favoured direction. When this easy direction $\mathbf{n}_0$ is $\mathbf{k}$, the normal to the substrate, or any direction inside its plane and within the obviously weak uniaxial surface order approximation, the first model proposed by Rapini and Papoular was to write the surface energy as $f_S = -\frac{1}{2} W_S(\mathbf{n} \cdot \mathbf{k})^2$. In fact, this expression for f_S was just the beginning of a series expansion in powers of $(\mathbf{n} \cdot \mathbf{k})^2$, the only angular term allowed by symmetry. Very soon afterwards, experimentalists noticed that higher powers such as $(\mathbf{n} \cdot \mathbf{k})^4$ were involved in f_S, especially for strong anchoring. As f_S is expected to arise from the anisotropy of van der Waals' forces between the nematic and the substrate, it was important to check the origin of these higher-power terms. In fact, it has been demonstrated that some of them were just originating from bulk effects like the flexoelectric effect for instance, which had been omitted in the first analysis. After these 'electrical' corrections, the Rapini–Papoular model seemed to hold correctly. In practice, many other macroscopic electrical effects should be taken into account, by adding new phenomenological terms to the surface energy. For instance, the easy direction $\mathbf{n}_0$ itself seemed to depend on the macroscopic distance between opposite boundary plates containing the nematic liquid crystal. Strange 'long-range forces', extending over the Debye screening length, were invoked. In fact, these changes of orientation can be explained by ion adsorption on surfaces: the surface electric field created, acting on the dielectric anisotropy, can induce surface transitions of orientation. More interesting, an intrinsic surface field appears each time the surface order parameter S_S is different from its bulk value S_b. Associated to the surface gradient of order $\nabla S \sim (S_S - S_b)/\xi$ is an 'order electric' polarization which generalizes the flexoelectric polarization for a system of space-varying quadrupoles. As this variation appears on the coherence length ξ, much shorter than the Debye screening length, the associated surface field is never screened. It gives rise also to a high-order surface term of $(\nabla S)^2[(\mathbf{n}\cdot\mathbf{k})^2 - \frac{1}{3}]^2$ which alone would tend to orient $\mathbf{n}_0$ at the 'magic' angle $(\mathbf{n}\cdot\mathbf{k})^2 = 1/3$. In practice, the surface gradient is strong for rough surfaces, which locally undulate at a scale ξ; rather than follow the surface undulation by expending a large elastic energy, the nematic decreases its surface order parameter, or even melts (Barbero and Durand, 1991). Obviously, one would also expect the surface order parameter to decrease when a strong external field locally curves the surface orientation, but what happens when an anchoring is completely 'broken' by such a strong field, so that no angular distortion remains at the surface? There exists now a more subtle effect, from the order–orientation coupling: the simplest way to generalize the Rapini–Papoular expression to take into account the surface order parameter is to write f_S as $\frac{1}{2}W(\mathbf{Q} - \mathbf{Q}_0)^2$, where $\mathbf{Q}$ is the surface tensor order parameter and $\mathbf{Q}_0$ its favoured value (Nobili and Durand, 1992). This expression shows that, even in the absence of surface angular distortion, the favoured surface scalar order parameter depends on the relative disorientation $\mathbf{n}\cdot\mathbf{n}_0$ as $S = \frac{3}{2}S_0[(\mathbf{n}\cdot\mathbf{n})^2 - \frac{1}{3}]$, with a tendency to negative (discotic-like) order $-S_0/2$ for total disorientation. Of course, a biaxial model would improve the description. This generalization shows that Rapini–Papoular model should work only for small surface distortions, or weak anchorings. The new tendency in surface anchoring phenomena is to take into account the order–orientation coupling. Under strong external fields, one expects even surface order S transitions for uniform orientation, rather than surface disorientations. New experiments are expected to demonstrate these predicted effects.

Inhomogeneous anchorings have been discussed in a stochastic model to consider the effect of Langmuir–Blodgett films on nematic orientation. This model shows that terms like $\sin^4\theta$ can result from the stochastic contribution of the direct Langmuir–Blodgett–substrate interaction. In fact, all the previous results could be generalized along these lines: all the electric polarization, for instance, connected with the state of the surface would give interesting effects. Disordered polarization components parallel to the surface would induce a spatial charge and field distribution along the surface, the energy of which should be taken into account. This could lead to a richer scenario than the one already discussed, and could be the subject of future work.

Finally, an old problem remains, generating passionate arguments and demonstrating the relative inefficiency of scientific meetings: it is the one concerning the so-called splay–bend elastic constant K_{13}. It comes from a divergence term in bulk curvature elasticity which integrates out as a surface term, depending on the normal derivative of the surface orientation. With this surface term, the minimization of the total free energy is an ill-posed problem, which leads to surface discontinuities of orientation. Some people (Pergamenshchik 1993) support another point of view. There must be an infinite number of k_{13}-like terms in the free energy; each one must be stabilized by higher-order terms, which means that the real total free energy is not known. Assuming that rapid distortions cannot exist in the nematic, they claim that the minimum of the unknown free energy is found as the solution suggested by Hinov (1990). In this way, torques calculated from the 'uncomplete' (but universally accepted) free energy are no longer conserved. Others suggest that to take care of these discontinuities a single higher-order elasticity should be included in the model, with more satisfactory solutions, but with the drawback that the rapid (and 'continuous') surface distortion should be mesoscopic, or even microscopic, for the free energy expansion to converge. On our part, it would be better to see the phenomenological model merge with microscopic models than to keep unbalanced torques. We hope that the predictions of both kinds of model will be seriously checked in the future to restore confidence in the continuum theory surface anchoring description of liquid crystals.

References

ALEXE-IONESCU, A. L. (1993) On the temperature dependence of the surface elastic constant K_{13}, *Phys. Lett.* **A175**, 352.

ALEXE-IONESCU, A. L., BARBERI, R., BARBERO, G., BEICA, T. and MOLDOVAN, R.(1992) Surface energy for nematic liquid crystals: A new point of view, *Z. Naturforsch.* **47a**, 1235.

ALEXE-IONESCU, A. L., BARBERO, G. and DURAND, G. (1993a) Temperature dependence of surface orientation of nematic liquid crystals, *J. Phys. II*, **3**, 1247.

ALEXE-IONESCU, A. L., BARBERO, G., IGNATOV, A. and MIRALDI, E. (1993b) Surface transitions in nematic liquid crystals oriented with Langmuir–Blodgett films, *Appl. Phys.* **56A**, 453.

ALEXE-IONESCU, A. L. BARBERO, G. and PETROV, A. G. (1993c) Gradient flexoelectric effect and thickness dependence of anchoring energy, *Phys. Rev. E* **48**, R1631.

ALEXE-IONESCU, A. L., BARBERO, G., GABBASOVA, Z., SAYKO, G. and ZVEZDIN, A. K. (1994a) Stochastic contribution to the anchoring energy: Deviation from the Rapini–Papoular expression, *Phys. Rev. E* **49**, 5354.

ALEXE-IONESCU, A. L., BARBERI, R., BARBERO, G. and GIOCONDO, M. (1994b) Anchoring energy for nematic liquid crystals: Contribution from the spatial variation of the elastic constants, *Phys. Rev. E* **49**, 5378.

BARBERI R. and DURAND, G. (1991) Electro-chirally controlled bistable surface switching in NLC, *Appl. Phys. Lett.* **58**, 2907.

BARBERI, R., GIOCONDO, M. and DURAND, G.(1992) A flexoelectricaly controlled surface bistable nematic switching, *Appl. Phys. Lett.* **60**, 1085.

BARBERI, R., GIOCONDO, M., MARTINOT-LAGARDE, Ph. and DURAND, G. (1993) Intrinsic multiplexability of surface bistable nematic displays, *J. Appl. Phys.* **62**, 3270.

BARBERO, G. (1984) Freedericks transition in hybrid aligned nematic liquid crystal cell, *Z. Naturforsch.* **39**, 575.

BARBERO, G., and BARBERI, R. (1983) Critical thickness of a hybrid aligned nematic liquid crystal cell, *J. Phys.* **44**, 609.

BARBERO, G. and BARBERI, R. (1991) *Physics of Liquid Crystalline Materials*, ed. F. Simoni and I.C. Khoo, Philadelphia: Gordon and Breach.

BARBERO, G. and DURAND, G. (1986) On the validity of the Rapini–Papoular surface anchoring energy form in nematic liquid crystals, *J. Phys.* **47**, 2129.

BARBERO, G. and DURAND, G. (1990a) Ions absorption and equilibrium distribution of charges in a cell of finite thickness, *J. Phys.* **51**, 281.

BARBERO, G. and DURAND, G. (1990b) Selective ions adsorption and non local anchoring energy in nematic liquid crystals, *J. Appl. Phys.* **67**, 2678.

BARBERO, G. and DURAND, G. (1991a) Order parameter spatial variation and anchoring energy for nematic liquid crystals, *J. Appl. Phys.* **69**, 6968.

BARBERO, G. and DURAND, G. (1991b) Curvature induced quasi-melting from rough surfaces in nematic liquid crystals, *J. Phys. II*, **1**, 651.

BARBERO, G. and DURAND, G. (1993) Splay-bend curvature and temperature-induced surface transitions in nematic liquid crystals, *Phys. Rev. E* **48**, 1942.

BARBERO, G. and OLDANO, C. (1985) Derivative-dependent surface-energy terms in nematic liquid crystals, *Il Nuovo Cimento* **6D**, 479.

BARBERO, G. and PETROV, A.G. (1994) Nematic liquid crystal anchoring on Langmuir-Blodgett films: Steric, biphilic, dielectric and flexoelectric aspects and instabilities, *J. Phys.: Condens. Matter* **6**, 2291.

BARBERO, G. and SIMONI, F. (1984) Non linear optical reorientation in hybrid aligned nematics, *J. Appl. Phys.* **55**, 304.

BARBERO, G. and STRIGAZZI, A. (1989) Second order elasticity in nematics, *Liq. Cryst.* **5**, 693.

BARBERO, D., MALVANO, R. and OMINI, M. (1977) On the refractive indexes of nematic liquid crystals, *Mol. Cryst. Liq. Cryst.* **39**, 69.

BARBERO, G., MADHUSUDANA, N. V. and DURAND, G. (1984a) Anchoring energy for nematic liquid crystals: An analysis of the proposed forms, *Z. Naturforsch.* **39**, 1066.

BARBERO, G., MADHUSUDANA, N. V., PALIERNE, J. F. and DURAND, G. (1984b) Optical determination of large distortion and surface anchoring torques in a nematic liquid crystals, *Phys. Lett.* **103A**, 385.

BARBERO, G., DOZOV, I., PALIERNE, J. F. and DURAND, G. (1986) Order electricity and surface orientation in nematic liquid crystals, *Phys. Rev. Lett.* **50**, 2056.

BARBERO, G., MADHUSUDANA, N. V. and OLDANO, C. (1989) Possibility of a deformed ground state in free standing nematic films, *J. Phys.* **5**, 2269.

BARBERO, G., GABBASOVA, Z. and OSIPOV, M. A. (1991) Surface order transition in nematic liquid crystals, *J. Phys. II*, **1**, 691.

BARBERO, G., BEICA, T., ALEXE-IONESCU, A. L. and MOLDOVAN, R. (1993) Influence of the surface topography on the surface transitions in nematics, *Liq. Cryst.* **14**, 1125.

BARBERO, G. EVANGELISTA, L. R. and PONTI, S. (1995) Surface deformations in nematics, *Phys. Rev. E* (to appear).

BLINOV, L. M., and SONIN, A. A. (1984) Determination of the binding energy of nematics with crystalline substrates from measurements of electro-optical effects, *Zh. Eksp. Teor. Fiz.* **87**, 476.

BLINOV, L. M. and KABAENKOV, A. Yu, (1987) Temperature dependence and the size-effect exhibited by the anchoring energy of a nematic with a planar orientation on a solid substrate, *Zh. Eksp. Teor. Fiz.* **93**, 1757.

BLINOV, L. M., RADZHABOV, D. Z., CHYUS, D. B. and IABLONSKII, S. V. (1991) Determination of the anisotropic part of the surface thermodynamic potential of a nematic liquid crystal, *JETP Lett.* **53**, 238.

BOUCHIAT, M. A. and LANGEVIN-CRUCHON, D. (1971) Molecular order at the free surface of a nematic from light reflectivity measurements, *Phys. Lett.* **34a**.

CHIARELLI, P., FAETTI, S. and FRONZINI, L. (1983) Structural transition at the free surface of the nematic liquid crystals MBBA and EBBA, *J. Phys.* **44**, 1061.

CHIARELLI, P., FAETTI, S. and FRONZONI, L. (1984) Critical behaviour of the anchoring energy of the director at the free surface of a nematic liquid crystal, *Phys. Lett.* **101a**, 31.

CHUVYROV, A. N. (1980) Influence of spontaneous polarization of mechanical stability of nematic liquid crystals, *Sov. Phys. Crystallogr.* **25**, 188.

CHUVYROV, A. N. and LACHIROV, A. V. (1978) Investigation of the nature of homeotropic orientation of nematic liquid crystal molecules, and the possibility of applying it in modulation spectroscopy, *Sov. Phys. JETP* **47**, 749.

CHUVYROV, A. N., SONIN, A. S. and ZAKIROV, A. D. (1976) Surface polarization of the nematic liquid crystal free surface, *Sov. Phys. Solid State* **18** 1797.

DEULING, H. J. (1974) On a method to measure the flexo-electric coefficients of nematic liquid crystals, *Solid State Commun.* **14**, 1073.

DIGUET, D., RONDELEZ, F. and DURAND, G. (1970) Anisotropie de la constante diélectrique et de la conductivité du p-métoxybenzidilène p-n-butylaniline en phase nématique *C.R. Hebdo. Sean. Acad. Sci.* **B271**, 954.

DOZOV, I., MARTINOT-LAGARDE, Ph. and DURAND, G. (1982) Flexoelectrically controlled twist of texture in a nematic liquid crystal, *J. Phys. Lett.* **43**, 365.

DOZOV, I., MARTINOT-LAGARDE, Ph. and DURAND, G. (1983) Conformational flexoelectricity in nematic liquid crystals, *J. Phys. Lett.* **44**, 817.

DOZOV, I., PENCHEV, I., MARTINOT−LAGARDE, Ph. and DURAND, G. (1984) On the sign of flexoelectric coefficients in nematic liquid crystals, *Ferroelec. Lett.* **2**, 135.

DURAND, G. (1976) *Electrohydrodynamics of Liquid Crystals,* Les Houches Lectures, Gordon and Breach: Philadelphia, p. 403.

DURAND, G. (1990) Order electricity in liquid crystals, *Physica A* **163**, 94.

ELSGOLTS, L. (1977) *Differential Equations and the Calculus of Variations,* Moscow: Mir.

EVANGELISTA, L. R. and BARBERO, G. (1984) On the electrostatic theory for quadrupolar materials, *Phys. Lett.* **185A**, 213.

FAETTI, S. (1994a) Resummation of higher-order terms in the free-energy density of nematic liquid crystals, *Phys. Rev. E* **49**, 5332

FAETTI, S. (1994b) Theory of surfacelike elastic contributions in nematic liquid crystals, *Phys. Rev. E* **49**, 4192.

FAETTI, S. (1991) *Physics of Liquid Crystalline Materials*, ed. F. Simoni and I.C. Khoo, Philadelphia: Gordon and Breach, p. 301.

FAETTI, S. and FRONZONI, L. (1978) Molecular orientation in nematic liquid crystal films with two free surfaces, *Solid State Commun.* **25**, 1087.

FLATISCHLER, L., KOMITOV, L., LAGERWALL, S. T., STEBLER, B. and STRIGAZZI, A. (1992) Surface induced alignment transition in a nematic layer with symmetrical boundary conditions, *Mol. Cryst. Liq. Cryst* **189**, 119.

FRANK, F. C. (1958) On the theory of liquid crystals, *Faraday Soc.* **25**, 19.

DE GENNES, P. G. (1969) Phenomenology of short-range-order effects in the isotropic phase of nematic materials, *Phys. Lett.* **30**, 454.

DE GENNES, P. G. and PROST, J. (1993) *The Physics of Liquid Crystals*, Oxford: Clarendon Press.

GOOSSEN, J. W. (1985) Bulk, interfacial and anchoring energies of liquid crystals, *Mol. Cryst. Liq. Cryst.* **124**, 305.

HELFRICH, W. (1974) Inherent bounds to the elasticity and flexoelectricity of liquid crystals, *Mol. Cryst. Liq. Cryst.* **26**, 1.

HILTROP. K and STEGEMEYER, H. (1978) On the orientation of liquid crystals by monolayers of amphiphilic molecules, *Mol. Cryst. Liq. Cryst.* **49**, 61.

HINOV, H. P. (1990) Further theoretical proofs for the existence of a relation between the elastic constants $K'_{11}(K_{11})$, $K'_{33}(K_{33})$ and K_{13} in nematics: Three- and one-dimensional cases, *Mol. Cryst. Liq. Cryst.* **178**, 53.

HOCHBAUM, A. and LABES, M. M. (1982) Alignment and texture of thin liquid crystal films on solid substrates, *J. Appl. Phys.* **53**, 2928.

ISRAELACHVILI, J. (1991) *Intermolecular and Surface Forces*, London: Academic Press.

JÉRÔME, B. (1991) Surface properties of nematics, *Rep. Prog. Phys.* **54**, 391.

KAZNACHEV, A. V. and SONIN, A. (1983) Spontaneous Freedericks transition, *Sov. Phys. Solid State* **25**, 528.

KOMITOV, L., LAGERWALL, S. T. SPARAVIGNA, A., STEBLER, B. and STRIGAZZI, A. (1992) Surface transition in a nematic layer with reverse pretilt, *Mol. Cryst. Liq. Cryst* **223**, 197.

KOMITOV, L., STEBLER, B., GABRIELLI, G., PUGELLI, M., SPARAVIGNA, A. and STRIGAZZI, A. (1994) Amphiphilic Langmuir–Blodgett films as a new tool for inducing alignment transitions in nematics, *Mol. Cryst. liq. Crys.* **243**, 107.

KUBO, R. (1967) *Statistical Mechanics*, Amsterdam: North-Holland.

LANDAU, L. D. and LIFSCHITZ, E. M. (1976) *Théorie de l'Élasticité*, Moscow: Mir.

DI LISI, G. A., ROSENBLATT, C., GRIFFIN, A. C. and HARI, UMA (1990) Behaviour of the anchoring strength coefficient near a structural transition at the nematic substrate interface, *Liq. Cryst.* **7**, 353.

MADA, H. (1979) Study on the surface alignment of nematic liquid crystals: Determination of the easy axis and temperature dependence of its field energy, *Mol. Cryst. Liq. Cryst.* **533**, 127.

MADHUSUDANA, N. V. and DURAND, G. (1985) Linear flexoelectrooptic effect in a hybrid aligned nematic liquid crystal cell, *J. Phys. Lett.* **46**, L195.

MADHUSUDANA, N. V. and PRATIBHA, R. (1982) Elasticity and orientational order in some cyanobiphenyls: Part IV Reanalysis of the data, *Mol. Cryst. Liq. Cryst.* **89**, 249.

MAHESWAR-MURTHY, P. R., RAGHUNATHAN, V. A. and MADHUSUDANA, N. V. (1993) Experimental determination of the flexoelectric coefficients of some nematic liquid crystals, *Liq. Cryst.* **14**, 483.

MARCEROU, J. P. and PROST, J. (1980) The different aspects of flexoelectricity in nematics, *Mol. Cryst. Liq. Cryst.* **58**, 259.

MEYER, R. B. (1969) Piezoelectric effects in liquid crystals, *Phys. Rev. Lett.* **22**, 918.

NEHRING, J. and SAUPE, A. (1971) Elastic theory of uniaxial liquid crystals, *J. Chem. Phys.* **54**, 337.

NEHRING, J. and SAUPE, A. (1972) Calculation of the elastic constants of nematics, *J. Chem. Phys.* **55**, 5527.

NOBILI, M. and DURAND, G. (1992) Disorientation induced disordering at a nematic liquid crystal solid interface, *Phys. Rev. A* **46**, R6174.

NYE, J. F. (1957) *Physical Properties of Crystals*, Oxford: Clarendon Press.

OLDANO, C. and BARBERO, G. (1985a) Ab-*initio* analysis of the second order elasticity effect on nematic configuration, *Phys. Lett.* **110A**, 273.

OLDANO, C. and BARBERO, G. (1985b) Possible boundary discontinuities of the tilt angle in nematic liquid crystals, *J. Phys. Lett.* **46**, L451.

OLDANO, C., MIRALDI, E., STRIGAZZI, A., TROSSI, L. and VALABREGA, P. (1984) Optical study of the molecular alignment in nematic liquid crystal in an oblique magnetic field, *J. Phys.* **45**, 355.

OSIPOV, M. A. and SLUCKIN, T. J. (1992) Statistical theory of order electric effect, *J. Phys. II* **97**, 1510.

PATEL, S. and YOKOYAMA, H. (1993) Continuous anchoring transition in liquid crystals, *Nature* **362**, 525.

PEREZ, E., PROUST, J. E. and TER-MINASSIAN-SARAGA, L. (1978) Films minces nématiques en alignement hybride, *Colloid. Polym. Sci.* **256**, 666.

PERGAMENSHCHICK, V. M. (1993) Phenomenological approach to the problem of the K_{13} surface-like elastic term in the nematic free energy, *Phys. Rev. E* **48**, 1954.

POINCARÉ, H. (1888) Sur la théorie analytique de la chaleur, *C. R. Acad Sci.* **107**, 961.

PRIESTLEY, E. B., WOJTOWICZ, P. J. and SHENG, PING (1976) *Introduction to Liquid Crystals*, New York: Plenum Press.

PROUST, J. E. and PEREZ, E. (1977) Films minces smectiques symétriques et asymétriques, *J. Phys. Lett.* **38**, L91.

PROUST, J. E. and TER-MINASSIAN-SARAGA, L. (1979) Films minces de cristaux liquides, *J Phys. Colloq.* **40**, C3-490.

RAPINI, A. and PAPOULAR. M. (1969) Distorsion d'une lamelle nématique sous champ magnétique. Condition d'ancrage aux parois, *J. Phys. Colloq.* **30**, C4-54.

SLUCKIN, T. J. (1995) Anchoring transitions at liquid crystal surfaces, *Physica* **213A**, 105.

SLUCKIN, T. J and PONIEWIERSKI, A. (1984) *Fluid and Interfacial Phenomena*, ed. C. A. Croxton, Chichester: John Wiley.

SONIN, A. A. and KAZNACHEEV, A. V. (1984) Theory of the spontaneous Freedericks transition, *Soc. Phys. Solid State* **26**, 486.

THURSTON, R. N., CHENG, J., MEYER, R. B. and BOYD, G. (1984) Physical mechanism of dc switching in a liquid crystal bistable boundary layer display, *J. Appl. Phys.* **56**, 263.

VERTOGEN, G. (1983) Elastic constants and the continuum theory of liquid crystals, *Physica* **117A**, 227.

VERTOGEN, G, FLAPPER, S. D. P. and DULLEMOND, C. (1982) Elastic constants of nematic and cholesteric liquid crystals and tensor fields, *J. Chem. Phys.* **76**, 616.

WARENGHEM, M. (1984) L'influence des substrats sur l'orientation du nématique dans une cellule nématique, *Mol. Phys.* **53**, 1381.

YANG, K. H. and ROSENBLATT, C. (1982) Determination of the anisotropic potential at the nematic liquid crystal-to-wall interface, *Appl. Phys. Lett.* **43**, 62.

YOKOYAMA, H. and VAN SPRANG, H. A. (1985) A novel method for determining the anchoring energy function at a nematic liquid crystal-wall interface from director distortions at high fields, *J. Appl. Phys.* **57**, 4520.

YOKOYAMA, H., KOBAYASHI, S. and KAMEI, H. (1987) Temperature dependence of the anchoring strength at a nematic liquid crystal-evaporated SiO interface, *J. Appl. Phys.* **61**, 4501.

3

Anisotropic Gels Obtained by Photopolymerization in the Liquid Crystal State

R. A. M. HIKMET

3.1 Introduction

Phase separation in liquid–polymer systems can lead to interesting morphologies. This intensively studied field has been subject to numerous publications and review articles where various ways of phase separation are described. The most common routes, however, involve polymerization (Dusek, 1971; Barret, 1975) and thermally (Hermans, 1978) induced phase separation. In the case of polymerization-induced phase separation, mixtures containing reactive and non-reactive molecules are used. Polymerization is induced thermally or photochemically leading to the phase separation of the mixture into polymer-rich and polymer-poor phases. In the case of thermally induced phase separation cooling a solution can also cause it to phase-separate into polymer-rich and polymer-poor phases. The thermodynamics of phase separation is also of great interest as it plays an important role in the structure formation within the two-phase system. In the case of binodal decomposition, the system finds itself in a metastable state where the nucleation and growth mechanism leads to the formation of polymer-rich and polymer-poor phases. In the case of spinodal decomposition the system is unstable and the fluctuations in the concentration grow with a dominant wavelength growing at the highest rate. In the case of binodal decomposition droplets of one phase embedded in a continuous phase are obtained whereas spinodal decomposition leads to two continuous interpenetrating phases. In such systems, however, the kinetics of phase separation is also influenced by other physical effects such as crystallization, glass transition and chemical cross-links. Therefore they also play an important role in the resultant morphology. In the case of high-molecular-weight polymers, different segments of the same molecule can take part in both phases of a two-phase system. If one of the phases is glassy or crystalline, acting as physical cross-links, the system is a thermoreversible gel (Miles, 1987). In the case where the cross-links are chemical then the system is not thermally reversible. Systems where the liquid phase is formed by LC molecules date back to the early 1970s when the polymerization of isotropic acrylates in mixtures of nematic LCs were investigated (Blumstein et al., 1971; Lecoin et al., 1975; Sanui et al., 1977). However, research into dispersions of LC molecules in isotropic polymer matrices (Fergason, 1985; Doane et al., 1986, 1988; Drzaic, 1986) (PDLCs) gained momentum owing to their applications using their electro-optical effects. A review of PDLCs can be found in Chapter 1. Anisotropic gel systems, which have been generating much attention, will be described further in this chapter.

Anisotropic gels are a new type of material obtained by *in situ* photopolymerization of mesogenic molecules in the presence of low-mass conventional LC molecules without reactive groups. In this way an anisotropic network containing LC molecules which are not chemically attached to the network is obtained. The network in the gels therefore possesses the structure of the initial LC mixture. These novel materials developed within Philips Research are of great academic as well as industrial interest. The industrial interest originates from the fact that they offer new possibilities for the manufacture and satisfactory functioning of existing LC devices (Hikmet and de Witz, 1991; Hasabe *et al.* 1994; Hikmet and Michielsen, 1995; Hikmet *et al.*, 1995a,b) while offering new effects (Hikmet 1990; Hikmet and Zwerver, 1992, 1993; Yang and Doane, 1992; Yang *et al.*, 1994) to be used in a new generation of LC devices (see also the chapters in this volume by Yang *et al.*, Kitzerow, Iwamoto *et al.* and Blinc *et al.*). At the same time the orientation and phase behaviour of LC molecules under restrictions which differ from that in the bulk are also of fundamental interest. In these systems the inclusion of a small amount of network (5%) has a drastic influence on the behaviour of the anisotropic properties of the LC molecules. For example, the threshold voltage for switching increases drastically (Hikmet, 1990; Hikmet and Zwerver, 1991b; Hikmet and Boots, 1995) with increasing network concentration and in the case of systems with a large birefringence, the application of an electric field can induce light scattering (Hikmet, 1990, 1992a; Hikmet and Zwerver, 1992; Hikmet and Boots, 1995). With increasing network concentration molecules change from showing first-order phase transition to a second-order phase transition material. Furthermore, as the anisotropic network has a profound effect on the orientational behaviour of the LC molecules manipulation of the network orientation directly affects the orientation of the LC molecules within the gel (Hikmet, 1991; Hikmet and Howard, 1993). It has been shown that the networks containing non-reactive molecules have been used in the production of orientation-inducing layers with adjustable tilt layers (Hikmet and de Witz, 1991), piezoelectric systems (Hikmet 1992c), and possible optical recording (Hikmet and Zwerver, 1993b). In the following sections some of the properties of the gels will be considered.

3.2 Production and Structure of Anisotropic Gels

Anisotropic gels were produced by using an LC mixture containing reactive and non-reactive molecules. Examples of various reactive LC molecules which can be used in the production of anisotropic gels are described in the chapter by Broer. In Figure 3.1 some examples of the chemical structures of molecules which have been successfully used in the production of anisotropic gels are shown. C6M is one of the most frequently used molecules in the production of gels as it mixes well with most LC molecules and does not crystallize readily. The LC mixture containing reactive and non-reactive molecules is also provided with a photoinitiator for inducing polymerization. The long-range orientation of the molecules can be chosen to be uniaxially planar, homeotropic or twisted and the polymerization is initiated using UV radiation freezing in the structure by the creation of a three-dimensional anisotropic network. Polymerization thus proceeds very fast and isothermally, avoiding polymerization taking place in an undesired phase or in a poorly oriented state. A review on photopolymerization of isotropic acrylates is given by Kloosterboer (1988). In most cases, during the polymerization of the gel mixture, the reaction is complete within a few minutes resulting in more than 70% conversion of the acrylate groups. The kinetics of polymerization of anisotropic gels has been described in an earlier article (Hikmet, 1991). The structure of the network in the anisotropic gels is determined to a large extent by the type of the reactive molecules used. Based on the type of network formed upon polymerization, two different types of gels have been characterized as schematically shown in Figure 3.2. In type 1 gel (Figure 3.2(a)) non-reactive LC molecules are confined between sheets of the anisotropic network. At higher concentrations of network the sheets join up to form cylindrical cavities as shown in Figure 3.2(a). The non-reactive

C6M

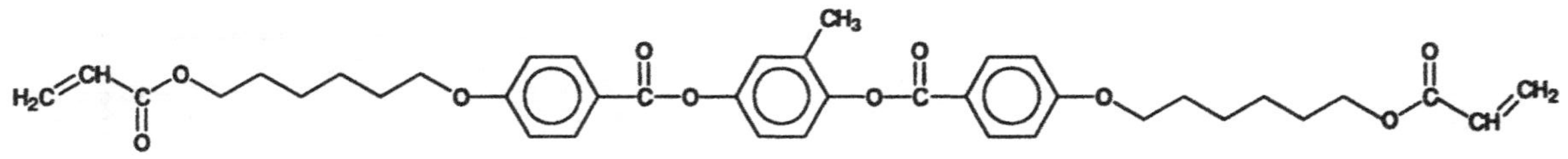

K   86ºC   N   116 ºC   I

(CₙH₂ₙ₊₁)

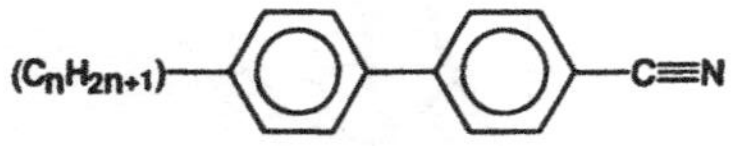

K15 n=5   K   240C   N   350C   I

K24 n=8   K   210C   SₐA   330C   N   400ºC   I

Figure 3.1 Examples of reactive and non-reactive LCs used in production of anisotropic gels.

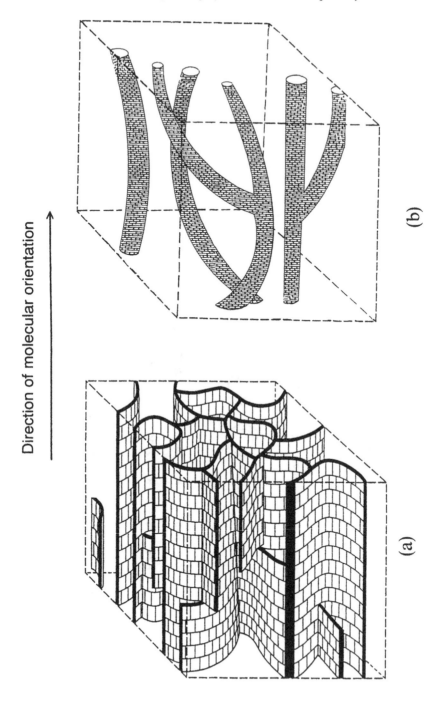

Direction of molecular orientation

(b)

(a)

Figure 3.2 Schematic drawing of the network structure in two types of gels: (a) type 1; (b) type 2.

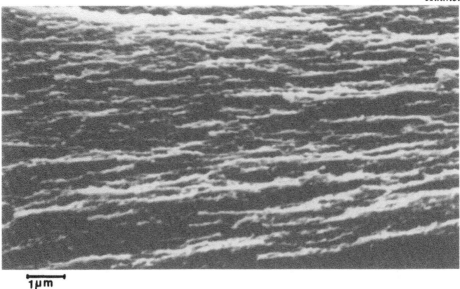

1μm

Figure 3.3 Scanning electron micrograph of a gel obtained after the extraction of the non-reactive LC (Hikmet 1991).

LC is contained in the cylindrical cavities. In type 2 gels, fibrils of anisotropic network run through the gel (Figure 3.2(b)). It has an open structure consisting of interconnecting phases of LC and the network. Owing to this structure in type 2 gels, the network has much less influence on the behaviour of the non-reactive LCs. In this chapter we shall mainly describe the properties of type 1 gels. An electron micrograph of a type 1 gel is shown in Figure 3.3 where a lamellar structure can clearly be seen. This photograph was obtained after the extraction of the non-reactive LC from the gel containing 30 % w/w network (polymerized C6M) and corresponds to the collapsed form of the structure schematically shown in Figure 3.2(a). The uniaxially oriented molecules of the network are confined in the plane of the lamellae. We shall return to a discussion of the structure within the gels and a comparison of the two different types of gels in the section on electrically controllable scattering.

Homogeneously oriented gels are highly transparent and highly birefringent. With increasing temperature the birefringence decreases continuously, eventually reaching a constant value above the temperature corresponding to the clearing point of the non-reactive LC molecules in the bulk. Birefringence as a function of temperature is shown in Figure 3.4 for gels containing various amount of network. The birefringence observed above the clearing temperature of the non-reactive LC is attributed to the network molecules which are thermally stable and remain oriented at elevated temperatures. However, this birefringence of the gels measured above the clearing temperature is higher than what is expected from the network alone. This indicates that part of the non-reactive LC molecules also remains oriented above their clearing point. It also indicates a high degree of interaction between the network and the non-reactive LC molecules. This large influence of the network on the behaviour of the LC molecules in fact forms the basis of some of the applications of the anisotropic gels. As the nature of this influence is of interest, the gels are being extensively studied (Hikmet 1991; Hikmet and Zwerver, 1991b, 1992; Hikmet and Howard, 1993; Jákli *et al.*, 1994; see also the chapters by Žumer and Crawford and Yang *et al.*).

The behaviour of the LC molecules in the gels was quantified using IR dichroism (Hikmet and Howard, 1993). The dichroic ratio $R = A_{\parallel}/A_{\perp}$) and the quantity S_0 is related to the order parameter S as (Kiefer and Bauer, 1989)

$$S_0 = \frac{R-1}{R+2} = S(1 - \tfrac{3}{2}\sin^2\theta) \tag{3.1}$$

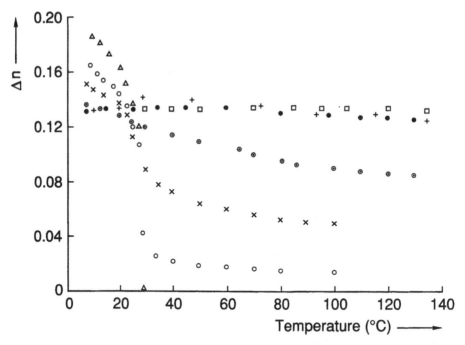

Figure 3.4 The birefringence of various gels as a function of temperature (Hikmet 1991): □ = 100% C6M; ○ = 90; + = 70; ⊙ = 50; × = 30; ● = 10; △ = 100% K15.

where θ is the angle between the direction of molecular orientation and the vibrational transition moment. We used characteristic absorption bands for the non-reactive LC molecules and for the network in order to estimate S_0 separately. Figure 3.5 shows S_0 as a function of temperature in a mixture containing 50% w/w C6M before and after polymerization. It can be seen that within experimental error, before polymerization S_0 for both non-reactive LC and C6M shows the same trend, decreasing with increasing temperature before discontinuously becoming zero at T_c of the mixture and showing typical behaviour for nematogens. After polymerization, the behaviours of the polymerized C6M (the network) and non-reactive LC molecules become different. For the network S_0 remains almost constant with increasing temperature, while S_0 for the non-reactive LC shows a strong decrease up to about 40°C above which the decrease becomes more gradual. The behaviour of non-reactive LC is, however significantly different from the behaviour shown by the bulk non-reactive LC which is also plotted in Figure 3.5 where the order parameter becomes zero at the clearing temperature of 40°C.

The behaviour of the non-reactive LC molecules in the presence of an anisotropic network can be explained in terms of a two-phase model (Hikmet, 1991; Hikmet and Zwerver, 1991b; Hikmet and Howard, 1993). According to the model within the gels two populations of non-reactive LC molecules are present. The behaviour of one of the populations (bound fraction) is determined to a large extent by the network and these molecules do not undergo the first-order nematic to isotropic whereas the other population (unbound fraction) behaves much like in the bulk, becoming isotropic at the clearing temperature. According to the two-phase model the shape of the S_0 curves for the whole population of non-reactive LC molecules depends on the fraction and the order parameter of the bound molecules. In order to simulate the effect of the amount of the bound fraction and its order parameter on the average S_0 measured using IR, we used the equation assuming a two-phase model (Hikmet and Howard, 1993):

$$R = \frac{A_{\parallel}}{A_{\perp}} = \frac{\varepsilon_{\parallel}(b)\phi(b) + \varepsilon_{\parallel}(u)\phi(u)}{\varepsilon_{\perp}(b)\phi(b) + \varepsilon_{\perp}(u)\phi(u)} \qquad (3.2)$$

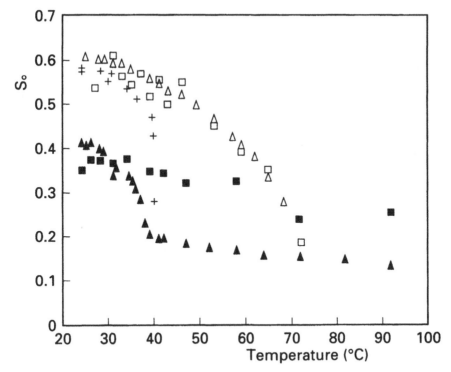

Figure 3.5 S_0 as a function of temperature for a mixture containing 50% C6M ($\square$), 8CB ($\triangle$). Open symbols denote before polymerization, solid symbols denote after polymerization; $+$ = pure 8CB in the bulk (Hikmet and Howard, 1993).

where $\varepsilon_{\|}$ and $\varepsilon_{\perp}$ are the extinction coefficients in the direction parallel and perpendicular to the director for bound (b) and unbound (u) fractions (ϕ), respectively. The order parameter of the components was modelled separately. The temperature dependence of the unbound fraction was modelled according to $(S_0)_u = (1 - T/T_c)^{0.2}$. Equations (3.1) and (3.2) were used to estimate the average S_0 for the two-phase system. In Figure 3.6, linearly decreasing values of S_0 with temperature, assumed to describe the behaviour of the bound fraction, are plotted together with the average S_0 calculated for the whole systems containing various bound and unbound fractions. As the bound fraction is strongly related to the network the order parameter of the bound fraction used in the model can be compared with the order parameter of the network. In Figure 3.7(a) S_0 obtained for the network in the gels and plasticized mixtures is plotted as a function of temperature. It can be seen that the order parameter of the network within samples of different compositions remains the same and shows only a slight temperature dependence. In the plots for the non-reactive LC molecules shown in Figure 3.7(b), however, it can be seen that at temperatures below 40°C the curves tend to merge, whereas at higher temperatures with increasing network concentration the S_0 at a given temperature increases. The behaviour shown in Figure 3.7(b) is very similar to what is observed in Figure 3.6 supporting the validity of the two-phase model. However, it is important to point out here that the very sharp discontinuous changes around the T_c shown by the calculated curves in Figure 3.6 are not observed in the case of experimental results. One explanation for this behaviour can be the broadening of the nematic to isotropic transition of the unbound fraction of non-reactive LC as compared with the behaviour shown in the bulk. The validity of this model was further checked by calculating the effect of keeping the fraction of unbound molecules constant while varying the order parameter of the unbound fraction at a given temperature. The simulated

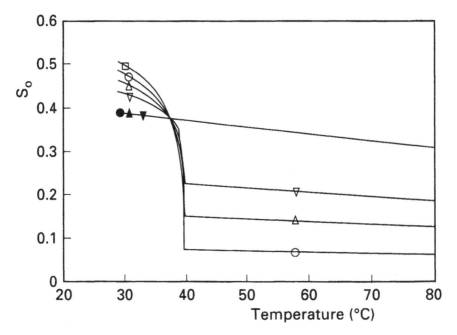

Figure 3.6 Simulated S_0 as a function of temperature for bound fractions (solid symbols) and average for both bound and unbound fractions (open symbols) for systems with various bound fractions: □ = 0; ○ = 20; △ = 40; ▽ = 60% (Hikmet and Howard, 1993).

behaviour was again in very good agreement with the measurements. This indicates that the observed behaviour within the gels is sufficiently described by the two-phase model.

Based on the two-phase model the fractions of bound and unbound non-reactive molecules were estimated from their various properties such as birefringence, heat of transition, IR dichrosim in the case of nematic gels and spontaneous polarization and the effective tilt angle in the case of ferroelectric gels. Figure 3.8 shows the estimated bound fractions of non-reactive LC molecules in anisotropic gels. In this figure it can be seen that the fraction of the bound fraction of non-reactive LC molecules increases almost linearly with increasing network fraction.

Even though the behaviour of the non-reactive LC molecules in the gels can be explained in terms of the two-phase model to get more insight into the bound fraction, it is interesting to consider the dielectric relaxation behaviour of the gels. Dielectric spectroscopy (Hikmet and Zwerver, 1991b) was used in order to investigate the relaxation of the non-reactive LC molecules about their short axis. It was found that at a given temperature with increasing network concentration the mean relaxation times (τ) shift to lower frequencies and the single mean relaxation time effective in the bulk is replaced by a distribution of relaxation times. Figure 3.9 shows Bauer plots for various gels. In these plots the other effect seen concerns the discontinuous jump in the mean relaxation time at the clearing temperature. It can be seen that the discontinuity become less defined with increasing network concentration and in networks containing more than 20 wt.% network no discontinuity could be observed. In the nematic phase the dispersion is best described by the Debye diffusion equation relating the relaxation time to the bulk viscosity (η) and the nematic potential as follows (de Jeu, 1980):

$$\tau \propto \eta G_{\parallel} \alpha \exp\left(\frac{\Delta H_N + \Delta H_{visc}}{RT}\right) \tag{3.3}$$

In Figure 3.9 it can be seen that the slope of the lines changes only slightly in the nematic phase with increasing network concentration. This indicates only a slight variation in the

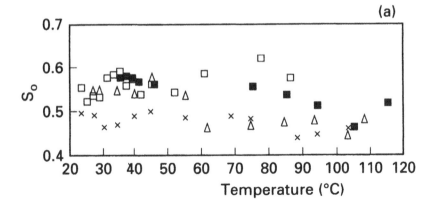

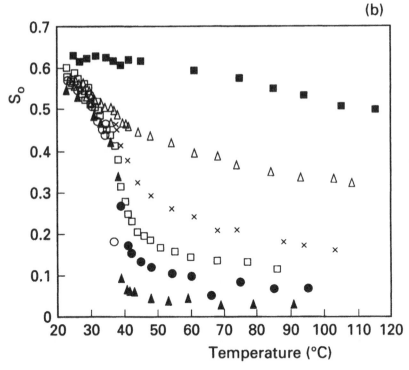

Figure 3.7 S_0 as a function of temperature for various gels (Hikmet and Howard, 1993): ■ = 80C6M; △ = 60C6M; × = 40% C6M; □ = 30% C6M; ● = 20% C6M; ▲ = 10% C6M, ○ = pure 8CB.

activation energy with increasing network concentration; hence ΔH_N and ΔH_{visc} for the non-reactive LC are almost the same as in the bulk. The increase in the mean relaxation times with increasing network concentration is associated with an increase in the viscosity caused by a decrease in the free volume. The effect of the free-volume change manifests itself in a decrease in the glass transition temperature with increasing network concentration. Indeed the thermomechanical measurements showed that the α relaxation peaks associated with the glass

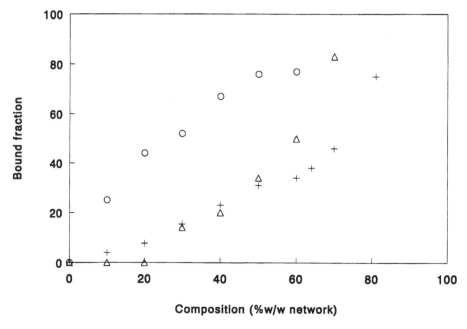

Figure 3.8 Estimated bound fraction for gels using various techniques (Hikmet *et al.*, 1995).

transition shifted to higher temperatures with increasing network concentration (Hikmet and Zwerver 1991b). The increase in the distribution of relaxation times therefore is associated with the inhomogeneity of the system and the existence of a distribution of domains containing the non-reactive LC. This distribution of domain sizes causes fluctuations in the free volume, thus broadening the distribution of relaxation times. The bound and unbound fractions can therefore be associated with the distribution of domain sizes and a critical domain size below which the non-reactive LC molecules do not show a first-order transition as theoretically predicted. With increasing network concentration the population of domains smaller than the critical size also increases and hence the bound fraction increases as experimentally observed.

It is also interesting to consider here the behaviour of the non-reactive LC in cholesteric gels. As is well known, cholesterics can reflect a band of circularly polarized light. The position (λ_m) of the reflection band is given by (de Vries, 1951)

$$\lambda_m = p\langle n \rangle \tag{3.4}$$

where $\langle n \rangle$ is the mean refractive index and p is the helical pitch. Chiral gels have been extensively studied regarding their structure and their switching properties (Hikmet and Zwerver, 1991a, 1992; Yang and Doane, 1992; Behrens and Kitzerow, 1994). In the case of TN cells they are also considered for improving the viewing angle dependence (Hasebe *et al.*, 1994). Figure 3.10(a) shows the transmission spectra for a mixture containing 20% C6M before and after polymerization together with the non-reactive chiral LC. It can be seen that after polymerization a second reflection band appears. One of the bands stays at the same position as before polymerization while the second peak appears at the position of a non-reactive mixture. In gels the secondary peak is associated with the unbounded fraction of the non-reactive LC while the main peak to a large extent is caused by the bound fraction. This can be better seen in Figure 3.10(b) where the behaviour of the gel is plotted as a function of temperature. It can be seen that the secondary peak disappears rapidly with increasing temperature. When higher network concentrations were used it was also found that the secondary peak did not appear.

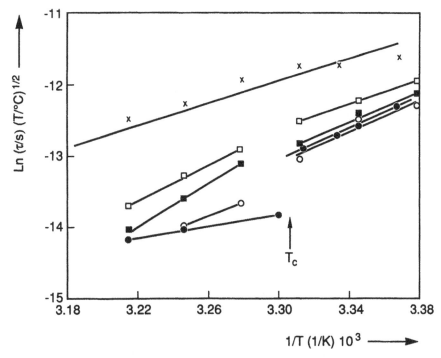

Figure 3.9 Bauer plots for the relaxation of K15 molecules in gels containing various amounts of K15 (Hikmet and Zwerver, 1991b): ● = 100%; ○ = 90%; ■ = 80%; □ = 70%; × = 60.

From the gap required for the suppression of the rotation of the director it was estimated that the bound fraction must have been confined in gaps thinner than 85 nm formed by layers of network.

3.3 Electrically Induced Light Scattering and Colour Changes

3.3.1 *Homogeneous Alignment*

As mentioned earlier, macroscopically oriented gels are transparent and cause no light scattering (Hikmet, 1990; Hikmet and Boots, 1995). However, application of an electric field can cause the transparent system to change its appearance and become increasingly translucent. The light scattering property of the gels is very much dependent on the initial macroscopic orientation of the system which can be disturbed by an electric field. As a result of the electric field, the LC molecules between the network become reoriented, causing 'large-scale' refractive index fluctuations within the gels giving rise to light scattering as schematically shown in Figure 3.11 for a gel with a negative dielectric anisotropy. The effect of polarization direction on scattering is shown in Figure 3.12(a). It shows that light polarized in the direction of molecular orientation is scattered to a much higher extent than the light polarized in the direction perpendicular to the molecular orientation. This behaviour is associated with the orientation of the molecules in the domains. In gels with a homogeneous orientation as the electric field is applied, the molecules tend to rotate about their short axis and become tilted with respect to the initial direction of molecular orientation. The ordinary index (n_0) of a uniaxially oriented system shows no dependence on the angle between the incident beam of light and the direction of molecular orientation whereas the effective extraordinary refractive index $(n_e)_{eff}$ is given by

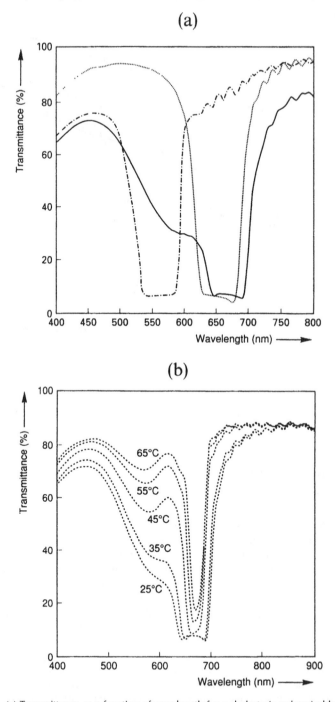

Figure 3.10 (a) Transmittance as a function of wavelength for a cholesteric polmerizable mixture containing 20% C6M (80% non-polymerizable cholesteric mixture): ·····, before polymerization; ———, after polymerization; ·······; non-polymerizable cholesteric mixture. (b) Transmittance as a function of temperature after polymerization (Hikmet and Zwerver, 1992).

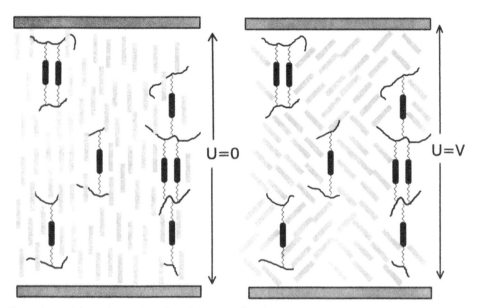

Figure 3.11 Schematic representation of scattering domains.

$$(n_e)_{eff} = \frac{n_e n_o}{(n_o^2 \sin^2\alpha + n_e^2 \cos^2\alpha)^{1/2}} \tag{3.5}$$

where α is the angle between the incident beam of light and the direction of molecular orientation. Electrically induced light scattering from gels with homogeneous orientation is found to be confined along a line perpendicular to the direction of molecular orientation. The scattering patterns obtained from such a gel and intensity profiles at various voltages are shown in Figures 3.12(b) and (c) respectively. The fact that the scattering is confined along a line perpendicular to the molecular orientation indicates that the cylindrical domains with a high aspect ratio are elongated in the direction of the molecular orientation as shown in Figure 3.2(a).

As the domain structure is a very important factor determining the switching behaviour of the gels, theoretical calculations were performed which predict two regimes influencing the switching behaviour of the LC molecules in the presence of the anisotropic network. These two regimes can be applied to two different types of gels shown in Figure 3.2. The threshold voltage for switching in these gels was described by an analytical model (Hikmet and Boots, 1955). In the model a cell $0 < \alpha < d$ with $\alpha = x, y, z$, where d_z is the cell thickness, is considered. The network is assumed to form N planes perpendicular to the three directions and the distance between the planes is $\delta_z = d_z/(N + 1)$. In the presence of the network during the application of the electric field in the z direction just above the threshold field the director will rotate over an infinitesimal angle of ϕ. The free-energy density is modelled as

$$f(\mathbf{r}) = \tfrac{1}{2}K[\nabla \phi(\mathbf{r})]^2 - \tfrac{1}{2}\varepsilon_0 \Delta\varepsilon E^2[\phi(\mathbf{r})]^2 + P(\mathbf{r})[\phi(\mathbf{r})]^2 \tag{3.6}$$

where $P(\mathbf{r})$ is the network contribution to the orientation free-energy density. It can be shown (Hikmet and Boots, 1995) that the threshold voltage is given by

$$V_{th} = \pi[(1 + m_z + m_y + m_z)K/(\varepsilon_0\Delta\varepsilon)]^{1/2} \tag{3.7}$$

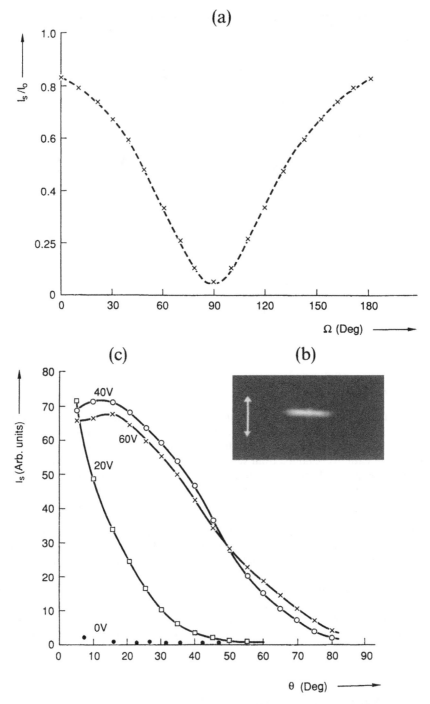

Figure 3.12 Scattering characteristics of a gel containing 5% network with a homogeneous orientation. (a) Scattered intensity as a function of the angle Ω which the director makes with respect to the plane of linearly polarized light (Hikmet 1990). (b) Photograph of the scattering pattern; the arrows indicate the direction of molecular orientation. (c) The scattered intensity as a function of angle along the line at various voltages.

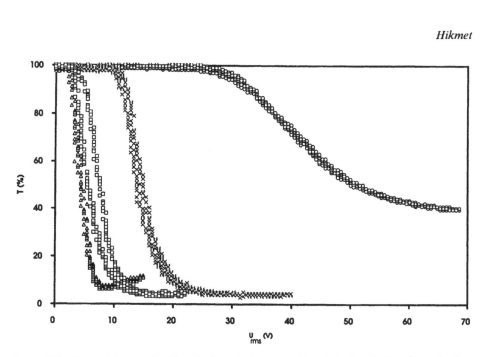

Figure 3.13 Transmission as a function of voltage for type 2 gel ($\triangle = 5\%$, $\square = 10\%$) and type 1 gel ($\times = 5\%$, $\bigcirc = 10\%$) polymer.

with

$$m_\alpha = d_z^2 \, \min[\delta_\alpha^{-2}, \, P_\alpha/(\delta_\alpha K \pi^2)] \qquad \alpha = x, y$$

$$m_z = d_z^2 \, \min[(\delta_z^{-2} - d_z^{-2}), \, P_z/(\delta_z K \pi^2)]$$

where K is the Frank elastic constant, ε_0 is the pemittivity of free space and $\Delta\varepsilon$ is the dielectric anisotropy of the LC. The minimum of the arguments of min should be taken. It can be seen that V_{th} is independent of cell thickness if m_x, m_y, $m_z \ll 1$. If m is not negligible but the average matrix influence P_α/δ_α is still small compared with $K\pi^2/\delta_\alpha^2$, V_{th} gives an upper limit for δ. In the case of gels with an interconnecting phase of polymer network and the LC, as the domain size is considered to be larger than the cell thickness, V_{th} is not expected to show cell thickness dependence and would be expected to be slightly higher than in the bulk. In the case of gels with a domain size smaller than the cell thickness a profound thickness dependence and a large increase in the threshold are expected.

In order to produce different types of gels two different types of LC diacrylate which can lead to two different types of networks were used. In Figure 3.13 the transmission–voltage behaviour of the gels is shown for two different types of gels in a 6 μm cell. It can be seen that in the case of the type 2 gel the threshold voltage (V_{th}) to cause scattering remains almost the same with increasing network concentration while the voltage to cause maximum scattering moves slightly to higher voltages. In the case of type 1 gels both the threshold voltage and the voltage to cause maximum scattering move rapidly to higher voltages with increasing network concentration.

It is interesting to consider here the two regimes predicted by theoretical considerations to describe the switching in two different types of gels. According to eqn (3.7) V_{th} can vary with the cell thickness depending on the domain size within the gels. The transmission–voltage curves for gels obtained using two different polymers are examined in various cell gaps. It was found that for type 2 gels V_{th} shows almost no dependence on the cell thickness indicating that the domain size within these gels is large compared with the cell thicknesses used. As opposed to type 2 gels the gels of type 1 show a strong thickness dependence and

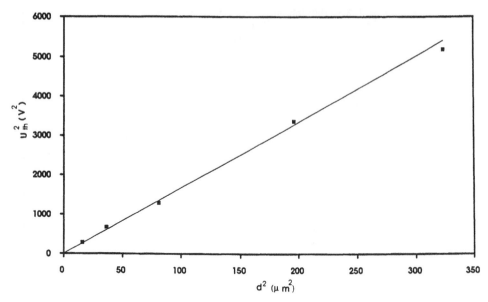

Figure 3.14 Squared threshold voltage as a function of squared cell thickness for type 1 gels; ■ = 10%.

with increasing cell thickness (d) V_{th} and the voltage to induce maximum scattering move to higher voltages. According to the theory there is a quadratic relation between the cell gap and V_{th} given by the eqn (3.6). In order to check the validity of this relationship we plotted in Figure 3.14 $(V_{\text{th}})^2$ against d^2 for type 1 gels. It can be seen that for both concentrations a linear relationship was obtained supporting the theory. It is important to point out here that the type 1 gel containing 10% C6M shows a higher slope. According to the theory the slope of the lines is related to the interaction parameter P and a characteristic dimension (δ_c) for the system defining the domain with the lowest threshold voltage. If the parameter P is known it would be possible to calculate δ_c. However, as it is not known it is only possible to make a relative comparison between the gels assuming an equal interaction parameter for type 1 gels at two concentrations. We determined the slopes of the lines to be about 3 and 16 $(V/\mu m)^2$. One can conclude that $\delta_c(5\% \text{ gel})/\delta_c(10\% \text{ gel}) = (16/3)^{0.5} = 2.3$, i.e. δ_c in the 5% gel is more than twice that in the 10% gel. Assuming a high P in all directions (a rigid network) and $\delta_y = \delta_z = \delta_c$ and $\delta_x = \infty$, the 5% gel was calculated to be about $\delta_c \leqslant 1.4 \ \mu m$.

The network structure has an important influence not only on the switching voltages but also on the switching times within the gels. We compared the switching speeds of two different types of gels in cells of various thicknesses. In all cases in order to obtain roughly the same rise time (t_r) for various cell gaps the applied voltage had to be increased with increasing cell gap. The decay time (t_d) on the other hand was measured after removing the applied voltage and short-circuiting the cell. In the case of type 1 gels no dependence of t_d on the cell thickness could be observed while the type 2 gels show a pronounced cell thickness dependence. The switching times can be expressed as (Goodman, 1974)

$$t_r^{-1} = \varepsilon_0 \Delta \varepsilon E^2 / \eta - t_d^{-1} \tag{3.8}$$

$$t_d^{-1} = K \pi^2 / \eta \delta_c^2 \tag{3.9}$$

In the type 1 gels the absence of thickness dependence on the decay time indicates that in these the characteristic distance influencing the behaviour of the LC molecules is the cell

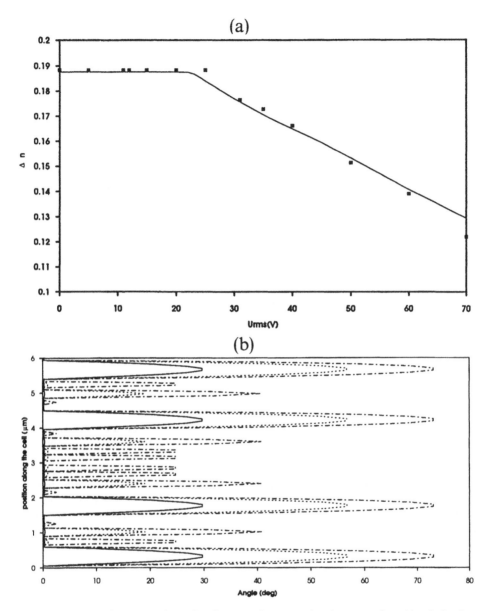

Figure 3.15 (a) Birefringence–voltage plot of type 1 gel (■ = 10%) polymer together with calculated line. (b) Calculated director angle along the cell thickness at various voltages: ———, 50 V; ····, 60 V; ·······, 70 V.

thickness rather than the distance in the polymer network. In the case of type 2 gels the fact that t_d is highly dependent of the cell thickness indicates that δ_c within these gels is much smaller than the cell thicknesses used and is defined by the network. However, this decrease is not as dramatic as the increase observed in the threshold voltage (Figure 3.14). The response of the LC molecules to the electric field was estimated by measuring the birefringence of the gel as a function of voltage as shown in Figure 3.15(a). In the same figure we also show the calculated line according to the theory. It was assumed that the network is rigid and does not change its orientation under the influence of the electric field and that

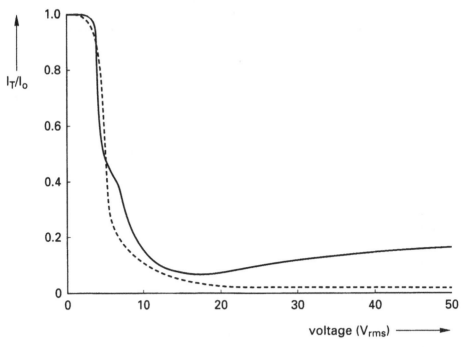

Figure 3.16 Transmitted intensity as a function of voltage for a cell with homeotropic orientation: ——, without dye; – – – –, with dye (Hikmet 1992a).

the LC domains contain any polymer. The observed behaviour could not be satisfactorily described by a single domain size. Therefore in order to get a good fit of the experimental data a distribution of domain sizes was assumed. Figure 3.15(b) shows the director profile along the cell thickness at various applied voltages. Points where the angle is zero at high voltages define the point where the polymer network is present. V_{th} is determined by the largest domain size within the system whereas the shape of the birefringence–voltage curve is determined by the distribution of the sizes. The results again show that the theory can be used in order to explain the observed behaviour.

3.3.2 *Homeotropic Gels*

Homeotropically aligned gels (Hikmet, 1992a) can be used to obtain isotropic scattering from the system, increasing the contrast. As opposed to the planar gels in homeotropically aligned gels, LC molecules with negative dielectric anisotropy need to be used. In such a system application of an electric field causes the molecules to align themselves in directions perpendicular to the electric field. Unlike the previous case, there is no defined direction for molecules to realign and therefore they are aligned randomly creating domains to scatter both polarizations of the incident light beam. In Figure 3.16 the intensity of light through such a homeotropically aligned gel is shown as a function of applied voltage. Another effect is obtained when the gels are provided with dichroic dyes. In the absence of the network LC material containing dichroic dye molecules with homeotropic orientation can only be switched between the transparent and the dark states without the grey levels. For this purpose homeotropically oriented gels containing dichroic molecules can be used so that grey

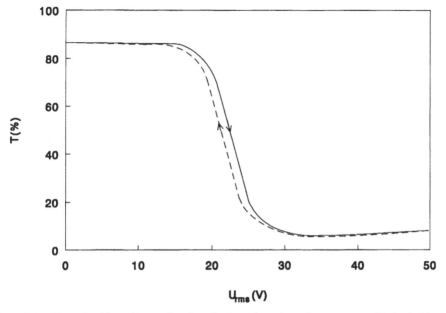

Figure 3.17 Transmitted intensity as a function of voltage through a cell containing a chiral gel with a Grandjean texture.

levels as also shown in Figure 3.16 are obtained. It can be seen that with increasing voltage transmission through the system decreases gradually, instead of a discontinuous decrease obtained for the LC system in the absence of a network.

3.3.3 Chiral Gels

Chiral gels (Hikmet and Zwerver, 1991a, 1992) are obtained by providing the system with chiral molecules inducing a helical twist within the system. The orientation of the helix within the gels is chosen to be perpendicular to the surfaces (Grandjean texture) or random (focal conic). Gels with a Grandjean texture giving no reflection band in the visible range are clear. As in the previous cases, the application of an electric field causes the gels to change their appearance and become translucent and revert to the transparent state upon removing the electric field as shown in Figure 3.17. The chiral gels, like uniaxially planar oriented gels, tend to scatter selectively a component of light which is circularly polarized and has the opposite sense of the helix within the gel. It is again for this reason that high scattering efficiencies were not obtained from these systems. However, the selectivity is not as strong as that shown by homogeneously aligned gels for linear polarization. On the other hand the cholesteric gels giving a reflection band in the visible region respond to electric fields in another way. The typical response of a gel to various applied voltages is shown in Figure 7.18. In this figure it can be seen that the reflection band shifts to lower wavelengths with increasing voltage. In the case of low-molecular-mass molecules, in the absence of a network the application of an electric field causes rotation of the cholesteric helix, causing the formation of a scattering texture. This scattering focal-conic texture remains even when the electric field is removed so that the system is not reversible. In the case of the gels due to the network, this rotation occurs in a controlled way causing a shift in the position of

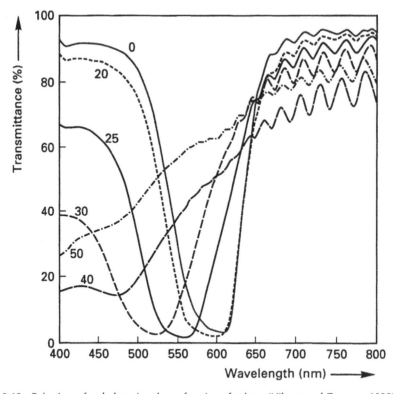

Figure 3.18 Behaviour of a cholesteric gel as a function of voltage (Hikmet and Zwerver, 1992).

the reflection band. This change in the position of the reflection band is caused by the rotation of the cholesteric helix within the cell. Eventually at higher voltages the helix unwinds and the cell becomes transparent. Upon removing the electric field the gel reverts back to its highly colourful initial Grandjean texture. The use of chiral gels is also described in the chapter by Yang *et al.*, and Kitzerow; matrix addressing for displays is discussed by Yang *et al.* In the chapters by Iwamoto *et al.* and Bos *et al.* the use of gels in the TN configuration is discussed.

3.4 Network-stabilized FLC–FLC Gels

In the ferroelectric liquid crystal (FLC) phase the director is at an angle with respect to the smectic layers. In the bulk the director rotates about a helix with the axis perpendicular to the smectic layers. In the case of surface-stabilized ferroelectric liquid crystal displays (SSFLCDs) (Clark and Lagerwall, 1980) the FLC is placed in a cell with a thin gap (of only a few micrometres). The pitch in the nematic phase is then suppressed and the LC molecules become uniaxially oriented in the nematic phase. Subsequently the FLC is cooled, preferably via the smectic A phase to the ferroelectric chiral smectic C phase (C*). Various devices, such as flat-panel screens for video and data graphic applications and high-speed shutters, have been developed on the basis of the surface-stabilized concept. The advantages of FLCs are the high speed of switching, the bistability in the case of passive addressed displays and the viewing angle independence of the contrast. However, the SSFLCD also has

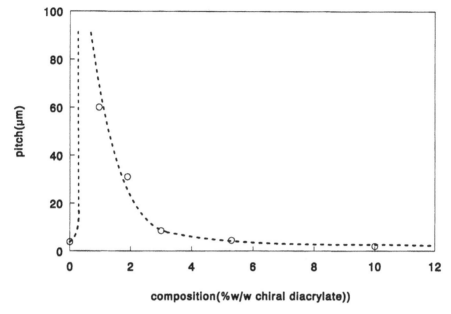

Figure 3.19 Effect of reactive chiral diacrylate on the pitch in the nematic phase of an FLC.

various problems, such as the difficulty of obtaining the desirable orientation and preserving it during the lifetime of the display, the poor shock sensitivity, the difficulty of obtaining grey levels and the small cell gap technology. The requirement of the small cell gap is due to the high birefringence of the FLC. In FLC displays use is made of the retardation mode, in which the incoming polarized light is rotated 90° as a result of the $\lambda/2$ retardation. However, owing to the dispersion in birefringence the retardation has to be of first order, which limits the cell thickness to the order of 2 μm. The possibility of using an FLC–PDLC has been suggested (Kitzerow *et al.*, 1992) (see also the chapter by Kitzerow) to solve these problems. However, owing to the shearing required during the polymerization to induce orientation this method has its limitations. The use of small amounts of isotropic acrylates in the production of networks has been also investigated (Pirš *et al.*, 1992; see also the chapter by Blinc *et al.*). Here we would like to present some experimental results of a new concept of network stabilization of the FLC using ferroelectric gels (Hikmet and Michielsen, 1995; Hikmet *et al.*, 1995a,b).

3.4.1 *Pitch Compensation and Stabilization of Orientation*

First it is interesting to consider Figure 3.19 which shows the possibility of compensating the pitch (Hikmet and Michielsen, 1995) of the non-reactive FLC in the nematic phase by adding chiral diacrylate to the mixtures containing 20% non-chiral LC diacrylate. The compensation of the chiral pitch is very important with respect to, for example, obtaining uniaxially oriented ferroelectric samples in thick cells. Figure 3.19 shows the effect of the addition of chiral diacrylate on the helical pitch within the mixture measured at 58 °C. It can be seen that when the chiral acrylate concentration increases the pitch first increases rapidly and then the helix changes its sense and the pitch starts to decrease. This figure clearly

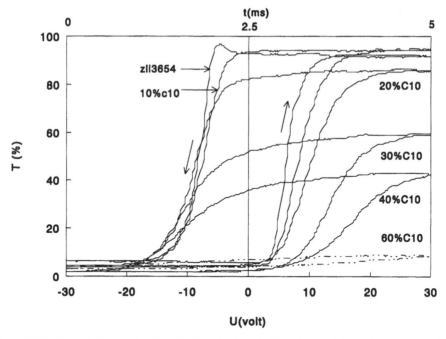

Figure 3.20 Transmission as a function of voltage for ferroelectric gels containing various amounts of network (Hikmet et al., 1995a).

shows that the pitch can be fully compensated by adding a small amount of chiral dia-crylate before polymerization. The mixture was put into a 7.7 μm cell provided with uniaxially rubbed nylon layers above transparent electrodes. After uniaxial orientation had been induced in the material in the cell at 58°C polymerization of the acrylates in the cell was initiated using a UV source. During the polymerization a three-dimensional anisotropic network containing the non-reactive FLC molecules (ferroelectric gels) was produced. After polymerization the cell was cooled to ambient temperature, at which the molecules (FLC and polymer) retained their uniaxial orientation. Other heating and cooling cycles did not change the uniaxial orientation of the molecules. This is different from the behaviour of pure non-reactive FLC placed into such a 7.7 μm cell, where the orientation was not homo-geneous but twisted at most temperatures in the N* phase. Cooling such an orientation to the ferroelectric phase led to a texture with defects. A much better orientation of the pure FLC could be obtained in much thinner (2 μm) cells in which surface stabilization could be induced. The fact that the non-reactive FLC molecules in the gel remained well oriented uniaxially throughout the entire nematic and ferroelectric ranges of the non-reactive FLC in the bulk indicates the great influence of the network molecules on the non-reactive FLC molecules which are not chemically attached to the network. For this reason we would like to refer to the FLC gels as 'anisotropic network-stabilized ferroelectric liquid crystal' (see the chapter by Blinc *et al.*).

3.4.2 *Optical Properties*

The transmission–voltage behaviour of the FLC gels was investigated by placing the cells between crossed polarizers so that a maximum dark state was obtained during switching.

Figure 3.20 shows the curves obtained for ferroelectric gels containing various amounts of network. It can be seen that the switching voltage increases with polymer concentration and the switching takes place over a wide voltage range. It can also be seen that the maximum transmitted intensity decreases with increasing polymer concentration. The transmission (T) of the wavelength λ through the crossed polarizers containing a material with a birefringence Δn and thickness d and the optic axis oriented at an angle ω with respect to one of the polarizers is given by

$$T = \sin^2(2\omega) \sin^2(\pi \Delta n d / \lambda) \tag{3.10}$$

The decrease observed in the maximum transmission with increasing network concentration is caused by (i) a change in the effective birefringence and (ii) a change in the effective tilt angle as a function of the composition during the application of the electric field. Both effects are caused by the network and have therefore been studied in detail.

3.4.2.1 *Effective birefringence and tilt angle of the gels*

Light propagation in the ferroelectric gels was simulated with the aid of a simple theoretical model. In this model the molecules were assumed to be contained in thin alternating layers containing the network and the ferroelectric molecules. The light propagating through such a system was calculated using the program developed by Mansuripur(1990). For the system to behave like a uniaxial crystal the layers have to be smaller than the wavelength of light; the results are then independent of the layer thickness. Therefore the total thickness of each polymer–FLC bilayer was kept constant in the calculations and the thicknesses of the components were varied in the same ratio as the composition. The direction of the average fast optic axis was determined by determining the direction along which the linearly polarized light remain linear. Figure 3.21(a) shows the calculated apparent tilt angle (the angle between the direction defined by the network and the average optic axis) as a function of the composition for various Δn (0.13 and 0.1) of the network. In the calculations the Δn of the non-reactive FLC was assumed to be 0.13 and the non-reactive FLC molecules were assumed to be oriented by $25°$ with respect to the uniaxially oriented network. It can be seen that in all cases the effective tilt angle decreases when the network concentration increases, as is also observed experimentally. It is also interesting to note here that the decrease in effective tilt angle is larger at higher Δn values of the polymer. In the same way we also calculated the effective birefringence as a function of the composition; the results are shown in Figure 3.21(b). Calculations were carried out for two orientations: (i) non-reactive FLC and the polymer oriented in the same direction: (ii) non-reactive FLC and the polymer oriented $25°$ with respect to each other. Here again it can be seen that in both cases Δn decreases with increasing network concentration. This shows the possibility of reducing effective birefringence using a network with a low birefringence. In Figure 3.21(a) and (b) theoretical calculations are compared with the experimental data. It can be seen that both the effective birefringence and the extinction angle decrease with increasing network concentration. The same trend is also shown by the experimental points. However, the behaviour of the gels is also influenced to a large extent by the bound fraction of the non-reactive molecules which remained oriented along the direction of the network. For this reason the measured data do not show a good correlation with the calculated lines.

The switching behaviour was investigated by measuring the extinction angle and the birefringence as a function of the voltage (Figure 3.22). In this figure it can be clearly seen that the birefringence first decreases with increasing voltage before increasing again. The switching of the molecules takes place in domains and the size of the domain determines the threshold voltage for the switching. Such behaviour also explains the behaviour observed for transmission/voltage (Figure 3.20) where the switching of non-reactive FLC molecules over a broad voltage range is suggested. It may be assumed that, as the concentration

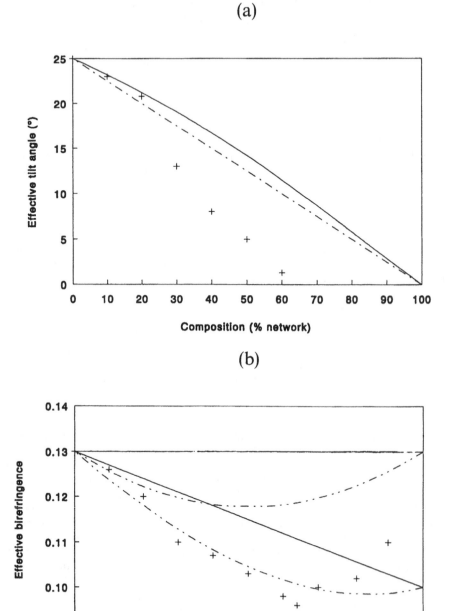

Figure 3.21 Calculated and measured extinction angle and birefringence as a function of composition. (a) Polymer $\Delta n = 0.13$ (-·-·-); $\Delta n = 0.1$ (——). The FLC and the polymer are oriented at $25°$ with respect to another. (b) Full lines correspond to the network and the FLC oriented in the same direction. Chain lines correspond to the FLC and the network oriented at $25°$ with respect to one another. For the FLC $\Delta n = 0.13$ and for the network $\Delta n = 0.13$, 0.1 (Hikmet *et al.*, 1995a).

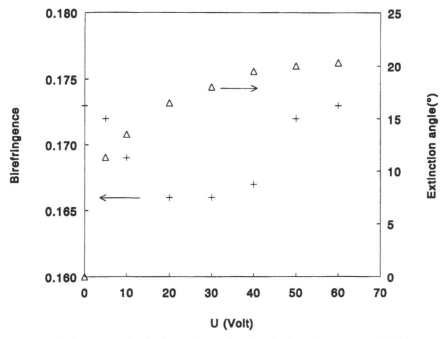

Figure 3.22 Birefringence and extinction angle as a function of voltage (Hikmet *et al.*, 1995a).

increases, different amounts of non-reactive FLC molecules switch from their initial orienta-tion to an angle of 25°C with respect to the initial orientation. This is also suggested by the results of the simulations shown in Figure 3.21(b), where the birefringence is plotted as a fraction of the molecules oriented at 25° with respect to each other. It can be seen that the minimum at 50% yields an effective birefringence of about 0.118. Likewise, in Figure 3.22 it may be assumed that at about 60 V 50% of the non-reactive FLC molecules are oriented at 25°. Above this voltage more of the non-reactive FLC molecules reorient and as a result the birefringence increases to a value very close to the initial value.

3.4.3 *Passive Addressing*

In order to reorient the chevrons to the upright position (bookshelf structure) the cells con-taining the gels were subjected to a 10 Hz peak-to-peak 50 V square wave. Subsequently the standard bistability test was carried out using a periodically inverted bipolar pulse set. The results of the bistability test obtained for various compositions are shown in Figure 3.23. It can be seen that all of the compositions used in the study showed bistable switching. Earlier in the text it was shown (Figure 3.20) that the gels switched over a wider voltage range. We therefore tried to use this effect to obtain stable grey levels. For this purpose we used two adjacent bipolar periodic (100 Hz) pulses (bipolar pulses were used to prevent the accumulation of ionic charges within the system). The first bipolar pulse served to reset (V_r) the system; its magnitude was kept constant. The second bipolar pulse served to switch (V_w) the system to various transmission levels; its magnitude was varied. Figure 3.23(a) shows the change in the transmission as a function of time during and after the application of such a pulse for various V_w values. It can be seen that after the reset pulse, which brought the

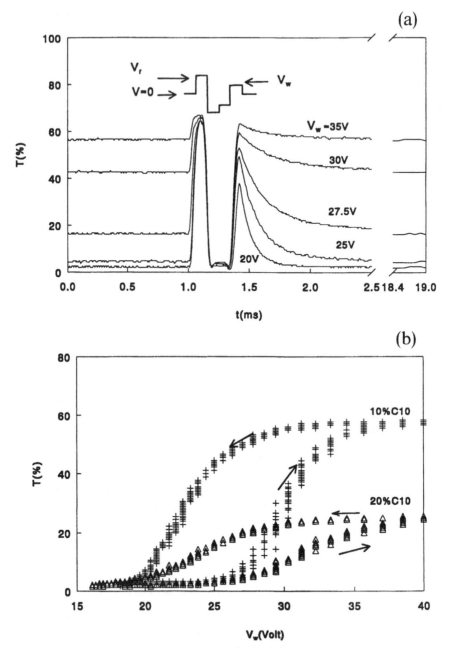

Figure 3.23 (a) Transmission through crossed polarizers during the application of a bipolar reset pulse (V_r) of ± 40 V followed by write voltage pulses (V_w) of various magnitudes for a gel containing 10% polymer. (b) Transmission as a function of a periodically varied write voltage (V_w) for two different gels: + = 10% network; $\triangle$ = 20% network.

transmission to zero, the second pulse raised the transmission. After the end of the pulse the intensity dropped to a lower value within about 0.5 ms, where it stayed until the following pulse. This shows that grey levels can be obtained in this way. The explanation for this

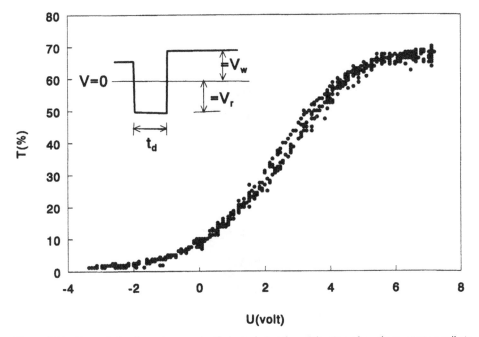

Figure 3.24 Transmission through crossed polarizers during the application of a voltage across a cell via the transistor. Voltages refer to the source voltage.

multistable switching may be switching in different domains at different voltages. The spontaneous polarization and the results of the birefringence measurements both also point to domain switching. Optical microscopy revealed that the texture was very fine and had almost the resolution of the optical microscope. It is very different from the texture obtained when pure substances are used in the texture method (Hartmann, 1991). In Figure 3.23(b) we have plotted the transmission as a function of V_w during the application of a periodically varied V_w. It can be seen that in this way smooth transmission–voltage curves can be obtained.

3.4.4 *Active Matrix Addressing*

For gels it has been shown that the molecules react to the amplitude as well as the polarity of the electric field. Depending on the sign of the electric field molecules rotate clockwise or anticlockwise to become tilted with respect to their initial direction of orientation defined by the rubbing direction. With increasing voltage the tilt angle also decreases. When such a cell is placed between crossed polarizers the transmission through the cell increases with increasing electric field. However, in a display a short pulse of a voltage is applied across such a cell by an electronic switch such as a transistor. In this way the cell is charged up by applying a certain voltage across and directly isolating it so that the field remains across the cell until the next pulse. In such an addressing scheme a reset pulse of 60 μs and −6 V was applied in order to bring the system to an identical starting situation followed 100 μs later by a 60 μs write pulse with a varying magnitude. This pulse sequence was repeated at 50 Hz. The voltages quoted here refer to the source voltage applied to the transistor. Figure 3.24 shows the transmission voltage characteristics of a gel. The effective voltage across the

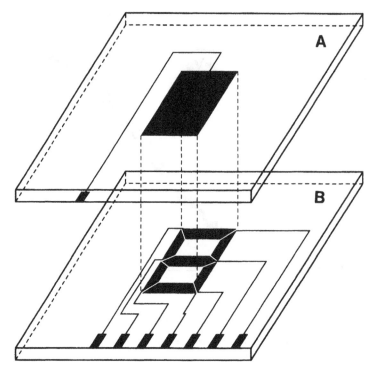

Figure 3.25 LC cell construction.

cell is lower since some of the charges are compensated by polarization reversal. However, it can be seen that by using active matrix addressing smooth transmission–voltage curves are obtained using gels, showing that they can also be used in active addressing to obtain continuous grey levels.

3.5 Structured Gels

In most display devices the contrast is produced by applying an electric field between electrodes and changing the orientation of the LC molecules. In order to make patterns appear in a display cell it is therefore necessary to apply the electric field locally, thus altering the orientation of the molecules. In most cases this is achieved by patterning electrodes on both surfaces in a display cell and aligning them precisely. An LC cell containing patterned electrodes is shown in Figure 3.25. It can be seen that in order to avoid the appearance of the connections leading to the display segments on the B surface, the indicated area of the electrode on the A surface needs to be patterned as well. Furthermore these patterned electrodes needed to be aligned precisely. Although this is done in the present LC cell technology, it is obviously interesting to be able to pattern only one of the electrodes, leaving the other unpatterned. As in such a cell undesired lines leading to the display elements will also appear, it is necessary to deactivate the LC material in some parts of the cell. This can be achieved by the creation of an anisotropic gel. The display cell with a transparent patterned and an unpatterned electrode is filled with an LC mixture containing reactive LC molecules. The cell is later irradiated through a mask so that only the areas desired to be activated by the electric field are not exposed to UV radiation. In the case of Figure 3.25 only the area marked on the A surface is thus masked off and the rest of the cell irradiated. In this way a patterned anisotropic gel containing areas without network molecules is formed. In these patterned gels when an electric field is applied to address an element an electric field also exists along the lines leading to the element since the other surface is totally covered with the electrode. However, owing to the polymer network the voltage required to change the orientation of the molecules in the unirradited areas (areas without the network) causes no displacement of the molecules. In this way, even though only one of the electrodes is patterned upon appliction of an electric field, only the elements required to be addressed appear, while the connections to the segments remain invisible.

3.6 Conclusions and Outlook

In conclusion it has been shown that anisotropic gels obtained by photopolymerization of reactive molecules in the LC state provide a unique method of stabilizing the orientation of the LC molecules in the bulk. This new method has already been used to produce new optical effects as well as improve the properties of existing devices. However, much research is still needed to build a better understanding of the gels and explore the possibilities offered by them.

References

BARRET, K. E. (1975) *Dispersion Polymerisation in Organic Media*, London: John Wiley.

BEHRENS, U. and KITZEROW, H.-S. (1994) Utility of liquid crystalline diacrylates for bistable cholesteric gel displays, *Pol. Adv. Technol.* **5**, 1.

BLUMSTEIN, A., KITAGAWA, N. and BLUMSTEIN, R. (1971) Polymerisation of p-(methacryloyloxy)- benzoic acid within liquid crystalline media, *Mol. Cryst. Liq. Cryst.* **12**, 215.

CLARK, N. A. and LAGERWALL, S. T. (1980) Submicrosecond bistable electro-optic switching in liquid crystals, *Appl. Phys. Lett.* **36**, 899.

DE JEU, W. H. (1980) *Physical Properties of Liquid Crystal Materials*, Gordon and Breach: Reading, MA.

DE VRIES, H. (1951) Rotary power and other optical properties of certain liquid crystals, *Acta Crystallogr.* **4**, 219.

DOANE, J. W., VAZ, N. A., WU, B. G. and ZUMER, S. (1986) Field controlled light scattering from nematic microdroplets, *Appl. Phys. Lett.* **48**, 296.

DOANE, J. W., GOLEMME, A., WEST, J. L., WHITEHEAD, J. B. and WU, B. G. (1988) Polymer dispersed liquid crystals for display applications, *Mol. Cryst. Liq. Cryst.* **165**, 511.

DRZAIC, P. (1986) Polymer dispersed nematic liquid crystal for large area display and light valves, *J. Appl. Phys.* **60**, 2142.

DUSEK, K. (1971) *Polymer Networks Structure and Properties*, ed. A.J. Chompff and S. Newmann, London: Plenum Press.

FERGASON, J. L. (1985) Encapsulated liquid crystal, *SiD Dig. Tech. Pap.* **16**, 68.

GOODMAN, L. A. (1974) *Introduction to Liquid Crystals*, ed. E.B. Priesly, P.J. Wojtowicz and P. Sheng, London: Plenum Press.

HARTMANN, W. J. A. M. (1991) Ferroelectric liquid crystal displays for television applications, *Ferroelectrics* **122**, 1.

HASEBE, H., TAKATSU, H., IIMURA, Y. and KOBAYASHI, S. (1994) Effect space of polymer network made of liquid crystalline diacrylate on characteristics of liquids crystal device, *Jpn. J. Appl. Phys.* **33**, 6245.

HERMANS, J. J. (1978) *Polymer Solution Properties*, Parts 1 & 2, Pennsylvania: Hutchinson and Ross.

HIKMET, R. A. M. (1990) Electrically induced light scattering from anisotropic gels, *J. Appl. Phys.* **68**, 4406.

HIKMET, R. A. M. (1991), Anisotropic gels and plasticised networks formed by liquid crystal molecules, *Liq. Cryst.* **9**, 405.

HIKMET, R. A. M. (1992a) Electrically induced light scattering from anisotropic gels with negative dielctric anisotropy, *Mol. Cryst. Liq. Cryst.* **213**, 117.

HIKMET, R. A. M. (1992b) Anisotropic gels in liquid crystal devices, *Adv. Mater.* **4**, 679.

HIKMET, R. A. M. (1992c) Piezoelectric networks obtained by photopolymerization of liquid crystal molecules, *Macromolecules*, **25**, 5759.

HIKMET, R. A. M. and BOOTS, H. M. J. (1995) Domain structure and switching behaviour of anisotropic gels, *Phys. Rev. E* **51**, 5824–5830.

HIKMET, R. A. M. and HOWARD, R. (1993) Structure and properties of anisotropic gels and plasticised networks containing molecules with a smectic-A phase, *Phys. Rev. E* **48**, 2752.

HIKMET, R. A. M. and MICHIELSEN, M. (1995) Anisotropic network stabilised ferroelectric gels, *Adv. Mater.* **7**, 300.

HIKMET, R. A. M. and DE WITZ, C. (1991) Gel layers for inducing adjustable pretilt angles in liquid crystal systems, *J. Appl. Phys.* **70**, 1265.

HIKMET, R. A. M. and ZWERVER, B. H. (1991a) Cholesteric networks containing free molecules, *Mol. Cryst. Liq. Cryst.* **200**, 197.

HIKMET, R. A. M. and ZWERVER, B. H. (1991b) Dielectric relaxation of liquid crystal molecules in anisotropic confinements, *Liq. Cryst.* **10**, 835.

HIKMET, R. A. M. and ZWERVER, B. H. (1992) Structure of cholesteric gels and their electrically induced light scattering, *Liq. Cryst.* **12**, 319.

HIKMET, R. A. M. and ZWERVER, B. H. (1993) Cholesteric gels formed by LC molecules and their use in optical storage, *Liq. Cryst.* **13**, 561.

HIKMET, R. A. M., BOOTS, H. M. J. and MICHIELSEN, M. (1995a) Ferroelectric liquid crystal gels – network stabilized ferroelectric display, *Liq. Cryst.* **19**, 65–74.

HIKMET, R. A. M., BOOTS, H. M. J. and MICHIELSEN, M. (1995b) Anisotropic network stabilized ferroelectric gels for active matrix addressing, *J. Appl. Phys.* (in press).

JÁKLI, A., BATA, K., FODOR-KSOBA, K., ROSTA, L. and NOIREZ, L. (1994) Structure of networks dispersed in liquid crystals: Small angle neutron study, *Liq. Cryst.* **17**, 227.

KIEFER, R. and BAUER, G. (1989) Molecular biaxiality in nematic liquid crystals as studied by infrared dichroism, *Mol. Cryst. Liq. Cryst.* **174**, 101.

KITZEROW, H.-S., MOLSEN, H., and HEPPKE, G. (1992) Helical unwinding in polymer-dispersed ferroelectric liquid crystals, *Pol. Adv. Technol.* **3**, 231.

KLOOSTERBOER, J. G. (1988) Network formation by chain crosslinking photopolymerisation and its applications in electronics, *Adv. Polym. Sci.* **84**, 1.

LECOIN, D., HOCHAPFEL, A. and VIOVYL, R. (1975) Comportement de la phase n dans des reactions de polymerisation, *J. Chim. Phys. Phys.-Chim. Biol.* **72**, 1029.

MANSURIPUR, M. (1990) Analysis of multilayer thin-film structures containing magneto-optic and anisotropic media at oblique incidence using 2×2 matrices, *J. Appl. Phys.* **67**, 6466.

MILES, M. J. (1987) *Developments in Crystalline Polymers*, ed. D.C. Basset, Vol. 2, Amsterdam: Elsevier.

PIRŠ, J., BLINC, R., MARTIN, B., MUŠEVIČ, I., PIRŠ, S., ŽUMER, S. and DOANE, J. W. (1992) *14th ILCC*, Poster C-P89.

SANUI, K., IMAMURA, H. and OGATA, N. (1977) Radical polymerisation of methacrylates in liquid crystalline solvents, *Chem. Abstr.* **86**, 171985e.

YANG, D. K. and DOANE, J. W. (1992) Cholesteric liquid crystal/polymer gel dispersion: Reflective display application, *SID 92 Dig.* p. 759.

YANG, D.-K., WEST, J. L., CHIEN, L.-C. and DOANE, J. W. (1994) Control of reflectivity and bistability in displays using cholesteric liquid crystals, *J. Appl. Phys.* **76**, 131.

4

Polymer Network Assemblies in Nematic Liquid Crystals

S. ŽUMER and G. P. CRAWFORD

4.1 Introduction

Low-concentration polymer networks assembled in an anisotropic liquid crystal environment have recently attracted a lot of attention because of their potential use in flat-panel displays (Hikmet, 1990; Yang *et al.*, 1992) and their interesting physical phenomena (Crawford *et al.*, 1994; Jákli *et al.*, 1994; Hikmet and Howard, 1993). These networks of very diluted polymers are capable of capturing the image of the orientational order of the nematic liquid crystal environment in which they are assembled (Crawford *et al.*, 1994). In the last few years polymer network assemblies have been successfully used in many new electro-optic technologies such as the normal and reverse mode light shutters (Yang *et al.*, 1992) and bistable reflective mode display (West *et al.*, 1993; Yang *et al.*, 1994a,b).

In addition, polymer network assemblies have ameliorated intrinsic problems which have hindered the development of several current liquid crystal display technologies; reduced the annoying off-axis haze in light-scattering-based devices (Yang *et al.*, 1992); alleviated mechanical shock failure and precision cell gap tolerances in ferroelectric and antiferroelectric liquid crystal displays via volume stabilization (Pirš *et al.*, 1995); eliminated the stripe deformations that occur in 270° super-twisted-nematic (STN) displays (Fredley *et al.*, 1994); allowed non-rubbed alignment techniques in nematic (Jain and Kitzerow, 1994a) and smectic (Jain and Kitzerow, 1994b) based displays; and modified twisted-nematic displays (Bos *et al.*, 1993; Hasebe *et al.*, 1994). (For more details see the chapters in this volume by Yang *et al.*, West, Yuan, Bos *et al.* and Iwamoto *et al.*) From the basic science perspective, these liquid crystal dispersion materials are soft condensed matter composites. They share many physical properties analogous to liquid crystals confined to non-planar, well-defined, sub-micrometre-sized spherical and cylindrical cavities (Golemme *et al.*, 1988; Crawford *et al.*, 1993) and random porous matrices (Bellini *et al.*, 1992; Wu *et al.*, 1992; Iannachione *et al.*, 1993; Clark *et al.*, 1993; Maritan *et al.*, 1994). (Short reviews on these related systems are given in the introduction and chapters by Kitzerow, Bellini and Clark, Finotello *et al.*, Tripathi and Rosenblatt, and Aliev.)

The success of polymer network assemblies in electro-optic and flat-panel display applications has been the principal driving force behind many of the basic studies on the structure and morphology of the polymer network. The clues used to identify these complex networks were inferred from studies of their physical properties: electron microscopy (Mariani *et al.*, 1986; Hikmet, 1991), birefringence (Hikmet, 1990, 1991), dichroism (Hikmet and Howard,

1993) and dielectric spectroscopy (Hikmet and Zwerver, 1991) in dispersions of high polymer concentration, and from magnetic resonance (Stannarius *et al.*, 1991), diamagnetic and viscosity measurements (Jákli *et al.*, 1992), small-angle neutron scattering measurements (Jákli *et al.*, 1994), birefringence (Crawford *et al.*, 1995) and nuclear magnetic relaxation (Vilfan *et al.*, 1995). (For more details see the chapters by Hikmet, Yang *et al.*, Jákli *et al.*, and Vilfan and Vrbančič.) Polymer-network-induced nematic ordering on micrometre and nanometre scales governs the optical properties of nematic and isotropic phases which is the focus of this contribution. The polymer network assemblies capture both simple director fields and defects. Simple phenomenological models allow the analysis of network-induced optical birefringence. Particular attention is focused on the structural information which can be obtained by a careful analysis of the experimental data.

4.2 Materials Preparation

Samples were prepared by dissolving 2–4 wt.% of the diacrylate monomer 4,4'-bis-acryloyl-biphenyl (BAB) and approximately 0.5 wt.% of the photoinitiator benzoin methyl ether (BME) into the room temperature nematic liquid crystal 4'-pentyl-4-cyanobiphenyl (5CB) (Stannarius *et al.*, 1991). The monomers are chemically similar to the 5CB so that if an oriented network assembles, it would have similar optical properties as the liquid crystal with similar degrees of orientational order. The nematic–isotropic (NI) transition temperature of the bulk 5CB is 35°C. The sample was heated to about 100°C and rigorously agitated to ensure a homogeneous mixture. From the mixture, two types of network assemblies in liquid crystals were prepared: one type formed in stable spherical nematic droplets and another in cylindrical capillary tubes. The droplet samples were prepared by emulsifying approximately 2 wt.% of the mixture into a glycerin matrix. By agitating the system, spherical droplets of 25–40 μm in diameter were formed. The glycerin matrix induces homogeneous anchoring conditions. For hometropic anchoring conditions, the glycerin matrix is doped with a 5 wt.% lecithin surfactant (Volovik and Lavrentovich, 1983). The samples were polymerized by exposing them to UV radiation emitted by a high-pressure mercury lamp at room temperature ($\sim$13° below the bulk NI transition for 5CB) for approximately 1 hour. Samples of droplets with aligned director fields were made by sandwiching the materials between transparent conducting substrates (coated with indium tin oxide (ITO)) and aligned by concurrently applying a 400 Hz AC voltage during the photopolymerization process.

For cylindrical capillary samples (diameters 100–900 μm), the cavity walls were treated with a lecithin surfactant prior to the introduction of liquid crystal (Crawford *et al.*, 1992b; Scharkowski *et al.*, 1993) to ensure homeotropic anchoring. The samples were then subjected to UV radiation for approximately 1 hour. Also, a strong electric field applied perpendicular to the cylindrical axis was used to induce a homogeneous director field during photopolymerization.

4.3 Measurements of Optical Anisotropy

Long-range orientational order of the liquid crystal molecules with anisotropic optical properties yields macroscopic birefringence of the material. In a composite system there are contributions arising from all components. In our system both comonents are relatively well dispersed on the scale comparable with the wavelength of visible light. Therefore it is natural to define the effective dielectric anisotropy of our medium by a spatial average over scales smaller than the wavelength of light. Introducing the local order parameter Q to describe the degree of molecur orientational order measured against the nematic director, the effective dielectric anisotropy is simply given by $\langle \Delta \varepsilon \rangle = \langle \Delta \varepsilon_0 Q \rangle$ where $\Delta \varepsilon_0$ is the aniso-

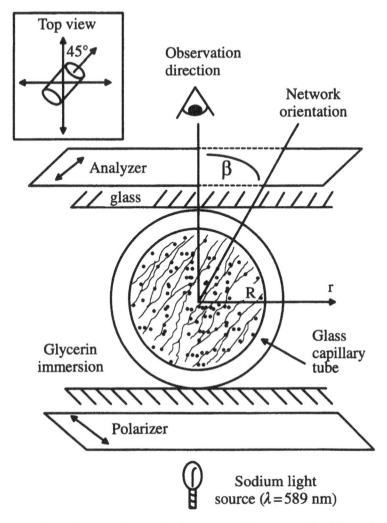

Figure 4.1 Experimental setup used to study polymer networks in capillary tubes (also used to study polymer networks in spherical droplets).

tropy in case of complete order. If $\Delta\varepsilon_o$ is much smaller than the dielectric constant of the medium, the effective optical birefringence is well described by $\langle\Delta n\rangle = \langle\Delta n_o Q\rangle$, where Δn_o is the local birefringence of the perfectly ordered material. The optical axis is everywhere parallel to the nematic director. If the nematic director field varies in space one can, at least in principle, by studying different birefringent textures obtained by optical polarizing microscopy, determine both the director field and the degree of orientational order.

The director field is typically identified from the general features of the textures (Ondris-Crawford *et al.*, 1991). As we will see later in these composite systems the presence of a substantial disorder can make this possible only in the high-temperature (low birefringent) phase.

Let us describe the details of the approach developed by Crawford *et al.* (1995) to investigate the low birefringence of polymer network systems in a liquid crystal solvent above the NI phase transition T_{NI}. Instead of preparing planar samples oriented by a treated surface (see the chapter by Yang and coworkers), samples were prepared in cylindrical capillary tubes. This approach enables large capillary tubes to be employed (up to 1 mm diameter) making these fragile, very-low-concentration samples robust and easily assessable. The experimental apparatus presented in Figure 4.1 allows one to record the interference pattern resulting after the ordinary and extraordinary beams experience different phase shifts traversing through the birefringent material. To measure the effective $\langle \Delta n \rangle$, we calculate the interference patterns for a cylindrical tube between crossed polarizers with a homogeneously aligned director field $\mathbf{n}$ (see Figure 4.1). The intensity of the transmitted light of wavelength λ at 45° with respect to the plane defined by the director field $\mathbf{n}$ (optic axis) and the incoming light direction $\mathbf{k}$ is given by the simple expression

$$I = I_0 \sin^2(\delta/2) \tag{4.1}$$

where δ is the phase shift between the ordinary and extraordinary components of light given by $2\pi \Delta n(\beta) d/\lambda$, $d = 2(R_0^2 - r^2)^{1/2}$ is the relevant thickness of the cylinder as a function of position and R_0 is the radius of the cylinder; r is the distance from the symmetry plane of the capillary tube, and β is the angle between the vectors $\mathbf{k}$ and $\mathbf{n}$ (see Figure 4.1). The effective birefringence in this configuration is

$$\langle \Delta n(\beta) \rangle = \langle [\sin^2\beta + (1 + \Delta n/n_o)^{-2} \cos^2\beta]^{-1/2} - 1 \rangle n_o \tag{4.2}$$

and simplifies to $\langle \Delta n(\beta) \rangle = \langle \Delta n \rangle \sin^2\beta$ when $\langle \Delta n \rangle \ll n_o$ (ordinary index of refraction) as in our case. The phase shift, δ, is thus given by the expression

$$\delta = 4\pi (R_o^2 - r^2)^{1/2} \langle \Delta n \rangle \sin^2\beta/\lambda \tag{4.3}$$

and substituted into eqn (4.1) generates the intensity distribution of the transmitted light $I(\rho)$. The values of $\langle \Delta n \rangle$ are determined by fitting the experimental intensity function $I(\rho)$. (For details and limitations see Scharkowski *et al.* (1993). A simulated intensity distribution is presented in Figure 4.2 and is directly compared with the experiment.

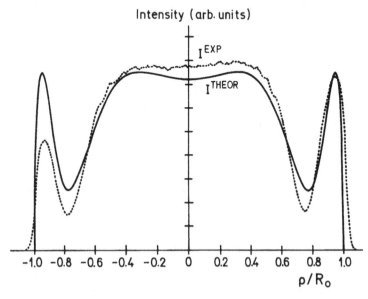

Figure 4.2 Calculated interference patterns of a homogeneously aligned polymer network in a capillary tube directly compared with experiment. The experimental intensity distributions were recorded at $T = 40$°C for the 2 wt.% BAB network in the 5CB liquid crystal. The capillary tube was oriented at 45° between crossed polarizers with $\beta = 90°$, $\lambda = 589$ nm (sodium line). The sample was oriented at $\beta = 90°$.

4.4 Oriental Order Captured by Polymer Networks

Nematic order in the prepolarization state is characterized by the director field and degree of order. Assembling a polymer network from monomers of liquid-crystalline-like molecules in a nematic solution certainly yields a trace of the initial structure. Here we describe a few simple examples where optical properties of the composite system can unveil the image memorized by polymer networks.

4.4.1 Nematic Director Fields

Before we consider particular examples, it is worth while to review the elastic continuum theory that predicts nematic director-field configurations in supramicrometre-size geometries under stable conditions (Kralj and Žumer, 1992; Crawford *et al.*, 1992a). The nematic director profile depends on the balance of elastic, surface and field forces. The total free energy of the liquid crystal needed to deduce equilibrium supramicrometre structures can be expressed as (Ondris-Crawford *et al.*, 1993)

$$F = \tfrac{1}{2} \int_V \, [K_{11}(\text{div } \mathbf{n})^2 + K_{22}(\mathbf{n} \cdot \text{curl } \mathbf{n})^2 + K_{33}(\mathbf{n} \times \text{curl } \mathbf{n})^2$$

$$- K_{24} \, \text{div}(\mathbf{n} \times \text{curl } \mathbf{n} + \mathbf{n} \, \text{div } \mathbf{n}) + K_{13} \, \text{div}(\mathbf{n} \, \text{div } \mathbf{n})$$

$$- \Delta\varepsilon\varepsilon_o(\mathbf{E} \cdot \mathbf{n})^2 - (\Delta\chi/\mu_o)(\mathbf{B} \cdot \mathbf{n})^2] \, \mathrm{d}V$$

$$+ \tfrac{1}{2} \int_S \, (W_\phi \sin^2 \phi + W_\theta \cos^2 \phi)\sin^2\theta \, \mathrm{d}S \qquad (4.4)$$

where K_{11}, K_{22} and K_{33} are the splay, twist and bend bulk elastic constants; K_{24} and K_{13} are the saddle–splay and mixed splay–bend surface elastic constants; W_ϕ and W_θ are the azimuthal and polar surface anchoring strengths; $\Delta\chi$ is diamagnetic anisotropy; and $\mathbf{E}$ and $\mathbf{B}$ are the magnitudes of the electric and magnetic applied fields, respectively. Furthermore, V and S denote the volume and surface of the system; the angles ϕ and θ represent the deviation of the director at the surface away from the preferred in-plane direction and the deviation away from the surface normal, respectively. (Specific details concerning the surface effects are discussed in the chapter by Barbero and Durand.) Minimization of this kind of free energy has been extensively used to predict stable nematic director-field configurations in planar (Lavrentovich and Pergamenshchik, 1994) (see also the chapter by Barbero and Durand), spherical (Kralj and Žumer, 1992) and cylindrical (Allender *et al.*, 1991) geometries.

4.4.1.1 Polymer networks in droplets

In the case of homogeneous planar anchoring characteristic of the glycerin matrix, the stable director field in spherical droplets is of the bipolar type (Volovik and Lavrentovich, 1983) presented schematically in Figure 4.3(a). The bipolar director field is a splay–bend-type deformation and is characterized by two point defects known as boojums at the poles of the droplets. Details of the structure depend on the ratio of the elastic constants, anchoring strengths, droplet size, and magnitude of the external field. After photopolymerization during which the assembled network memorizes the original director field one would expect the texture of the nematic structure, but this is not the case. The birefringent texture of the droplets in the nematic phase is practically obscured by strong light scattering (see Figure 4.3(a)). Even though the network on the average grows along the director direction a certain

87

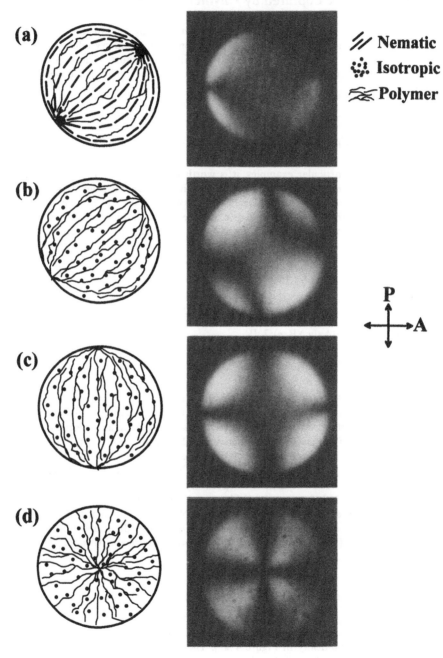

Figure 4.3 Capturing the image of the orientational order in nematic liquid crystal droplets; schematic presentations alongside the optical polarizing microscope photographs. The pure glycerin binder favours homogeneous alignment and results in a bipolar configuration of droplet diameter of approximately diameter of approximately 25 μm: the polymerized polymer network (2 wt.%) in the nematic phase (a), and in the isotropic phase at various orientations of the symmetry axis of the droplet (b) and (c). The doped lecithin binder favours homeotropic anchoring and results in 40 μm droplets with radial director field (d).

amount of cross-linking takes place. The scattering indicates that the resulting inhomogeneities in the orientational ordering, although not directly visible under the microscope, are certainly on the scale of the wavelength of light. Therefore from the nematic textures it is impossible to determine how well the network memorizes the original director field.

Above the NI transition of the bulk 5CB a different situation occurs. The scattering is negligible but the residual optical anisotropy indicates the presence of the oriented polymer network. The birefringent textures (see Figures 4.3(b) and (c)) demonstrate that the polymer network has completely captured the image of the director field of the bipolar droplet during photopolymerization. To understand the difference in behaviour between the network dispersed in the nematic and isotropic liquid crystalline phases, we estimate the phase shift $\delta = \langle \Delta n \rangle d/\lambda$ between ordinary and extraordinary light beams traversing a distance d through a droplet. For $\delta \sim \pi$, the microscope texture has full intensity and will only have fringes if $\delta > 2\pi$. The absence of fringes and high intensity of the texture above T_{NI} indicates that $\langle \Delta n \rangle \sim 0.05$. To estimate the scattering we assume that the sample consists of domains with size $D < \lambda$. Using the anomalous diffraction approximation for scattering (Žumer, 1988) one can estimate the cross-section of such a domain: $\sigma \sim (\Delta n_o Q)^2 D^4/\lambda^2$. Furthermore, following the work of Žumer et al. (1989) and taking D^{-3} for the density of randomly oriented domains, one can estimate the extinction coefficient as $\sigma/D^3 \sim (\Delta n_o Q)^2 D/\lambda^2$. For a droplet of diameter d with inhomogeneities on the length scale $D \sim \lambda$, the relative amount of the transmitted light, I_{trans}, is approximately given by $\exp[-(\Delta n_o Q)^2 d/\lambda]$. For a droplet of diameter 20 μm on crossing the T_{NI} transition point the exponent is reduced from approximately 1 to 10^{-2}. This estimate explains strong light scattering in the nematic phase and no scattering in the isotropic phase.

In large droplets when the matrix surrounding the droplets induces homeotropic anchoring the radial nematic director field (Figure 4.3(d)) is stable (Kralj and Žumer, 1992). Such a structure has only a splay deformation with a point defect in the centre of the droplet topologically classified as a hedgehog. Prior to photopolymerization in the nematic phase, the microscope textures of all droplets in the sample were the typical star pattern and invariant with respect to rotations of the microscope stage, thus proving that the director field was radial (Ondris-Crawford et al., 1991; Xu et al., 1994). The radial droplets were also strongly scattering light in the nematic phase after photopolymerization, similar to the bipolar droplets shown in Figure 4.3(a). Above the NI transition, the star texture was apparent (see Figure 4.3(d)), indicating that the polymer network has captured the image of the orientational order of the radial droplet.

Another interesting feature of polymer network assemblies is their ability to capture metastable or even unstable defect structures. In large droplets with homeotropic boundary conditions, an unstable hedgehog defect can occur in a non-symmetric position inside the spherical droplet. Volovik and Lavrentovich (1983) have extensively studied the topological dynamics of similar defects in liquid crystal droplets by varying the surface interaction energies of the liquid crystal molecules at the droplet wall. By doping the liquid crystal droplet/glycerin suspension with a lecithin surfactant, the bipolar droplet was shown to transform continuously into the radial droplet (Volovik and Lavrentvich, 1983). We have captured indefinitely two of the intermediate steps in the transformation process that are shown in Figures 4.4(a) and (b).

The radial droplet can also be transformed into the axial structure (splay–bend deformation) via an applied external electric field. Erdmann et al. (1990) extensively studied this transformation. We applied 1.5 V/μm and 3.5 V/μm electric fields to the radial droplet to induce the axial configuration. The photopolymerization was carried out concurrently with the application of the electric field. After photopolymerization was complete, the electric field was removed. The textures above the NI transition presented in Figures 4.4(c) and (d) show that the axial transition is captured indefinitely. The large black region in the microscope texture in Figure 4.4(d) corresponds to the low curvature of the director field in the centre of the droplet caused by the large electric field applied during photopolymerization.

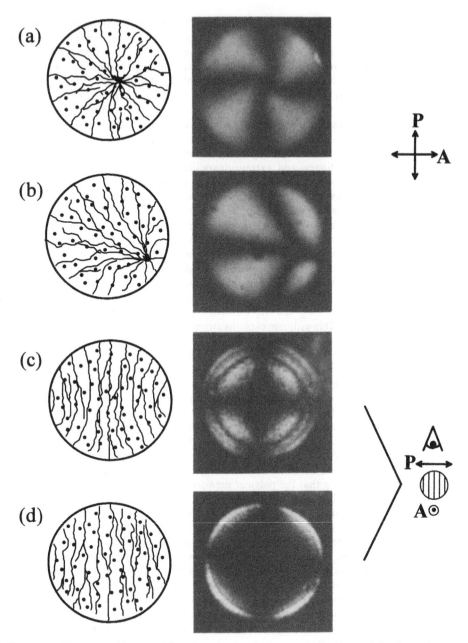

Figure 4.4 The captured images of the orientational order in nematic liquid crystal droplets with homeotropic anchoring; schematic presentations alongside the optical polarizing microscopy photographs. The point defect is outside its centre symmetric position (a) and (b). The axial director field configuration after being polymerized in the presence of a 1.5 V/μm field (c) and (d) 3.5 V/μm electric field; the droplets viewed parallel to the symmetry axis are shown in the isotropic phase and are approximately 24 μm in diameter.

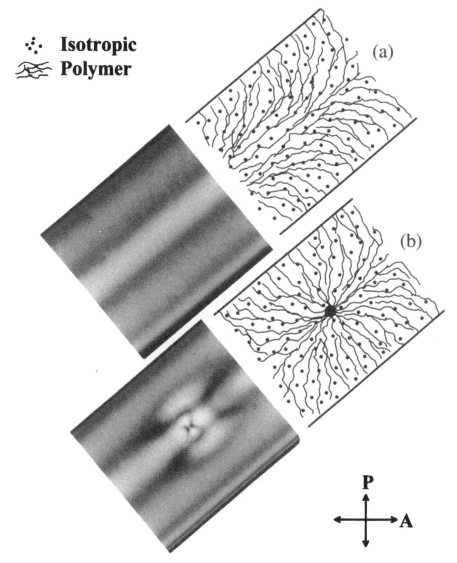

Isotropic

Polymer

Figure 4.5 The captured image of the orientational ordering of the escaped-radial director field stable in 150 μm capillary tubes with lecithin surface treaments. The optical polarizing microscope photographs are recorded at 25 K above the nematic–isotropic transition temperature. The photograph and schematic presentation are shown in (a) and those with a point defect are shown in (b).

4.4.1.2 *Polymer networks in capillary tubes*

The polymer network inside a capillary tube was assembled in the escaped-radial (splay–bend deformation) nematic director field which is radial at the surface and escapes along the axial direction through the central region of the tube (Cladis and Kléman, 1972; Crawford *et al.*, 1992a). In the nematic phase after photopolymerization, the birefringent texture shows a significant incease in light scattering. At temperatures above the bulk NI transition, a weak optical anisotropy remains similarly as in the droplet case. The textures above the

NI transition shown in Figure 4.5 reveal that the polymer network assembly has captured and stored the details of the escaped-radial director field. The interference colours observed with white light in a cylinder with $R_o = 150$ μm indicate low birefringence.

Crawford *et al.* (1994) also studied the stability of the network by substituting the liquid crystal with another solvent. The polymer network assembly in the liquid crystal was extracted from the capillary tube into a dish of hexane. The sample was left in the solvent for several days to ensure that all of the liquid crystal was dissolved out of the network. The sample was then placed between crossed polarizers of an optical microscope (see book cover); the resulting texture reveals that the polymer network assembly has retained the orientational order of the escaped-radial director field. An interesting aspect of these materials is their stability. After extraction and solvent evaporation, the network has shrunk to approximately one-third its original size. The network can then be permeated with either a liquid crystal or an isotropic fluid and will approximately resume its original size. Further inspection of the network revealed that it retained its original configuration prior to processing.

4.4.2 Order Parameter

The effective degree of order of a composite polymer network–liquid crystal system depends on both local (molecular) and mesoscopic (submicrometre to micrometre) scale ordering. To obtain quantitative information about different aspects of the ordering Crawford *et al.* (1995) performed several measurements on the birefringence of the liquid crystal–polymer network assembly in the isotropic liquid crystal phase after the sample was formed in the initially homogeneous state in the nematic phase.

4.4.2.1 Isotropic network in a liquid crystal

The UV photopolymerization performed in the isotropic phase results in a polymer network assembly which, between crossed polarizers in the nematic phase, was highly light scattering and completely black when viewed above the NI transition temperature. This experiment reveals that the polymer network assembled in the isotropic phase does not possess any long-range order. Such a situation prevents us from using birefringence measurements for a local-order study. If there was any residual local order, it was completely averaged out by a beam sampling randomly oriented areas of local order.

4.4.2.2 Ordered network in a liquid crystal

The pretransitional increase of the birefringence shown in Figure 4.6(a) as a function of temperature for the 4 wt.% BAB sample suggests that in addition to the contribution of the polymer network assembly, there is also a contribution from the paranematic order induced in the isotropic liquid crystal phase by the internal surfaces of the polymer network assembly. The paranematic order induced by different surfaces far above T_{NI} has been observed in many liquid crystal systems: for example, Chen *et al.* (1989) and Moses and Shen (1991) on planar surfaces, Golemme *et al.* (1988) in polymer-dispersed liquid crystal systems, and Crawford *et al.* (1993) in cylindrical cavities. The variation of the birefringence as a function of the initial concentration of the BAB is shown in Figure 4.6(b).

4.4.2.3 Ordered network in an isotropic solvent

To decipher the relative contributions to the net birefringence of the polymer network and the surface-induced order, we replaced the liquid crystal which surrounds the network with

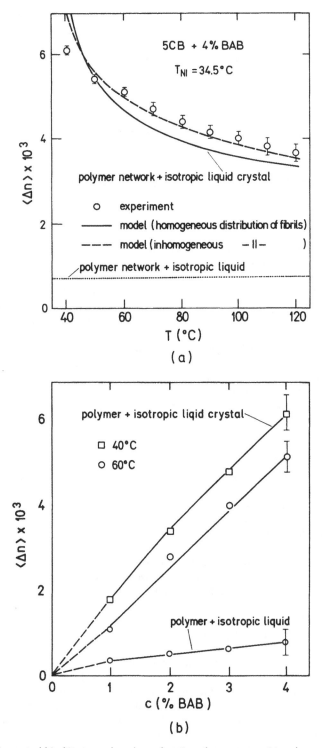

Figure 4.6 The optical birefringence plotted as a function of temperature (a) and concentration (b). The fits to two theoretical models are shown as well.

the isotropic fluid chlorobenzene. Since the chlorobenzene molecules are too spherical to expect any ordering at the polymer network surfaces, the birefringence can solely be attributed to the polymer network assembly. Applying the same method to measure the birefringence of the polymer network assembly in the isotropic solvent, $\langle \Delta n \rangle_{pil}$, as shown in Figure 4.2, was estimated to be between 3 and 8×10^{-4} depending on the network concentration c (see Figure 4.6(b)) but independent of temperature as shown in Figure 4.6(a). In not detecting any appreciable change in the volume of the network when the liquid crystal was replaced with the chlorobenzene, one can assume that the effective order parameter of the network is also the same in the liquid crystal matrix.

4.5 Modelling Polymer-induced Paranematic Ordering

4.5.1 *Uniformly Distributed Polymer Network*

Let us now first follow a simple analysis performed by Crawford *et al.* (1995). They went beyond the *two-fraction model* proposed by Hikmet and Howard (1993) (see also the chapter by Hikmet) that cannot be used to explain pretransitional ordering, by resorting to phenomenology employed to describe the surface-induced ordering in confined liquid crystals (Crawford *et al.*, 1993).Taking into account that the birefringence of both the liquid crystal and the polymer has its origin in the oriented biphenyl molecular core, one can simply assume that in a completely ordered state the optical anisotropy Δn_o is the same. Therefore from now on we use $\Delta n_o = 0.35$ obtained from data taken on the 5CB compound (Horn, 1978). As stated above, the lack of scattering above the T_{NI} transition ensures that the birefringence in all three examples is simply given by $\langle \Delta n \rangle \sim \Delta n_o \langle Q \rangle$ where only the effective $\langle Q \rangle$ depends on the sample. By decomposing $\langle Q \rangle$ into two contributions, the direct polymer network contribution and the fluid contribution, where order is induced by the presence of the oriented polymer, can be expressed as

$$\langle \Delta n \rangle = \Delta n_o [cQ_p + (1 - c)Q_f] \tag{4.5}$$

Here c is the concentration of the polymer, Q_p is the spatial average order parameter of the polymer network and Q_f is the average order parameter of the fluid.

In the case where the network is dissolved in the isotropic fluid, one finds that Q_p is between 0.08 and 0.05 for various values of c. The value of Q_p is substantially less than that of the liquid crystal where it was formed (~ 0.45), which indicates that a substantial amount of cross-linking has occurred.

In the case where the network is dissolved in the isotropic liquid crystal, a simple model is instructive; for example, assume that the network is covered by a layer of partially oriented liquid crystal. It is reasonable to assume that the average induced order in this layer is simply equal to Q_p; this is known as the strong coupling regime. To estimate the thickness of the ordered liquid crystalline layer, the nematic correlation length, ζ, which describes how far the orientational perturbation extends, can be used. According to Landau theory, ζ is given by $\zeta_o(T/T^* - 1)^{-1/2}$ with $\zeta_o = 0.65$ nm and $T^* = 306$ K for the 5CB compound. (For more details see de Gennes and Prost (1993)). Crawford *et al.* (1991, 1993) have shown that a better description of the surface layer is obtained if a layer of constant order of molecular thickness ζ_m is incorporated into the conventional profile. In the case of 5CB, it was found that ζ_m was approximately 1 nm. Introducing the surface of the polymer network per unit of its volume as α, the volume of the ordered liquid crystalline layer per unit of the polymer volume is simply $(\zeta + \zeta_m)\alpha$. So the birefringence $\langle \Delta n \rangle_{pilc}$ of the ordered polymer network in the isotropic liquid crystal can be (using eqn (4.5)) expressed in the following form:

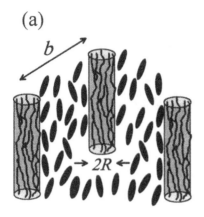
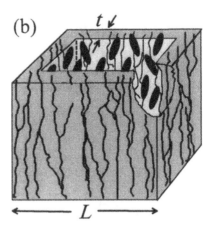

Figure 4.7 Simple model structures of the polymer network. (a) The fibril model, where parallel cylindrical fibrils of diameter $2R$ uniformly distributed form a square lattice with spacing b. (b) The sheet model, where polymer sheets of width L and thickness t form a square lattice.

$$\langle \Delta n \rangle_{\text{pilc}} = \Delta n_o c Q_p [1 + (\zeta + \zeta_m)\alpha] \tag{4.6}$$

Using eqn (4.6) to fit our experimental data in Figure 4.6 and adjusting only α, we determined $\alpha = 1 \pm 0.5$ nm^{-1}. Although the result is qualitative when the surface is not planar on the length scale ζ, one can use it to envision the polymer network distribution. Two very simple model structures (see Figures 4.7(a) and (b)) were used. (For related model structures see also the chapters by Hikmet and Yang *et al.*) The first model structure is a regular square array of cylindrical fibrils with radius R and interfibril distance b which yields the relative internal surface area as $\alpha = 2/R$ and the network concentration as $c = \pi R^2/b^2$. Consequently, a radius of the fibrils of $R \sim 2$ nm and an interfibril distance between $b = 16$ and 34 nm as c goes from 4% to 1% were obtained. The second model structure is a square lattice of polymer sheets of thickness t and intersheet distance L. The internal surface area of the sheet model is $\alpha = 1/t$ and the concentration is $c = 2t/L$. We can therefore estimate the sheet thickness to be $t \sim 1$ nm and the lattice distance to be between $L = 50$ and 200 nm depending whether c is 4% or 1%. Because of the symmetry of the nematic liquid crystal phase where the polymer network was assembled, the actual structure of the network is probably closer to the cylindrical fibril model. The conclusion of this very simple analysis is that both the fibril and sheet model structures indicate that at least one dimension of the network, either the fibril or sheet thickness, is comparable with molecular dimensions. In the following section we describe a more realistic model.

4.5.2 *Non-uniformly Distributed Polymer Network*

The oversimplified model presented above has two weak points: (i) the estimated order parameter is much lower than that of the nematic matrix where the network was formed and (ii) the rather poor description of the pretransitional temperature dependence of the birefringence. On the other hand, enhanced light scattering introduced by the polymer network in the nematic phase and scanning electron microscopy images (Crawford *et al.*, 1995; the chapter by Yang *et al.*) both indicate structures on the submicrometre scale. Recent small-angle neutron scattering studies by Jákli *et al.* (1994) using the same materials are consistent with bundles of mean diameter of about 60 nm. Their studies, however, are not sensitive to the pretransitional surface-induced order that the birefringence measurements reveal. All these inconsistencies can be summarized in a conclusion that the non-uniform network

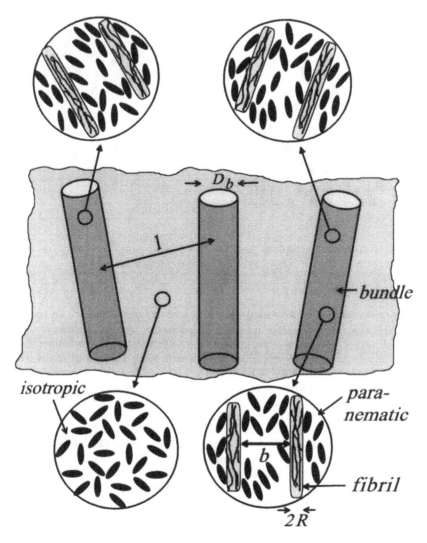

Figure 4.8 The model structure of an inhomogeneous network with bundles of polymer fibrils that are *polymer rich* in a *polymer-poor* environment. Within each bundle of average diameter $D_b \sim 60$ nm there are nanometre fibrils with radius R forming a regular net with interfibril distance b. The distance between bundles is l. The non-perfect alignment of fibrils and bundles is schematically shown as well.

distribution and the distinction between ordering on the molecular and mesoscopic scale have been neglected. This is also true of the approach followed by Yang *et al.* (this volume) where the induced order parameter profile around polymer fibre is better modelled than in our previous section.

We, therefore, follow modelling by Crawford *et al.* (1995) introducing a model structure (Figure 4.8) where molecular fibrils with nanometre-size radii surrounded by relatively thin layers of liquid crystal form bundles of *polymer-rich* material separated by a rather pure liquid crystal (*polymer-poor*) background. The order parameter of the network is described by $Q_p = Q_l Q_m$ where Q_l and Q_m represent the order on the local and mesoscopic scale, respectively. As thin fibrils grow in the oriented nematic phase one can expect that the local order parameter of the network is comparable with that of the matrix where it was assembled. With a lack of information we simply assume the equality yielding $Q_1 \sim 0.45$ for our situation. The disorder resulting from the cross-linking is assumed to be important on mesoscopic scales partially within a bundle and partially from one bundle to another bundle. Using $Q_p \sim 0.08$ to 0.05 one can conclude that $Q_m \sim 0.15$ to 0.1. Within this model structure the interfibril distances are much smaller than in the case of uniform distribution of fibrils; therefore saturation effects in the surface-induced ordering are more pronounced. To overcome this problem we follow Fung *et al.* (1995) and solve the problem of para-nematic ordering induced by an array of fibrils. To make the approach as simple as possible, within each bundle parallel fibrils of equal radius R are assumed to form a rectangular array with interfibril spacing b (Figure 4.8). The surface of the fibrils is assumed to induce nematic ordering along the fibril direction. Following the standard approach the Landau–de Gennes free energy

$$f = f_0 + \tfrac{1}{2}a(T - T^*)Q^2(\mathbf{r}) - \tfrac{1}{3}BQ^3(r) + \tfrac{1}{4}CQ^4(\mathbf{r}) + \tfrac{1}{2}L[\nabla Q]^2 + G[Q(\mathbf{r}) - Q_1]^2\delta(|\mathbf{r}| - R)$$

(4.7)

is used to describe paranematic ordering in the confined space. a, B, C, T^* and L are phenomenological material parameters of the bulk liquid crystal (here we use 5CB data (Coles, 1978), G is the surface coupling constant, and Q_1 is the preferred orientational order parameter induced in the liquid crystal at the polymer surface. This type of free energy has been successfully employed in describing the ordering in well-defined cylindrical geometries (Crawford *et al.*, 1993). The minimization yields a differential equation for the order parameter profile given by

$$a(T - T^*)Q(\mathbf{r}) - BQ^2(\mathbf{r}) + CQ^3(\mathbf{r}) - L\Delta Q = 0$$

(4.8)

and the surface condition is expressed by

$$L(\nabla Q)_\perp - 2G(Q - Q_1) = 0$$

(4.9)

which provides the boundary value of the solution. The profiles of numerical solutions are presented in Figure 4.9 and, using a three-dimensional visualization, in Figure 4.10. When

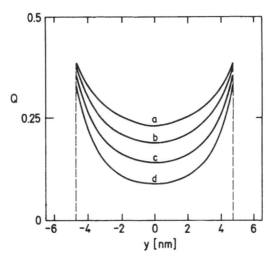

Figure 4.9 Paranematic order parameter profiles between two neighbouring fibrils as a function of the position (y) on a line along the diagonal direction in the rectangular lattice of fibrils. Curves a–d corresponds to $\Delta T = 12$, 15, 20 and 30 K respectively. The surface coupling constant $G = 1.2 \times 10^{-2}$ J/m^2, the lattice constant $b = 8$ nm and a fibril radius of 1.7 nm are used.

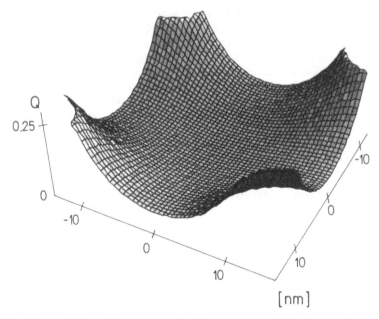

Figure 4.10 3D presentation of the paranematic order parameter profile in the space between the four closest fibrils. The parameters of Figure 4.9 are used.

interfibril distances are large, which seems not to be the real case, these solutions approach the analytical ones described in the chapter by Yang and coworkers.

Using the above solutions and implementing the values determined by Jákli *et al.* (1994) of 60 nm for the bundle diameter D_b, there are three parameters to be determined by the fitting procedure: fibril radius R; interfibril distance b; and surface coupling strength G. The fit of the experimental temperature dependence of the birefringence shown in Figure 4.6 yields $R \sim 1.7$ nm, $b \sim 8$ nm and $G \sim 1.2 \times 10^{-2}$ J/m^2. The model reasonably describes the saturation of the surface-induced ordering on approaching the T_{NI} transition which we were unable to describe with the simple model used in section 4.5.1. The value of 4 wt.% for the network concentration allows us to estimate interbundle distance l to be approximately 0.11 μm.

The values obtained for the fibril radius and interfibril distance are close to our first estimates and yield only a slightly different relative internal surface $\alpha = 1.2$ nm^{-1}. In addition the large polymer–liquid crystal coupling constant G (for details see Crawford *et al.* (1993) indicates that the induced liquid crystalline order parameter at the polymer surface is weakly temperature dependent, which is consistent with the findings of the NMR relaxation study (see the chapter by Vilfan and Vrbančič-Kopač; further the estimated interbundle distance seems to be reasonably consistent with the scanning electron microscope studies (see the chapter by Yang *et al.*). Altogether we can conclude that our visualization of the polymer network morphology describes reasonably well all experimental evidence.

4.6 Summary

We have presented a short review illustrating how simple birefringence measurements can be used to model the ordering, internal surface and distribution of polymer networks assembled in a liquid crystal environment. Independent of the details of the model structure used, the optical anisotropy data indicate that the network is finely distributed on the molecular level. Combining this fact with the evidence of substantial disorder on the submicrometre scale, a model structure is envisioned where molecular fibrils with a thickness of a few nanometres form bundles rich in polymer. We conclude that these bundles are responsible for the inhomogeneities on the mesoscopic scale.

Acknowledgements

We wish to acknowledge the NSF ALCOM Center at Kent State University for their support. We also wish to thank Professor Bill Doane for his pioneering work in this area and congratulate him on his 60th birthday.

References

ALLENDER, D. W., CRAWFORD, G. O. and DOANE, J. W. (1991) Determination of the liquid crystal surface elastic constant k_{24}, *Phys. Rev. Lett.* **67**, 1442–1445.

BELLINI, T., CLARK, N. A., MUZNY, D. D., WU, L., GARLAND, C. W., SCHAEFER, D. W. and OLIVER, B. J. (1992) Phase behavior of the liquid crystal 8CB in silica aerogel, *Phys. Rev. Lett.* **69**, 788–791.

BOS, P. J., RAHMAN, J. A. and DOANE, J. W. (1993) A low threshold voltage polymer network TN device, *Soc. Inf. Disp. Dig.* **XXIV**, 877–880.

CHEN, W., MARTINEZ-MIRANDA, L. J., HSIUNG, H. and SHEN, Y. R. (1989) Orientational wetting behavior of a liquid crystal homologous series, *Phys. Rev. Lett.* **62**, 1860–1863.

CLADIS, P. E. and KLÉMAN, M. (1972) Non-singular disclinations of strength $S = +1$ in nematics, *J. Phys.* **40**, 591–598.

CLARK, N. A., BELLINI, T., MALZBENDER, R. M., THOMAS, B. N., RAPPAPORT, A. G., MUZNY, C. D., SCHAEFER, D. W. and HRUBESH, L. (1993) X-ray scattering study of smectic ordering in a silica aerogel, *Phys. Rev. Lett.* **71**, 3505–3508.

COLES, H. J. (1978) Laser and electronic field induced birefringence studies on the cyanobiphenyl homologues, *Mol. Crys. Liq. Cryst.* **49**, 67–74.

CRAWFORD, G. P., YANG, D.-K., ŽUMER, S., FINOTELLO, D. and DOANE, J. W. (1991) Ordering and self diffusion in the first molecular layer of a liquid crystal–polymer interface, *Phys. Rev. Lett.* **66**, 723–726.

CRAWFORD, G. P., ALLENDER, D. W. and DOANE, J. W. (1992a) Surface elastic and molecular anchoring properties of nematic liquid crystals, *Phys. Rev. A* **45**, 8693–8708.

CRAWFORD, G. P., MITCHELTREE, J. A., BOYKO, E. P., FRITZ, W., ŽUMER, S. and DOANE, J. W. (1992b) K_{33}/K_{11} determination in nematic liquid crystals, *Appl. Phys. Lett.* **60**, 3226–3228.

CRAWFORD, G. P., ONDRIS-CRAWFORD, R., ŽUMER, S. and DOANE, J. W. (1993) Anchoring and orientational wetting transitions of confined liquid crystals, *Phys. Rev. Lett.* **70**, 1838–1841.

CRAWFORD, G. P., POLAK, R. D., SCHARKOWSKI, A., CHIEN, L.-C., ŽUMER, S. and DOANE, J. W. (1994) Nematic director fields captured in polymer networks cofined to spherical droplets, *J. Appl. Phys.* **75**, 1968–1971.

CRAWFORD, G. P., SCHARKOWSKI, A., FUNG, Y. K., DOANE, J. W. and ŽUMER, S. (1995) Nematic internal surface, orientational order, and distribution of a polymer network in a liquid crystal matrix, *Phys. Rev.* (to be published).

DE GENNES, P. G. and PROST, J. (1993) *The Physics of Liquid Crystals*, Oxford: Clarendon Press.

ERDMANN, J. H., ŽUMER, S. and DOANE, J. W. (1990) Configuration transition in nematic liquid crystals to a small spherical cavity, *Phys. Rev. Lett.* **64**, 1907–1910.

FREDLEY, D. S., QUINN, B. M. and BOS, P. J. (1994) Polymer stabilized SBE devices, *Conference Record of the 1994 Int. Display Research Conf.*, pp. 480–483.

FUNG, Y., YANG, D.-K., BORŠTNIK, A. and ŽUMER, S. (1995) Pretransitional ordering in isotropic liquid crystals with dispersed polymer networks, *Phys. Rev. E* (submitted).

GOLEMME, A., ŽUMER, S., ALLENDER, D. W. and DOANE, J. W. (1988) Continuous nematic–isotropic transition in submicrometer-size liquid crystal droplets, *Phys. Rev. Lett.* **61**, 1937–1940.

HASEBE, H., TAKATSU, H., IIMURA, Y. and KOBAYASHI, S. (1994) Effect of a polymer network made of diacrylate on the characteristics of a liquid crystal display, *Jpn. J. Appl. Phys.* **33**, 6245–6248.

HIKMET, R. A. M. (1990) Electrically induced light scattering from anisotropic gels, *J. Appl. Phys.* **68**, 4406–44012.

HIKMET, R. A. M. (1991) Anisotropic gels and plasticized networks formed by liquid crystal molecules, *Liq. Cryst.* **9**, 405–416.

HIKMET, R. A. M. and HOWARD, R. (1993) Structure and properties of anisotropic gels and plasticized networks containing molecules with a smectic A phase, *Phys. Rev. E* **45**, 2752–2759.

HIKMET, R. A. M. and ZWERVER, B. H. (1992) Dielectric relaxation of liquid crystal molecules in anisotropic confinements, *Liq. Cryst.* **10**, 835–847.

HORN, R. G. (1978) Refractive indices and order parameters of two liquid crystals, *J. Phys.* **39**, 105–109.

IANNACCHIONE, G. S., CRAWFORD, G. P., ŽUMER, S., DOANE, J. W. and FINOTELLO, D. (1993) Randomly constrained orientational order in porous glass, *Phys. Rev. Lett.* **71**, 2595–2598.

JAIN, S. C. and KITZEROW, H.-S. (1994a) Bulk induced alignment of nematic liquid crystals by photopolymerization, *Appl. Phys. Lett.* **64**, 2946–2948.

JAIN, S. C. and KITZEROW, H.-S. (1994b) A new method to align smectic liquid crystals by photopolymerization, *Jpn. J. Appl. Phys.* **33**, Part 2, L656–L659.

JÁKLI, A., KIM, D. R., CHIEN, L.-C. and SAUPE, A. (1992) Effect of a polymer network on the alignment and rotational viscosity of a nematic liquid crystal, *J. Appl. Phys.* **72**, 3161–3164.

JÁKLI, A., BATA, L., FODOR-CSORBA, K., ROSTA, L. and NOIREZ, L. (1994) Structure of polymer networks dispersed in liquid crystals: Small angle neutron scattering study, *Liq. Cryst.* **17**, 227–234

KRALJ, S. and ŽUMER, S. (1992) Freedericks transitions in supra-nematic droplets, *Phys. Rev. A* **45**, 2461–2470.

KRALJ, S. and ZUMER, S. (1993) The stability diagram of a nematic liquid crystal confined to a cylindrical cavity, *Liq. Cryst.* **15**, 591–598.

LAVRENTOVICH, O. D. and PERGAMENSHCHIK, V. M. (1994) Stripe domain phase of a thin nematic film and the K_{13} elasticity, *Phys. Rev. Lett.* **73**, 979–982.

MARIANI, P., SAMORI, B., ANGELONI, A. S. and FERRUTI, P. (1986) Polymerization of bisacrylic monomers within a liquid crystalline smectic B solvent, *Liq. Cryst.* **5**, 1477–1487.

MARITAN, A., CIEPLAK, M., BELLINI, T. and BANAVAR, J. R. (1994) Nematic-isotropic transition in porous media, *Phys. Rev. Lett.* **72**, 4113–4116.

MOSES, T. and SHEN, Y. R. (1991) Pretransitional surface ordering and disordering of a liquid crystal, *Phys. Rev. Lett.* **67**, 2033–2036.

ONDRIS-CRAWFORD, R. J., BOYKO, E. P., WAGNER, B. G., ERDMANN, J. H., ŽUMER, S. and DOANE, J. W. (1991) Microscope textures of nematic droplets in polymer dispersed liquid crystals. *J. Appl. Phys.* **69**, 6380–6386.

ONDRIS-CRAWFORD, R. J., CRAWFORD, G. P., ŽUMER, S. and DOANE, J. W. (1993) Curvature induced configuration transition in confined liquid crystals, *Phys. Rev. Lett.* **70**, 194–197.

PIRŠ, J., MARIN, B., PIRŠ, S. and DOANE, J. W. (1995) Polymer network stabilized ferroelectric liquid crystal display, *Mol. Cryst. Liq. Cryst.* (to appear).

SCHARKOWSKI, A., CRAWFORD, G. P., ŽUMER, S. and DOANE, J. W. (1993) A method for the determination of the elastic constant ratio K_{33}/K_{11} in nematic liquid crystals, *J. Appl. Phys.* **73**, 7280–7287.

STANNARIUS, R., CRAWFORD, G. P., CHIEN, L.-C. and DOANE, J. W. (1991) Nematic director orientation in a liquid crystal dispersed polymer: A DNMR approach, *J. Appl. Phys.* **70**, 135–143.

VILFAN, M., LAHAJNAR, G., ZUPANČIČ, I., ŽUMER, S., BLINC, R., CRAWFORD, G. P. and DOANE, J. W. (1995) Dynamics of nematic liquid crystal constrained by the polymer network: A proton NMR study, *J. Chem. Phys.* (in press).

VOLOVIK, G. E. and LAVRENTOVICH, O. D. (1983) Topological dynamics of defects: Boojums in nematic droplets, *Sov. Phys. JETP* **58**, 1159–1166 [*Zh. Eksp. Teor. Fiz* **85**, 1997–2010].

WEST, J. L., AKINS, R. B., FRANCL, J. and DOANE, J. W. (1993) Cholesteric/polymer dispersed light shutter, *Appl. Phys. Lett.* **63**, 1471–1473.

WU, X. L., GOLDBURG, W. I., LIU, M. X. and XUE, J. Z. (1992) Slow dynamics of isotropic–nematic phase transitions in silica aerogels, *Phys. Rev. Lett.* **69**, 470–473.

XU, F., KITZEROW, H. S. and CROOKER, P. P. (1994) Director configurations of nematic liquid crystal droplets: Negative dielectric anisotropy and parallel surface anchoring, *Phys. Rev. E* **49**, 3061–3068.

YANG, D.-K., CHIEN, L.-C. and DOANE, J. W. (1992) Cholesteric liquid crystal/polymer dispersion for haze free light shutters, *Appl. Phys. Lett.* **60**, 3102–3104.

YANG, D.-K., WEST, J. L., CHIEN, L.-C. and DOANE, J. W. (1994a) Control of reflectivity and bistability in displays using cholesteric liquid crystals, *J. Appl. Phys.* **76**, 1331–1333.

YANG, D.-K., DOANE, J. W., YANIV, Z. and GLASSER, J. (1994b) Cholesteric reflective display: Drive scheme and contrast, *Appl. Phys. Lett.* **64**, 1905–1907.

ŽUMER, S. (1988) Light scattering from nematic droplets: Anomalous-diffraction approach, *Phys. Rev. A* **37**, 4006–4015.

ŽUMER, S., GOLEMME, A. and DOANE, J. W. (1989) Light extinction in a dispersion of small nematic droplets, *J. Opt. Soc. Am. A* **6**, 403–411.

5

Polymer-stabilized Cholesteric Textures

Materials and applications

D.-K. YANG, L.-C. CHIEN and Y. K. FUNG

5.1 Introduction

5.1.1 Cholesteric Liquid Crystal

Cholesteric liquid crystals are chiral nematics. Locally, a cholesteric liquid crystal is very similar to a nematic liquid crystal: the elongated molecules are aligned, on the average, along the nematic director **n**. Cholesterics, however, have a helical structure: the director is uniformly twisted along a perpendicular axis called the helical axis. The distance along the helical axis for the director to twist 2π is called the pitch P. The period of the material is $P/2$ because **n** and $-$**n** are equivalent. If the helical axis is chosen to be parallel to the z axis of a Cartesian coordinate, then the director **n** is in the x–y plane with the components given by (cos ϕ, sin ϕ, 0), where $\phi = (2\pi/P)z + \phi_0$. The helical twist is defined as $\phi' = \partial\phi/\partial z = 2\pi/P = q_0$.

Cholesteric liquid crystals consist of either chiral molecules which are optically active molecules or nematic molecules with chiral dopants. When a chiral agent is mixed with a nematic liquid crystal, the pitch of the mixture is given by (de Gennes and Prost, 1993a)

$$\frac{1}{P} = \frac{1}{(\text{HTP})X_c} \tag{5.1}$$

where HTP and X_c are the helical twisting power and concentration of the chiral agent, respectively. The pitch can be easily adjusted by varying the concentration of the chiral agent. HTP is mainly a characteristic parameter of the chiral dopant and depends slightly on the nematic host.

Cholesteric liquid crystals have been utilized in many applications such as twisted-nematic liquid crystal displays (TN-LCDs), phase change effect dichroic liquid crystal displays, cholesteric–nematic phase change liquid crystal displays and thermochromic liquid crystal devices. In this chapter, we will discuss the displays utilizing the reflective and scattering effects of cholesteric liquid crystals.

When a cholesteric liquid crystal is sandwiched between two parallel plates, at zero field the liquid crystal is in the planar state (also called planar texture or Grandjean texture), shown in Figure 5.1(a), where the helical axes are perpendicular to the plates. In this state, the refractive index of the material has a periodic structure along the z direction (normal to the cell surface) and the material Bragg reflects light peaked at wavelength $\lambda = \bar{n}P$, where $\bar{n}$

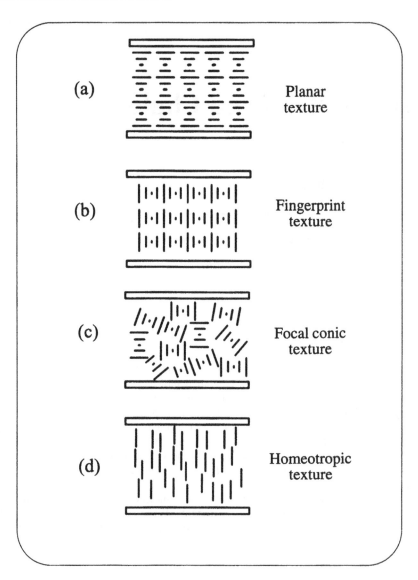

Figure 5.1 Textures of a cholesteric liquid crystal.

is the average refractive index. The reflection peak width is determined by $\Delta\lambda = \Delta nP$, where Δn is the birefringence of the material. If $\bar{n}P$ is in the visible light region, the cell reflects brilliant coloured light; if $\bar{n}P$ is in the infrared or ultraviolet regions, the cell is transparent for visible light. When an electric field is applied normal to the plates, for a positive cholesteric liquid crystal ($\Delta\varepsilon > 0$), the applied field exerts a torque on the molecules, which tends to align the director along the field. As a result of the competition between the intermolecular interaction and the applied field, a small field will switch the material into the fingerprint state (also called fingerprint texture), shown in Figure 5.1(b), where the helical axes are parallel to the plates. In practice, because of the anchoring effect of the surfaces of the cell, polydomains of fingerprint texture are formed, as shown in Figure 5.1(c). The helical axes of the domains are more or less randomly oriented throughout the cell. This state is referred to as the focal-conic state and the material is scattering because of the abrupt change of the refractive indices at the domain boundaries (Blinov, 1983). When the applied field is increased above the threshold given by (de Gennes and Prost, 1993b)

$$E_c = \frac{2\pi^2}{P} \sqrt{\frac{\pi K_{22}}{\Delta\varepsilon}} \tag{5.2}$$

the liquid crystal is switched into the homeotropic state (also called homeotropic texture) where the helical structure of the material is untwisted with the director perpendicular to the plates, shown in Figure 5.1(d), and the material becomes transparent.

In order to avoid confusion, two points need to be clarified. First, throughout this chapter, for historical reasons, the terminology 'texture' is used, which means 'state'. The second concerns the focal-conic texture. When a polymer is dispersed in cholesteric liquid crystal, the liquid crystal is in the polydomain state which is different from the focal-conic texture. We, however, still call this polydomain state the focal-conic texture (Blinov, 1983).

5.1.2 *Polymer-dispersed Liquid Crystals*

Liquid crystal/polymer composites have been studied intensively for both fundamental science and applications. Fergason developed a system in which a liquid crystal is dispersed in a waterborne polymer to form an emulsion (Fergason, 1985; Drzaic, 1988)). The liquid crystal is encapsulated by the polymer to form micrometre-size droplets. The emulsion is then sandwiched between two parallel plates. In the literature this system is called NCAP (Nematic Curvilinearly Aligned Phase). A more popularly used system, polymer dispersed liquid crystal (PDLC) (Doane *et al.*, 1986; Doane, 1991), was invented by J. W. Doane's group at Kent State University. In PDLC micrometre-size droplets of the liquid crystal are dispersed in the polymer binder. In the fabrication of PDLC, a liquid crystal is mixed with a prepolymer, or a polymer or a mixture of polymer and solvent to form a homogeneous solution. The liquid crystal is then phase separated from the polymer by one of the following methods (West, 1988): (i) polymerization-induced phase separation (PIPS); (ii) thermally induced phase separation (TIPS); or (iii) solvent-induced phase separation (SIPS). The operating mechanisms of NCAP and PDLC are the same. The director of the liquid crystal inside a droplet is more or less along one direction (referred to as the director of the droplet). In a field-on condition, the droplet directors are aligned along the applied field; for normally incident light the liquid crystal has the ordinary refractive index $n_\perp$ which is matched to the refractive index of the polymer n_p; therefore the light goes through the material without being scattered (Žumer, 1988; Kelly and Palffy-Mohuray, 1994). In a field-off condition, the droplet directors are randomly oriented throughout the sample; the refractive indices of the liquid crystal and the polymer are not matched and the material becomes optically scattering.

One problem with PDLCs is that in a field-on condition, the material is transparent only for a light beam incident normally and becomes hazy for obliquely incident light (Doane *et al.*, 1988). Doane *et al.* (1989) have suggested an intriguing modification. Instead of an isotropic polymer, a liquid crystal polymer is used in PDLC. The refractive indices, $n_{p\parallel}$ and $n_{p\perp}$, of the liquid crystal polymer match those of the liquid crystal, respectively. In the field-on condition, the liquid crystal polymer and the liquid crystal are all aligned along the applied field; the refractive indices are matched for light incident at all angles and therefore the material is transparent in all directions. However, a different problem arises: in order for the refractive indices of the liquid crystal polymer and the liquid crystal to be matched, the mesogenic moieties of the liquid crystal polymer must have a structure similar to that of the liquid crystal. For such a liquid crystal and liquid crystal polymer system, phase separation becomes difficult.

5.1.3 *Polymer-stabilized or Polymer-modified Liquid Crystals*

In order to solve the problem of refractive index mismatching discussed previously, we introduce a new method: dispersing a small amount of polymer in liquid crystals (Yang *et al.*, 1992; Doane *et al.*, 1991). Polymer is in the minority and typically its weight concentration is less than 10%. We call this new system polymer-stabilized liquid crystal (PSLC) or polymer-modified liquid crystal (PMLC). In PSLCs, liquid crystal is mixed with a monomer, which is then polymerized. The monomer can be either mesogenic, i.e. have a rigid core as liquid crystal molecules, or isotropic, i.e. consist of a flexible hydrocarbon chain. But better results can be achieved by using a mesogenic monomer. Because of the low concentration of polymer, the refractive index mismatch problem is eliminated. Particularly when the polymer is a mesogenic network, the scattering caused by the refractive index mismatch between the liquid crystal and the polymer network is very small and negligible. In PSLCs the monomer can be either thermal polymerizable or photopolymerizable, but thermal curable monomers are not good candidates because polymerization of a low-concentration monomer takes a very long time.

Because the concentration of the monomer is low, the mixture of the liquid crystal and the monomer is usually in a liquid crystal phase. When the monomer is being polymerized, a surface alignment layer or external field can be used to control the orientation of the liquid crystal, and consequently the direction of the polymer network. In this way the polymer formed is anisotropic and has an aligning effect on the liquid crystal (Hikmet, 1991; Broer *et al.*, 1990; Fung *et al.*, 1995). After polymerization, the polymer network can be used in turn to control the orientation of the liquid crystal (Yang *et al.*, 1992). Now the aligning effect of the polymer network is a bulk effect and more efficient than a surface alignment layer. Bos *et al.* (1993; see also their chapter in this volume) have used a polymer network to modify TN-LCD and STN-LCD. Blinc *et al.* (this volume) have used a polymer network to stabilize FLC. In this chapter we discuss the work we did on polymer-stabilized cholesteric texture (PSCT) displays. For additional details see also the chapters by Hikmet and Žumer and Crawford.

For a cholesteric liquid crystal sandwiched between two parallel plates, polymers can be used to stabilize the planar texture and/or the focal-conic texture. When the pitch of a cholesteric liquid crystal is in the region from ultraviolet to near-infrared light, both the planar and focal-conic texture can be stabilized at zero field (Yang *et al.*, 1994a,b). In this region the most prominent application is the bistable reflective cholesteric display (BRCD). The material can be switched between the reflective planar texture and the scattering focal-conic texture. With the pitch of a cholesteric liquid crystal in the infrared wavelength region, we made light valves of two different modes: normal mode and reverse mode (Yang *et al.*, 1992). In the normal mode light valve, the scattering focal-conic texture is stabilized at zero field and the material is opaque; in a field-on condition, the material is switched into the homeotropic texture and becomes transparent. In the reverse mode, the planar texture is stabilized at zero field and the material is transparent for visible light because the reflection peak of the material is in the infrared region; in a field-on condition, the material is switched into the focal-conic texture and becomes scattering.

5.2 **Materials**

This section describes the molecular engineering of network-forming materials in order to understand the orientation of dispersed polymer networks which are formed from photopolymerizations in different liquid crystal solvents.

5.2.1 *Molecular Engineering of Polymerizable Monomers*

The introduction of polymer network/liquid crystal dispersions, in name, was due to the indigent solubility of one diacrylate monomer in a cholesteric mixture. It is necessary to perform the molecular engineering of the diacrylate monomers, in turn, to vary the flexibility of the molecule, the length of the rigid core and the polymerizable functional group, and then to improve the electro-optic properties.

The diacrylate monomers, **1** to **4**, designed and synthesized for PSCT applications are listed in Scheme 1. These monomers were used for the photopolymerizations in liquid crystals to induce phase separations. The design of the monomer stems from the building block, biphenylene. In this way, the rigidity of the resultant network is tuned by varying the constitution of a molecule. The detail of synthesis will be reported elsewhere (Chien, 1995). The reactive monomer 4,4'-bisacryloyloxy-1,1'-biphenylene (BAB) **1**, consisting of a biphenylene core and the diacrylate reactive groups directly linked to the rigid core, forms a rigid and densely cross-linked network when photopolymerized in liquid crystals. The 4,4'-bis[6-(acryloyloxy)- hexyloxy]-1,1'-biphenylene (BAB-6) **2**, with the incorporation of hexamethylene spacer groups into the BAB, forms a less rigid network when polymerized in liquid crystals. The flexible spacer gave the monomer molecules more conformational freedom to organize themselves in the liquid crystal during polymerization. The BAB-6 was found to provide better alignment, high contrast ratio, and hysteresis for the normal mode devices with grey scale for multiplexing capability, giving rise to its flexible spacers. The 4,4'-bis-{4-[6-(acryloyloxy)-hexyloxy]benzoate}-1,1'-biphenylene (BABB-6) **3**, with an extended rigid core (two extra benzene rings and two ester linking groups), was found to provide high contrast ratio, low voltage driving, better alignment and freedom from haze for liquid crystals in reverse mode devices. By replacing the diacrylate of BABB-6 with a dimethacrylate, the 4,4'-bis-{4-[6-(methacryloyloxy)-hexyloxy]benzoate}-1,1'-biphenylene (BMBB-6) **4** was found to provide little or no hysteresis and fast switching in reverse mode cholesteric displays.

The mesomorphic behaviour of these polymerizable monomers is studied by both polarizing optical microscopy (POM) and differential scanning calorimetry (DSC). The transition temperatures were obtained from DSC and confirmed with POM. Monomer BAB melts directly from crystal to isotropic liquid at a temperature of 150°C. The monomer BAB-6 has a much lower transition temperature of 80.6°C than that of BAB owing to the incorporation of the flexible spacers. This monomer exhibits a supercooled phenomenon and a high ordered smectic phase was observed but not identified on cooling. As the length of rigid core increases by two aromatic rings and two ester linkages, BABB-6 exhibits several higher-ordered smectic phases (S × 1 108.5°C; S × 2 112.3°C), a smectic C phase at 119.2°C and a nematic phase at 124.9°C. Monomer BABB-6 undergoes thermal polymerization when heated to above 180°C. Nevertheless, there is no distinguishable texture change observed by POM. By substituting the hydrogen with a methyl group on the acrylates, the BMBB-6 exhibits two crystal–crystal transitions and an enantiotropic smectic C phase at a temperature of 113.8°C. Thermal polymerization occurs before the temperature reaches 180°C on heating the BMBB-6 and no texture change is observed up to 275°C, indicating the formation of an anisotropic network.

5.2.2 *Photopolymerizations*

The formulations for polymerizations in PSCT display cells include a nematic mixture (E48) and/or a cholesteric mixture (combinations of E48, R1011, CB15, CE1, etc.), a polymerizable monomer or a mixture of polymerizable monomers, and a photoinitiator. The photo-initiated polymerization of diacrylate monomers is illustrated in Scheme 2. The photo-initiator benzoin methyl ether (BME, Polysciences Inc.) upon absorbing the UV radiation diacrylate monomers:

1. BAB

$CH_2=CHCO_2$—⬡—⬡—$O_2CCH=CH_2$

2. BAB-6

$CH_2=CHCO_2(CH_2)_6O$—⬡—⬡—$O(CH_2)_6O_2CCH=CH_2$

3. BABB-6

$CH_2=CHCO_2(CH_2)_6O$—⬡—CO_2—⬡—⬡—O_2C—⬡—$O(CH_2)_6O_2CCH=CH_2$

4. BMBB-6

$CH_2=CCH_3CO_2(CH_2)_6O$—⬡—CO_2—⬡—⬡—O_2C—⬡—$O(CH_2)_6O_2CCCH_3=CH_2$

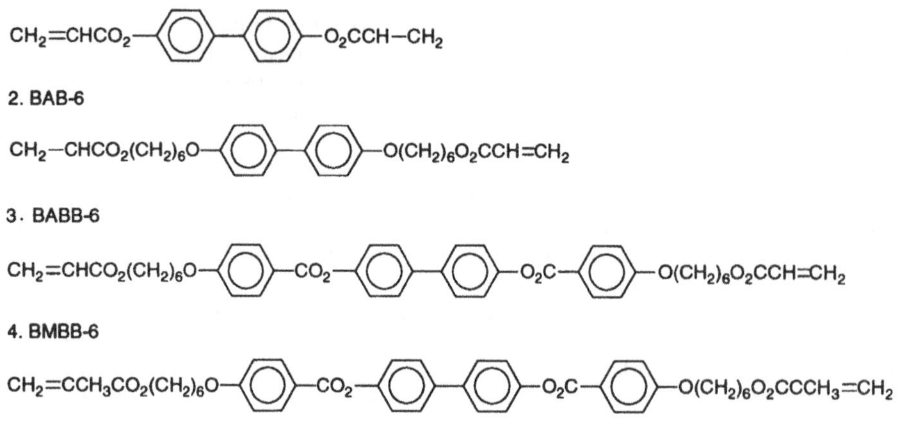

Scheme 1

becomes electronically excited, subsequently decomposes into free radicals, and then undergoes an extremely fast photocleaving process. The resulting benzoyl radicals react with the double bonds of the diacrylate monomer to form a three-dimensional polymer network by a chain reaction. The growth of the polymer network relies on the presence of a propagating radical chain end and the monomer molecules in the vicinity. As the polymerization proceeds, the viscosity of the whole mixture increases. Therefore at the final stage of polymerization, the completion of polymerization is a diffusion control process. As long as there exist radicals the polymerization will proceed even in the dark.

The polymer network morphology can be varied, depending on the types of monomer and photoinitiator, the monomer/photoinitiator ratio, the intensity of UV, the oxygen quenching and chain transfer quenching of polymerization, the application of an external field during polymerizations, the different surfaces during polymerizations as well as the different liquid crystal formulations in a polymerization mixture.

When a diacrylate monomer is polymerized in a liquid crystal solvent, the orientation and order of the resultant network depends on the orientation and order of the liquid crystal. The polymerizations of reactive monomers in liquid crystals were examined using a differential photocalorimeter. Sample B (bistable mode) was prepared with the standard chiral materials at a concentration of 28% which resulted in a green reflected colour. Sample N (normal mode) was prepared with the standard chiral materials at a concentration of 8.5% and a scattering milky material resulted. The BAB concentration of 2.5% and BME concentration of 0.2% by the weight of liquid crystal were added to the cholesteric mixtures. The kinetics of the polymerizations in these liquid crystal solvents was studied by exposing the resulting cholesteric mixture to the UV radiation (40 mW/cm^2) of a Perkin-Elmer Differential Photo Calorimeter DPC 7. The results indicate that at the early stage of polymerization, the polymerization proceeds faster in sample B than in sample N, and as the polymerization goes on, that of sample N reaches completion faster than that of sample B giving rise to the higher viscosity of sample B. The results clearly reflect the diffusion-controlled polymerization process at the final stage of polymerization. Completion of the polymerizations is within 5 minutes in the bulk samples. The polymer networks formed are anisotropic, depending on the orientation of the liquid crystals during polymerization. After polymerization, these polymer networks were found to affect the orientation of the liquid crystals according to microscopy observation.

In conclusion, the contribution of materials chemistry to the molecular engineering of polymerizable monomers led to the successful development of several PSCT display devices. An exploration of the new materials for new or improved electro-optics is necessary.

Photochemistry of benzoin methyl ether (Norish Type I cleavage) initiated polymerization:

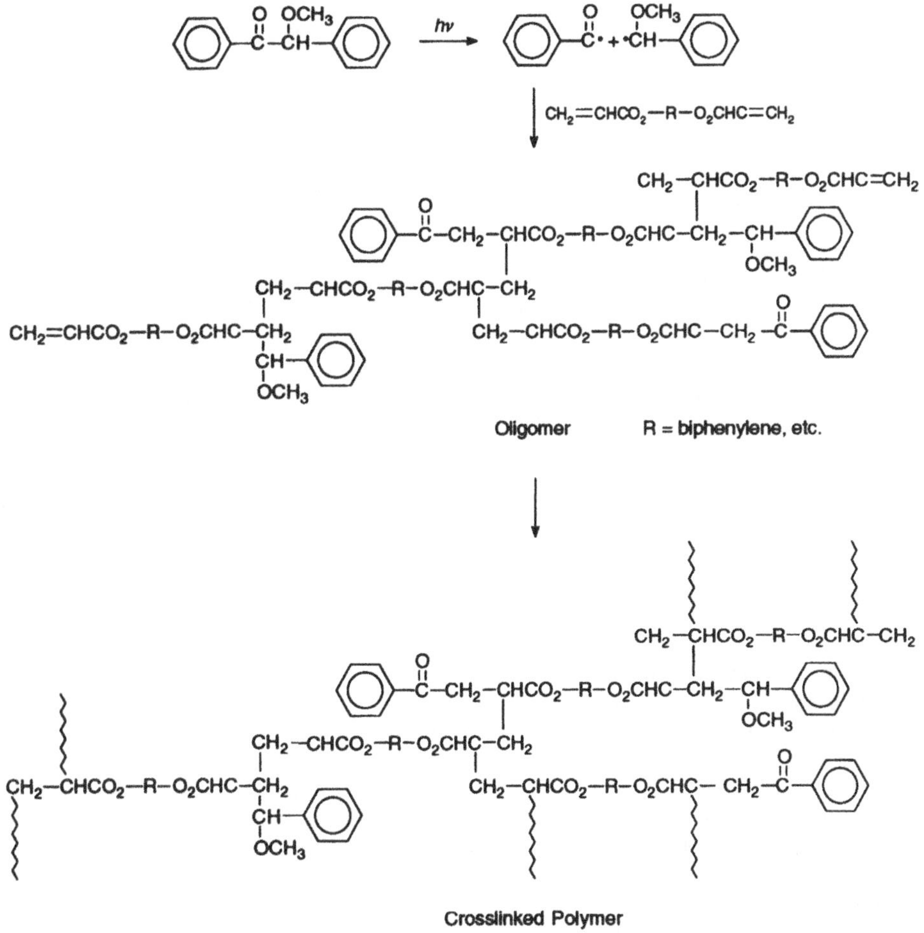

Oligomer   R = biphenylene, etc.

Crosslinked Polymer

Scheme 2

5.2.3 *SEM Study of Polymer Network*

When the liquid crystal and monomer mixture is irradiated by UV light, the monomer is polymerized to form an anisotropic network because of the aligning effect and the anisotropic diffusion properties of the liquid crystal. Although the concentration of the monomer is low, the formed polymer network phase separates from the liquid crystal, which is confirmed by neutron scattering and scanning electron microscopy (SEM) (Fung *et al.*, 1995; Jákli *et al.*, 1994; Rajaram and Hudson, 1994). We used SEM to study the morphology of polymer networks formed in liquid crystals. The results indicate that the orientational order and the direction of the polymer network depend on the order and direction of the liquid crystal during the polymerization.

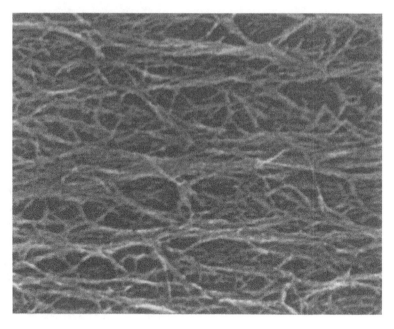

$\overline{1\,\mu m}$

Figure 5.2 Top view SEM picture of homogeneously aligned polymer network.

Samples for SEM study were prepared in the following way. First the monomer was mixed with the low-molecular-mass liquid crystal to form a homogeneous solution. In a vacuum chamber the sample cell was filled with the mixture. While the mixture was in the liquid crystal phase, it was irradiated by UV light for polymerization. After polarization, the sealant material on opposite edges of the cell was removed to allow hexane to diffuse into the cell when it was immersed. It took a few days for the solvent to diffuse into a cell of area 1 square inch (6.45 cm²). When the solvent diffused into the cell, it replaced the liquid crystal. The cell was taken out of the solvent and left to dry in a vacuum oven for a few hours. Once the evaporation was completed, the cell was opened with caution. In this way the low-molecular-mass liquid crystal was extracted from the sample and only the polymer network was left on the substrates of the cell. The substrates with the polymer were then sputtered with a thin layer of palladium for SEM study.

5.2.3.1 Polymer network formed in nematic liquid crystal

We prepared a homogeneously aligned nematic cell with buffed polyimide alignment layers on the inner surfaces of the substrates. The buffing directions of the top and bottom plates were antiparallel. The cell gap was controlled by 15 μm glass fibre spacers. The mixture consisted of 97.0% nematic liquid crystal ZLI4389 (from E. Merck), 2.7% monomer BMBB-6 and 0.3% photoinitiator BME. Uniform homogeneous alignment of the mixture was observed by optical microscopy. At ambient temperature, the mixture was in the nematic phase and the monomer was polymerized. The cell was prepared for SEM study as already described. The SEM picture of the polymer network is shown in Figure 5.2; the net-

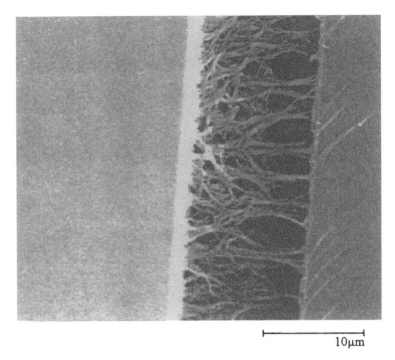

10μm

Figure 5.3 Edge view SEM picture of homeotropically aligned polymer network.

work is fibre-like, anisotropic and aligned along the buffing direction. The anisotropy of the network is believed to be created by the aligning effect and anisotropic diffusion properties of the liquid crystal during polymerization. When the liquid crystal was extracted, the polymer network collapsed to some extent in the direction perpendicular to the substrate. Furthermore there may be some residual liquid crystal around the network.

Homeotropically aligned nematic samples can be obtained using either homeotropic alignment layers, such as silane, on the inner cell surfaces or by applying an electric field normal to the cell. We made a 15 μm thick cell without any surface treatment for alignment except cleaning. The cell was irradiated by UV light with the material in the homeotropic state in the presence of an AC field of 20 V (r.m.s.). The frequency of the field was 500 Hz. The SEM picture of the polymer network shown in Figure 5.3 is the edge view. The network is fibre-like and perpendicular to the substrates, and the ends of the network are attached to the substrates. The top view SEM picture of the polymer network is shown in Figure 5.4; the network looks like a honeycomb which is believed to be formed by retraction of the polymer fibres in the direction parallel to the substrate during solvent evaporation. We made a second cell which was cured in the presence of an electric field of 100 V. The SEM picture of the network, taken with the normal of the substrate tilted 45° away from incident electron beam, is shown in Figure 5.5. One end of the network is attached to the substrate like tree roots while the other end shows a sharp fracture formed when the cell was split. Homeotropic aligned nematic samples with homeotropic alignment layers were also investigated and the SEM pictures of the polymer networks formed in these cells are quite similar to those of the polymer networks formed in the homeotropic state with an applied field.

Polymer networks formed from different monomers were investigated. When monomer BABB-6 or BMBB-6 was used, the polymer networks formed in the nematic phase were

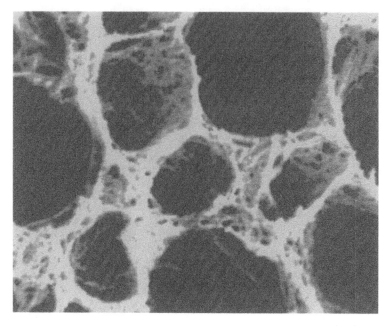

$\vdash\!\!\!\dashv$
1 μm

Figure 5.4   Top view SEM picture of homeotropically aligned polymer network.

fibre-like and anisotropic, very similar to those of BAB-6. However, the morphology of the polymer formed by BAB is quite different. A sample of 96.2% nematic liquid crystal ZLI4389 (from E. Merck), 3.5% monomer BAB and 0.3% photoinitiator BME was prepared. The monomer was polymerized in the homogeneous nematic state as in the case of BAB-6. The polymer looks like an aggregate of beads as shown in Figure 5.6. We believe that the difference between the polymer formed by BAB and that by BAB-6 is caused by the difference in the structure; BAB does not have flexible tails as BAB-6 does. It is impossible for BAB monomers to be linked together with the backbones parallel to each other.

The orientational order of the polymer network depends on that of the liquid crystal during the polymerization. We prepared a sample using the same mixture as the sample shown in Figure 5.2, except the monomer was polymerized at 80 °C at which the material was in the isotropic phase. Since the molecules were randomly oriented, the polymer network was also randomly oriented and formed globules as shown in Figure 5.7. Experiments are under way to study quantitatively the relation between the order parameter of the liquid crystal during the polymerization and that of the polymer network.

5.2.3.2   *Polymer network formed in cholesteric liquid crystal*

We studied the polymer networks formed in cholesteric liquid crystals. The cholesteric liquid crystal was a mixture of 99.5% nematic E48 and 0.5% chiral agent R1011 and had a pitch of 6 μm. The cholesteric liquid crystal, the monomer BABM6 and the photoinitiator BME were mixed in the weight ratio of 93 : 7 : 0.7. The cell was made from glass plates with buffed polyimide for homogeneous alignment and the cell gap was 15 μm. The material was in the planar state when irradiated by UV for polymerization. The liquid crystal director

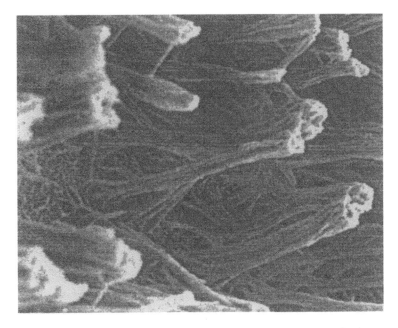

1 μm

Figure 5.5 45° view SEM picture of homeotropically aligned polymer network.

was parallel to the cell surface and the helical axis was perpendicular to the cell surface. The resulting polymer network mimicked the configuration of the liquid crystal director: the polymer networks were parallel to the cell surface and twisted around an axis perpendicular to the surface, as shown by the top view SEM picture in Figure 5.8.

A cholesteric sample cured in the presence of an electric field was also prepared. The liquid crystal was a mixture of 97.8% nematic E48 and 2.2% chiral agent R1011. The liquid crystal was mixed with the monomer BMBB-6 and the photoinitiator BME in the ratio 95:4.5:0.5. The sample was polymerized in an applied electric field with the material aligned in the homeotropic state. Now the environment was the same as in the case of the nematic liquid crystal. The polymer networks were examined by SEM and found to be perpendicular to the cell surface as shown in Figure 5.9. This SEM picture was taken with the normal of the substrate titled 45° away from the incident electron beam. The large void in the middle of the picture was formed when the two glass plates were taken apart; the polymer in that region adhered to the other glass plate.

The effect of UV intensity on the formation of the polymer network was studied. The same mixture as in the previous sample shown in Figure 5.8 was used. The samples were cured in the planar state. The SEM pictures of the polymer networks formed under different UV curing intensities are shown in Figures 5.10(a)–(c). The UV intensities are: (a) 0.04, (b) 0.4 and (c) 4.0 mW/cm². The SEM pictures show that under a lower UV curing intensity laterally large but less dense fibrils are formed. As the UV intensity increases, the fibrils become thinner and more dense. In general, at the lower UV intensity, fewer free radicals are produced and hence fewer propagating chains ends, resulting in polymer networks with high molecular weights of cross-linking. All the polymer networks shown in Figure 5.10 are long and the effect of UV intensity on the longitudinal size of the fibrils is not easy to see. The UV intensity effect on the lateral size of the fibrils is significant and shows clearly. The

⊢──⊣
1 μm

Figure 5.6 45° view SEM pictue of polymer formed by monomer BAB in the homogeneously aligned nematic liquid crystal.

density and size of the fibrils of the polymer network are very important in polymer-stabilized cholesteric texture displays; the size of the fibrils affects the stability of the display, and the density of the fibrils controls the focal-conic domain size. These effects will be discussed in detail in section 5.3.

5.2.4 Size of Polymer Networks

In the previous section, we have demonstrated that SEM is a useful technique to study the morphology of polymer networks. The SEM pictures show that the polymer networks formed in liquid crystals are fibre-like, anisotropic and mimic the structure of the liquid crystals; however, SEM cannot provide accurate information on the size of the fibrils because of the following difficulties. First, the liquid crystal has to be extracted and the polymer network may collapse; what is seen by SEM may not be exactly the same as the polymer network before the liquid crystal was extracted. Second, there may be residual liquid crystals on the polymer network surface. Finally, the polymer networks have to be coated with palladium. These three factors cause a discrepancy between the real size and the observed size of the fibrils of the polymer network under SEM.

Jákli *et al.* (1994) have employed a small-angle neutron scattering technique to study the size of the fibrils of the polymer networks in liquid crystals. This technique can provide information on the shape, size and surface roughness of the polymer network. Their results show that the fibril of the polymer network is rod-like and the surface of the polymer network is rough. The scattered intensity of rod-like particles is given by

$$I(q) = I_0 q^{-1} \exp(-q^2 R_c/2) \tag{5.3}$$

⊢—⊣
1 μm

Figure 5.7 Top view SEM picture of polymer formed in the isotropic phase.

where q is the scattering vector and R_c is the radius of gyration of the cross-section of the polymer network. When $\ln(I(q)q)$ is plotted against q^2, the slope is given by $-R_c/2$. In their experiment monomers BAB and 4,4'-bis(2-methylpropenoyloxy)-biphenyl (BMB) were used. By fitting their experimental data they obtained a value of 300 Å for the radius of the fibrils of the polymer. However, the monomers used in their experiment do not have flexible alkyl chains, and our SEM experiments show that monomers without flexible alkyl chains do not form fibre-like networks (Fung *et al.*, 1995; Rajaram and Hudson, 1994).

Following the birefringence study by Crawford *et al.* (1995) we performed a detailed investigation of polymer networks in liquid crystals, which happened to be able to provide information on the size of polymer networks. When the material is heated to the isotropic phase, the total birefringence observed is contributed by the birefringence of the polymer network and the polymer-induced birefringence of the nearby liquid crystals. The latter depends on the size of the polymer network.

We made a homogeneously aligned nematic sample with two glass plates with antiparallel rubbed polyimide coating. The mixture was made of liquid crystal 5CB (97.8 wt.%), the monomer BMBB-6 (2.0 wt.%) and the photoinitiator BME (0.20 wt.%). The mixture had an NI transition temperature of 34.7°C. The cell thickness was controlled by 29 μm mylar spacers. The material was irradiated by UV light of intensity 4 mW/cm^2 for polymerization. The liquid crystal director was unidirectionally aligned along the rubbing direction during the polymerization. The formed polymer network was also aligned along the rubbing direction as shown in the SEM study.

In the experiment, He–Ne laser light was used to study the birefringence. The sample was placed between the polarizer and analyzer, which were perpendicular to each other, and the rubbing direction was oriented at 45° to both the polarizer and analyzer. The temperature of the sample was controlled by a hot stage with a precision of 5 mK. The intensity of the

115

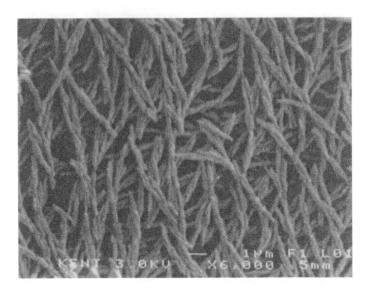

├──────┤
1μm

Figure 5.8 Top view SEM picture of polymer network formed in the planar texture of the cholesteric liquid crystal.

transmitted light was detected by a photodiode. The intensity I of the transmitted light is related to the birefringence, Δn, of the sample by

$$I = I_0 \sin^2\left(\frac{\pi d \Delta n}{\lambda}\right) \tag{5.4}$$

where I_0 is the intensity of incident light, λ is the wavelength of the light and d is the sample thickness. I was measured at different temperatures and we used the above formula to calculate the birefringence Δn at those temperatures. Δn against temperature T is shown by the open circles in Figure 5.11.

We developed a theoretical model to calculate the birefringence. We used a cylinder to represent the fibril, as shown in Figure 5.12. The polymer chains inside the cylinder have an orientational order parameter S_p and the radius of the cylinder is R. The fibril consists of cross-linked molecular chains and therefore the order parameter S_p of the chains can be assumed to be temperature independent. The liquid crystal molecules on the surface of the cylinder have an orientational order parameter S_o. The liquid crystal molecules in the bulk have an order parameter $S(\mathbf{r})$, which is a function of position $\mathbf{r}$. The total birefringence is given by

$$\Delta n = \Delta n_p + \Delta n_{lc} \tag{5.5}$$

where Δn_p and Δn_{lc} are the birefringences of the polymer and liquid crystal, respectively. First, let us discuss Δn_p which is given by

$$\Delta n_p = d\rho_p \Delta n_{po} \int_0^R S_p 2\pi r \, dr = d\rho_p \Delta n_{po} S_p \pi R^2 \tag{5.6}$$

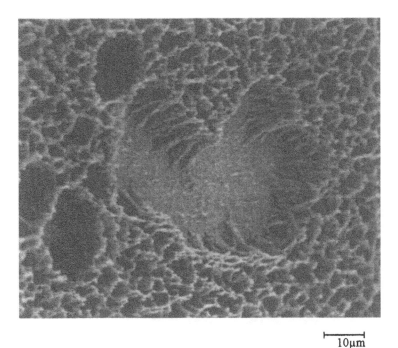

$\overline{\qquad}$
10μm

Figure 5.9 45° view SEM picture of polymer network formed in the homeotropic texture of the cholesteric liquid crystal.

where d is the cell thickness and p_p is the number of fibrils (the cylinders) per unit area perpendicular to the networks. There are a few unknown parameters in eqn (5.6). Fortunately, we were able to measure Δn_p experimentally. After we studied the birefringence of the polymer network and the polymer-induced birefringence of the liquid crystal, we put the sample in the solvent octane which dissolved the liquid crystal but did not change the structure of the polymer network. Then we again measured the birefringence of the sample, which was contributed solely by the polymer network. Δn_p was found to be 1.26×10^{-3} and approximately temperature independent.

Now let us consider the polymer-induced birefringence of the liquid crystal, which is given by

$$\Delta n_{lc} = d\rho_p \Delta n_{lco} \int_0^{2\pi} \int_R^\infty S(\mathbf{r}) d\phi r \, dr \tag{5.7}$$

where Δn_{lco} is the birefringence of the perfectly ordered 5CB. In order to calculate Δn_{lc}, we have to find $S(\mathbf{r})$. In the isotropic phase, the free-energy density of the liquid crystal is given by the Landau–de Gennes theory (de Gennes, 1971)

$$f = f_0 + \frac{1}{2} a(T - T^*)S^2 + \frac{1}{2} L(\nabla S)^2 \tag{5.8}$$

where f_0 is the free energy independent of the order parameter, T^* is the virtual second-order phase transition temperature and L is the elastic constant. It is assumed that the order parameter S_0 on the cylinder surface is temperature independent. The order parameter S is invariant along the z axis (which is parallel to the cylinder). Furthermore, S is independent of the azimuthal angle ϕ, and therefore $S(\mathbf{r}) = S(r)$. The total free energy is given by

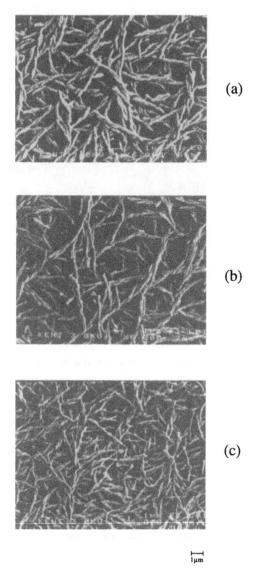

H
1μm

Figure 5.10 Top view SEM pictures of polymer network formed in the planar texture of the cholesteric liquid crystal under three different UV intensities: (a) 0.04 mW/cm²; (b) 0.4 mW/cm²; and (c) 4.0 mW/cm².

$$F = 2\pi \int_R^\infty \left[\frac{1}{2} a(T - T^*)S^2 + \frac{1}{2} L(\delta S/\delta r)^2 \right] r \, dr \tag{5.9}$$

Minimizing the free energy we get

$$\frac{d^2Q}{du^2} + \frac{I}{u}\frac{dQ}{dr} - u = 0 \tag{5.10}$$

where $Q = S/S_0$ and $u = r/\xi$, and

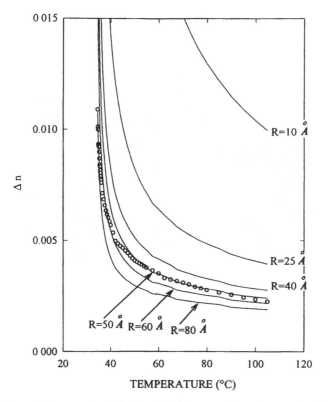

Figure 5.11 Open circles: experimental data of birefringence of the polymer network and the liquid crystal at various temperatures. Full lines: theoretical fitting with various radii of the polymer network.

$$\xi = \sqrt{\frac{L}{T^*a}} \sqrt{\frac{T^*}{T - T^*}}$$

is the correlation length. Equation (5.10) is the modified Bessel equation. The boundary conditions are

$Q = 1$ when $u = R/\xi$

$Q = 0$ when $u \to \infty$ ⠀⠀⠀⠀⠀⠀⠀⠀⠀⠀⠀⠀⠀⠀⠀⠀⠀⠀⠀⠀⠀⠀⠀⠀⠀⠀⠀⠀⠀⠀⠀(5.11)

The solution of eqn (5.10) is

$$Q(u) = \frac{K_0(u)}{K_0(R/\xi)} \tag{5.12}$$

where K_0 is the zero-order modified Bessel function (Abramowitz and Stegun, 1972). The birefringence can be calculated from eqn (5.7):

$$\Delta n_{lc} = d\rho_p \Delta n_{lco} 2\pi S_o \xi^2 \frac{\int_{R/\xi}^{\infty} K_0(u)u \; du}{K_0(R/\xi)} \tag{5.13}$$

A review of the literature reveals the following parameters (Sheng, 1982; Coles, 1978; Horn, 1978): $T^* = 307$ K, $\Delta n_{lco} = 0.35$ and $S_o = 0.3$. Therefore the only fitting parameter is the radius R of the fibril cylinder. The fittings with various values of R are shown by the full lines in Figure 5.11; $R = 50$ Å fits the data best. Because the diameter of a polymer

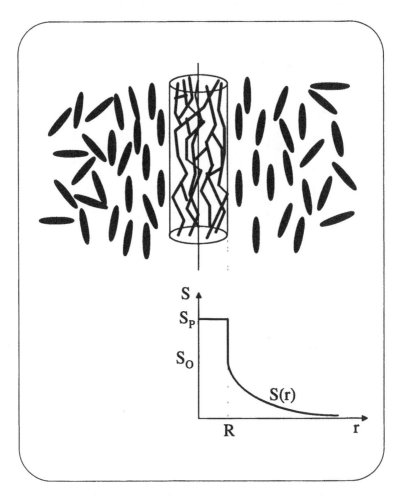

Figure 5.12 Top: schematic diagram of the fibril of the polymer network and the nearby nematic liquid crystal molecules in the isotropic phase. Bottom: the order parameter of the polymer chains and liquid crystal as a function of position.

molecular chain is about 10 Å, there are about 100 molecular chains inside the cylinder, and because the polymer concentration is 2 wt.%, the interfibril distance is about $\sqrt{\pi/0.02}R \approx 10R = 500$ Å, which differs from what was observed under SEM. This discrepancy may be caused by the collapse when the liquid crystal was extracted in the SEM study. The data presented here are from a preliminary study, and systematic and precise study has to be carried out. What we would like to say here is that a birefringence study is a useful technique to investigate the structure of the polymer network. The aligning effect of the polymer network on the liquid crystal molecules is an indication that the network is anisotropic.

Crawford *et al.* (1995) have also performed a birefringence study of polymer networks formed by the monomer BAB. In their model, the surface of the fibril of the polymer is treated as a flat surface and the order parameter S depends only on the coordinate x, which is perpendicular to the polymer network surface. By fitting their data the diameter of the fibril was found to be 20 Å, i.e. the polymer network consisted of nearly single molecular chains. Not much different are the conclusions of the chapter by Žumer and Crawford based on numerical calculations of the order parameter profile for an array of fibrils.

5.3 Polymer-stabilized Cholesteric Texture Displays

Cholesteric liquid crystals have unique optical properties because of their helical structure. When a cholesteric liquid crystal has a planar texture, it exhibits Bragg reflection, when it has a focal-conic texture, it scatters light, and when it has a homeotropic texture where the helical structure is unwound, it is transparent. The stability of these three textures depends on the pitch of the material, boundary conditions, field conditions and dispersed polymer. When the pitch of a cholesteric liquid crystal is comparable with the wavelength of visible or UV light, both the planar and focal-conic textures are stable at zero field in cells with proper surface anchoring condition or dispersed polymer. The display utilizing this bistability is called bistable cholesteric display (BCD). When the pitch of a cholesteric liquid crystal is in the infrared light region, the bistability disappears. The planar or the focal-conic texture can be stabilized at zero field by polymer networks. This display is referred to as the polymer-stabilized cholesteric texture (PSCT) light shutter.

5.3.1 *Bistable Cholesteric Display*

When the pitch of a cholesteric liquid crystal is short, bistability of the planar and focal-conic texture can be obtained (Yang *et al.*, 1991, 1994a,b); Dir *et al.*, 1972; Greubel *et al.*, 1973). This bistability is determined by the free energy of the planar and focal-conic textures as well as the transitions among the planar, focal-conic and homeotropic textures. Both the energy of the defects in the focal-conic texture and the transitions depend on the pitch. In this section we will consider that the material has a short pitch.

First, we discuss the transitions or the switching mechanism (Yang *et al.*, 1995). When a field higher than E_c is applied normal to a cholesteric cell, the material will be switched into the homeotropic texture in which the liquid crystal director is aligned perpendicular to the cell surface. We choose the z direction of a Cartesian coordinate perpendicular to the cell surface. When the applied field is reduced below E_c, the homeotropic texture is unstable and the liquid crystal will relax. In this relaxation, there are two competing modes: (i) the HF mode in which the liquid crystal relaxes into the focal-conic texture (precisely speaking, this texture is the fingerprint texture) where the helical axis is parallel to the cell surface; (ii) the HP mode in which the liquid crystal relaxes into the planar texture where the helical axis is perpendicular to the cell surface. The dynamics of these two modes is a complicated subject and we will discuss it only briefly here.

Let us consider the HF mode. In this mode only one parameter is needed to describe the transition. This parameter is the twist angle ϕ as shown in Figure 5.13. The y axis is chosen to be parallel to the helical axis. The dynamic equation can be derived from the Landau–Khalatnikov equation

$$\gamma \frac{\partial \phi}{\partial t} = K_{22} \frac{\partial^2 \phi}{\partial y^2} - \frac{\Delta \varepsilon}{4\pi} E^2 \sin \phi \cos \phi \qquad (5.14)$$

where γ is the rotational viscosity coefficient, K_{22} is the twist elastic constant, $\Delta \varepsilon$ is the dielectric anisotropy and E is the bias field. Even when $E < E_c$, there is a large energy barrier between the homeotropic state and the focal-conic state (de Gennes and Frost, 1993c). Thermal fluctuations themselves are not able to overcome this energy barrier. Irregularities such as defect, surface roughness and guest particle are needed to initiate the transition.

121

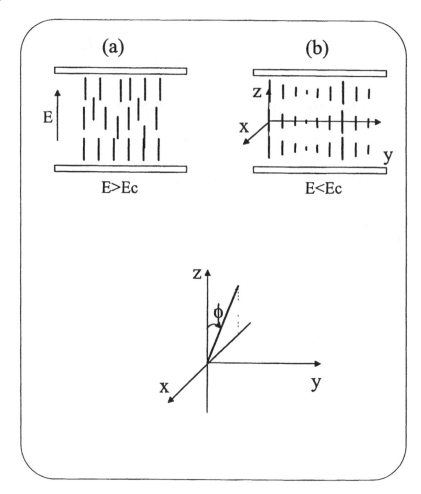

Figure 5.13 Schematic diagrams of the configuration of the liquid crystal director in the homeotropic–focal-conic transition.

Therefore the HF mode is a heterogeneous nucleation transition. When $E = 0$, eqn (5.14) becomes a diffusion equation; the transition time is given by

$$\tau_{HF} = \gamma L^2 / K_{22} \tag{5.15}$$

where L is the distance between two neighbouring nucleation sites. Experimentally we observed that the average value of L depends on the bias field and is typically longer than 5 μm. For a material with the following parameters, $\gamma = 5 \times 10^{-2}$ NS/m^2, $K_{22} = 5 \times 10^{-12}$ N, $L = 5 \times 10^{-6}$ m, then the transition time $\tau_{HF} = 5 \times 10^{-2} \times (5 \times 10^{-6})^2/5 \times 10^{-12} = 250$ ms.

Kawachi and Kogure (1977) suggested that the homeotropic–planar transition can be achieved through a conic helical process as shown in Figure 5.14. When the applied field is reduced, the polar angle θ will increase from 0° to 90° and the azimuthal angle ϕ will twist to form a spiral structure with the helical axis along z. In this conic helical structure, the splay and bend as well as the twist elastic energies are involved. The dynamic equation of θ and ϕ can be derived from the Landau–Khalatnikov equation:

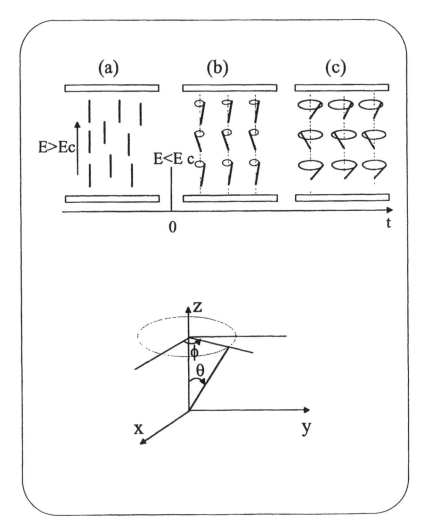

Figure 5.14 Schematic diagrams of the configuration of the liquid crystal director in the homeotropic–planar transition. During the transition the polar angle increases with time.

$$\gamma \frac{\partial \theta}{\partial t} = K_{33} \frac{\partial^2 \theta}{\partial z^2} + \left[2(K_{33} - K_{22})\left(\frac{\partial \phi}{\partial z} \sin \theta\right)^2 - K_{33}\left(\frac{\partial \phi}{\partial z}\right)^2 \right.$$

$$\left. + 2K_{22}q_0 \frac{\partial \phi}{\partial z} - \frac{\Delta \varepsilon}{4\pi} E^2 \right] \sin \theta \cos \theta \tag{5.16}$$

$$\gamma \frac{\partial \phi}{\partial t} = [K_{33} + (K_{22} - K_{33}) \sin^2\theta] \frac{\partial^2 \phi}{\partial z^2} + 2 \frac{\partial \theta}{\partial z} \frac{\cos \theta}{\sin \theta}$$

$$\times \left(2(K_{22} - K_{33}) \sin^2\theta \frac{\partial \phi}{\partial z} + K_{33} \frac{\partial \phi}{\partial z} - K_{22}q_0 \right) \tag{5.17}$$

Here we assume $K_{11} = K_{33}$ for simplicity. At the beginning of the transition, θ is very small. Because $\sin \theta$ is in the denominator of the non-linear term in eqn (5.17), ϕ changes very quickly until $\partial \phi / \partial z$ reaches a value such that the following condition is satisfied:

$$2(K_{22} - K_{33}) \sin^2\theta \frac{\partial\phi}{\partial z} + K_{33} \frac{\partial\phi}{\partial z} - K_{22}q_0 = 0$$

Because θ is very small,

$$\frac{\partial\phi}{\partial z} = \frac{K_{22}}{K_{33}} q_0 \tag{5.18}$$

Physically the azimuthal rotation of the liquid crystal molecules is very quick when θ is very small. The helical twist $\partial\phi/\partial z$ will take the value given by eqn (5.18) in order to minimize the free energy.

Now let us consider the critical field for the homeotropic–planar transition. In order for the polar angle to change from $0°$ to $90°$, the non-linear term in eqn (5.16) has to be positive. Since $\sin\theta\cos\theta$ is always positive ($0° \leqslant \theta \leqslant 90°$), it is required that

$$2(K_{33} - K_{22})\left(\sin\theta \frac{\partial\phi}{\partial z}\right)^2 - K_{33}\left(\frac{\partial\phi}{\partial z}\right)^2 + 2K_{22}q_0 \frac{\partial\phi}{\partial z} - \frac{\Delta\varepsilon}{4\pi} E^2 \geqslant 0 \tag{5.19}$$

Putting eqn (5.18) into (5.19) we will get

$$E \leqslant \left(\frac{K_{22}}{K_{33}} \frac{4\pi}{\Delta\varepsilon} q_0^2\right)^{1/2} = \frac{2}{\pi} \sqrt{\frac{K_{22}}{K_{33}}} E_c = E_{HP} \tag{5.20}$$

When the applied field is reduced below E_{HP}, the HP mode becomes possible. Numerical calculations show that the polar angle θ changes from $0°$ to $90°$ in a time interval of

$$\tau_{HP} \approx \frac{\gamma P_0^2}{K_{22}} \tag{5.21}$$

Within this time interval, $\partial/\partial z$ almost remains at the value of $(K_{22}/K_{33})q_0$. This state with the pitch

$$P = 2\pi \Big/ \sqrt{\left(\frac{K_{22}}{K_{33}} q_0\right)} = \frac{K_{33}}{K_{22}} P$$

and polar angle $\theta = 90°$ is referred as the transient planar state. The homeotropic–transient planar transition time is given by eqn (5.21). This transient planar state is not stable because the free energy has not yet reached its minimum value. The liquid crystal will relax further into the stable planar state with the intrinsic pitch P_0. The transition from the transient planar texture to the stable planar texture is again a slow diffusion process with defects involved. The important point is that once the liquid crystal is in the transient planar state, if the applied field is sufficiently low, the liquid crystal will relax into the stable planar state but not the focal-conic state.

As discussed above, there are two possible relaxation modes in a cholesteric liquid crystal when the applied field is reduced. These two modes compete with each other. Under a given bias field, the mode with the shorter transition time wins. For a cholesteric liquid crystal reflecting visible light, the pitch $P_0 \sim 0.5$ μm. When the applied field is removed suddenly, the HP mode will win because $\tau_{HP} = (P_0/L)^2\tau_{HF} = \tau_{HF}/100$. Therefore the reflecting planar texture is obtained. In this homeotropic–planar transition, first the liquid crystal transforms very quickly into the transient planar texture with the pitch $(K_{33}/K_{22})P_0$, and then the liquid crystal relaxes slowly into the planar texture with the intrinsic pitch P_0 (Saupe and Lu, 1980; Kawachi et al., 1975). Experimentally we observed the reflection of the transient planar texture at the wavelength $(K_{33}/K_{22})\bar{n}P_0$ (Yang et al., 1995). If the applied field is decreased slowly (in a time interval longer than 200 ms), at the beginning the field is in the region $E_{HP} < E < E_c$, only the HF mode can be realized and the liquid crystal relaxes into the focal-conic texture. Once the liquid crystal is in the focal-conic texture, it will remain in this texture even if the applied field is decreased further to zero.

124

When the pitch of a cholesteric liquid crystal is increased, the homeotropic–transient planar transition time τ_{HP} becomes longer. When P_0 becomes comparable with the distance between two neighbouring nucleation sites of the homeotropic–focal-conic transition, no mode will dominate even if the field is removed suddenly. Therefore the bistable switching mechanism disappears for long-pitch cholesteric liquid crystal.

Another subject which should be addressed is the energy of defects in cholesteric liquid crystals. There are many defects in the focal-conic texture. At zero field, the free energy of the focal-conic texture is higher than that of the planar texture. However, there is an energy barrier between them. When the pitch is short, the energy barrier cannot be overcome by thermal fluctuations. Therefore both the planar and focal-conic textures are stable at zero field (physicists may think that the focal-conic texture is metastable). As suggested by P. H. Keys, the energy of a defect in cholesteric liquid crystals is pitch dependent. The energies of the defects increase with increasing pitch. When the pitch is long enough, the free energy of the focal-conic texture, due to the defects, becomes too high, and the focal-conic texture becomes unstable. This explains why the bistability disappears in cholesteric liquid crystals with a long pitch. So far the energy of defects in cholesteric liquid crystals is not well understood. Much more research should be done on this subject.

Bistability is an intrinsic property of short-pitch cholesteric liquid crystals. We have observed bistability when a short-pitch cholesteric liquid crystal is sandwiched between plates with the following surface treatments: rough surface, homeotropic anchoring surface, weak homogeneous anchoring surface, and so forth (Lu *et al.*, 1955). However, when a cell with a rubbed polyimide surface is used, the bistability disappears because the strong homogeneous anchoring of the polyimide favours the planar texture and the focal-conic texture is unstable at zero field (Hulin, 1972). However, when a small amount of polymer is dispersed in the cholesteric liquid crystal in this type of cell, then both the planar and focal-conic textures become stable at zero field (Yang *et al.*, 1994a). The resulting system is called the polymer-stabilized cholesteric texture (PSCT) display.

Now we discuss the electro-optical properties of the PSCT display (Yang and Doane, 1992; Doane *et al.*, 1992). The cell was constructed with two glass plates with a rubbed polyimide alignment layer and had a cell gap of 5 μm; 1.2 wt.% monomer BAB and 0.2 wt.% photoinitiator BME were added to a cholesteric liquid crystals which reflected green light. The mixture was poured into the cell in a vacuum chamber. The monomer was polymerized under UV light. During the polymerization the material had the homeotropic texture in the presence of an electric field. The reflection spectra of the planar and focal-conic textures at zero field are shown in Figure 5.15. The reflection spectrum of the planar texture, indicated by R, peaks at the wavelength given by the Bragg formula $\lambda_0 = \bar{n}P_0$ and the width of the peak is governed by $(\Delta n/n)\lambda_0$. The reflection of the focal-conic texture, indicated by S, is much weaker; the measured reflection is mainly caused by reflection from the glass–air interface and the scattering is caused by focal-conic domains. High contrasts can be achieved with large focal-conic domains which cause less scattering. If the surface of the back plate is painted black, the planar texture appears brilliantly reflective while the focal-conic texture appears black.

We studied the response of the material to voltage pulses and the result is shown in Figure 5.16 in which the vertical axis is the reflectance of the material and the horizontal axis the r.m.s. value of the pulse. The details of the curves depend on the width of the pulse. The pulse width used here is 20 ms. Curve a is for the case where the initial state of the material has the planar texture and curve b is for the case where the initial state has the focal-conic texture. Two things should be emphasized here: first, the reflectance is measured *a few seconds after the pulse*, and during the measurement the voltage applied is *zero*; second, initially the material has either completely the planar texture or completely the focal-conic texture. If the initial state has the planar texture, grey-scale memory can be easily obtained with the voltage of the pulse in the region 18–34 V. We think it is better to call this display a bistable display instead of multistable, because the state with intermediate reflection is a

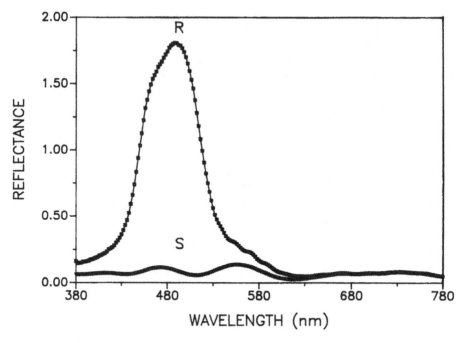

Figure 5.15 Reflection spectra of the PSCT bistable reflective cholesteric display. R: planar texture, X: focal-conic texture.

combination of the planar and focal-conic domains. The switching between the planar and the focal-conic texture is shown in Figure 5.17. If we want to switch the material from the planar to the focal-conic texture, we apply a low-voltage pulse V_L. After the pulse, the material will remain in the focal-conic texture. If we want to switch the material from the focal-conic to the planar texture, we apply a high-voltage pulse V_H. During the pulse, the material is switched into the homeotropic texture. After the pulse the material relaxes into the planar texture. For the material whose response is shown in Figure 5.16, we choose $V_L = 34$ V and $V_H = 50$ V.

West *et al.* (1993) reported a bistable reflective cholesteric display with high-concentration (~ 20 wt.%) isotropic polymer. The liquid crystal is dispersed in the polymer and exists in the form of droplets a few micrometres in diameter. The bistability is preserved, but reflectivity is lower in the planar state due to the smaller liquid crystal volume. In this material the helical axes have a wide distribution in orientation and result in a broad reflection peak of the planar texture. Good black and white displays can be made from this dispersion.

Bistable cholesteric material (BCM) is very useful for display applications. It can be used to make a highly multiplexed display on a passive matrix because of its bistability. The manufacture of a passive matrix is much cheaper than an active matrix. In addition, a bistable cholesteric display does not need polarizers, resulting in improved light efficiency. The display appears bright even at room light conditions without power-hungry backlights. Finally, the reflection is from the liquid crystal. Plastic substrates with non-uniform birefringence can be used to construct displays which are much lighter in weight and mechanically shockproof. Making use of the short transition time of the homeotropic–transient planar transition we have designed a dynamic drive scheme with which the display can be addressed at a speed of 0.5 ms per line (Huang *et al.*, 1995). Figure 5.18 is a picture of an 80 dpi 4×4 in² (103 cm²) bistable reflective cholesteric display. Bistable reflective cholesteric displays can be used for electronic viewers, signs and supermarket labels.

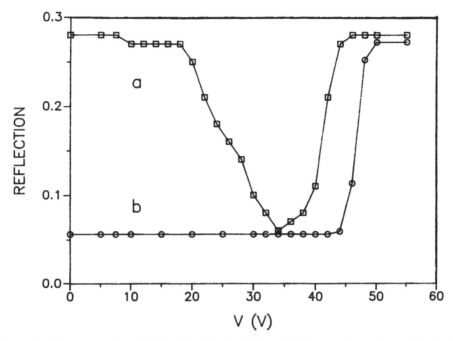

Figure 5.16 Response of the PSCT bistable reflective cholesteric display to voltage pulses: a, the liquid crystal was initially in the planar texture; b, the liquid crystal was initially in the focal-conic texture.

5.3.2 *Polymer-stabilized Cholesteric Texture (PSCT) Normal Mode Light Shutter*

When the pitch of a cholesteric liquid crystal is long, the bistable effect disappears. Between the planar and focal-conic textures, only one of them can be stabilized at zero field. A PSCT normal mode light shutter is achieved by stabilizing the focal-conic texture at zero field with dispersed polymer networks (Yang *et al.*, 1992). In this shutter, the polymer network is perpendicular to the cell surface as shown in Figure 5.3. At zero field (the off-state), as a result of the competition between the aligning effect of the polymer network and the intermolecular interaction, the liquid crystal is in the focal-conic texture, as shown in Figure 5.19(a). In each domain the liquid crystals still have the helical structure. However, the helical axes of the domains are oriented more or less randomly throughout the cell. At the domain boundaries, the refractive indices change abruptly, and the material scatters light. When a field is applied normal to the cell (the on-state), the helical structure is unwound, as shown in Figure 5.19(b), and the liquid crystal is aligned along the field which is parallel to the polymer network. In this homeotropic texture the liquid crystal becomes single domain and is transparent.

In our electro-optical measurements, an He–Ne laser was used. The incident light was a collimated beam 3 mm in diameter. The transmitted light was detected by a photodiode with a collection angle of 5°. The applied voltage had a frequency of 1 kHz. In the transmittance–voltage measurement the voltage was typically varied at the rate of 1 V/s. The display cell used did not have any surface alignment layers and the cell gap was controlled by glass fibre spacers. Typically a few weight per cent monomer was added to the liquid crystal and the mixture was in a homogeneous solution. The viscosity of the mixture was low. The mixture was poured into the liquid crystal cell in a vacuum chamber and the cell was irradiated by UV light for polymerization. During polymerization, the material had a

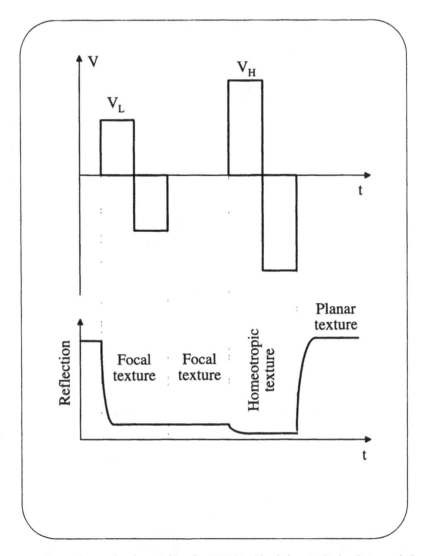

Figure 5.17 Top: voltage pulses for switching the PSCT bistable cholesteric display during and after the pulses.

homeotropic texture in the presence of an externally applied electric field. The polymer network formed was perpendicular to the cell surface. The intensity of the UV light for curing did not have a great effect on the electro-optical properties of the PSCT normal mode shutter. Typically the UV intensity was about 5 mW/cm² and the irradiation time was half an hour.

We made a display with the following combination: 88.6% nematic liquid crystal ZLI4389, 8.5% chiral agent CB15, 2.7% monomer BMBB-6 and 0.2% photoinitiator BME. The cell gap was 10 μm. The transmittance–voltage curve is shown in Figure 5.20(a). In the off-state, the material has the focal-conic texture and the transmittance is less than 1%. As the applied voltage is increased, the transmittance is low until close to the focal-conic–homeotropic transition. When the voltage is high enough, the material is switched into the homeotropic state and the transmittance reaches the maximum which is close to 90%. The

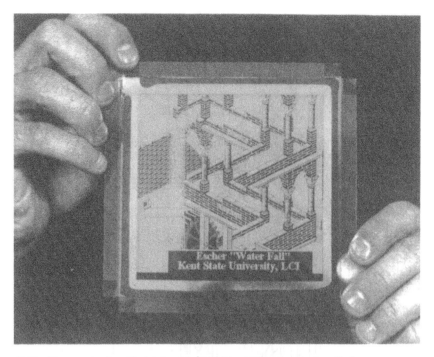

Figure 5.18 Photogragh of an 80 dpi 4 × 4 in² bistable cholesteric reflective display.

main cause of light loss is the reflection at the glass–air interface. When the voltage is decreased, the material relaxes back to the focal-conic texture and becomes scattering. A hysteresis of a few volts is observed in the focal-conic–homeotropic transition (also called the cholesteric–nematic transition). The contrast ratio is more than 100. The response of the material to a 100 ms wide 20 V pulse is shown in Figure 5.20(b). The turn-off time is about 20 ms. The turn-on time is about 50 ms which is long. In order to achieve a fast turn-on time, a higher voltage is required.

In PSCT shutters, the polymer concentration is low and the scattering caused by refractive index mismatching is negligible. The transmission spectra of the PSCT normal mode shutter are shown in Figure 5.21. In the on-state, the material is transparent for all light in the visible region. In the off-state, the focal-conic domain size has a wide distribution. The scattering is also almost wavelength independent in the visible region. This wavelength-independent contrast is an advantage of PSCT over PDLC where colour dispersion is a problem.

PSCT shutters have very wide viewing angles. In the experiment studying viewing angle, the cell was immersed in glycerine contained in a transparent plastic cylinder to prevent reflection and refraction at the glass–air interface. The cell was rotated such that the normal of the cell made various angles with the incident laser light beam. The on-state transmittance against the angle of the incident light is shown by the open squares in Figure 5.22. The transmittance of PDLC at various angles is represented by the open circles in the figure for comparison.

We have systematically studied the effects of polymer networks as well as the pitch of the cholesteric liquid crystal on the electro-optical properties of the PSCT normal mode light valve. In this study the nematic liquid crystal was ZLI4389, the monomer was BMBB-6 and the chiral agent was R1011. First we produced a polymer–chirality phase diagram shown in Figure 5.23. The diagram is divided into three regions: (1) unstable, (2) stable and scattering,

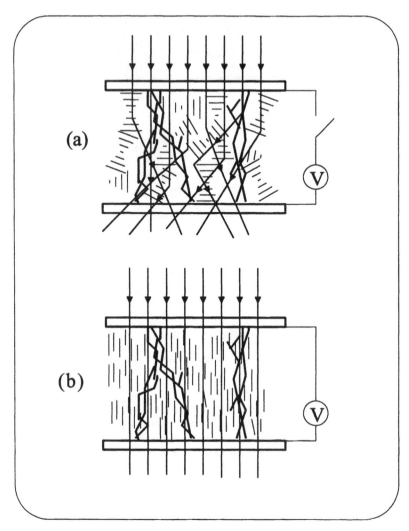

Figure 5.19 Schematic diagam of the configurations of the liquid crystal director of the PSCT normal mode light valve in the off-state and on-state. The thick dark lines represent the polymer network.

and (3) stable and clear. In region (1), the polymer concentration is not high enough and the focal-conic texture is not completely stabilized. After being switched on and off, the focal-conic domain size is not stable and changes with time. The domain size is too large and scattering in the off-state is low. Region (2) is the operation region of PSCT normal mode shutters. In this region the focal-conic texture is stabilized; in the off-state the domain size does not change with time and the material is very scattering. In region (3) the polymer concentration is too high and its aligning effect is too strong, so the homeotropic texture is stabilized and the material is transparent. The boundary between regions (1) and (2) is approximately independent of the chiral concentration. However, the boundary between regions (2) and (3) has approximately a linear dependence on the chiral concentration. The higher the chiral concentration, the shorter the pitch, and more polymer is required to unwind the helical structure.

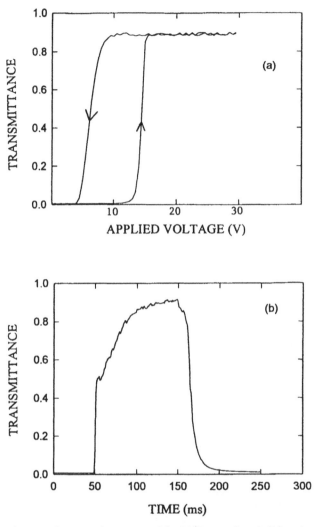

Figure 5.20 (a) The transmittance–voltage curve of the PSCT normal mode light valve. (b) The response of the PSCT normal mode light valve to a 100 ms long voltage pulse.

In the experiment studying the effects of the pitch of the cholesteric liquid crystal, we used chiral agent CB15 which has a helical twisting power HTP $\approx 8 \ \mu m^{-1}$. The pitch is given by $P = 1/(\text{HTP})C_c$, where C_c is the chiral concentration. The monomer was BAB-6 and had a concentration of 2.7%. Figure 5.24(a) shows the contrast which peaks at 9% CB15. The pitch is one of the factors determining the focal-conic domain size. Roughly speaking, 1–2 μm focal-conic domains strongly scatter visible light. When the pitch is long, the domains are too large and do not scatter light strongly; when the pitch is short, the domains are too small and do not scatter light strongly either. Figure 5.24(b) shows the drive voltage which increases approximately linearly with the chiral concentration. This is just what we would expect from the critical field given by eqn (5.2), i.e. $V_d \propto 1/P \propto C_c$. The turn-on and turn-off times are shown in Figure 5.24(c). The turn-off time decreases with increasing chiral concentration. This is consistent with $\tau_{\text{off}} \propto 1/q_o^2 \propto 1/C_c^2$. The turn-on time seems to be independent of the chiral concentration.

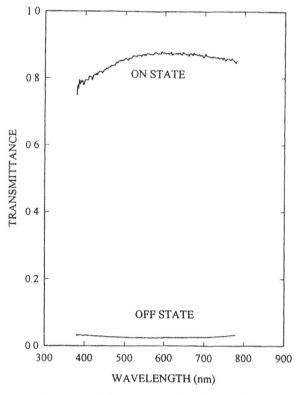

Figure 5.21 The transmission spectra of the PSCT normal mode light valve in the off-state and on-state.

We then studied the effects of polymer concentration. The nematic E48 and chiral agent CB15 were mixed in the ratio of 91:9. Monomer BAB-6 was used. Polymer concentration is another factor controlling the focal-conic domain size. Figure 5.25(a) shows the contrast which peaks at 2.7% polymer. Figure 5.25(b) shows the drive voltage which is independent of the polymer concentration. The drive voltage shown here is the one with increasing voltage. When the liquid crystal has the focal-conic texture, it is not affected by the aligning force of the polymer. However, if the liquid crystal initially has the homeotropic texture, it is affected by the aligning force of the polymer. This can be seen from the turn-off time shown in Figure 5.25(c) and will be discussed further when we consider the hysteresis of PSCT normal mode shutters. The higher the polymer concentration, the stronger its aligning effect, and therefore the longer is the turn-off time. The turn-on time seems to be independent of the polymer-concentration.

There is always a hysteresis in the focal-conic–homeotropic transition, which makes it very difficult to achieve grey scale in a PSCT normal mode display. However, the hysteresis can be used to obtain bistability under a bias field (Greubel, 1974). In Figure 5.20(a), if the bias voltage of $V_0 = 10$ V is applied, the material can have either the homeotropic texture with high transmittance or the focal-conic texture with low transmittance. If the initial state has the focal-conic texture, the material will have the focal-conic texture at V_0; if the initial state has the homeotropic texture, the material will have the homeotropic texture at V_0. Bistable displays using this hysteresis have been studied for two decades. For bistable operation large hysteresis is desired. Mochizuki and coworkers at Fujitsu have successfully made a 5 million pixel projection display (Mochizuki *et al.*, 1990; Yabe *et al.*, 1991). In order to obtain large hysteresis their display requires a homeotropic surface alignment layer and thin

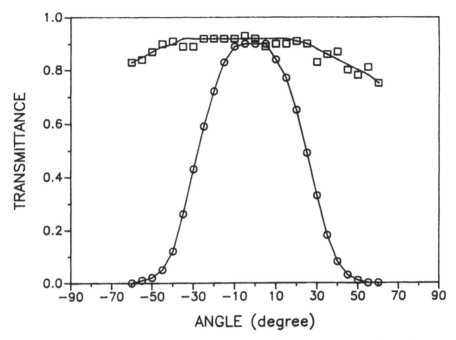

Figure 5.22 Transmittance of the PSCT normal mode light valve and a PDLC light valve in the on-state at various viewing angles: □, PSCT normal mode light valve; ○, PDLC light valve.

cell thickness. The second requirement limits the contrast of the display. This problem can be solved by dispersing polymer networks in the liquid crystal (Fung *et al.*, 1993). We found that the hysteresis can be enhanced with anisotropic polymer networks along the normal of the cell as in PSCT normal mode shutters. Now the hysteresis is independent of the cell thickness because the aligning effect of the polymer network is a bulk effect. The hysteresis of PSCT normal mode material can be characterized by ΔV defined as the voltage difference between the two points with 50% of the maximum transmittance. In the experiment the liquid crystal used was 97.8% nematic ZLI4389 and 2.2% chiral agent R1011. The hysteresis ΔV against the concentration of monomer BMBB-6 is plotted in Figure 5.26. The hysteresis also depends on the variation speed of the applied voltage. The voltage variation speeds for the two curves are 0.25 V/s and 0.0083 V/s, respectively. The hysteresis is larger at a faster varying field because the homeotropic texture is only metastable at a field below E_c. However, the hysteresis does not vanish even at a practically infinitely slow voltage-varying speed. With carefully chosen bias voltage, contrasts higher than 100 can be achieved.

The PSCT normal mode light shutter is suitable for small-size switchable window applications. It can also be used to make high-resolution projection displays on a passive matrix because of its bistability under a bias voltage. Manufacture is simple and the cost is low; however, this display does not have grey scale and cannot be operated at video speed.

5.3.3 *Polymer-stabilized Cholesteric Texture (PSCT) Reverse Mode Liquid Light Shutter*

In a PSCT reverse mode light shutter, the pitch of the cholesteric liquid crystal is long and the planar texture is stabilized at zero field by the polymer network (Yang *et al.*, 1992). The

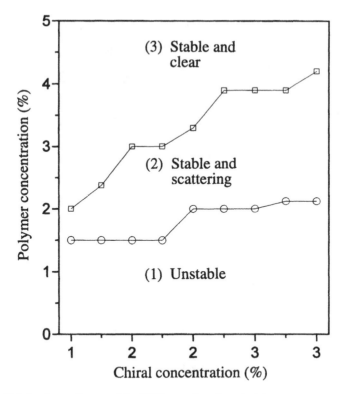

Figure 5.23   Chiral–polymer diagram of the PSCT normal mode material.

cell was constructed from two glass plates with a rubbed polyimide alignment layer. The liquid crystal, monomer and photoinitiator mixture were poured into the cell in a vacuum chamber. The material was irradiated by UV light for polymerization. For PSCT reverse mode shutters, the UV intensity was typically about 0.05 mW/cm^2 and the curing time was about 5 hours. During polymerization, the material had the planar texture. The polymer network formed is parallel to the cell surface as shown in Figure 5.8 and stabilizes the planar texture. At zero field, the material has the planar texture as schematically shown in Figure 5.27(a). Now because the pitch of the cholesteric liquid crystal is in the infrared region, the material is transparent for visible light. When a sufficiently high field is applied, as a result of the competition between the aligning effect of the field and the intermolecular interaction, the material is switched into the focal-conic texture, as shown by Figure 5.27(b), and becomes scattering. When the applied field is turned off, the material relaxes back to the planar texture. With appropriate polymer network structure, the polymer network is quite stable and can withstand voltages more than twice the drive voltage. Of course, when a very high voltage is applied, the material will be switched into the homeotropic texture and the polymer will be distorted. When the very high voltage is turned off the material will not relax back to the transparent planar texture.

In the experiment studying the PSCT reverse mode light shutter, the setup was the same as that used for PSCT normal mode light shutters. We made a PSCT reverse mode shutter with the following combination: 92.4% nematic E48, 0.6% chiral agent R1011 and 7% monomer BMBB-6. The photoinitiator used was always BME and its concentration was controlled at 1.5% of the monomer. The cell gap was 15 μm. The transmittance–voltage curve is shown in Figure 5.28(a). In the off-state, the material has the planar texture and the

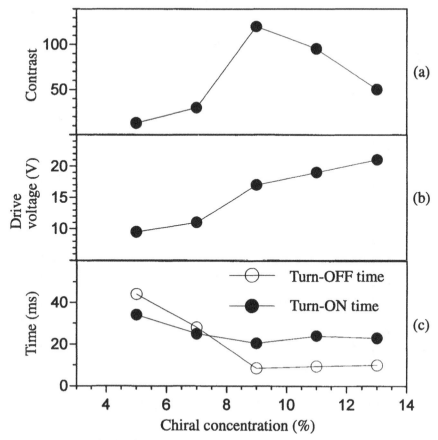

Figure 5.24 Effects of the chiral concentration on the electro-optical properties of the PSCT normal mode material.

transmittance is about 85%. The main light loss is due to reflection at the glass–air interface. Now the polymer concentration is 7%, which is higher than that in the PSCT normal mode shutter. The polymer causes a few per cent of the light to be scattered. As the applied voltage is increased, the material is gradually switched into the focal-conic texture and becomes light scattering. When the voltage is turned down, the material relaxes back to the planar texture. In this planar–focal-conic transition, there is no hysteresis if the aligning effect of the polymer network is strong enough. The response of the shutter to a 100 ms long 25 V pulse is shown in Figure 5.28(b). The turn-on and turn-off times are 4 ms and 10 ms, respectively. These response times are short and suitable for video rate operation.

The stability of PSCT reverse mode shutters has to be examined closely. In these shutters, the applied field is perpendicular to the cell surface and tends to distort the structure of the polymer network which is parallel to the cell surface. The situation is different in PSCT normal mode shutters where the applied field is parallel to the polymer network. In PSCT reverse mode shutters, the polymer network has to be able to withstand application of the drive voltage over a long time period.

The UV intensity for polymerization is crucial to PSCT reverse mode shutters. Stable polymer networks, which have to be large, can be achieved only at very slow polymerization rate. The transmittance–voltage curves of three cells cured at different UV intensities are shown in Figure 5.29. The materials used here were 92.4% nematic E48, 0.6% chiral

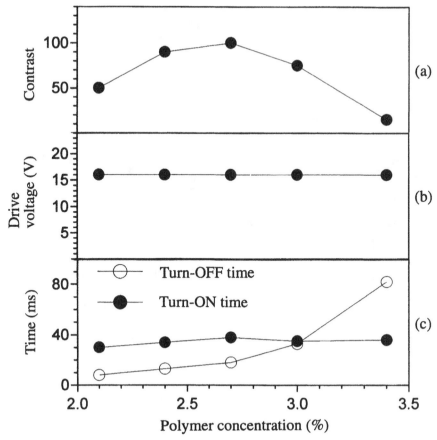

Figure 5.25 Effects of the polymer concentration on the electro-optical properties of the PSCT normal mode material.

agent R1011 and 7% monomer BMBB-6. For the cell cured at 4.0 mW/cm², the polymer network was thin and dense; the domain size was too small, which resulted in a high drive voltage and low contrast ratio. As a matter of fact, this cell was not stable and the off-state transmittance decreased after a long period of application of the voltage pulse. For the cell cured at 0.04 mW/cm², the polymer network was thick and stable. The cell had a much better performance, i.e. low drive voltage and high contrast.

Polymer concentration affects the stability of the polymer network and the domain size in the on-state. In the experiment studying the effect of polymer concentration, we used E48, R1011 and BMBB-6. The concentration of R1011 was 0.5% and the UV intensity was 0.04 mW/cm². The contrast is shown in Figure 5.30(a). The maximum contrast was achieved with 7% BMBB-6. Figure 5.30(b) shows the drive voltage which increases linearly with polymer concentration. Figure 5.30(c) shows the turn-off time which is approximately inversely proportional to the polymer concentration. The full line is the fitting by $\tau_{off} = a/C_p^2$, where a is the fitting parameter. This suggests that the aligning field is linearly proportional to the polymer concentration C_p.

The pitch of the cholesteric liquid crystal is another factor affecting the electro-optical properties of PSCT reverse mode shutters. The chiral concentration should be adjusted such that the pitch P and the cell thickness d satisfy the relation $P = 2d/m$, where m is a positive integer. Our experimental results showed that the electro-optical performance of the PSCT

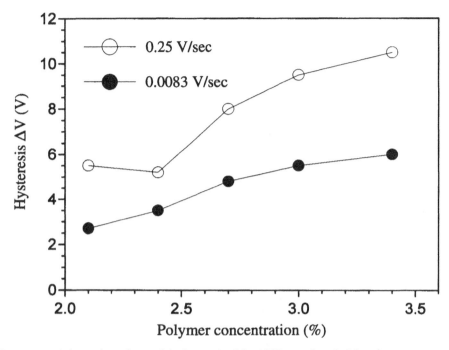

Figure 5.26 Polymer dependence of the hysteresis of the PSCT normal mode light valve.

reverse mode shutter was optimized when the pitch was one-third of the cell thickness. The scattering of the material became polarization dependent if the pitch was longer than the cell thickness.

PSCT reverse mode shutters, like PSCT normal mode shutters, have a wide viewing angle because the polymer concentration is still low. The transmission spectra of both the on-state and off-state are almost wavelength independent. PSCT reverse mode material scatters more light in the backward direction than PSCT normal mode material and PDLCs, and therefore is suitable for black-and-white reflective displays. We have made a prototype projection of PSCT reverse mode material on an MIM matrix using fluorinated liquid crystal TL203. The charge hold time (90%) was more than 100 ms. For a 10 μm cell, the drive voltage was 15 V and contrast was 50. The response times of the PSCT reverse mode material were about 10 ms, and there was no hysteresis. Therefore PSCT reverse mode material is suitable for video rate grey-scale active matrix displays.

5.4 Conclusion

Polymer-stabilized and polymer-modified liquid crystals are a new and exciting area. When a small amount of monomer, especially a mesogenic monomer, is mixed with a liquid crystal, the mixture is in a liquid crystal mesophase. When the monomer is polymerized in a liquid crystal environment, the resulting polymer network mimics the order and orientation of the liquid crystal. Desired structures of polymer networks can be achieved by using alignment layers and externally applied fields, which align the liquid crystal, during polymerization. After polymerization, polymer networks can be used in turn to align the liquid

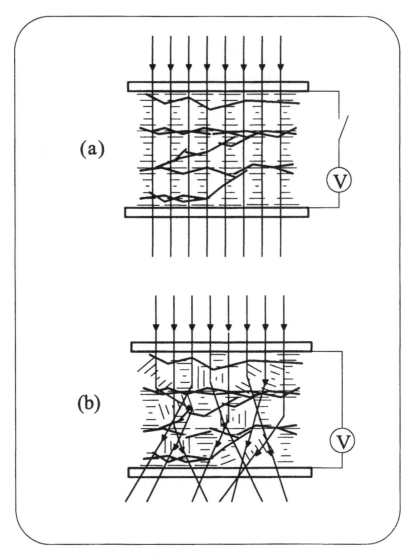

Figure 5.27 Schematic diagram of the configurations of the liquid crystal director of the PSCT reverse mode light valve in the off-state and on-state. The thick dark lines represent the polymer network.

crystal. The aligning effect of the polymer network is a bulk effect; it can be used to achieve the desired director configuration of liquid crystals, which is sometimes impossible with surface alignments.

Morphology and birefringence studies prove our expectation that polymer networks formed in liquid crystals are anisotropic. Homogeneous alignment of polymer networks is achieved by using a homogeneous alignment layer and homeotropic alignment of polymer networks is obtained by using either a homeotropic alignment layer or an electric field applied across the cell. Fibre-like polymer networks can be obtained with mesogenic monomers with flexible tails and flexible monomers. Study of polymer-induced birefringence has proven to be a useful tool in investigating the structure of polymer networks.

Low-concentration polymer network dispersion is a very useful technique in liquid crystal

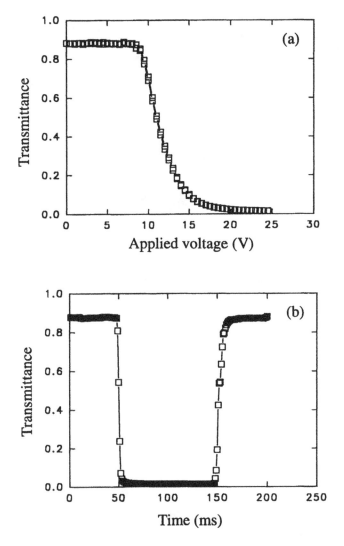

Figure 5.28 (a) The transmittance–voltage curve of the PSCT reverse mode light valve. (b) The response of the PSCT reverse mode light valve to a 100 ms long voltage pulse.

displays. Polymer networks can be used to stabilize and modify liquid crystals. In this dispersion, the optical scattering effect is negligible since the polymer concentration is low. Using polymer networks we are able to stabilize one or two textures of cholesteric liquid crystals at zero field. With the bistable reflective cholesteric material, a highly multiplexed display can be made on a passive matrix. The display does not need polarizers and backlight. The contrast is high and the viewing angle is wide. With the polymer-stabilized cholesteric texture scattering material, either normal mode or reverse mode light valves can be constructed. These valves have a wide viewing angle. The viscosity of the material is low and the material can be poured into a cell in a vacuum chamber. With high-resistivity liquid crystals, the PSCT material can be used for active matrix displays.

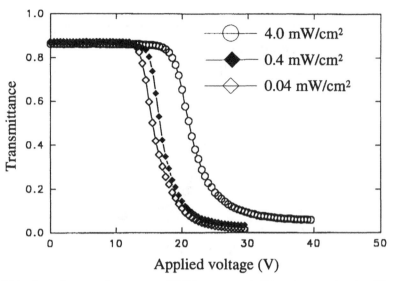

Figure 5.29 Transmittance–voltage curves of PSCT reverse mode light valves cured under different UV intensities.

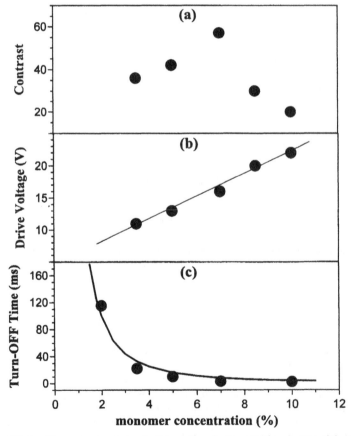

Figure 5.30 Effects of the monomer concentration on the electro-optical properties of the PSCT reverse mode material.

Acknowledgements

Much of the work presented in this chapter was done in collaboration with Professors J. W. Doane and P. Palffy-Muhoray. It would have been impossible to write this chapter without the help of Y. Sun. We would like to thank D. Baker, M. Boyden, C. Citano, Don Davis, W. Fritz, X. Huang, B. Jones, A. Kolosovkaya, Z. Lu, M. Pfeiffer, R. Ma, I. Shenouda, D. St John and G. Ventouris, who have contributed greatly to the polymer-stabilized cholesteric texture (PSCT) project. We would also like to thank Drs J. Kelly, O. Lavrentorvich, P. Bos, J. West, S. Žumer and G. Crawford for stimulating discussions. This work was partially supported by NSF under ALCOM-STC Grant DMR 89-20147, ARPA/High Definition Display Technology Contract #DMR-972-91-J-1020 and Hughes Research Lab.

References

ABRAMOWITZ, M. and STEGUN, I. A. (1972) *Handbook of Mathematical Functions*, 10th edn, New York: Dover, p. 374.

BLINOV, L. (1983) *Electro- and Magneto-optical Properties of Liquid Crystals*, Chichester: John Wiley, p. 213.

BOS, P., RAHMAN, J. and DOANE, J. W. (1993) A low-threshold-voltage polymer network TN device, *SID Dig. Tech. Pap.* **24**, 877–880.

BROER, D., GOSSINK, R. and HIKMET, R. (1990) Oriented polymer networks obtained by photopolymerization of liquid-crystalline monomers, *Ang. Makromol. Chem.* **183**, 45–65.

CHIEN, L.-C. (1995) *Recent Advances in Liquid Crystal Polymers*, American Chemical Society Book Series.

COLES, H. J. (1978) Laser and electric field induced birefringence studies of the cynaobiphenyl homologues, *Mol. Cryst. Liq. Cryst.* **49** (Letters), 67–74.

CRAWFORD, G., SCHARKOWSKI, A., FUNG, Y., DOANE, J. W. and ZUMER, S. (1995) *Phys. Rev. E* **52**, 1273–1276.

DE GENNES, P. G. (1971) Short range order effects in the isotropic phase of nematic and cholesterics, *Mol. Cryst. Liq. Cryst.* **12**, 1993.

DE GENNES, P. and PROST, J. (1993a) *The Physics of Liquid Crystals*, 2nd edn, Oxford: Clarendon Press, p. 282.

DE GENNES, P. and PROST, J. (1993b) *The Physics of Liquid Crystals*, 2nd edn, Oxford: Clarendon Press, pp. 288–291.

DE GENNES, P. and PROST, J. (1993c) *The Physics of Liquid Crystals*, 2nd edn, Oxford: Clarendon Press, p. 292.

DIR, G. A. *et al. (1972)* Cholesteric liquid crystal texture change displays, *SID Proc.* **13**, 106–113.

DOANE, J. W. (1991) Polymer-dispersed liquid crystals: Boojums at work, *MRS Bull.* **16**, 22–28.

DOANE, J. W., VAZ, N., WU, B.-G. and ŽUMER, S. (1986) Field controlled scattering from nematic microdroplets, *Appl. Phys. Lett.* **48**, 269–271.

DOANE, J. W., GOLEMME, A., WEST, J., WHITEHEAD, J. and WU, B.-G. (1988) Polymer dispersed liquid crystals for display application, *Mol. Cryst. Liq. Cryst.* **165**, 511–532.

DOANE, J. W., WEST, J. L., TAMURA-LIS, W. and WHITEHEAD, J. (1989) Wide-angle view PDLC displays, *Pacific Polymer Preprint* pp. 245–247.

DOANE, J. W., YANG, D.-K. and CHIEN, L.-C. (1991) Current trends in polymer dispersed liquid crystals, *Conference Record of Int. Display Research Conf.*, SID, pp. 175–178.

DOANE, J. W., YANG, D-K. and YANIV, Z. (1992) Front-lit panel display from polymer stabilized cholesteric textures, *Proc. SID, Japan Display '92*, pp. 73–76.

DRZAIC, P. (1988) A new director alignment for droplets of nematic liquid crystals with low bend to splay ratio, *Mol. Cryst. Liq. Cryst.* **154**, 289–296.

FERGASON, J. (1985) *SID Dig. Tech. Pap.* **16**, p. 68.

FUNG, Y., YANG, D.-K., DOANE, J. W. and YANIV, Z. (1993) Projection display from polymer stabilized cholesteric textures, *Proc. Int. Display Research Conf.*, *EuroDisplay '93*, pp. 157–160.

FUNG, Y., YANG, D.-K., SUN, Y., CHIEN, L.-C., ŽUMER, S. and DOANE, J. W. (1995) *Mol. Cryst. Liq. Cryst* (submitted).

GREUBEL, W. (1974) Bistability behavior of texture in cholesteric liquid crystals in an electric field, *Appl. Phys. Lett.* **25**, 5–7.

GREUBEL, W., WOLFF, U. and KRUGER, H. (1973) Electric field induced texture changes in certain nematic/cholesteric liquid crystals, *Mol. Cryst. Liq. Cryst.* **24**, 103–111.

HIKMET, R. (1991) Anisotropic gels and plasticized networks formed by liquid crystal molecules, *Liq. Cryst.* **9**, 405–416.

HORN, R. G. (1978) Refractive indices and order parameters of two liquid crystals, *J. Phys.* **39**, 105–109.

HUANG, X.-Y., YANG, D.-K., BOS, P. and DOANE, J. W. (1995) *SID Dig. Tech. Pap.*, pp. 347–350.

HULIN, J. P. (1972) Parametric study of the optical storage effect in mixed liquid-crystal systems, *Appl. Phys. Lett.* **21**, 455–457.

JÁKLI, A., BATA, L., FODOR-CSORBA, K., ROSTAS, L. and NOIREZ, L. (1994) Structure of polymer networks dispersed in liquid crystals: small angle neutron scattering study, *Liq. Cryst.* **17**, 227–234.

KAWACHI, M. and KOGURE, O. (1977) Hysteresis behavior of texture in the field-induced nematic-cholesteric relaxation, *Jpn. J. Appl. Phys.* **16**, 1673–1778.

KAWACHI, M. *et al.* (1975) Field-induced nematic-cholesteric relaxation in a small angle wedge, *Jpn. J. Appl. Phys.* **14**, 1063–1064.

KELLY, J. and PALFFY-MOHURAY, P. (1994) The optical response of polymer dispersed liquid crystals, *Mol. Cryst. Liq. Cryst.* **243**, 11–29.

KEYS, P. H. Private communication.

LU, Z.-J., YANG, D.-K. and DOANE, J. W. (1995) *SID Dig. Tech. Pap.*, pp. 172–175.

MOCHIZUKI, A. *et al.* (1990) A 1120 × 768 pixel four-color double layer liquid crystal projection display, *Proc. SID* **31**, 155–161.

RAJARAM, C. and HUDSON, S. (1994) Morphology of polymer stabilized liquid crystals, *ALCOM Proc.* **6**, 1–6.

SAUPE, A. and LU, L. (1980) On the dynamics of the nematic-cholesteric transition and undulation instabilities, *Freiburger Arbeitstag. Flussigkrist.* **3**, 26–28.

SHENG, P. (1982) Boundary-layer phase transition in nematic liquid crystals, *Phys. Rev. A* **26**, 1610–1617.

WEST, J. (1988) Phase separation of liquid crystals in polymers, *Mol. Cryst. Liq. Cryst.* **157**, 427–441.

WEST, J. L., AKINS, R. B., FRANCL, J. and DOANE, J. W. (1993) Cholesteric/polymer dispersed light-shutter, *Appl. Phys. Lett.* **63**, 1471–1473.

YABE, Y., YAMADA, H. and HOSHI, T. (1991) A 5-M pixel overhead projection display utilizing a nematic-cholesteric phase-transition liquid crystal, *SID 91 Dig.* pp. 261–264.

YANG, D.-K. and DOANE, J. W. (1992) Cholesteric liquid crystal/polymer gel dispersion: Reflective display application, *SID Dig. Tech. Pap.* pp. 759–761.

YANG, D,-K. and LU, Z.-J. (1995) Switching mechanism of bistable cholesteric reflective displays, *SID Dig. Tech. Pap.*, pp. 351–354.

YANG, D.-K. and PALFFY-MOHURAY, P. (to be published).

YANG, D.-K., CHIEN, L.-C. and DOANE, J. W. (1991) Cholesteric liquid crystal/polymer gel dispersion bistable at zero field, *Conference Record. of Int. Display Research Conf. SID* pp. 49–51.

YANG, D.-K., CHIEN, L.-C. and DOANE, J. W. (1992) Cholesteric liquid crystal/polymer dispersion for haze-free light shutters, *Appl. Phys. Lett.* **60**, 3102–3104.

YANG, D.-K., WEST, J., CHIEN, L.-C. and DOANE, J. W. (1994a) Control of reflectivity and bistability in displays using cholesteric liquid crystals, *J. Appl. Phys.* **76**, 1331–1333.

YANG, D.-K., DOANE, J. W., YANIV, V. and GLASSER, J. (1994b) Cholesteric reflective display: Drive scheme and contrast, *Appl. Phys. Lett.* **64**, 1905–1907.

ŽUMER, S. (1988) Light scattering from nematic droplets: Anomalous-diffraction approach, *Phys. Rev. A*, **37**, 4006–4015.

6

Liquid Crystal–Gel Dispersions Prepared in the Isotropic Phase

A. JÁKLI, K. FODOR-CSORBA and A. VAJDA

6.1 Introduction

Polymerization and gelation processes in solvents have been studied for a long time. For one of the first reviews we just refer here to Flory's famous book of polymer chemistry (Flory, 1953). Even the idea of performing chemical reactions in anistropic liquids is about 80 years old (Svedberg, 1916). It was suggested that the anisotropic media can play a role of a 'catalyst' in certain reactions. For a review see Percec *et al.* (1992). These reactions have attracted significant interest from 1982, when interesting electro-optical properties of liquid crystal–polymer composite systems were realized (Crighead *et al.*, 1982). The so-called polymer-dispersed liquid crystals (PDLCs) are one type of composite (Fergason, 1985; Doane *et al.*, 1986; Drzaic, 1986). They contain about the same amount of polymer and liquid crystal, and consist of liquid crystal droplets in a continuous polymer matrix. PDLCs can be switched between opaque and transparent states by relatively high voltages, but it is difficult to achieve a haze-free transparent state. For more details see the chapter by Crawford and Žumer in this book.

Another type of liquid crystal–polymer composites are obtained when relatively small amounts of reactive monomer are dissolved and polymerized in the liquid crystal medium. In analogy to PDLCs sometimes they are called liquid crystal dispersed polymers (LCDPs) Here we call them liquid crystal–gel dispersions, since the polymer forms a network in the continuous liquid crystal matrix. In such systems the polymer imposes an alignment of the liquid crystal molecules, but it can be altered by external fields. Similar to PDLCs they can be used for light shutters, but with principally a haze-free transparent state. This idea has been studied in detail using mesogenic reactive monomer (Hikmet, 1990, 1991a,b; Hikmet and Zwerver, 1991, 1992; Hikmet and Higgins, 1992), with non-mesogenic monomer for cholesterics (Yang *et al.*, 1992) and nematic droplets suspended in a glycerine matrix (Crawford *et al.*, 1994). In these cases the polymerization was carried out in the liquid crystal phase usually in thin films. These studies are summarized by Hikmet, by Žumer and Crawford, Yang *et al.* and Broer in other chapters of this book.

Liquid crystal–gel dispersions can also be prepared in bulk, in the isotropic phase of the liquid crystal (Stannarius *et al.*, 1991; Jákli *et al.*, 1992, 1994, 1995; Jákli, 1994). In such cases the polymer network has an isotropic distribution and imposes a misalignment on the liquid crystal. In an external field, however, the liquid crystal can be oriented. Consequently this polymer becomes somewhat distorted resulting in interesting storage and electro-optic

143

effects (Jákli, 1994). The properties of the gel dispersions can be very different depending on whether the liquid crystal is a good or a poor solvent for the polymers. Accordingly the materials have paste- or glass-like consistency with unchanged or strongly depressed phase transitions, respectively. In this chapter we summarize recent studies and present new results regarding such systems.

6.2 Studied Systems

Small amounts (0.5–3wt.%) of bifunctional monomers and 0.5–1.5% photoinitiator were dissolved and UV polymerized in the isotropic phase of four different liquid crystals. The experiments were carried out using two bifunctional reactive monomers: (i) 4,4′-bisacryloyloxybiphenyl (BAB) (Mariani *et al.*, 1986), this was synthesized in the Liquid Crystal Institute of the Kent State University by L. C. Chien; (ii) 4,4′-bis(2-methylpropenoyloxy)-biphenyl (BMB), this was synthesized in the Liquid Crystal Group of the Research Institute for Solid Physics in Budapest by the following procedure. Methacrylic acid (38 mmol) was dissolved in carbon tetrachloride (30 ml) and was added to a well-stirred solution of dicyclohexyl carbodiimide (38 mmol) and 4,4′-biphenol (19 mmol) in 30 ml carbon tetrachloride. The reaction was catalysed by *p*-dimethylamino pyridine. The slightly exothermic reaction proceeded over 4 hours. The precipitate was filtered off and the solvent evaporated. The residue was crystallized from ethanol several times. The purity of the product was checked by thin-layer chromatography on Kieselgel 60 F_{254} plates eluted with *n*-hexane and ethyl acetate in the ratio 2:1, melting point = 151 °C.

The reactive monomers BAB and BMB have the following structures:

Here *R* stands for *H* (BAB) or CH_3(BMB). The two double bonds are opened by photoinduced free-radical reactions. They participate in the construction of the two distinct chains, and

becomes a cross-linking bridge of the structure. Owing to this chemical process branched polymer chains form. They can be characterized by the parameters *N* and *P*. *N* is the degree of polymerization, i.e. the number of monomer units in a chain, and *P* is the polydispersity defined as the ratio of the weight and number-averaged molecular weight. In the present case the distribution of the molecular weight is probably very broad, i.e. the polydispersity *P* is large.

Benzoin methyl ether (BME) was used as a photoinitiator. Under UV light it generates radicals by intramolecular cleavage of a chemical bond which has a lower dissociation energy than that of the excitation energy of the reactive excited state. The following radical formation is created by the UV light:

If the concentration of the radical is small compared with the monomer the B and ME radicals have comparable efficiency as initiators (Pappas and Asmus, 1982). At high radical and low monomer concentrations the B radical is primarily responsible for initiation and the ME radical participates, predominantly as a chain-terminating agent (Hageman *et al.*, 1979). In our experiments the ratio of the photoinitiator to the monomer is about one-half. This seems to correspond to high radical concentration since, according to infrared spectroscopy (Fodor *et al.*, unpublished) the B radicals are responsible for the initiation. For larger amounts of photoinitiator the reaction starts simultaneously at more points and one can expect a lower degree of polymerization and shorter polymerization time. Small amounts of photoinitiator result in a higher degree of polymerization and a slow polymerization process. The following four liquid crystal compounds were used as solvents:

1 4-cyano-4'-*n*-pentylbiphenyl (5CB, from Merck Ltd). It has a nematic phase below 35.3°C and crystallizes below room temperature.

2 4-cyano-4'-*n*-octylbiphenyl (8CB, from Merck Ltd). It has a phase sequence of isotropic–nematic–smectic A–crystal with the transition temperatures of 40.5, 33.5 and 21.5°C respectively.

3 (S)-4'-(2-methylbutyloxy)phenyl 4-undecyloxybenzoate (MBOPE110BA) (Bata *et al.*, (1987). Cooling from the isotropic phase it goes to the SmA phase at 66 °C. It has a monotropic SmC* phase between 46 and 37 °C and is crystalline below it.

4 Cholesteryl propionate (CHP, from Thermax Thermographic Measurement Limited). Its phase sequence can be seen in the first row of Table 6.2.

In Table 6.1 we list the main parameters of the studied samples.

Table 6.1 List of main parameters of the studied samples

Liquid crystal	Monomer	UV curing
5CB	BAB 0.5%, 1%, 1.5%	50°C, 40, 30, 20 min
8CB	BMB 1.5%	50°C, 30 min
MPOBE11OBA	BMB 3%	75°C, 30 min
CHP	BMB 1%, 2%, 3%	120°C, 1 h, 2 h, 3 h (each)

On dissolving the reactive monomers and the photoinitiator the clearing point shifted down by a few degrees. The polymerizations were carried out in bulk (sample volumes were larger than 0.2 cm^3) in the isotropic phase of the liquid crystal solvent. To ensure complete mixing the samples were kept at elevated temperatures for at least 2 hours prior to UV irradiation. In all cases the UV light was provided by mercury-arc lamps that produced an intensity of 1–3 mW/cm^2. The polymerization typically took an hour (or even more at higher temperatures). This time is much larger than that needed in the experiments of Hikmet (minutes) with similar light intensities (Hikmet, 1991). The reasons for the difference could be the following. Owing to the higher degree of ordering the rate of polymerization is higher in the liquid crystal state (especially in the nematic and cholesteric phases) than in the isotropic state (Hoyle *et al.*, 1988). Furthermore, Hikmet used more than 10 wt.% solute monomers, and we used less than 3 wt.%.

6.3 Experimental Results

The results for the 5CB (Jákli *et al.*, 1992), 8CB (Jákli *et al.*, 1995) and MBOPE110BA (Jákli, 1994) are similar to each other, but very different from that obtained for the CHP samples. In section 6.3.1 we summarize the main results for the first three samples, and in section 6.3.2 we present new results regarding the behaviour of the CHP cells. From their appearance and consistency we shall refer to them as *pastes* and *glasses*, respectively.

6.3.1 *Liquid Crystal Pastes*

Samples containing 0.5 % and 1 % of reactive monomers remained macroscopically a fluid after 25 min of UV exposure, while samples containing 1.5 % BAB or BMB could be smeared like a paste or cream. The rotational viscosities of 5CB containing a small amount of BAB were determined by suspending cylindrical samples in a magnetic field on a thin tungsten wire and measuring the torque at constant angular velocities (Jákli *et al.*, 1992). The rotational viscosities of the fluid samples deviate only moderately (increased by less than 30 %) from pure 5CB. The samples with paste-like consistency became anisotropic in the measuring field and the viscosities could not be determined. It was observed that samples macroscopically separated after about 1 week into fractions of paste-like consistency, usually at the bottom or the side walls of the containers, and fluid fractions of lower density. The ratio of the paste to the fluid parts was approximately 0.1 and 0.9 in the case of 0.5 % and 1.5 % polymers, respectively. Studies on several liquid crystal–polymer dispersions showed that samples containing only 2 % of polymers stay uniformly paste-like for longer periods. Pastes are white and strongly light scattering in the liquid crystal phase and remain slightly turbid even in the isotropic phase. This indicates that the polymer aggregates and separates from the liquid crystal even in the isotropic phase. It was suggested that in separating the polymer aggregates to fibrils which are physically interconnected (Jákli *et al.*, 1992).

From a practical point of view it can be important that the pastes can be smeared on a substrate. By adding a small number of spacer balls to the bulk paste one can easily prepare thin films from them (Saupe and Jákli, 1994). The films can be switched between scattering and haze-free transparent states similar to the polymer-dispersed liquid crystal (PDLC) films. The voltage range between the misaligned scattering state and the aligned transparent state is broad. Good transparency requires typically 50 V (10–20 μm film thicknesses). The contrasts are good: for films of 5CB + 1.5 % BAB values as high as 500 could be achieved (Jákli *et al.*, 1992). The relaxation times are of the order of a few milliseconds: for 5CB containing polymerized BAB network they are 7 ms and 15 ms for 1.5 % and 0.5 % concentrations, respectively. The characteristics are stable provided we stay in the nematic phase.

The mesh size of the network determines the minimum field strength required to compete with the alignment by the network. The coherence length of the fields must be comparable with the mesh size, i.e. the distance between the fibrils. Analysing the behaviour of the samples in electric and magnetic fields, it was estimated that the mesh size is in the 0.1–1 μm range if the polymer concentration is a few per cent. Given the mesh size and the polymer concentration one can estimate the radius of the fibril (Jákli *et al.*, 1992). Assuming complete separation with a regular lattice of mesh size 1 μm and a polymer concentration of 1 %, the radius of the fibril is in the range of 600 Å. This means that the average strand would consist of a bundle of approximately 10^4 polymer chains. These predictions were checked experimentally by small-angle neutron scattering measurements (Jákli, 1994). The measurements revealed that the characteristic size of the separated phase is indeed in the range of a few hundred angstroms but somewhat smaller than estimated. The size slightly decreases on heating. For example, the cross-section radius of the separated polymer fibres of the 5CB + 1.5wt.% BAB system is 200 Å at 60 °C and 270 Å at room temperature. This indicates that in practice some part of the polymer is dissolved in the liquid crystal; it can also be concluded from the shift of the phase transition temperatures. Neglecting confinement effects a fully separated system should have the same transition temperature as the pure liquid crystal. In a real system after polymerization the phase transition temperature is typically less than a degree below that of pure liquid crystal. This is much smaller than the shift prior to polymerization on dissolving the monomer and phoinitiator (3–5 °C). The interfaces between the different phases seem to be diffuse. The roughness, i.e. the surface area of the polymer–liquid crystal interface, decreases with temperature.

Pastes show strong memory effects. If the sample is cooled in the presence of magnetic fields it becomes aligned. The alignment stays after removal of the field indicating that the alignment is fixed by the dispersed network. Diamagnetic anisotropy measurements on 5CB (Jákli *et al.*, 1992) and the Bragg reflection of the smectic layers of 8CB detected by neutron scattering (Jákli *et al.*, 1995) show that the alignment is essentially as perfect as the field-induced alignment of the pure liquid crystal samples. On heating the pastes to the isotropic phase and cooling without the magnetic field, the alignment recovers to almost the same quality. After several heating–cooling cycles without the external field the quality of the resulting alignment begins to fade. The alignments of both the liquid crystal 8CB and the polymer network were monitored by neutron scattering measurements (Jákli *et al.*, 1995). On repeating the heating-cooling cycles it was found that, in the range of a few hundred ångströms, the orientation of the polymer network stays random while the alignment of the liquid crystal varies between random and perfectly aligned. As a measure of the alignment of the smectic layers the half width of the Bragg reflection (w_b) can be used. It would be $0°$ for perfect alignment and $180°$ for non-aligned layers. The layers of a virgin 8CB + 1.5% BMB sample were randomly aligned. By cooling it in the presence of a magnetic field of 1.4 T the alignment can be characterized by $w_b = 4.5°$. On heating above the clearing point by 5°C and cooling after an hour without the magnetic field we find that $w_b = 11°$.

6.3.2 *Polymer-induced Glasses*

In cholesteryl propionate mixtures, contrary to liquid crystal pastes, the phase transition temperatures considerably shifted downwards after polymerization. The phase transition temperatures, as determined by the observed texture changes, are listed in Table 6.2. The mixtures with different BMB concentrations and illumination times are denoted by the ratio of the BMB concentration in wt.% and the illumination time in hours. The phase transition temperatures are different in heating and cooling. The hysteresis increases with larger polymer concentration and illumination time. In the mixtures, where we do not indicate crystallization, the cholesteric phase is stable at room temperature. Keeping the sample at around 50 °C for hours, crystalline nuclei appear and grow slowly. On heating the sample from such a mixed state we observe that the crystals melt only at the temperature corresponding to the melting of the non-illuminated mixtures.

The UV light causes slight changes even in the pure cholesteryl propionate, and after 3 hours of irradiation the temperature of the phase transition decreases by 5 °C. Thin-layer chromatography and infrared spectroscopy (S. Holly, private communication) indicated the appearance of some new molecules, but their content is surely less than 1%. Cholesteryl derivatives are known to be sensitive to UV light in the presence of oxygen (W. Weissflog, private communication) and the degradation is probably due to this effect. All the irradiated pure CHP samples show textures similar to the non-irradiated films. The phase transitions decreased up to 10 °C in the non-irradiated mixtures too, but they also present the same textures as the pure, non-irradiated CHP. This decrease is typical for solutions but are much smaller than after polymerization. It is remarkable that only a few per cent of polymer depresses the isotropic cholesteric phase transition by about 50 °C and suppresses the crystallization completely.

Table 6.2 Phase transitions of cholesteryl propionate gel dispersions

Concentration (%)/ curing time (h)	Phase transitions/°C heating (cooling)	
	Cr–Ch	Ch–I
0/0	95 (82)	110 (110)
0/1	93 (78)	109 (109)
0/2	92 (74)	107 (109)
0/3	92 (68)	104 (105)
1/0	90 (83)	109 (107)
1/1	83 (74)	90 (88)
1/2	78 (60)^a	87 (78)
1/3	–(–)	78 (68)
2/0	87 (78)	107 (105)
2/1	83 (60)	90 (88)
2/2	–(–)	76 (73)
2/3	–(–)	76 (46)
3/0	80 (70)	99 (94)
3/1	78 (52)^a	82 (80)
3/2	–(–)	72 (70)
3/3	–(–)	71 (57)

^{a} Crystallization happens only for slow cooling ($<0.5°C$/min).
– No crystallization until room temperature.

CHP+1%BMB+0.5%BME

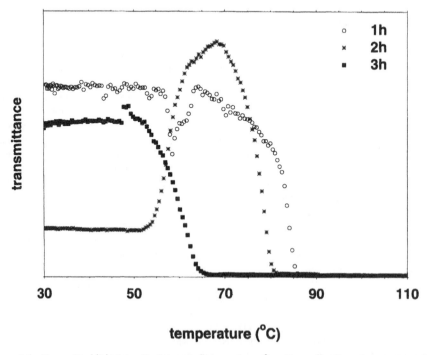

Figure 6.1 Transmitted light intensity (integrated over a 1 mm² area) as a function of temperature for a 6 μm thick film containing 1 wt.% of reactive monomer, BMB, and 0.5 wt.% of photoinitiator, BME.

CHP+2%BMB+1%BME

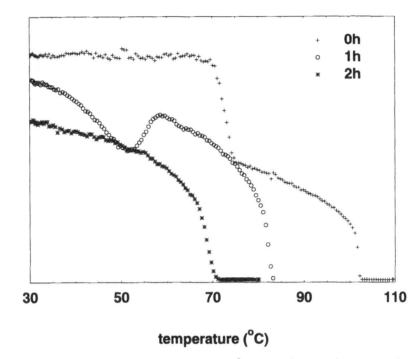

Figure 6.2 Transmitted light intensity (integrated over a 1 mm² area) as a function of temperature for a 6μm thick film containing 2 wt.% of reactive monomer, BMB, and 1.0 wt.% of photoinitiator, BME.

The phase transitions under a cooling rate of 2°C/min are illustrated in Figures 6.1–6.3 where we plotted the temperature dependencies of the light intensities transmitted through 6 μm films placed in a polarizing microscope. The transmittance was measured by replacing one of the oculars by a Mettler photomonitor. During the measurements the textures were also observed. In the figures zero transmittance corresponds to the isotropic phase. On cooling to the liquid crystal phase the transmittance increases. Although the steepness of the curves decreases for higher concentrations and illumination times, the phase transition always looks sharp. The coexistence of the isotropic and cholesteric phases are observable only in a narrow (< 1 °C) range. The transmittance is plotted in arbitrary units, but in the same figure the units are the same. For samples containing 2% and 3% of polymers the transmittance at low temperatures decreases with the illumination time. This does not hold for the 1% solution: after 3 hours of illumination the transmittance is larger than after 2 hours. In this case, however, the phases are different. After 1 and 2 hours of illumination the samples are crystalline (with different crystal structures), but after 3 hours they are cholesteric. Comparing the same phases, therefore, we can say that between crossed polarizers the textures are darker for higher polymer concentration and longer polymerization time.

The pure cholesteryl propionate in the cholesteric phase show a bright blue colour in reflection. At lower temperatures the reflected colour has a slight greenish tone. The polymer dispersions also show reflected colour in their cholesteric phase, even at room temperature. In time the colour fades slowly, but for the mixtures 2/3, 3/2 and 3/3 it stays for several weeks. The reflected colour is green and almost independent of the temperature. The films show grainy domain textures, but without considerable light scattering. The textures consist of densely packed defect lines and are not differently oriented domains as in pastes.

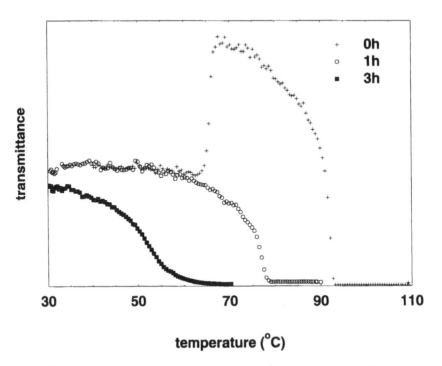

CHP+3%BMB+1.5%BME

Figure 6.3 Transmitted light intensity (integrated over a 1 mm² area) as a function of temperature for a 6 μm thick film containing 3 wt.% of reactive monomer, BMB, and 1.5 wt.% of photoinitiator, BME.

Contrary to the paste systems, between crossed polarizers the films are completely black in the isotropic phase.

The consistency of the mixtures is also different from the paste systems. On cooling they become very viscous and below 30–40 °C the mixtures 2/3, and 3/2 and 3/3 are rigid, like glasses. In the bulk their free surfaces are smooth. On heating, at the top of the cholesteric phase, the sample becomes fluid (in the film flows appear). The glassy to fluid transitions are completely reversible.

We prepared 6 μm thick films sandwiching the pure CHP (0/0) and the mixture 3/3 between glass plates coated with indium tin oxide and studied their electro-optical behaviour. The pure CHP spontaneously assumes planar alignment and (except around defect lines and knots) the films could not be switched because of their negative dielectric constant (de Gennes, 1974). At higher temperatures (⩾40 °C) the films containing mixture 3/3 can be switched by strong fields (10 V/μm) between greenish and transparent states. The textures between crossed polarizers, without and with fields, are shown in Figures 6.4(a) and (b) respectively. The switching is very different from that of the pastes. Owing to the large viscosity the switching time is slow, almost a second. The texture is stable when the field is turned off. In this respect it resembles the behaviour of the polymer-stabilized cholesteric display (Yang *et al.*, 1992) (see also the chapters by Yang *et al.* and Yuan and West in this volume). When the field is turned on again it takes a few seconds until a saturated contrast level is reached. At room temperature the contrast of the switching is very small.

We also studied the stability of the textures of the mixture 3/3. At room temperature the texture is stable at rest for weeks or so. If, however, a strong electric field is applied, the original transparent domain texture transforms to a green planar texture. The textures

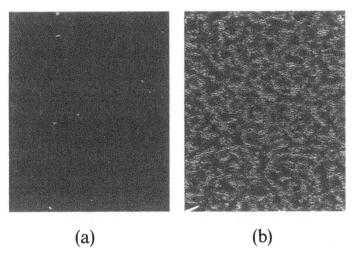

(a)   (b)

Figure 6.4   Textures of a 6 μm film of mixture 3/3 (CHP + 3 wt.% BMB + 1.5 wt.% BME, UV cured for 3 hours) between crossed polarizers at $T = 52\,^{\circ}$C; magnification $N = 160$: (a) at $U = 100$ V; (b) at $U = 0$ V.

between crossed polarizers before and after 40 min of voltage treatment are shown in Figures 6.5(a) and (b), respectively. Once the sample is heated to the isotropic phase and cooled back again without the field the original grainy texture is recovered. It is, however, slightly brighter resembling the previous texture.

At higher temperatures (40–50 $^{\circ}$C) the texture is stable only for a few hours even at rest. When the film is kept a few degrees below the isotropic–cholesteric transition darker parts appear and grow slowly. Later, crystals nucleate typically in the dark spots. The decomposed texture after 12 hours can be seen in Figure 6.6. Under strong fields a spinodal-type decomposition starts after a few minutes. The textures at different durations of the voltage treatments are shown in Figure 6.7. The decomposed structure is almost completely forgotten once the film is reheated to the isotropic phase. On cooling the texture resumed its original, grainy texture.

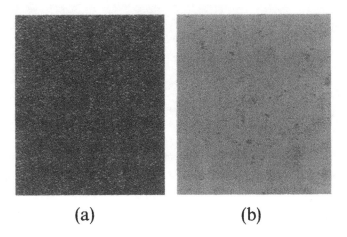

(a)   (b)

Figure 6.5   Textures of a 6 μm film of mixture 3/3 (CHP + 3 wt.% BMB + 1.5 wt.% BME, UV cured for 3 hours) between crossed polarizers at $T = 25\,^{\circ}$C; magnification $N = 160$: (a) before voltage treatment; (b) after 40 minutes of voltage treatment ($U = 200$ V_{p-p}, $f = 0.5$ Hz).

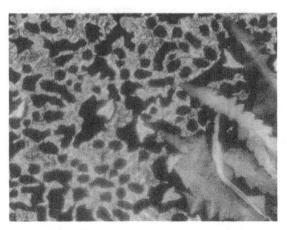

Figure 6.6 Decomposition of a 6 μm film of mixture 3/3 (CHP + 3 wt.% BMB + 1.5 wt.% BME, UV cured for 3 hours) between crossed polarizers at = 52°C 12 hours after it was cooled to the cholesteric phase; magnification $N = 160$. Black spots are either isotropic or homeotropic cholesteric; the dendrites are in the crystal phase.

6.4 Discussion

The properties of liquid crystal polymer dispersions prepared by similar polymerization techniques can differ considerably for different solvents. According to our observations one type of mixture shows practically no change of phase sequence and presents strong memory effects. In the mixtures of cholesteryl propionate the phase sequence changed dramatically, and they showed only weak memory effects. In the polymer science literature there are a number of examples for both types of behaviour. Although the anisotropy of the solvent molecules can result in additional peculiarities, the main features can be understood on the

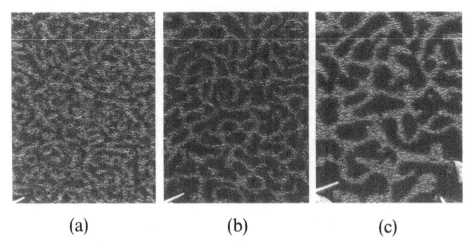

(a)   (b)   (c)

Figure 6.7 Spinodal-type decomposition of a 6 μm film of mixture 3/3 (CHP + 3 wt.% BMB + 1.5 wt.% BME, UV cured for 3 hours) between crossed polarizers at $T = 52°C$. Textures after (a) 5, (b) 15 and (c) 90 minutes of application of a square-wave voltage, $U = 200$ V_{p-p}, $f = 0.5$ Hz.

basis of the results for normal liquid solvents. For this reason we first invoke the relevant results and terminology available from the literature.

6.4.1 *Normal Liquid Solvent*

If not stated otherwise, in this section we mainly use the statements that can be found in de Gennes' book on polymers (de Gennes, 1979).

A polymer gel is a network of flexible chains that can be obtained by chemical or physical processes. At high temperatures the system can be a simple solution of polymers (so called *sol*). On cooling we get a *gel*. If we raise the temperature again we recover the sol, but the transition temperatures measured on heating are often higher than those measured on cooling.

For polymers that are in a solvent of small molecules the chain–solvent interactions can be characterized by a Flory parameter, χ, which depends on the temperature. In most systems $\chi(T)$ is a decreasing function of temperature. At lower χ values steric repulsion dominates and the chain tends to swell. When $\chi > 1/2$ we have a poor solvent and a phase separation phenomenon occurs. Because solvent expulsion is slow in the gel phase, the separation often does not take place on a macroscopic scale. Small 'pockets' with different polymer concentrations can appear. The apearance of ribbons, fibrils and nodules with sizes in the range of 200 to 1000 Å is very usual. These effects are called 'microsyneresis'. They can also occur in good solvents by slightly lowering the solvent quality (e.g. by cooling).

In Figure 6.8 (de Gennes, 1979) the possible gelation processes are illustrated in a phase diagram for gelating systems. ϕ is the concentration of the bifunctional monomers and T_{eq} is an equivalent temperature which decreases from infinity to low values as the chemical reaction progresses. The diagram shows a one-phase range limited by a coexistence curve at lower temperatures. The one-phase region itself is divided into two parts by a sol–gel transition–line. At an intermediate concentration, on decreasing the temperature one can cross the sol–gel line and make a one-phase sol–gel transition. On further cooling we cross the one-phase–two-phase line and eventually phase separation occurs. It initially proceeds via nucleation or spinodal decomposition. On slightly entering the two-phase region, demixing can take place only by nucleation of a droplet of one phase inside the other. This implies interfacial energy at the droplet surface and is slow. Going more deeply into the two-phase region, we reach a state where the interfacial energy vanishes and the system breaks up spontaneously into small domains. Spinodal decomposition involves the formation of a connected, bicontinuous structure and the concentration difference gradually evolves to the equilibrium concentration difference.

6.4.2 *Liquid Crystal Paste Systems*

The phase behaviour of the pastes can be explained by the phase diagram in Figure 6.8. The turbidity observed in their isotropic phase indicates that the solvent becomes poor even during the polymerization process. This means that the equivalent temperature falls in the two-phase region ($T_{eq} = T_C < T_B$) and results in a phase separation process in the isotropic phase. For simplicity in the following we discuss only polymers and solutions, but actually we mean dense and dilute polymer concentrations. The sizes of the separated parts observed by small-angle neutron scattering (Jákli, 1994) have the same order of magnitude that was estimated supposing 100% and 0% polymer concentrations, for the separated phases. This indicates that the system is not far from the 'collapse transition' where the gel expels most of the solvent. The extent of the equilibrium segregation depends on the temperature. Usually longer polymer chains have a stronger tendency towards separation, so the two

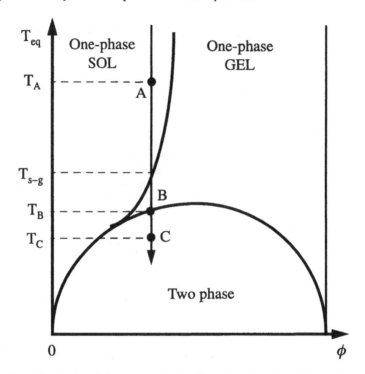

Figure 6.8 Phase diagram for gelating systems (for details see de Gennes (1979)). ϕ is the concentration of the bifunctional monomers and T_{eq} is an equivalent temperature which decreases from infinity to low values as the chemical reaction progresses. The diagram shows the one-phase region limited by a coexistence curve at lower temperatures. The one-phase region itself is divided into two parts by a sol–gel transition line. At an intermediate concentration, by decreasing the temperature along the arrow one can cross the sol–gel line and make a one-phase sol–gel transition. On further cooling at B we eventually enter the two-phase region.

phases in equilibrium do not have the same weight distribution for the containing polymer chains (Flory, 1953). Accordingly, on cooling first only the larger polymers separate out and the smaller branched chains are still dissolved. On entering the liquid crystal phase the majority of the polymers have already separated, so the isotropic–nematic transition is not depressed significantly. On decreasing the temperature further the solvent becomes poorer facilitating the separation of the small chains, too. We suppose that the newly separated parts stick to the already-present polymer stem and become more diffuse on the polymer–liquid crystal interface. This scenario is confirmed by small-angle neutron scattering (Jákli, 1994) where it was found that the characteristic size of the separated polymer increases on cooling.

The liquid crystal alignment memorized by the polymer (Jákli *et al.*, 1995) (see also the chapter by Žumer and Crawford) is characteristic only of anisotropic solvents. In length scales accessible by the small-angle neutron scattering technique the polymer network looks stable memorizing the isotropic condition during preparation. It was therefore proposed (Jákli *et al.*, 1995) that the alignment is determined in smaller length scales by polymer parts separated only in the nematic phase. When the polymer is dissolved, it is flexible and its shape reflects the symmetry of the system. Consequently the polymers that segregated after the phase transition are oriented by the director of the liquid crystal (which is parallel to the magnetic field). On heating to the isotropic phase the separated polymers do not dissolve readily and memorize the initial orientation. Consequently, when the sample is again cooled to the nematic phase the polymer network still contains some smaller oriented, more diffuse parts.

6.4.3 Polymer-induced Glasses

The cholesteryl propionate mixtures do not become turbid, either during the UV curing, or when they are cooled to the liquid crystal phase. This indicates that the cholesteryl propionate is a good solvent for the branched polymer chains. The phase behaviour of the cholesteryl propionate mixtures can be illustrated on the gelation phase diagram (Figure 6.8) following the vertical arrow over the range of intermediate concentrations. After the chemical reaction we reach a value of $T_{eq} = T_A$ in the one-phase sol region (A). After cooling the sample through the sol–gel transition (T_{s-g}) we enter the one-phase gel regime. It is a swollen gel that behaves elastically and can be rigid as a glass. On further cooling we reach the two-phase range. In this state the material decomposes into two states as indeed was observed experimentally (Figure 6.6). With increasing temperature, and with some hysteresis, we again reach the sol state, i.e. the material can flow. In our case the actual phase diagram is more complicated, since there are isotropic–cholesteric and cholesteric–crystal phase transitions. During the isotropic–cholesteric transition the quality of the solvent decreases. The spinodal-type decomposition facilitated by the electric field (Figure 6.7) is probably present only in anisotropic solvents. The crystallization (in the case of the low polymer concentrations, like sample 1/1) drastically modifies the separation process. These features are unique for liquid crystal solvents and deserve more study in the future.

The unusually large ($\sim 50\,°C$) depression of the isotropic–cholesteric transition is probably due to the fact that the polymer chains are dissolved in the isotropic phase. In the liquid crystal phase the Flory parameter, χ, can be approximated as (Brochard, 1979)

$$\chi \cong \chi_0 + \chi_1 S^2 \tag{6.1}$$

Here S is the order parameter of the liquid crystal. Positive χ_1 means that the solvent quality increases during the isotropic–liquid crystal transition, while the solvent is poorer (in the liquid crystal phase if $\chi_1 > 0$. In the first case the clearing point increases owing to the addition of the polymer. In the latter (and most frequent) case the clearing point is depressed. More than $10\,°C$ clearing point depressions are not unusual when a few per cent of polymer is dissolved in a nematic liquid crystal (Kronberg *et al.*, 1978). The depression of the phase transition is connected to the appearance of a two-phase region, where the isotropic and liquid crystal state coexist. During the cooling and heating processes we do not see the two-phase boundaries, probably because they had no time to reach their equilibrium structure. We must mention other factors that could also contribute to the observed depression. There is a double bond in the cholesteryl part of the cholesteryl propionate, which may chemically interact with the radicals of the benzoin methyl ether. The chemical networking of the liquid crystal molecules, however, would result in an increase, instead of a decrease, of the phase transition temperatures (Broer *et al.*, 1988). Confinement effects in the case of pore (or mesh) sizes of the order of few nanometres can result in supercooling of almost 20% below the bulk liquid freezing point (Awschalom and Warnock, 1989). Since the gels have reflected colour in the visible light range, in our case the size of the confinement cannot be smaller than a few hundred angstroms.

The large depression of the isotropic–cholesteric transition and the formation of the glassy state can have practical importance and deserve further study.

Acknowledgements

This work was supported in part by the Hungarian National Science Foundation (OTKA) under contract number T 7409. The authors are grateful to Profesor A. Saupe, Dr S. Holli, Profesor S. Žumer, Dr W. Weissflog and Dr J. Szabon for helpful discussions.

References

AWSCHALOM, D. D. and WARNOCK, J. (1989) Orientational dynamics of supercooled liquids in restricted geometries, *Molecular Dynamics in Restricted Geometries*, ed. J. Klafter and J. M. Drake New York: John Wiley, pp. 351–370.

BATA, L., BUKA, Á., ÉBER, N., JÁKLI, A., PINTÉR, K., SZABON, J. and VAJDA, A. (1987) Properties of homolous series of ferroelectric liquid crystals, *Mol. Cryst. Liq. Cryst.* **151**, 47–68.

BROCHARD, F. (1979) Solutions de polymères flexibles dans un liquide nématique, *C.R. Acad. Sci. Paris* **289**, Serie B, 229–232.

BROER, D. J. Finkelmann, H. KONDO K. (1988) *In situ* polymerization of an orientated liquid crystalline acrylate, *Makromol. Chem.* **189**, 185–194.

CRAWFORD, G. P., POLAK, R. D., SCHARKOWSKI, A., CHIEN, L.-C., DOANE, J. W. and ŽUMER, S. (1994) Nematic director fields captured in polymer networks confined to spherical droplets, *J. Appl. Phys.* **75**, 1968–1971.

CRIGHEAD, H. G., CHENG, J. and HACKWOOD, S. A. (1982) New display based on electrically induced index matching in an inhomogeneous medium, *Appl. Phys. Lett.* **40**, 22–24.

DOANE, J. W., WAZ, N. A., WU B. G. and ŽUMER, S., (1986) Field controlled light scattering from nematic microdoplets, *Appl. Phys. Lett.* **48**, 269–271.

DRZAIC, P. S. (1986) Polymer dispersed nematic liquid crystal for large area displays and light valves, *J. Appl. Phys.* **60**, 2142–2148.

DE GENNES, P. G. (1974) *The Physics of Liquid Crystals*, Oxford: Clarendon Press.

DE GENNES, P. G. (1979) *Scaling Concepts in Polymer Physics*, Ithaca, NY: Cornell University Press.

FERGASON, J. L. (1985) *Dig. SID'85* **16**, 68.

FLORY, P. (1953) *Principles of Polymer Chemistry*, Ithaca, NY: Cornell University Press.

FODOR-CSORBA, K., HOLLI, S., VAJDA, A. and JÁKLI, A. Unpublished.

HAGEMAN, H. J., VAN DER MAADEN, F. P. B. and JANSSEN, P. C. G. M. (1979) Photoinitiators and photoinitiation, 1. Vinyl polymerization photoinitiated by benzoin methyl ether, *Makromol. Chem.* **180**, 2531–2537.

HIKMET, R. A. M. (1990) Electrically induced light scattering from anisotropic gels *J. Appl. Phys.* **68**, 4406–4412.

HIKMET, R. A. M. (1991a) Anisotropic gels and plasticized networks formed by liquid crystal molecules, *Liq. Crys.* **9**, 405–416.

HIKMET, R. A. M. (1991b) From liquid crystalline molecules to anisotropic gels, *Mol. Cryst. Liq. Cryst.* **198**, 357–370.

HIKMET, R. A. M. and HIGGINS, J. A. (1992) Fast switching anisotropic networks obtained by *in situ* photopolymerization of liquid crystal molecules, *Liq. Cryst.* **12**, 831–845.

HIKMET, R. A. M. and ZWERVER, B. H., (1991) Cholesteric networks containing free molecules, *Mol. Cryst. Liq. Cryst.* **200**, 197–204.

HIKMET, R. A. M. and ZWERVER, B. H. (1992) Structure of cholesteric gels and their electrically induced light scattering and colour changes, *Liq. Cryst.* **12**, 319–336.

HOYLE, C. E., CHAWLA, C. P. and GRIFFIN, A. C. (1988) Photopolymerization of a liquid crystalline monomer, *Mol Cryst. Liq. Cryst.* **157**, 639–650.

JÁKLI, A. (1994) Structure and optical properties of liquid crystal dispersed polymers, *Mol. Cryst. Liq. Cryst.* **25**, 289–301.

JÁKLI, A., KIM, D.-R., CHIEN, L.-C. and SAUPE, A., Effect of a polymer network on the alignment and the rotational viscosity of a nematic liquid crystal, *J. Appl. Phys.* **72**, 3161–3164.

JÁKLI, A., BATA, L., FODOR-CSORBA, K., ROSTA, L. and NOIREZ, L. (1994) Structure of polymer networks dispersed in liquid crystals: Small angle neutron scattering study, *Liq. Cryst.* **17**, 227–234.

JÁKLI, A., ROSTA, L. and NOIREZ, L. (1995) Anistropy of non-mesogenic polymer networks dispersed in liquid crystal matrix, *Liq. Cryst.* (in press).

KRONBERG, B., BASSIGNANA, I. and PATTERSON, D. (1978) Phase diagrams of liquid crystal + polymer systems, *J. Phys. Chem.* **82**, 1714–1719.

MARIANI, P., SAMORI, B., ANGELONI, A. S. and FERUTTI, P. E. (1986) Polymerization of bisacrylic monomers within a liquid crystalline smectic B solvent, *Liq. Cryst.* **1**, 327–336.

PAPPAS, S. P. and ASMUS, R. A. (1982) Photoinitiated polymerization of methyl methacrylate with benzoin methyl ether. III. Independent photogeneration of the ether radical, *J. Polym. Sci. Polym. Chem.* **20**, 2643–2653.

PERCEC, V. JONSSON, H. and TOMAZOS, D. (1992) Reactions and interactions in liquid crystalline media, *Polymerization in Organized Media*, ed. C. M. Paleos, Philadelphia: Gordon & Breach, pp. 1–103.

SAUPE, A. and JÁKLI, A. (1994) Method of preparing LC films, US Patent 5,368,770.

STANNARIUS, R., CRAWFORD, G. P., CHIEN, L.-C. and DOANE, J. W. (1991) Nematic director orientation in a liquid crystal dispersed polymer: A deuterium NMR approach, *J. Appl. Phys.* **70**, 135–143.

SVEDBERG, T. (1916) *Kolloid-Z.* **18**, 54.

YANG, D.-K., CHIEN, L.-C and DOANE, J. W. (1992) *Appl. Phys. Lett.* **60**, 3102–3104.

Nuclear Magnetic Resonance of Liquid Crystals with an Embedded Polymer Network

M. VILFAN and N. VRBANČIČ-KOPAČ

7.1 Introduction

Since the discovery of polymer-dispersed liquid crystals (PDLCs) (Doane *et al.*, 1986), nuclear magnetic resonance (NMR) has been intensively used to study director configurations, molecular dynamics, and the nematic–isotropic transition in microconfined liquid crystals. The size of the cavities accessible to this method is in the submicrometre range where the large surface-to-volume ratio creates a number of interesting effects. NMR is a technique which is extremely sensitive to the orientational order of molecules (NMR lineshape studies), and which reveals molecular dynamics in a broad frequency range (nuclear spin relaxation). An *ab initio* interpretation of NMR data in liquid crystals is generally a complicated task in view of the large number of parameters involved. However, when one is interested in the effects of the confinement, only the *difference* in the NMR response between the constrained and bulk systems has to be investigated. This makes the interpretation easier and more reliable.

In the study of PDLC materials, NMR was first applied to confirm the bipolar director-field configuration in droplets with parallel anchoring at the wall (Golemme *et al.*, 1988). This gave direct experimental evidence for the theoretical prediction that the isotropic–nematic first-order phase transition is replaced by a continuous evolvement or orientational order in droplets below the critical size. The critical diameter was found to be about 0.13 μm under strong parallel anchoring conditions (Golemme *et al.*, 1988; Vilfan *et al.*, 1989). Both static and dynamic NMR methods, applied above the nematic–isotropic transition, proved the existence of a weakly orientationally ordered surface layer at the cavity wall.

In 1989, the confined liquid crystals in cylindrical cavities became the focus of NMR research. NMR of deuterons was used to discriminate unambiguously between various nematic structures resulting from an interplay between elastic forces, morphology and size of the cavity, and surface interactions in cylindrical cavities (Crawford *et al.*, 1991a,b, 1992; Allender *et al.* 1991; Ondris-Crawford *et al.*, 1993a,b, 1994). The most intensively studied were the escaped-radial and planar–polar configurations for the homeotropic surface anchoring. NMR spectra provide information on the director field, on the density of singular point defects, on the strength of surface anchoring, and on the surface elastic constants. The orientationally ordered surface layer in the isotropic phase has been investigated in detail by means of deuteron resonance lineshapes (Crawford *et al.*, 1991c,d). Furthermore, NMR in these systems allowed for the observation of the homeotropic-to-parallel anchoring transition as the molecular length of the surfactant at the interface is decreased (Crawford *et al.*, 1993).

Recently, the NMR studies extended from the cavities with well-defined shapes to irregular porous systems with randomly constrained orientational order. Kralj *et al.* (1993) and Zidanšek *et al.* (1995) investigated a liquid crystal captured in the silica aerogel matrix with a wide distribution of pore sizes. NMR yields here a continuous increase in the linewidth as the temperature is lowered from the isotropic into the nematic phase. Similarly, a gradual increase of the local orientational order with decreasing temperature has been observed for a nematic confined to the random network of pores with average diameter of approximately 7 nm in Vycor glass (Iannacchione *et al.*, 1993). It was shown – on the basis of the NMR linewidth – that such a severe constraint yields a suppressed and inhomogeneous nematic ordering.

In this chapter we report on the NMR studies of liquid crystals with an embedded low-concentration polymer network, which are of great current interest (see the chapters by Žumer and Crawford, Yang *et al.* and Blinc *et al.* in this volume). In these composite systems, the polymer only constitutes a few per cent of the dispersion in contrast to PDLC materials with a polymer content of about 70% and anisotropic gels with intermediate concentrations (Hikmet, 1990, 1991). The polymer network, assembled in the liquid crystal matrix, acquires partly the orientational order of the liquid crystal and retains the same orientation upon heating into the isotropic phase. Owing to the delicate balance of intermolecular interactions in the nematic phase, the polymer network – even in the low concentration of only a few per cent – changes dramatically the optical and mechanical properties of the system. The liquid crystal loses its fluidity to a large extent and does not easily align along an external field mismatching the direction of the network. The NMR studies were made in order to get an insight into the effect of the polymer network on the orientational order and dynamics of liquid crystal molecules. In addition, they also provide information on the morphology of the polymer network assembly.

Section 7.2 is devoted to NMR lineshape studies. It begins with a note on the NMR spectra in macroscopic bulk nematic and isotropic phases (section 7.2.1), intended for non-specialists in the field. Section 7.2.2 contains a description of the so far available – rather scarce – NMR spectra of liquid crystals with the polymer network. In section 7.2.3, the effects of the constraint on the NMR spectra in the nematic phase are discussed and compared with experiment. The advantages of the NMR method in studying the competing effects between the polymer network and magnetic field are demonstrated. Section 7.2.4 deals with the constrained isotropic phase.

Section 7.3 comprises nuclear spin relaxation studies. In the first part (section 7.3.1), the combined effect of motions in the kilohertz and megahertz frequency range on different relaxation rates is evaluated for a model system. Experimental results for the deuteron spin–lattice and spin–spin relaxation rates for a liquid crystal-polymer network dispersion are presented and compared with the related bulk in section 7.3.2. Different relaxation mechanisms, which are not present in the bulk but which appear in the composite system, are described. The section concludes with a discussion on dynamic processes as deduced from deuteron relaxation data. Complementary proton relaxation data are the topic of section 7.3.3.

In section 7.4, preliminary measurements of the translational self-diffusion coefficient for the constrained liquid crystal are presented. The last section summarizes conclusions on the structure of the polymer network and on the order and dynamics of liquid crystal molecules obtained from NMR studies. A comparison with related confined systems like PDLC materials and cylindrical cavities is present throughout the chapter.

7.2 NMR Lineshape Studies

7.2.1 *A Note on NMR Lineshape for Macroscopic Bulk Liquid Crystals*

In this section a short and simplified introduction to the liquid crystal NMR lineshape is given. It is intended for the non-specialist in the field. For an exhaustive review of the topic see Doane (1979) and Dong (1994).

In an NMR experiment, the nuclei with magnetic moment different from zero serve as the probe. The interaction of the nuclear magnetic moment with the external magnetic field of the NMR apparatus results in several equidistant Zeeman energy levels. The application of a perpendicular rotating magnetic field induces transitions among them when the frequency of the rotating field matches the Larmor frequency ω_o of the nuclei. The NMR spectrum shows at which frequency the matching occurs, i.e. the NMR line appears. For the magnetic fields which are commonly used in NMR spectrometers, the resonant frequency belongs to the megahertz range: it is 42.6 MHz for protons and 6.5 MHz for deuterons in a field of 1 T. In NMR studies of liquid crystals with embedded polymer networks only these two kinds of nuclei have so far been used as a probe. Therefore we will discuss here only the NMR spectra of protons and deuterons.

The Zeeman energy levels of nuclei in solids and liquid crystals are perturbed by the interactions of the nuclear spins with their intermediate surroundings. In the case of protons with spin $\frac{1}{2}$ magnetic dipole–dipole interactions between the neighbouring spins are the main perturbation. For deuterons with spin 1 the electric quadrupole interaction of the nuclei with the electric field gradient (EFG) tensor is dominating. These interactions cause a non-uniform shift of the Zeeman energy levels. The NMR line is therefore either broadened or split into several distinct lines.

To illustrate the effect of the dipole–dipole interaction on the NMR spectrum, let us consider first an isolated rigid proton pair. Its spectrum consists of two lines separated by

$$\Delta\nu_d = \frac{3}{2\pi} \frac{\gamma^2 \hbar}{r^3} \frac{1}{2} (3 \cos^2\Theta_0 - 1) \tag{7.1}$$

where γ denotes the proton gyromagnetic ratio, r the interproton distance, and Θ_0 the angle of the interproton vector with respect to the external magnetic field. Deuterons in rigid molecules give a similar splitting of the NMR line

$$\Delta\nu_q = \frac{3}{2} \frac{e^2 qQ}{h} \frac{1}{2} (3 \cos^2\Theta_0 - 1) \tag{7.2}$$

where $e^2 qQ/h$ stands for the static quadrupole coupling constant (approximately 165 kHz for deuterons in the alkyl chain and 185 kHz for aromatic deuterons, for example), and Θ_0 for the angle between the EFG tensor symmetry axis (usually parallel to the C–D bond in liquid crystals) and the magnetic field. In eqn (7.2), the axial symmetry of the EFG tensor has been assumed.

In a *nematic liquid crystal*, the factor $\frac{1}{2}(3 \cos^2\Theta_0 - 1)$ is averaged over molecular motions which are faster than $(2\pi\Delta\nu_d)^{-1}$ or $(2\pi\Delta\nu_q)^{-1}$, respectively. These are mainly conformational changes of the molecule, its rotation around the long axis and fluctuations of the long axis about the local preferred direction, i.e. the local director **n**. Taking into account the uniaxiality of the bulk nematic phase, the actual splitting of the NMR line for a proton pair is given by (Doane, 1979):

$$\Delta\nu_d = \frac{3}{2\pi} \frac{\gamma^2 \hbar}{r^3} \frac{1}{2} (3 \cos^2\Theta_B - 1)[S\langle\tfrac{1}{2}(3 \cos^2\beta - 1)\rangle + \tfrac{1}{2}(S_{xx} - S_{yy})\langle\sin^2\beta \cos 2\alpha\rangle] \tag{7.3}$$

and for deuterons by

$$\Delta\nu_q = \frac{3}{2} \frac{e^2 qQ}{h} \frac{1}{2} (3 \cos^2\Theta_B - 1)[S\langle\tfrac{1}{2}(3 \cos^2\beta - 1)\rangle + \tfrac{1}{2}(S_{xx} - S_{yy})\langle\sin^2\beta \cos 2\alpha\rangle] \tag{7.4}$$

α and β denote the azimuthal and polar angles of the interproton vector (in eqn (7.3)) or of the EFG tensor symmetry axis (in eqn (7.4)) with respect to the long molecular axis. The averaging in eqns (7.3) and (7.4) is performed over the conformational changes of the molecule. The factor S (sometimes named S_{zz}) represents the well-known orientational order parameter, $S = \langle\frac{1}{2}(3 \cos^2\Theta - 1)\rangle$, where Θ stands for the angle between the instantaneous direction of

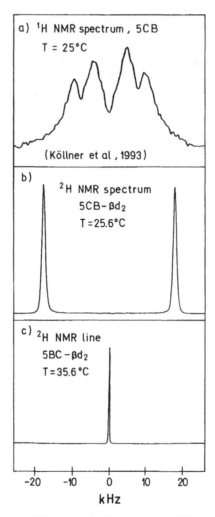

Figure 7.1 Proton and deuteron NMR spectra of 5CB: (a) proton NMR spectrum of 5CB in the nematic phase (Köllner *et al.*, 1993); (b) deuteron NMR spectrum of 5CB-βd_2 in the nematic phase (Crawford *et al.*, 1991a); (c) deuteron NMR spectrum of 5CB-βd_2 in the isotropic phase.

the long molecular axis and the local director, and $\langle \; \rangle$ means the ensemble average. The effect of order director fluctuations has been neglected in view of their small amplitude. $(S_{xx} - S_{yy})$ is the molecular biaxiality order parameter and describes the deviation of the molecular shape from the exact cylindrical symmetry. It is usually almost one order of magnitude smaller than S. The second terms in eqns. (7.3) and (7.4) are therefore often neglected. Finally, Θ_B describes the orientation of the director in the magnetic field. In the bulk nematic liquid crystal, Θ_B is spatially uniform and equals zero for compounds with positive anisotropy of the magnetic susceptibility.

The most important aspect concerning eqns (7.3) and (7.4) is that the *splitting of the NMR line is related directly to the orientational order parameter S of the molecules and to the orientation of the director.* For the cylindrically symmetric molecules it is proportional to S and is maximal for the director parallel to the magnetic field **B**.

In Figure 7.1(a) the proton spectrum of 4'-n-pentyl-4-cyanobiphenyl (5CB) is presented. The deuteron spectrum of the same compound, selectively deuterated in the second position of the hydrocarbon chain, is shown in Figure 7.1(b). The deuteron spectrum shows two well-resolved lines. A broader spectral distribution, observed for protons, results from a number of non-equivalent proton pairs. There are, however, some pairs of strongly interacting spins that give resolved lines superimposed on a broad background. Two sets of well-resolved lines are clearly seen in Figure 7.1(a). One set can be roughly ascribed to the proton pairs in the aromatic rings and the other to the pairs in the CH_2 groups. Though the proton spectra are useful in liquid crystal studies, the order parameter investigations have been made mostly on selectively deuterated compounds because of their better resolution. For example, using deuterium NMR, the decrease of the orientational order along the hydrocarbon chain in thermotropic liquid crystals (Bos *et al.*, 1977) and in lyotropic membranes (Charvolin and Hendrikx, 1984) was found, and the even–odd effect was observed. NMR measurements also allowed for the study of biaxiality in cholesteric liquid crystals (Doane *et al.*, 1982). Another example is the gradual onset of orientational order when the substance is cooled down through various smectic phases (Figueirinhas *et al.*, 1987).

When the liquid crystal is melted into *the isotropic phase*, the phase transition is accompanied by a collapse of the multiple NMR line into a single line (Figure 7.1(c)). Here the spin interactions are averaged out by fast and isotropic molecular reorientations. The width of the NMR line shrinks by several orders of magnitude well below 100 Hz. The measured width is usually determined by the non-homogeneity of the magnetic field and not by the intrinsic interactions of the spin system.

7.2.2 *Experimental Results*

The samples of the liquid crystal–polymer network dispersions, used so far in the NMR studies, are prepared (Stannarius *et al.*, 1991; Crawford *et al.*, 1994) by dissolving a small amount (1–4 wt.%) of the monomeric 4,4'-bis-acryloylbiphenyl (BAB) and about 0.5 wt.% of the photoinitiator benzoin methyl ether (BME) in the liquid crystal 5CB-αd_2 (4'-n-pentyl-4-cyanobiphenyl, deuterated in the first position of the hydrocarbon chain).

5CB-αd_2: $\quad$ CN—◯—◯—C $^2H_2$—CH_2—CH_2—CH_2—CH_3

BAB: $\quad$ CH_2=CH—CO_2—◯—◯—CO_2—CH=CH_2

The sample is heated to a temperature far above the nematic clearing point ($T_{NI} \simeq 308$ K for pure 5CB), homogenized, and then exposed to the UV irradiation at ambient temperature. During the UV exposure the monomer molecules of BAB are activated and linked together to form a polymer network. Special tricks are needed to control the orientation of the nematic phase during the polymerization process. For this purpose, an external electric or magnetic field acting on the liquid crystal between the glassy plates is applied. The samples with the network have a well-defined but slightly lower nematic–isotropic transition temperature than the bulk (by 1–2 K).

The optical textures of the dispersion show in a straightforward way that the polymer network predominantly acquires the alignment of the nematic director field (Hikmet, 1990, 1991; Stannarius *et al.*, 1991). Birefringence measurements (Žumer and Crawford, this volume), complemented by other experimental methods (Jákli *et al.*, 1992, 1994), indicate that the network of low-concentration liquid crystal–polymer dispersions can be envisioned in the form of fibres or bundles of fibres that arrange fairly parallel during the growth in the nematic environment. The network elements adjust parallel to the nematic director not only in a uniform nematic phase but also in cylindrical or spherical cavities with rather complicated director-field configurations (Crawford *et al.*, 1994). When the liquid crystal is heated into the isotropic phase, the network retains its oriented structure and the system exhibits a weak optical birefringence.

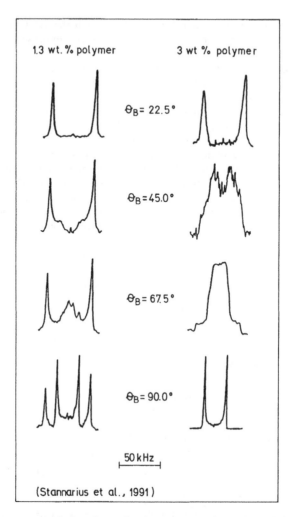

Figure 7.2 Deuteron spectra of a liquid crystal–polymer network dispersion in the nematic phase for different orientations of the network with respect to the magnetic field (Stannarius et al., 1991). At 3% polymer concentration, the orientation of the liquid crystal is dominated by the network, whereas the competing effect of the magnetic field can be observed for the 1.3% polymer dispersion.

The first NMR studies of these systems (Stannarius *et al.*, 1991) checked the orientational order parameter of the liquid crystal with the embedded network and studied the competition between the network and magnetic field in orienting the nematic phase. In Figure 7.2, the deuteron NMR spectra for the nematic phase of the mixture of 50% 5CB and 50% 5OCB-αd$_2$, constrained by the BAB network, are presented (Stannarius *et al.*, 1991). The liquid crystal in a 350 μm thick cell was homogeneously oriented by means of an electric field applied during the polymerization. The NMR spectra for two polymer concentrations (about 1 wt.% and 3 wt.%) give exactly the same splitting of the two deuteron lines as the pure nematic at the same temperature when the network orientation matches the direction of the magnetic field. A slight broadening of the individual lines can be observed. The shape of the spectra changes drastically on rotating the cell in the magnetic field (Figure 7.2).

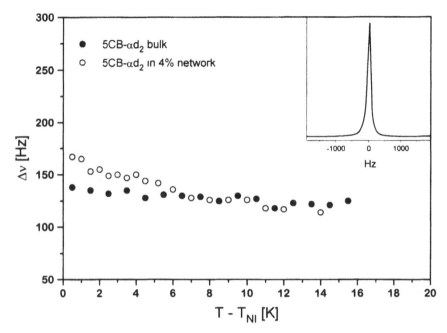

Figure 7.3 Temperature dependence of the deuteron NMR linewidth of 5CB-αd₂ with 4% polymer network and of the macroscopic bulk 5CB in the isotropic phase. The linewidths are strongly influenced by the inhomogeneity of the magnetic field. The inset shows a typical spectrum of the dispersion.

In the polymer-constrained isotropic phase the deuteron spectrum consists of only one, relatively narrow line. It has been investigated for the oriented network in the cell, and for the samples in the NMR tubes where no special precaution has been made to prevent the domain structure during the network formation. The latter data (4% BAB network in 5CB-αd₂) are presented in Figure 7.3. The linewidth of the constrained isotropic 5CB-αd₂ almost equals that of the pure liquid crystal and exceeds its width only in the vicinity of T_{NI}.

In order to understand the above experiments, the effect of constraining surfaces on the liquid crystal NMR lineshape will be discussed in the next section.

7.2.3 *Effect of Constraint on the Nematic Phase*

In a spatially constrained liquid crystal the director field $\mathbf{n}(\mathbf{r})$ is space and – in view of the molecular translational diffusion – also time dependent. Additionally, the order parameter S can vary within the sample, $S = S(\mathbf{r})$. If the biaxiality is neglected, the deuterons in a local region, represented by the positional vector $\mathbf{r}$, yield a spectrum of two lines split by

$$\Delta v_q(\mathbf{r}) = \Delta v_{q0} \tfrac{1}{2}[3 \cos^2\Theta_B(\mathbf{r}) - 1]S(\mathbf{r}) \tag{7.5}$$

Here, according to eqn (7.4), Δv_{q0} denotes the splitting of the perfectly oriented nematic phase ($S = 1$) along the magnetic field. The whole spectrum sums up the contributions from different parts of the sample. Its *frequency distribution* thus reflects the whole *director field* if the lineshape is not distorted by the spatial variation of S, magnetic field, or molecular self-diffusion. Their influence, however, is negligible in many cases as shown below.

165

1 The variation of $S(\mathbf{r})$ throughout the sample is important only for very small cavities (size of the order of about 10 nm), or at temperatures close to the phase transition (within about 1 K). It changes considerably also in a narrow region in the intermediate vicinity of structural defects. In most experiments, however, cavities with a radius between 0.1 and 1 μm are used. The order parameter S is thus to a good approximation constant throughout the cavity for temperatures well below T_{NI}.

2 The strong magnetic field of the NMR spectrometer might distort the surface-induced alignment in large cavities. The range of the surface influence is determined by the magnetic coherence length

$$\xi_B = (K/\Delta\chi)^{1/2}B^{-1} \tag{7.6}$$

where K is one of the liquid crystal elastic constants and $\Delta\chi$ the anisotropy of the magnetic susceptibility. For 5CB in a magnetic field of 1 T, the magnetic coherence length is about 8 μm at room temperature (Vilfan *et al.*, 1988). Hence, the distortions due to the magnetic field prevail only in cavities with a diameter larger than ξ_B or at distances larger than ξ_B from the surface.

3 The translational diffusion of molecules might partly smear the spectral distribution provided that the molecules diffuse to the regions with different director orientation within the NMR time scale:

$$t_{NMR} = (2\pi\Delta\nu_{q,\,bulk})^{-1} \tag{7.7}$$

is approximately 4×10^{-6} s for a typical splitting of 40 kHz. An estimate of the corresponding diffusion length gives $\langle r^2 \rangle^{1/2} \sim (6Dt_{NMR})^{1/2} \sim 30$ nm for $D \sim 4 \times 10^{-11}$ m^2/s (bulk 5CB at room temperature (Noack, 1993)). As the changes in director orientation in the confined nematic phase occur mostly at larger distances, the translational diffusion practically does not affect the spectral distribution except in very small cavities ($\lesssim 100$ nm).

The deuteron NMR spectra are thus extremely useful in determining the director field for the constrained liquid crystals. In a cavity with the director everywhere parallel to the magnetic field $\mathbf{B}$, the spectrum consists of two single lines at $\pm\nu_{q,\,bulk}$, split by $\Delta\nu_{q,\,bulk}$ (eqn (7.4), Figure 7.4(a)). When the director orientations are randomly distributed in space, the corresponding lineshape is a Pake-type powder pattern (Figure 7.4(b)). It displays two singularities at $\pm\frac{1}{2}\nu_{q,\,bulk}$ and two shoulders at $\pm\nu_{q,\,bulk}$. The singularities arise from the regions with the director $\mathbf{n}$ perpendicular to $\mathbf{B}$ and the shoulders from $\mathbf{n}$ parallel to $\mathbf{B}$. The planar isotropic distribution of directors gives four singularities at $\pm\frac{1}{2}\nu_{q,\,bulk}$ and $\pm\nu_{q,\,bulk}$ for the magnetic field in the plane of directors (Figure 7.4(d)). On the other hand, this plane perpendicular to the magnetic field yields a spectrum of only two sharp lines split by $\frac{1}{2}\Delta\nu_{q,\,bulk}$ (Figure 7.4(c)).

In more complicated cases of confinement (Figure 7.4(e)) it is not possible to deduce the director–field configuration directly. The spectrum should be first calculated on the basis of a given model. The comparison with experimental data can then either confirm or reject the proposed structure. But even for the systems where knowledge of the structure is very scarce, the spatial distribution of directors can be extracted from the deuteron spectrum provided that it is not smeared out by translational self-diffusion, magnetic field, or spatial dependence of the order parameter.

The deuteron NMR spectrum of the 5CB–5OCB-αd$_2$ mixture constrained by the 1 wt.% and 3 wt.% BAB networks, which is oriented along the magnetic field ($\Theta_B = 0°$), gives exactly the same splitting of the two deuteron lines as the pure nematic at the same temperature (Stannarius *et al.*, 1991). According to eqns (7.4) and (7.5), this shows that the order parameter S is not affected by the presence of the network and that the orientation of the liquid crystal is along the network fibres. However, a slight broadening of the lines indicates a non-perfect sample alignment. The estimated width of the angular distribution of directors is about $\pm 10°$. The spectra change drastically upon turning the direction of the network away from the

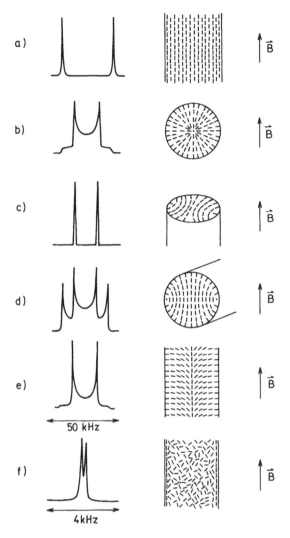

Figure 7.4 Calculated deuteron NMR spectra for different nematic director fields (Crawford *et al.*, 1991a): (a) nematic phase with parallel–axial structure in the cylindrical cavity and splitting $\Delta v_{q,bulk}$; (b) Pake-type powder spectrum for isotropic spatial distribution of directors in a spherical cavity with radial structure; (c) spectrum of a planar–polar structure in cylindrical cavities with all directors perpendicular to the magnetic field and splitting $\frac{1}{2}\Delta v_{q,bulk}$; (d) spectral pattern for the isotropic distribution of nematic directors in a plane parallel to the magnetic field; (e) escaped-radial structure in a cylindrical cavity. The calculated spectra reproduce excellently the experimental ones. The splitting of the deuteron NMR line in the isotropic phase in a cylindrical cavity is shown in (f). It reflects the residual nematic ordering at the cavity wall. The splitting here is considerably smaller than in the nematic phase.

magnetic field of the spectrometer (4.7 T). A pure nematic would adjust itself in a short time to the direction of the magnetic field and no angular dependence would be observed. A polymer network, however, imposes its orientation on a major portion of the liquid crystal and overcomes the effect of the magnetic field. The shape of the spectra for the sample with 3 wt.% network changes with increasing angle Θ_B (Figure 7.2) according to eqn (7.5). At $\Theta_B = 90°$, it shows the same pattern as presented in Figure 7.4(c), indicating that the major part of the liquid crystal follows the orientation of the network and makes an angle of 90° with the magnetic field. This strong effect of the network is rather astonishing as it represents only 3% of the sample weight. When the polymer content is lowered to about 1%, the effect of the network becomes comparable with that of the magnetic field. The $\Theta_B = 90°$ spectrum (Figure 7.2) shows that a part of the liquid crystal is aligned along **B** (outer peaks) and a part perpendicular to it (inner peaks). A detailed analysis demonstrates that the fixed and realigned parts of the nematic are almost equal in this case.

The first NMR experiments on liquid crystals, interwreathed by a low-concentration polymer network, lead to the conclusion that the deuteron lineshape studies provide an excellent insight into the orientational effects of the network in the nematic phase (similarly as they do for PDLC materials and cylindrical cavities). The competition between the magnetic field and polymer network and, consequently, the polymer network stability can be studied in this way.

7.2.4 *Effect of Constraint on the Isotropic Phase*

In the constrained isotropic phase, the confining surfaces can induce a weak orientational order. The phase is therefore sometimes named paranematic instead of isotropic. We will use the term isotropic when the transition between the high- and the low-temperature phase is clearly first order. Such is the case in liquid crystals with an embedded low-concentration polymer network. Recent optical birefringence measurements indicate that there is a weakly orientationally ordered liquid crystal layer close to the fibres (Žumer and Crawford, this volume). The preferred orientation of liquid crystal molecules at the interface is expected to be parallel to the fibre. As the distance from the fibre is increased, the order decays exponentially with the correlation length ξ. The phase is truly isotropic only at distances of several correlation lengths from the internal polymer surfaces.

The NMR spectrum of the constrained isotropic phase is motionally narrowed, but the surface-induced orientational order results in a detectable line broadening with respect to the bulk. If there were no translational self-diffusion, the NMR spectrum would consist of a sharp central line – corresponding to the true isotropic regions away from the surface – and a superimposed nematic-like spectrum (though narrower) corresponding to the surface director field. The experiment shows that, in contrast to the nematic phase, the effect of molecular self-diffusion cannot be avoided in the constrained isotropic phase. The time of the NMR measurement, t_{NMR}, is here longer than in the nematic phase in view of the smaller residual interaction and the diffusion is faster. The migration of molecules thus affects the lineshape essentially. The deuteron NMR line broadening depends on the magnitude and spatial distribution of the orientational order parameter $S(\mathbf{r})$, on the average time spent by molecules in the oriented region and on the surface orientation with respect to the magnetic field.

To estimate the broadening, we use a simple, two-site model with the order parameter equal to S_0 in the surface layer and zero elsewhere. The thickness l_0 of the surface layer is of the order of one molecular length. Assuming that the vicinity of internal surfaces allows for all molecules in the sample an exchange between the ordered and disordered regions within t_{NMR}, the deuteron NMR linewidth is given by

$$\Delta v_q \simeq \Delta v_{q0} \tfrac{1}{2}(3 \cos^2 \Theta_B - 1) S_0 \eta \qquad \eta \simeq A l_0 / V \tag{7.8}$$

Θ_B denotes the angle between the molecular preferred orientation at the surface and the magnetic field, and η the fraction of liquid crystal molecules in the surface layer. η is roughly equal to $A l_0 / V$ for the layer thickness l_0 considerably smaller than the curvature of the internal surface area A. V here stands for the volume of the liquid crystal. The above model is used to a good approximation in the whole temperature range of the isotropic phase except close to T_{NI}.

Not only a broadening, but also a well-resolved splitting of the deuteron NMR line has been observed for 5CB-βd_2 confined in cylindrical cavities in inorganic Anopore membranes (~ 0.2 μm in diameter). A typical spectrum is shown in Figure 7.4(f). The absence of a central peak and the angular dependence definitely demonstrate fast averaging due to diffusion over the whole cavity (Crawford *et al.*, 1991c,d, 1993). A detailed study of the temperature dependence of the splitting allowed an accurate determination of S_0. It was found that S_0 exhibits – depending on the type of the surface – either a pretransitional increase or a weak temperature dependence.

Analysis of the NMR lineshape in the constrained isotropic phase is considerably more complicated in systems with a non-uniform orientation of the surface with respect to the magnetic field. Such systems include liquid crystals with an embedded polymer network and those constrained into irregular cavities of gels and glasses. Owing to translational diffusion within the NMR time, the molecules experience surface regions with different director orientation with respect to the magnetic field. As a consequence, the factor $\frac{1}{2}(3\cos^2\Theta_B - 1)$ in eqn (7.8) is partially averaged. The evaluation of the *local order parameter* S_0 is hence not straightforward and its value can be easily underestimated.

The narrow deuteron NMR line of the isotropic 5CB-αd_2, constrained by the 4 wt.% BAB network, only slightly exceeds in width that of the pure liquid crystal (Figure 7.3). However, in view of eqn (7.8), an interpretation in terms of vanishing orientational order above T_{NI} would be misleading. Although the small increase close to the phase transition represents a poor proof of the orientational order, it should be noted that even such a narrow line allows for the existence of a local S_0 of the order of about 0.1. According to eqn (7.8), the linewidth depends on S_0, η and orientation of the surface director with respect to the magnetic field. In a network which is finely distributed down to the nanometre scale, A/V is approximately 0.04 mm^{-1} (Žumer and Crawford, this volume); the corresponding $\eta \simeq 0.06$ for the 4% polymer concentration and $l_0 \sim 1.2$ nm (Crawford *et al.*, 1993). The lower limit for the local surface order parameter S_0 can be calculated using $\Delta\nu_q \approx 170$ Hz, $\Delta\nu_{q0} \simeq 110$ kHz, and $\Theta_B = 0$ throughout the sample. This estimate gives $S_0 > 0.03$. However, because of cross-linking of the fibrils and domain structure of the sample, the partially averaged term $\frac{1}{2}(3\cos^2\Theta_B - 1)$ is smaller than one. The local S_0 is therefore considerably larger than its lower limit. The lack of a more obvious difference in the linewidths between the bulk and network dispersion is due to the masking of intrinsic linewidths, which are smaller than about 100 Hz, by the non-homogeneity of the magnetic field. Intrinsic NMR linewidths are related to the spin–spin relaxation rate, where a large difference between the bulk and constrained sample has been observed. It will be presented in the next section.

7.3 Nuclear Spin Relaxation

7.3.1 *Nuclear Spin Relaxation for a Model System and Bulk Liquid Crystals*

In the previous section, the first-order effect of local nuclear spin interactions, resulting in the splitting of the NMR line, was discussed. It was shown that the largest effect is generally observed in static systems which are not subjected to motional narrowing. The main second-order effect of dipole–dipole or quadrupolar spin interactions is to induce spin relaxation that is essentially a dynamic process. It is produced by time-dependent local magnetic or electric fields and directly related to the intensity and frequency of the thermal motion of spin-bearing molecules. Spin relaxation thus provides valuable information about molecular dynamics.

Most commonly measured are the *spin–lattice relaxation time* T_1 and the *spin–spin relaxation time* T_2. The former one is the characteristic time for the exponential increase of the longitudinal spin magnetization (parallel to the static magnetic field) if it was initially turned away from the equilibrium value. In this process, the spin system exchanges energy with the surroundings, which are called the 'lattice'. The second characteristic relaxation time, T_2, describes the decay of magnetization transverse to the static field. Sometimes the spin–lattice relaxation time is measured in the rotating frame resulting in the time constant $T_{1\rho}$.

The spin relaxation behaviour can be expressed for a proton pair in terms of three spectral densities $J_k(\omega)$ for $k = 0, 1, 2$ (Abragam, 1961; Fukushima and Roeder, 1981):

$$T_1^{-1} = \frac{3}{2} \frac{\gamma^4 \hbar^2}{r_{ij}^6} [J_1(\omega_0) + 4J_2(2\omega_0)] \tag{7.9a}$$

$$T_{1\rho}^{-1} = \frac{3}{4} \frac{\gamma^4 \hbar^2}{r_{ij}^6} [3J_0(2\omega_1) + 5J_1(\omega_0) + 2J_2(2\omega_0)] \tag{7.9b}$$

$$T_2^{-1} = \frac{3}{4} \frac{\gamma^4 \hbar^2}{r_{ij}^6} [3J_0(0) + 5J_1(\omega_0) + 2J_2(2\omega_0)] \tag{7.9c}$$

with

$$J_k(\omega) = \mathrm{Re} \int_{-\infty}^{\infty} \langle F_k(0) F_k^*(t) \rangle \exp(i\omega t) \, dt \tag{7.10}$$

The spin–lattice relaxation rate T_1^{-1} is determined by the spectral densities at Larmor frequency ω_0 and at double Larmor frequency $2\omega_0$. For $T_{1\rho}^{-1}$ and T_2^{-1}, additional terms at $2\omega_1$ and at zero frequency, respectively, contribute to the relaxation. The frequency ω_1 in the rotating magnetic field is usually in the kilohertz range and two to three orders of magnitude smaller than the Larmor frequency ω_0. $F_k(t)$ are time-dependent spherical harmonics describing the well-known angular part of the dipole–dipole interaction between two spin $\frac{1}{2}$ nuclei separated by r_{ij}:

$$F_0(t) = \frac{1}{\sqrt{8}} [1 - 3\cos^2\Theta_0(t)] \tag{7.11a}$$

$$F_1(t) = \frac{\sqrt{3}}{2} \cos\Theta_0(t) \sin\Theta_0(t) \exp(i\phi_0(t)) \tag{7.11b}$$

$$F_2(t) = \frac{\sqrt{3}}{4} \sin^2\Theta_0(t) \exp(2i\phi_0(t)) \tag{7.11c}$$

$\Theta_0(t)$ and $\phi_0(t)$ denote here the polar and azimuthal angles between the internuclear vector and the magnetic field. If the dominant time-dependent interaction is electric quadrupolar instead of magnetic dipole–dipole – as in the case of deuterons – eqns (7.9)–(7.11) are still valid if the factor $\gamma^4\hbar^2/r_{ij}^6$ is replaced by $\pi^2(e^2qQ/h)^2$ to account for a different type of dominant interaction.

Expressions (7.9) can be straightforwardly evaluated for an isolated proton pair undergoing random isotropic motion which is characterized by a single correlation time τ_c. One gets (Abragam, 1961; Fukushima and Roeder, 1981)

$$T_1^{-1} = \frac{3}{2} \frac{\gamma^4 \hbar^2}{r_{ij}^6} \frac{1}{5} \left(\frac{\tau_c}{1 + \omega_0^2\tau_c^2} + \frac{4\tau_c}{1 + 4\omega_0^2\tau_c^2} \right) \tag{7.12a}$$

$$T_{1\rho}^{-1} = \frac{3}{4} \frac{\gamma^4 \hbar^2}{r_{ij}^6} \frac{1}{5} \left(\frac{3\tau_c}{1 + 4\omega_1^2\tau_c^2} + \frac{5\tau_c}{1 + \omega_0^2\tau_c^2} + \frac{2\tau_c}{1 + 4\omega_0^2\tau_c^2} \right) \tag{7.12b}$$

$$T_2^{-1} = \frac{3}{4} \frac{\gamma^4 \hbar^2}{r_{ij}^6} \frac{1}{5} \left(3\tau_c + \frac{5\tau_c}{1 + \omega_0^2\tau_c^2} + \frac{2\tau_c}{1 + 4\omega_0^2\tau_c^2} \right) \tag{7.12c}$$

Though referring to a simple motion, this model shows the basic features of the relaxation behaviour. The relaxation rates against the correlation time τ_c are plotted in Figure 7.5 for

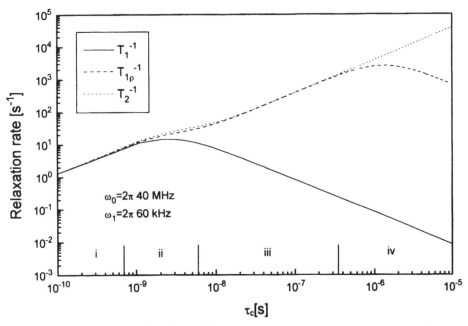

Figure 7.5 Relaxation rates T_1^{-1}, $T_{1\rho}^{-1}$ and T_2^{-1} calculated for different correlation times τ_c according to eqns (7.12) for $\omega_0 = 2\pi \times 40$ MHz, $\omega_1 = 2\pi \times 60$ kHz and $r_{ij} = 0.2$ nm.

the parameters $\omega_0 = 2\pi \times 40$ MHz, $\omega_1 = 2\pi \times 60$ kHz and $r_{ij} = 0.2$ nm. The calculated relaxation times exceed by up to 10 orders of magnitude the correlation time of molecular motion. For example, the correlation time $\tau_c = 10^{-10}$ s results in the spin relaxation time of about 1 s. In relation to Figure 7.5 it is important to note:

i   A motion with a short correlation time which fulfils the conditions $\omega_0\tau_c \ll 1$, $\omega_1\tau_c \ll 1$ (meaning in practice $\tau_c \lesssim 10^{-9}$ s), produces equal relaxation rates, which are in this limit independent of ω_0 and ω_1:

$$T_1^{-1} \simeq T_{1\rho}^{-1} \simeq T_2^{-1} \simeq \frac{3}{2}\frac{\gamma^4\hbar^2}{r_{ij}^6}\tau_c \tag{7.13}$$

Such a situation is often realized in simple isotropic liquids.

ii A motion with τ_c comparable with the inverse Larmor frequency, $\omega_0\tau_c \sim 1$, $\omega_1\tau_c \ll 1$ (this happens usually for $\tau_c \sim 10^{-9}$ to 10^{-8} s), causes a difference between the relaxation rate in the rotating frame and the spin–spin relaxation rate compared with the spin–lattice relaxation rate. The former two exceed T_1^{-1} by a factor of about three though T_1^{-1} is here maximal (region ii in Figure 7.5).

iii Correlation times of the order of 10^{-8} or 10^{-7} s, which lead to $\omega_0\tau_c > 1$ and $\omega_1\tau_c < 1$, produce a frequency-dependent relaxation rate T_1^{-1} which is proportional to ω_0^{-2}. It decreases with increasing τ_c. On the other hand, $T_{1\rho}^{-1}$ and T_2^{-1} are still roughly equal in magnitude and independent of ω_0 since the terms $J_0(2\omega_1)$ and $J_0(0)$ prevail in eqns (7.12b) and (7.12c). The ratios $T_{1\rho}^{-1}/T_1^{-1}$ and T_2^{-1}/T_1^{-1} are increasing.

iv Finally, slow motions with $\omega_0\tau_c \gg 1$ and $\omega_1\tau_c \sim 1$ or >1 (i.e. $\tau_c \gtrsim 10^{-6}$ s) provide a negligible contribution to T_1^{-1} but reflect themselves in $T_{1\rho}^{-1}$, and even more significantly in T_2^{-1}. It should be stressed, however, that eqns (7.12) are valid only as long as the correlation time does not exceed the inverse strength of the interaction which is responsible for the relaxation, i.e. for $\tau_c < (\gamma^4 \hbar^2 / r_{ij}^6)^{-1/2}$.

Suppose now that the molecules undergo a fast but anisotropic motion with the correlation time τ_{fast}, which averages the dipole–dipole or quadrupole interaction only partly. The remaining interaction may be modulated by another, slower motion with the correlation time τ_{slow}. If the faster and slower motions occur on different time scales they can be treated as statistically independent. The total spin–lattice relaxation rate thus consists of two contributions

$$T_1^{-1} = (T_1^{-1})_{fast} + (T_1^{-1})_{slow} \qquad (7.14)$$

and analogously $T_{1\rho}^{-1}$ and T_2^{-1}. The contributions originating from the fast and slow motions are given roughly by eqns (7.12) by inserting two different correlation times, τ_{fast} and τ_{slow}, respectively. Besides, $(T_1^{-1})_{fast}$ should be reduced by a factor due to the anisotropy of the motion (it is of the order $(1 - S)$ in liquid crystals). The residual interaction, which enters into the expression for $(T_1^{-1})_{slow}$, should be averaged over the faster motion. Comparing both contributions to the relaxation rates, it is rather obvious that the slower motion considerably enhances $T_{1\rho}^{-1}$ and T_2^{-1}, but adds a completely negligible contribution to T_1^{-1} because of the extremely large term $(\omega_0 \tau_{slow})^2$ in the denominator of spectral densities J_1 and J_2. For this reason the measurements of T_1 provide information on the *faster motion* (megahertz frequencies) and the measurements of $T_{1\rho}$ and T_2 on the *slower motion* (kilohertz frequencies).

It should be emphasized that, instead of measuring $T_{1\rho}$ or T_2, there is another possibility to scan a large frequency range from 10^3 up to 10^8 Hz by measuring only T_1. A special field-cycling technique, which measures low field relaxation while preserving the sensitivity of the high field, has been developed for this purpose (Noack, 1986, and references therein). It requires, however, special experimental equipment, which is not commercially available.

In *bulk nematic liquid crystals* there are more than two dynamic processes that contribute to the spin relaxation. These include:

- individual molecular reorientations (around the long axis and restricted fluctuations of this axis (R)),
- internal conformational changes of the molecule (R),
- translational self-diffusion (SD),
- order director fluctuations, which are collective motions of a large number of molecules (ODF).

Fortunately, the time scales on which they occur are sufficiently different to treat their contributions to the spin relaxation rate separately, like

$$T_1^{-1} = (T_1^{-1})_R + (T_1^{-1})_{ODF} + (T_1^{-1})_{SD} \qquad (7.15)$$

The contribution of local molecular reorientations, $(T_1^{-1})_R$, dominates in the upper megahertz range. Molecular self-diffusion influences relaxation in the same frequency range. It is effective for protons by modulating the interactions between nuclei in different molecules, but not for deuterons where the intramolecular quadrupole interaction prevails. By comparing experimental relaxation data with theoretical predictions, it was possible to determine the correlation times for molecular conformational changes, for rotation around the long axis and for its orientational fluctuations (Kohlhammer *et al.*, 1989). The contribution of order director fluctuations, $(T_1^{-1})_{ODF}$, is the result of fluctuations of the director relative to its time-average orientation. Order director fluctuations are characterized by a broad frequency distribution of thermally activated modes which depend on the viscoelastic properties of the liquid crystal. The resulting $(T_1^{-1})_{ODF}$ obeys the well-known square root dispersion law, i.e. $(T_1^{-1})_{ODF} \propto \omega_0^{-1/2}$. Order director fluctuations constitute the dominant relaxation process in the kilohertz region in the nematic phase, as shown first by Wölfel *et al.* (1975) (see also Wölfel (1978) and Noack (1986)).

The spin relaxation in the *bulk isotropic phase* is governed by individual molecular reorientations, conformational changes, and – only for protons – by molecular translational self-diffusion. The absence of order director fluctuations results in a smaller and frequency-independent relaxation rate in the kilohertz regime. Far above the nematic–isotropic transition (~ 20 K above T_{NI}), the conditions for the extreme narrowing limit are fulfilled and the relaxation rates behave according to eqn (7.13), i.e. $T_1^{-1} \simeq T_{1\rho}^{-1} \simeq T_2^{-1}$. A faster increase of T_2^{-1} compared with T_1^{-1} as the temperature is lowered indicates anisotropic local molecular reorientations. A few degrees above T_{NI}, there is a pronounced increasing in the relaxation rates $T_{1\rho}^{-1}$ and T_2^{-1} owing to nematic fluctuations within the isotropic phase.

7.3.2 *Deuteron Spin Relaxation*

7.3.2.1 *Experimental results*

Here we report on the deuteron spin–lattice relaxation (T_1) and spin–spin relaxation times (T_2) of 5CB–polymer network dispersions. The NMR samples, prepared as described in section 7.2.2, have a macroscopic multidomain structure as no orienting attempt has been made during the network formation in the nematic phase. The spin–lattice relaxation time T_1 was measured by the π–τ–$(\pi/2)_x$–δ–$(\pi/2)_y$–δ–echo pulse sequence, and T_2 by the solid echo sequence, $(\pi/2)_x$–τ–$(\pi/2)_y$–τ–echo, which eliminates the static quadrupole interaction and magnetic field inhomogeneities. The observed relaxation processes were purely monoexponential over at least one decade. The experimental errors were less than $\pm 10\%$ in T_1 and less than $\pm 15\%$ in T_2 measurements. We also present the results for a liquid crystal confined to cylindrical cavities to compare the effect of constraint in regular and irregular geometry.

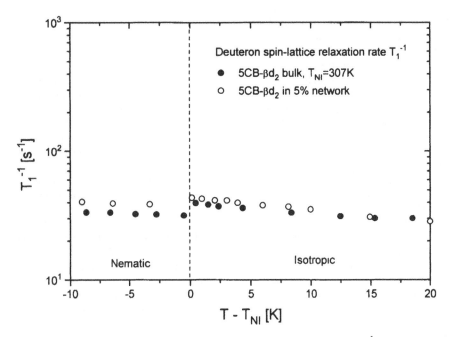

Figure 7.6 Temperature dependence of the deuteron spin–lattice relaxation rate T_1^{-1} for 5CB-βd$_2$ with the embedded 5% polymer network and for the pure, bulk liquid crystal 5CB-βd$_2$.

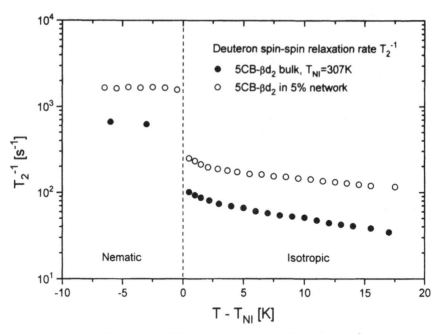

Figure 7.7 Temperature dependence of the deuteron spin–spin relaxation rate T_2^{-1} for 5CB-βd$_2$ with the embedded 5% polymer network and for the pure, bulk liquid crystal.

Figure 7.6 shows the temperature dependence of the deuteron spin–lattice relaxation rate T_1^{-1} as obtained at $\nu_0 = 13.6$ MHz upon cooling. In the high-temperature phase, there is no difference between the dispersion and pure bulk sample, which would seriously exceed the experimental error. In the nematic phase, T_1^{-1} of the dispersion is increased by about 30% compared with the pure liquid crystal. In contrast to a rather modest difference in T_1^{-1}, a huge increase upon constraint is observed in the spin–spin relaxation rate T_2^{-1} (Figures 7.7 and 7.8). The significant effect of the network is clearly visible here over the whole temperature range studied. In the isotropic phase, about 15 K above T_{NI} for example, the spin–spin relaxation rate is 120 s^{-1} for 5CB-βd$_2$ with the 5% polymer network, and only 40 s^{-1} for the bulk. In the nematic phase, $T_2^{-1} \sim 1700$ s^{-1} of the dispersion compares with about 600 s^{-1} of pure 5CB-βd$_2$. It is interesting to note that, in spite of a large difference related to the magnitudes of T_2^{-1}, the temperature dependences are rather similar in both cases. There is a smooth increase in the isotropic phase with decreasing temperature, and an increase in the curvature just above T_{NI}. The phase transition is accompanied by a jump for about one order of magnitude whereas both relaxation rates are almost independent of temperature in the nematic phase.

The effect of the constraint upon relaxation is most interesting in the *isotropic phase*. Similar values of T_1^{-1} for dispersion and pure liquid crystal on one hand, and significant differences in T_2^{-1} on the other, indicate a situation similar to that discussed in section 7.3.1. The liquid crystal molecules in the dispersion are obviously subjected – in addition to the fast local reorientations – to a slower dynamic process in the kilohertz frequency range.

The network directly influences the fraction η of molecules which are *bound temporarily to the internal surface of the polymer*. If the exchange of the bound and free molecules is fast enough, the magnetization decays nearly monoexponentially as was actually observed. The fast exchange limit is justified when

$$\tau_D, \tau_s/\eta \ll (T_1, T_2)_{\text{free}} \tag{7.16}$$

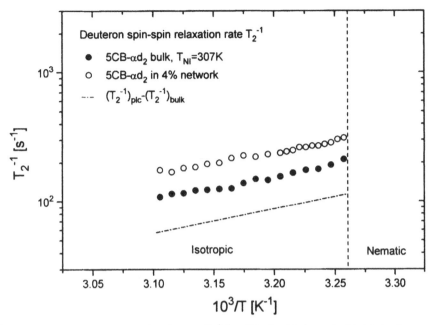

Figure 7.8 Deuteron spin–spin relaxation rate T_2^{-1} for 5CB-αd$_2$ with the embedded 4% polymer network in the isotropic phase. The chain line shows the difference between the dispersion and bulk values, $(T_2^{-1})_{plc} - (T_2^{-1})_{bulk}$.

where τ_D is a characteristic time needed for a molecule to reach the polymer surface by diffusion, and τ_s the average lifetime of molecules at the surface. Within this limit, the measured spin–spin relaxation rate of the liquid crystal–polymer network dispersion, $(T_2^{-1})_{plc}$, is given by

$$(T_2^{-1})_{plc} = \eta(T_2^{-1})_{bound} + (1 - \eta)(T_2^{-1})_{free} \tag{7.17}$$

$(T_2^{-1})_{free}$ is the spin–spin relaxation rate of deuterons well away from the fibre surface, and equals the relaxation rate of the pure isotropic phase. The bound molecules experience a faster relaxation. The mechanisms that account for this effect in the isotropic phase are described separately although they might be interconnected in view of similar time scales.

7.3.2.2 Relaxation mechanisms

(i) *Exchange of liquid crystal molecules between the orientationally ordered surface layers and isotropic regions* The effect of molecular exchange between ordered and disordered regions on the spin–spin relaxation is not specific to constrained liquid crystals. It was earlier observed in the bulk isotropic phase and convincingly explained by Martin *et al.* (1984). They found that the deuteron T_2^{-1} increases as $(T - T^*)^{-1}$ in a narrow temperature interval of a few kelvin above the nematic–isotropic transition (T^* denotes the supercooling limit, $T^* \simeq (T_{NI} - 1 \text{ K})$ for 5CB). This effect was interpreted in terms of fluctuations in the local nematic order which appear in the isotropic phase. Such ordered regions are of the size of the correlation length $\xi \simeq \xi_0[T^*/(T - T^*)]^{1/2}$ (for 5CB, $\xi_0 \simeq 0.65$ nm). As a molecule diffuses from the ordered into the disordered region, the quadrupole interaction of deuterons is modulated not only by fast local reorientations, but also on a much slower time scale. In fact, the quadrupole interaction, averaged over fast molecular reorientations, is zero in the disordered region and small but non-zero in the ordered region. This slow modulation is characterized by the exchange time τ_{exch} of molecules, which is proportional to ξ^2 and consequently to $(T - T^*)^{-1}$.

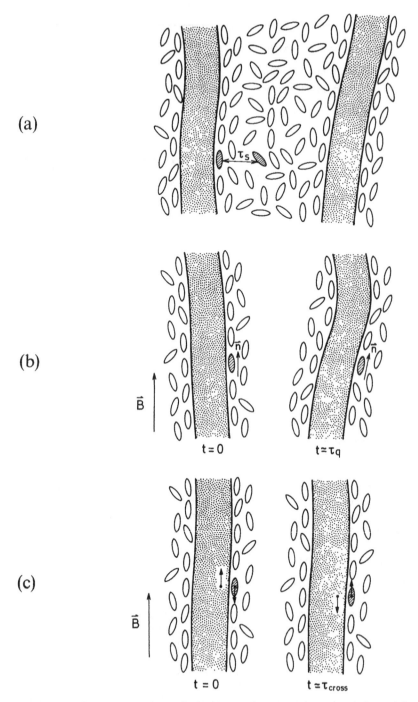

Figure 7.9 Schematic presentation of relaxation mechanisms in the isotropic phase of the dispersion: (a) exchange of molecules between the ordered layer at the surface and the disordered region away from it; (b) fluctuations of the local director at the interface, caused by fluctuations of the network; (c) cross-relaxation, i.e. direct exchange of the Zeeman energy between the liquid crystal protons and polymer protons at the interface. The cross-linking of the network is not shown and the degree of surface order is exaggerated.

In liquid crystals with an embedded polymer network, the orientational order in thin surface layers persists far above the transition temperature. The exchange-induced spin–spin relaxation rate, $(T_2^{-1})_{\text{exch}}$, is therefore effective in a much broader temperature range (Figure 7.9(a)). The same mechanism was found to account for the increase in the deuteron spin–spin relaxation rate up to about 25 K above T_{NI} in cylindrical microcavities (Vrbančič *et al.*, 1993) and PDLC materials (Dolinšek *et al.*, 1991).

Quantitatively, the contribution of molecular exchange between the ordered and disordered regions to the deuteron relaxation rate is given by

$$(T_2^{-1})_{\text{exch}} = \eta \, \tfrac{9}{16}(2\pi)^2 \left(\frac{e^2 qQ}{h}\right)^2 \langle \tfrac{1}{2}(3\cos^2\beta - 1)\rangle^2 S_0^2 [\tfrac{1}{2}(3\cos^2\Theta_B - 1)]^2 \tau_s \tag{7.18}$$

as derived by Halle and Wennerström (1981) and Burnell *et al.* (1981) for a two-site exchange model. The term

$$\left(\frac{e^2 qQ}{h}\right)^2 \langle \tfrac{1}{2}(3\cos^2\beta - 1)\rangle^2 S_0^2$$

represents the residual quadrupole interaction in the surface layer averaged over the faster individual molecular reorientations. Θ_B denotes the angle between the director of the oriented region and the external agnetic field. In evaluating eqn (7.18), it was assumed that the average lifetime of molecules at the surface, $\tau_s \sim \tau_{\text{exch}}$, exceeds $1/\omega_0$ considerably. In this case, the term with $J_0(0)$ in eqn (7.9c) dominates the terms with $J_1(\omega_0)$ and $J_2(2\omega_0)$, which can be omitted.

Equation (7.18) can be simplified by relating it to the splitting of the deuterium NMR doublet in the bulk nematic phase, $\Delta\nu_{\text{q, nem}}$ (given by eqn (7.4)), which can be easily measured. If the biaxiality effects are neglected,

$$(T_2^{-1})_{\text{exch}} = \eta \, \frac{1}{4} \left(\frac{2\pi\Delta\nu_{\text{q, nem}}}{S_{\text{nem}}} S_0\right)^2 [\tfrac{1}{2}(3\cos^2\Theta_B - 1)]^2 \tau_s \tag{7.19}$$

In a liquid crystal sample with polymer network, the directors of the oriented regions assume different directions with respect to the magnetic field. Therefore the spatial average of $\langle \tfrac{1}{4}(3\cos^2\Theta_B - 1)^2 \rangle$ enters into eqn (7.19) in the first approximation. For the isotropic distribution of directors, $\langle \tfrac{1}{4}(3\cos^2\Theta_B - 1)^2 \rangle = \tfrac{1}{5}$ and

$$(T_2^{-1})_{\text{exch}} \simeq \eta \, \frac{1}{20} \left(\frac{2\pi\Delta\nu_{\text{q, nem}}}{S_{\text{nem}}} S_0\right)^2 \tau_s \tag{7.20}$$

The increase in T_2^{-1} in the constrained liquid crystals thus gives information on the lifetime of molecules in the surface layer τ_s and on the surface order parameter S_0. Both quantities are related to the strength of molecular anchoring at the polymer surface.

(ii) *Fluctuations of the polymer network* Another possible mechanism to increase the relaxation rate of bound liquid crystal molecules are orientational fluctuations of the network, similar to order director fluctuations in a nematic liquid crystal (Figure 7.9(b)). Their characteristic feature is a wide distribution of correlation times connected to different eigenmodes of the system. An excitation mode of the wavevector $\mathbf{q}$ decays with the correlation time $\tau_q = \beta q^{-2}$ (de Gennes, 1979), where β is a material-dependent constant. The wavevectors are limited on one side by the largest wavelength of fluctuations, and on the other side by the size of a single polymer segment (Rorschach and Hazlewood, 1986).

Network fluctuations are effective in relaxing the deuterons of bound molecules if their preferential orientation, i.e. local director, adapts to the instantaneous direction of the fibre (Figure 7.9(b)). However, a liquid crystal molecule, which is on the average bound to the network for a time τ_s, experiences only the orientational fluctuations of the fibres with correlation times τ_q shorter than τ_s. Besides, the amplitudes of director fluctuations are strongly

177

restricted by the network structure. The modulation of residual quadrupole interaction by network fluctuations is therefore much weaker than in the exchange process where the quadrupole interaction varies between zero and its maximal value at the surface.

$(T_2^{-1})_{fluct}$ can be described by an expression similar to eqn (7.20) with τ_q replacing τ_s and multiplied by a reduction factor $\varepsilon(q)$ to account for the reduced amplitude of the fluctuations:

$$(T_2^{-1})_{fluct} \simeq \eta \frac{1}{20} \sum_q \varepsilon^2(q) \left(\frac{2\pi\Delta\nu_{q,\,nem}}{S_{nem}} S_0 \right)^2 \tau_q \tag{7.21}$$

In view of $\varepsilon(q) \ll 1$ as well as $\tau_q < \tau_s$, it is reasonable to assume that $(T_2^{-1})_{fluct}$ is generally overwhelmed by the larger $(T_2^{-1})_{exch}$. Nevertheless, the importance of the network fluctuations cannot be ruled out completely without performing additional – frequency dispersion measurements.

The situation would be essentially different if the network were subjected to isotropic reorientations instead of fluctuations restricted in direction. In this case a correlation function $\langle F_k(0)F_k^*(t) \rangle$ (eqn (7.10)) would decay as $\exp[-t(1/\tau_s + 1/\tau_q)]$, and the effect of the shorter correlation time, i.e. τ_q, would prevail (Halle and Wennerström, 1981; Halle *et al.*, 1993; Furo and Halle, 1995).

It should be noted that the fluctuations of the network influence, although indirectly, also $(T_2^{-1})_{exch}$ by diminishing the residual quadrupole interaction. An estimate shows that the corresponding decrease in $(T_2^{-1})_{exch}$ is smaller than about 30% for fluctuations as large as about 20°.

(iii) *Molecular reorientation due to the translational displacement along the surface* This mechanism was introduced by Schauer *et al.* (1988) and by Kimmich and Weber (1993) in their studies of the increase of water relaxation rates in solutions of macromolecules. They showed that the diffusion of water molecules along the macromolecular surface changes their preferential orientation due to the curvature or roughness of the surface. However, the liquid crystal molecules are rather large to be affected by the roughness of the fibre surface in this way. Besides, the intermolecular interactions between the fibre and the liquid crystal – being of the same type as the interactions between the liquid crystal molecules themselves – do not suggest that the diffusion along the surface is faster than away from it.

7.3.2.3 *Discussion and comparison with related systems*

As shown above, the increase in deuteron T_2^{-1} for the constrained isotropic phase originates most probably from the exchange of molecules between the surface layer and isotropic regions. The whole difference $(T_2^{-1})_{plc} - (T_2^{-1})_{bulk}$ can be thus ascribed to $(T_2^{-1})_{exch}$ to a good approximation as given by eqn (7.20). The product of three quantities in eqn (7.20), $S_0^2\eta\tau_s$, is the only adjustable parameter to be calculated from experimental data, as the splitting $\Delta\nu_{q,\,nem}$ of α-deuterons in the nematic phase is known ($\simeq 51$ kHz at 28 °C) as well as the corresponding molecular order parameter, $S_{nem} \simeq 0.45$. The difference $(T_2^{-1})_{plc} - (T_2^{-1})_{bulk}$ ranges in the isotropic phase from 60 to 90 s^{-1} as shown by the broken line in Figure 7.8 for the 4% dispersion. The product $S_0^2\eta\tau_s$ is thus, for example, approximately 30×10^{-10} s at $T - T_{NI} \sim 10$ K.

To get knowledge of τ_s, the values for S_0 and η are taken in the first approximation from the recent optical birefringence studies of the same dispersion (Žumer and Crawford, this volume): $S_0 \simeq 0.05$ and $\eta \simeq 0.06$. The evaluated τ_s is approximately 2×10^{-5} s. This value represents in fact the upper limit. A larger τ_s would prevent the averaging during the NMR time and two superimposed lines would appear in the spectrum presented in Figure 7.3. On the other hand, τ_s can be shorter if the values for S_0 or η have been underestimated. The order parameter $S_0 \sim 0.05$, as detected by optical measurements, is averaged over distances which are of the order of the wavelength of light. On the other hand, S_0 in the NMR relaxation process described by eqn (7.20), is the local order parameter, which might therefore be larger

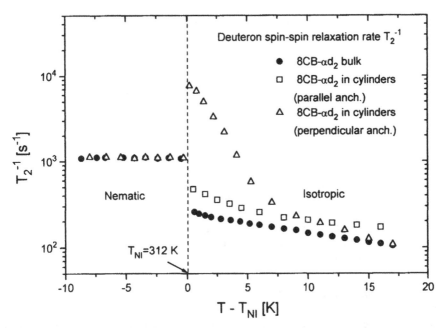

Figure 7.10 Temperature dependence of the deuteron spin–spin relaxation rate T_2^{-1} for 8CB-αd_2 in the cylindrical Anopore cavities with radius of about 0.1 μm: non-treated walls, parallel anchoring (squares); walls with organic coating, perpendicular anchoring (triangles); and bulk liquid crystal (solid circles).

than 0.05. The corresponding τ_s would be shorter. Proton relaxation data in the next section give arguments in the favour of this conjecture.

It is important to stress that – independent of the exact value of τ_s – only a large value of η (of the order of a few per cent) can account for the huge increase in the spin–spin relaxation rate for the dispersion in the isotropic phase. This requires a fine distribution of network on the nanometre scale to provide a large enough surface-to-volume ratio at the fixed content of the polymer. A smaller internal surface of the network would not account for the NMR relaxation data.

There is another important point resulting from the spin–spin relaxation of deuterons in the isotropic phase which should be noted. The difference between the isotropic constrained and bulk 5CB-αd_2 is only slightly temperature dependent in the whole range of the isotropic phase (Figure 7.8). The linear dependence of $(T_2^{-1})_{\text{plc}} - (T_2^{-1})_{\text{bulk}}$ against $1/T$ (broken line in Figure 7.8) over a broad temperature interval indicates that the slow motion, responsible for the relaxation, is thermally activated with the activation energy 30 kJ/mol. $(T_2^{-1})_{\text{plc}} - (T_2^{-1})_{\text{bulk}}$ does not show any pretransitional increase. This leads straightforwardly to the conclusion that S_0 in the simple two-site model is only weakly temperature dependent.

It is interesting to compare the above behaviour with the spin–spin relaxation rate observed for 5CB and 8CB in cylindrical cavities of Anopore membranes. The results for 8CB are presented in Figure 7.10. The spin–spin relaxation rate of the confined liquid crystal in the isotropic phase shows two completely different patterns of behaviour. The sample with non-treated walls, which promote parallel anchoring, yields a weakly temperature-dependent T_2^{-1}, similar to that presented for the liquid crystal with the low-concentration polymer network (Figures 7.7 and 7.8). The basic reason for the smooth temperature dependence of T_2^{-1} is the temperature invariance of S_0, which indicates strong surface coupling in these two cases. In contrast, the liquid crystal confined to cavities with organic coating $(CH_2)_{15}COOH$ on the walls shows a strong pretransitional increase in T_2^{-1}. It is correlated to a similar increase in

S_0, which was attributed to diminished strength of the interfacial molecular interaction due to the coating on the walls (Crawford et al., 1993).

We conclude this section with a short comment on nuclear spin relaxation in the *polymer-constrained nematic phase*. Here the pure orienting effect of the network, which has been described in section 7.2.3, increases both relaxation rates T_1^{-1} and T_2^{-1} when measured in multidomain samples. The spin-lattice relaxation rate of the uniformly oriented nematic increases as the director is turned away from the magnetic field. The increase amounts to about 30% at $\Theta_B = 90°$. The effect of director orientation is much stronger for T_2^{-1}. Its angular dependence in the nematic phase is proportional to $(\sin^2\Theta_B \cos^2\Theta_B)$, giving the smallest T_2^{-1} for the director parallel or perpendicular to the magnetic field, and maximal for $\Theta_B = 45°$. In our experiment, the magnetic field of 2.1 T is not strong enough to overcome the effect of the network in aligning the liquid crystal. The difference in T_1^{-1} and T_2^{-1} between the constrained and bulk nematic phase can thus be explained to a large extent by the orientational effects of the network. T_2^{-1} can be additionally affected by the diffusion of molecules between differently oriented regions, which causes a slow effective reorientation of molecules (TR mechanism (Žumer et al., 1989)). The exchange of the bound and free molecules at the surface is not effective in the nematic phase as the order parameter retains roughly the same value in the surface layer as away from the fibres.

7.3.3 Proton Spin Relaxation

Complementary information on molecular dynamics and order in the liquid crystal–polymer network dispersions can be obtained from proton relaxation data. Similar to the deuteron relaxation, no particular impact of the network has been observed in proton T_1 measurements at 46 MHz (Lahajnar et al., 1995; Vilfan et al., 1995). A small difference observed in the nematic phase can be ascribed to the non-uniform orientation of the dispersion in the magnetic field. A significant effect of the network is reflected, however, in the *spin–lattice relaxation in the rotating frame*, $T_{1\rho}^{-1}$, which is a sensitive tool to probe slower motions. The measurements of proton $T_{1\rho}$ for 5CB with 4% network have been performed in the isotropic phase for different frequencies ω_1 between 30 kHz and 200 kHz (Lahajnar et al., 1995). A striking difference between the relaxation rates of the dispersion and pure 5CB was observed (Figure 7.11). At $\omega_1 = 40$ kHz, the relaxation rate $T_{1\rho}^{-1}$ in the sample with 4% network is approximately 35 s^{-1} and exceeds that of the pure 5CB, which is approximately 5 to 10 s^{-1}, by about 30 s^{-1}. Besides, $T_{1\rho}^{-1}$ of the dispersion was found to depend strongly on the frequency ω_1 in contrast to the bulk $T_{1\rho}^{-1}$ which is frequency independent (inset in Figure 7.11).

Different behaviour and magnitudes of $T_{1\rho}^{-1}$ in the dispersion and pure isotropic phase can be explained by the influence of a dynamic process in the kilohertz frequency range. According to section 7.3.2.2, the most effective relaxation mechanism is the exchange of molecules between the ordered and disordered regions. Equations (7.18) and (7.20), adapted for proton spin–lattice relaxation in the rotating frame, give for a proton pair

$$(T_{1\rho}^{-1})_{\text{exch}} \simeq \eta \frac{9}{20} \frac{\gamma^4 \hbar^2}{r_{ij}^6} \langle \tfrac{1}{2}(3\cos^2\beta - 1) \rangle^2 S_0^2 \frac{\tau_s}{1 + 4\omega_1^2 \tau_s^2} \tag{7.22}$$

For an actual liquid crystal molecule, the summation and averaging over different proton pairs increase the prefactor. Fortunately, the measured $T_{1\rho}^{-1}$ dependence on the frequency ω_1 of the rotating field gives a direct estimate of the characteristic correlation time without a detailed knowledge of the prefactors. The fit of eqn (7.22) to $(T_{1\rho}^{-1})_{\text{plc}} - (T_{1\rho}^{-1})_{\text{bulk}}$, measured as a function of ω_1, shows that the relaxation process between 30 kHz and 200 kHz can be described to a good approximation by a single correlation time. The value of τ_s providing the best fit is $(4 \pm 1) \times 10^{-6}$ s. Its variation with temperature in the isotropic phase is smaller than the experimental error. As expected, the value of τ_s obtained from proton relaxation data is shorter than the upper limit evaluated in section 7.3.2.3. It is interesting to note that for $\tau_s \sim 4 \times$

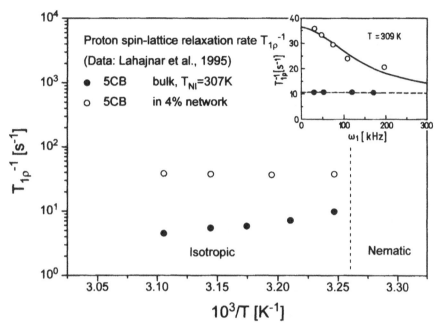

Figure 7.11 Temperature dependence of the proton spin–lattice relaxation rate in the rotating frame $T_{1\rho}^{-1}$ at $\omega_1 \sim 40$ kHz for the isotropic phase of 5CB constrained by the 4% polymer network, and for the isotropic phase of the pure, bulk 5CB (from Lahajnar *et al.* (1995) and Vilfan *et al.* (1995)).

10^{-6} s, $(T_{1\rho}^{-1})_{exch}$ assumes its largest possible value close to the maximum in Figure 7.5. The strength of the dipolar interaction required to account for the difference $(T_{1\rho}^{-1})_{plc} - (T_{1\rho}^{-1})_{bulk} \sim 30\,s^{-1}$ is rather large. It includes certainly – in addition to the interaction of each proton with its nearest neighbour – the contributions of more distant protons in the same molecule and of the polymer protons. Besides, the large value of proton $T_{1\rho}^{-1}$ in the dispersion requires for its explanation, as T_2^{-1} in the case of deuterons, a large internal surface area per unit volume of the network and a relatively high value of the local surface order parameter. By inserting τ_s, estimated from proton relaxation data, into the product $S_0^2 \eta \tau_s \sim 30 \times 10^{-10}$ s from deuteron relaxation data (Section 7.3.2.3), and using $\eta \simeq 0.06$ from optical measurements (Žumer and Crawford, this volume), we get $S_0 \simeq 0.11$. The estimated error of this result is rather large ($\sim 40\%$) in view of the uncertainty in the values of τ_s and η.

It should be mentioned that there is another mechanism which could enhance the spin–lattice relaxation of protons and which is not effective for deuterons. This is the *cross-relaxation*, i.e. direct transfer of Zeeman energy between the liquid crystal and polymer protons at the interface of two proton phases (Edzes and Samulski, 1978). The transfer is caused by mutual spin flips of bound liquid crystal protons and polymer protons (Figure 7.9(c)). The rate at which the transfer of the Zeeman energy takes place, τ_{cross}^{-1}, depends on the strength of the dipolar interaction between the two kinds of spins and on the average lifetime of liquid crystal molecules in the surface layer τ_s. The corresponding increase in the relaxation rate of bound molecules, $(T_{1\rho}^{-1})_{cross}$, is estimated to be at most $10\,s^{-1}$ in our case (Vilfan *et al.*, 1995). We conclude that the cross-relaxation could increase $T_{1\rho}^{-1}$ in the isotropic phase of the dispersion, but to a smaller extent than the exchange relaxation mechanism (assisted probably by network fluctuations). In systems with larger polymer content (PDLC materials with approximately 70% polymer) the cross-relaxation was found to dominate the proton relaxation. It was also measured directly (Jarh *et al.*, 1991; Cross and Fung, 1992, 1993) and related to the strength of surface interaction.

7.4 Translational Self-diffusion in the Isotropic Phase

The molecular self-diffusion coefficient in the isotropic phase can be measured directly by the pulsed gradient spin echo (PGSE) NMR technique (Callaghan, 1991). The diffusion coefficient – obtained from the slope of the NMR echo attenuation – does not depend on the time between gradient pulses Δ if the medium is spatially non-restricted over distances covered by diffusion of molecules during the measurement. For a typical liquid crystal in the isotropic phase, the diffusion coefficient is about 10^{-10} m^2/s (Noack, 1993) and the distance covered within $\Delta \sim 10$ ms is about 2.5 μm. The preliminary measurements show that the self-diffusion coefficient of 5CB constrained by the 4% network (Lahajnar *et al.*, 1995; Vilfan *et al.*, 1995) does not differ from the bulk coefficient if the diffusion time Δ is less than 10 ms. This gives evidence that the network elements on the molecular scale are in the form of fibres which allow a relatively non-hindered passage. A model of the small cages on the nanometre scale must be definitely excluded. However, at larger diffusion times, corresponding to micrometre distances, a decrease in the effective diffusion constant, not observed in the bulk, takes place (Figure 7.12). The decrease in the effective diffusion constant of the dispersion is similar to that in PDLC droplets with a diameter of about 1 μm (Lahajnar *et al.*, 1995) and to the decrease in cholesteric liquid crystals with a pitch of about 1.5 μm (Blinc *et al.*, 1985). This implies that the polymer network,

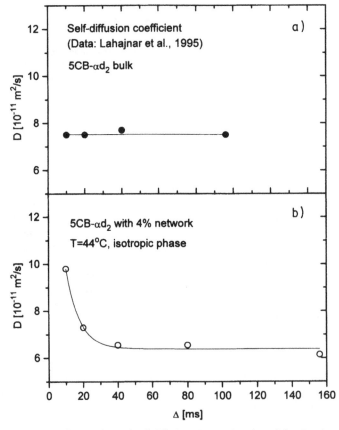

Figure 7.12 Coefficient of the translational self-diffusion, D, as a function of the time interval Δ between gradient pulses in the isotropic phase of: (a) bulk 5CB at $T - T_{NI} \sim 8$ K, and (b) 5CB with the embedded 4% polymer network at $T - T_{NI} \sim 11$ K (from Lahajnar *et al.*, 1995). The thin line is only a guide to the eye.

though finely distributed on the nanometre scale, also exhibits effective inhomogeneities on the submicrometre or micrometre scale. In our case, the inhomogeneities may be partly related to domain boundaries which have been present during the polymerization.

7.5 Conclusions

The described NMR studies provide new information on the structure and dynamics of liquid crystals with an embedded low-concentration polymer network. The NMR spectra of deuterons in the nematic phase are an excellent tool to study the orientational distribution of directors and the stability of the orientation imposed by the network. It was shown, for example, that the 3% network almost completely stabilizes the liquid crystal structure in a magnetic field as high as 4.7 T.

The proton and deuteron spin relaxations in the megahertz regime (T_1 data) differ only modestly in the dispersion and in pure liquid crystal. The difference is explained by the orienting effects of the network imposed on the nematic phase. On the other hand, the spin relaxation in the kilohertz region, studied by T_2 and $T_{1\rho}$ measurements, shows a huge increase in the relaxation rates upon constraint. This indicates the presence of one or more dynamic processes, specific for the dispersion. The experimental data for the isotropic phase are analyzed in terms of additional relaxation mechanisms with characteristic frequencies in the kilohertz range: the exchange of molecules between the ordered surface layer and disordered regions away from it, network fluctuations, and direct cross-relaxation between the polymer and liquid crystal protons. All these mechanisms, involving dynamic processes on the micro- and millisecond scale, are not effective in the spin relaxation at megahertz frequencies. The impact of the network is observed also in the translational self-diffusion of molecules at large diffusion times.

The proton relaxation data show that the dominant relaxation dynamic process in the range between 30 kHz and 200 kHz can be described to a good approximation by a single correlation time. Its value is approximately $(4 \pm 1) \times 10^{-6}$ s, independently of the strength of interaction and negligibly varying with temperature in the isotropic phase. The correlation time is most probably related to the average lifetime of molecules in the surface layer τ_s although a partial influence of network fluctuations and cross-relaxation cannot be excluded in the interpretation without additional deuteron T_2 dispersion measurements. The NMR relaxation data point out further that the local order parameter in the surface layer, $S_0 \sim 0.1$, is relatively large. This value has been estimated from the product $S_0^2 \eta \tau_s \sim 30 \times 10^{-10}$ s, which is the only unknown parameter in the deuteron T_2 relaxation data, and by using τ_s from proton data and $\eta \sim 0.06$ from optical measurements. The estimated value of $S_0 \sim 0.1$ exceeds that obtained from birefringence measurements, interpreted in terms of a simple, one-parameter model ($S_0 \sim 0.05$). The difference might be ascribed to the fact that the spin relaxation rates depend on the square of $(3 \cos^2 \Theta_B - 1)$ and are therefore less affected by the inhomogeneities in the network orientation and cross-linking than other experimental techniques. The surface order parameter, as seen by NMR, does not show any pretransitional increase on approaching the nematic phase. This indicates that the polymer–liquid crystal interactions at the interface are rather strong. It should be stressed that the NMR relaxation and diffusion measurements can only be explained if a finely distributed network of polymer fibres with a large internal surface-to-volume ratio is assumed. Such fibres bind temporarily to the surface a rather large fraction of molecules ($\eta \sim 0.06$ for the 4% polymer content). The smallest value of η, which can still account for the NMR relaxation data, available so far, is about 0.01 corresponding to the internal surface area per unit volume of the polymer network of 0.2 nm^{-1}. The diffusion measurements confirm that the dispersed polymer is in the form of fibres and exclude the model of closed cavities. There are, however, inhomogeneities in the polymer distribution on the submicrometre or micrometre scale as well.

So far, only a few liquid crystal–polymer network dispersions have been investigated by NMR. We expect that future research will include other compounds, in particular their twisted and/or smectic phases. In all NMR studies of the dispersions, deuterons and protons were used as the probe. Carbon-13 and xenon NMR could be used complementarily in the future. These nuclei recently gave promising results in cylindrical cavities (Schmiedel *et al.*, 1994; Long *et al.*, 1995). Besides, more sophisticated NMR techniques like field-cycling relaxation measurements, T_2 frequency dispersion studies, two-dimensional exchange NMR, and NMR imaging are expected to give valuable new information on the structure and dynamics of liquid crystals with an embedded polymer network. The end of this review means in our opinion only the beginning of research in the field.

Acknowledgements

M.V. would like to thank first of all Professor J. W. Doane for introducing her to relaxation in liquid crystals many years ago. The authors are indebted to Professor S. Žumer for many helpful discussions on the topics of this review, and to Professor R. Blinc for support. They would like to thank further Dr G. Crawford and Professor J. W. Doane for the samples of microconfined liquid crystals, which were made at the Liquid Crystal Institute in Kent, Ohio, and Professors G. Lahajnar and I. Zupančič for some of their data prior to publication.

References

ABRAGAM, A. (1961) *The Principles of Nuclear Magnetism*, Oxford: Clarendon Press.

ALLENDER, D. W., CRAWFORD, G. P. and DOANE, J. W. (1991) Determination of the liquid crystal surface elastic constant K_{24}, *Phys. Rev. Lett.* **67**, 1442–1445.

BLINC, R., MARIN, B., PIRŠ, J. and DOANE, J. W. (1985) Time-resolved NMR study of self-diffusion in the cholesteric phase, *Phys. Rev. Lett.* **54**, 438–440.

BOS, P. J., PIRŠ, J., UKLEJA, P. and DOANE, J. W. (1977) Molecular orientational order and NMR in uniaxial and biaxial phases, *Mol. Cryst. Liq. Cryst.* **40**, 59–77.

BURNELL, E. E., CLARK, M. E., HINKE, J. A. M. and CHAPMAN, N. R. (1981) Water in barnacle muscle. III. NMR studies of fresh fibers and membrane-damaged fibers equilibrated with selected solutes, *Biophys. J.* **33**, 1–26.

CALLAGHAN, P. T. (1991) *Principles of Nuclear Magnetic Resonance Microscopy*, Oxford: Oxford University Press.

CHARVOLIN, J. and HENDRIKX, Y. (1984) Amphiphilic molecules in lyotropic liquid crystals and micellar phases, *NMR in Liquid Crystals*, ed. J. W. Emsley, Amsterdam: Reidel Publishing Company.

CRAWFORD, G. P., VILFAN, M., DOANE, J. W. and VILFAN, I. (1991a) Escaped-radial nematic configuration in submicrometer size cylindrical cavities: Deuterium nuclear magnetic resonance, *Phys. Rev. A*, **43**, 835–842.

CRAWFORD, G. P., ALLENDER, D. W., VILFAN, M., VILFAN, I. and DOANE, J. W. (1991b) Finite molecular anchoring in the escaped-radial nematic configuration: A ^{2}H NMR study, *Phys. Rev. A*, **44**, 2570–2577.

CRAWFORD, G. P., YANG, D.-K., ŽUMER, S., FINOTELLO, D. and DOANE, J. W. (1991c) Ordering and self-diffusion in the first molecular layer at a liquid crystal–polymer interface, *Phys. Rev. Lett.* **66**, 723–726.

CRAWFORD, G. P., STANNARIUS, R. and DOANE, J. W. (1991d) Surface induced orientational order in the isotropic phase of a liquid crystal material, *Phys. Rev. A* **44**, 2558–2569.

CRAWFORD, G. P., ALLENDER, D. W. and DOANE, J. W. (1992) Surface elastic and molecular anchoring properties of nematic liquid crystals confined to cylindrical cavities, *Phys. Rev. A*, **45**, 8693–8708.

CRAWFORD, G. P., ONDRIS-CRAWFORD, R., ŽUMER, S. and DOANE, J. W. (1993) Anchoring and orientational wetting transition of confined liquid crystals, *Phys. Rev. Lett.* **70**, 1838–1841.

CRAWFORD, G. P., POLAK, R. D., SCHARKOWSKI, A., CHIEN, L.-C., DOANE, J. W. and ŽUMER, S. (1994) Nematic director fields captured in polymer networks confined to spherical droplets, *J. Appl. Phys.* **75**, 1968–1971.

CROSS, C. W. and FUNG, B. M. (1992) Cross relaxation of polymer dispersed liquid crystal droplets, *J. Chem. Phys.* **96**, 7086–7091.

CROSS, C. W. and FUNG, B. M. (1993) Measurement of the cross relaxation rate for a polymer dispersed liquid crystal system, *J. Chem. Phys.* **99**, 1425–1428.

DE GENNES, P. G. (1979) *Scaling Concepts in Polymer Physics*, Ithaca, NY: Cornell University Press.

DOANE, J. W. (1979) NMR of liquid crystals, *Magnetic Resonance of Phase Transitions*, ed. F. J. Owens *et al.*, New York: Academic Press, Ch. 4.

DOANE, J. W., YANIV, Z., CHIDICHIMO, G. and VAZ, N. (1982) NMR of the cholesteric and blue liquid crystalline phases, in R. Blinc and M. Vilfan (eds) *Proc. 7th AMPERE Int. Summer School, Portorož*, Ljubljana: J. Stefan Institute, pp. 181–198.

DOANE, J. W., VAZ, N. A., WU, B. G. and ŽUMER, S. (1986) Field controlled light scattering from nematic droplets, *Appl. Phys. Lett.* **48**, 269–271.

DOLINŠEK, J., JARH, O., VILFAN, M., ŽUMER, S., BLINC, R., DOANE, J. W. and CRAWFORD, G. P. (1991) Two-dimensional deuteron nuclear magnetic resonance of a polymer dispersed nematic liquid crystal, *J. Chem. Phys.* **95**, 2154–2161.

DONG, R. Y. (1994) *Nuclear Magnetic Resonance of Liquid Crystals*, Berlin: Springer-Verlag.

EDZES, H. T. and SAMULSKI, E. T. (1978) The measurement of cross relaxation effects in the proton NMR spin-lattice relaxation of water in biological systems: Hydrated collagen and muscle, *J. Magn. Reson.* **31**, 207–229.

FIGUEIRINHAS, J., ŽUMER, S. and DOANE, J. W. (1987) Orientational flow of monodomain samples of the smectic F phase: Deuterium nuclear magnetic resonance, *Phys. Rev. A* **35**, 4389–4396.

FUKUSHIMA, E. and ROEDER, S. B. W. (1981) *Experimental Pulse NMR – a Nuts and Bolts Approach*, Reading, MA: Addison-Wesley.

FURO, I. and HALLE, B. (1995) Micelle size and orientational order across the nematic isotropic transition: A field-dependent nuclear-spin-relaxation study, *Phys. Rev. E*, **51**, 466.

GOLEMME, A., ŽUMER, S., ALLENDER, D. W. and DOANE, J. W. (1988) Continuous nematic-isotropic transition in submicron size liquid crystal droplets, *Phys. Rev. Lett.* **61**, 2937–2940.

HALLE, B. and WENNERSTRÖM, H. (1981) Interpretation of magnetic resonance data from water nuclei in heterogeneous systems, *J. Chem. Phys.* **75**, 1928–1943.

HALLE, B., QUIST, P. O. and FURO, I. (1993) Microstructure and dynamics in lyotropic liquid crystals: Principles and applications of nuclear spin relaxation, *Liq. Cryst.* **14**, 227–263.

HIKMET, R. A. M. (1990) Electrically induced light scattering from anisotropic gels, *J. Appl. Phys.* **68**, 4406–4412.

HIKMET, R. A. M. (1991) Anisotropic gels and plasticized networks formed by liquid crystal molecules, *Liq. Cryst.* **9**, 405–416.

IANNACCHIONE, G., CRAWFORD, G. P., ŽUMER, S., DOANE, J. W., and FINOTELLO, D. (1993) Randomly constrained orientational order in porous glass, *Phys. Rev. Lett.* **71**, 2595–2598.

JÁKLI, A., KIM, D. R., CHIEN, L.-C. and SAUPE, A. (1992) Effect of a polymer network on the alignment and the rotational viscosity of a nematic liquid crystal, *J. Appl. Phys.* **72**, 3161–3164.

JÁKLI, A., BATA, L., FODOR-CSORBA, K., ROSTA, L. and NOIREZ, L. (1994) Structure of polymer networks dispersed in liquid crystals: Small angle neutron scattering study, *Liq. Cryst.* **17**, 227–234.

JARH, O., SEPE, A. and VILFAN, M. (1991) A direct NMR study of cross relaxation in confined liquid crystal, *Abstracts of 10th Eur. Experimental NMR Conf., Veldhoven*, Veldhoven: Center for Biomolecular Research.

KIMMICH, R. and WEBER, H. W. (1993) Nuclear magnetic relaxation spectroscopy of polymers and anomalous segment diffusion. Reorientations mediated by translational displacements, *J. Chem. Phys.* **98**, 5847–5854.

KOHLHAMMER, K., MÜLLER, K. and KOTHE, G. (1989) Molecular order and motion in nematogens from pulsed dynamic NMR: A comparative study of model compounds and parent liquid crystal polymers, *Liq. Cryst.* **5**, 1525–1547.

KÖLLNER, R., SCHWEIKERT, K. H., NOACK, F. and ZIMMERMANN, H. (1993) NMR field cycling study of proton and deuteron spin relaxation in the nematic liquid crystal 4-n-pentyl-4'-cyanobiphenyl, *Liq. Cryst.* **13**, 483–498.

KRALJ, S., LAHAJNAR, G., ZIDANŠEK, A., VRBANČIČ-KOPAČ, N., VILFAN, M., BLINC,

185

R. and KOSEC, M. (1993) Deuterium NMR of a pentylcyanobiphenyl liquid crystal confined in a silica aerogel matrix, *Phys. Rev. E* **48**, 340–349.

LAHAJNAR, G., ZUPANČIČ, I., VILFAN, M., ŽUMER, S., BLINC, R., CRAWFORD, G. P. and DOANE, J. W. (1995) A proton NMR study of the nematic liquid crystal 5CB constrained by the polymer network, *Abstracts of the Eur. Conf. on Liquid Crystals, Bovec*, ed. A. Mertelj and I. Muševič, Ljubljana: J. Stefan Institute.

LONG, H. W., LUZAR, M., GAEDE, H. C., LARSEN, R. G., KRITZENBEGER, J., PINES, A., and CRAWFORD, G. P. (1995) Xenon NMR study of a nematic liquid crystal centred to cylindrical submicron centres, *J. Phys. Chem.* **99**, 11 989–11 983.

MARTIN, J. F., VOLD, R. R. and VOLD, R. L. (1984) Pretransitional effects in nuclear spin-relaxation of p-methoxybenzylidene-p-n-butylaniline, *J. Chem. Phys.* **80**, 2237–2238.

NOACK, F. (1986) Field-cycling NMR spectroscopy, *Prog. NMR Spectrosc.* **18**, 171–208.

NOACK, F. (1993) Basic and novel aspects of NMR field-cycling spectroscopy, in R. Blinc *et al.* (eds) *Proc. AMPERE Summer Inst. on Advanced Techniques in Experimental Magnetic Resonance, Portorož*, Ljubljana: J. Stefan Institute, pp. 18–35.

ONDRIS-CRAWFORD, R. J., CRAWFORD, G. P., ŽUMER, S. and DOANE, J. W. (1993a) Curvature induced configuration transition in confined nematic liquid crystal, *Phys. Rev. Lett.* **70**, 194–217.

ONDRIS-CRAWFORD, R. J., CRAWFORD, G. P., DOANE, J. W., ŽUMER, S., VILFAN, M. and VILFAN, I. (1993b) Surface molecular anchoring in microconfined liquid crystals near the nematic-smectic A transition, *Phys. Rev. E* **48**, 1998–2005.

ONDRIS-CRAWFORD, R. J., AMBROŽIČ, M., DOANE, J. W. and ŽUMER, S. (1994) Pitch induced transition of chiral nematic liquid crystals in submicrometer cylindrical cavities, *Phys. Rev. E* **50**, 4773–4779.

RORSCHACH, H. E. and HAZLEWOOD, C. F. (1986) Protein dynamics and the NMR relaxation time T_1 of water in biological systems, *J. Magn. Reson.* **70**, 79–88.

SCHAUER, G., KIMMICH, R., and NUSSER, W. (1988) Deuteron field-cycling relaxation spectroscopy and translational water diffusion in protein hydration shells, *Biophys. J.* **53**, 397–404.

SCHMIEDEL, H., STANNARIUS, R., FELLER, G. and CRAMER, C. (1994) Experimental evidence of a conic helical liquid crystalline structure in cylindrical microcavities, *Liq. Cryst.* **17**, 323–332.

STANNARIUS, R., CRAWFORD, G. P., CHIEN, L.-C. and DOANE, J. W. (1991) Nematic director orientation in a liquid crystal dispersed polymer: A deuterium nuclear magnetic resonance approach, *J. Appl. Phys.* **70**, 135–143.

VILFAN, M., RUTAR, V., ŽUMER, S., LAHAJNAR, G., BLINC, R., DOANE, J. W. and GOLEMME, A. (1988) Proton spin-lattice relaxation in nematic microdroplets, *J. Chem. Phys.* **89**, 597–604.

VILFAN, I., VILFAN, M. and ŽUMER, S. (1989) Orientational order in bipolar nematic microdroplets close to the phase transition, *Phys. Rev. A* **40**, 4724–4730.

VILFAN, M., LAHAJNAR, G., ZUPANČIČ, I., ŽUMER, S., BLINC, R., CRAWFORD, G. P. and DOANE, J. W. (1995) Dynamics of a nematic liquid crystal constrained by the polymer network: a proton NMR study, *J. Chem. Phys.* **103**, 8726–8733.

VRBANČIČ, N., VILFAN, M., BLINC, R., DOLINŠEK, J., CRAWFORD, G. P. and DOANE, J. W. (1993) Deuteron spin relaxation and molecular dynamics of a nematic liquid crystal (5CB) in cylindrical microcavities, *J. Chem. Phys.* **98**, 3540–3557.

WÖLFEL, W. (1978) *Kernrelaxationsuntersuchungen der Molekülbewegung im Flüssigkristal PAA*, München: Minerva.

WÖLFEL, W., NOACK, F. and STOHRER, M. (1975) Frequency dependence of proton spin relaxation in liquid crystalline PAA, *Z. Naturforsch.* **30a**, 437–441.

ZIDANŠEK, A., KRALJ, S., LAHAJNAR, G. and BLINC, R. (1995) Deuteron NMR study of liquid crystals confined in aerogel matrices, *Phys. Rev. E* **51**, 3332–3340.

ŽUMER, S., KRALJ, S. and VILFAN, M. (1989) Nuclear magnetic relaxation in small nematic droplets induced by molecular self-diffusion, *J. Chem. Phys.* **91**, 6411–6420.

8

Polymer-dispersed and Polymer-stabilized Chiral Liquid Crystals

H.-S. KITZEROW

8.1 Introduction

During the last decade, the research on polymer-dispersed liquid crystals (PDLCs) has emerged into a rapidly growing field due to the pioneering work of Fergason (1984) and Doane *et al.* (1986, 1987). In the classic PDLC system, droplets of a nematic liquid crystal with positive dielectric anisotropy, dispersed in a polymer, are used to produce thin polymer films which can be switched from a scattering, translucent state to a transparent state by applying an electric field. This electro-optic effect is due to refractive index matching between the liquid crystal and the polymer in the field-on state, and due to mismatching of the refractive indices in the field-off state. PDLC systems are suitable for application in large-area light shutters (smart windows), flexible displays (e.g. for touch switches), projection displays with high brightness, storage devices, and non-linear optical elements. PDLC films can be prepared by encapsulation, starting from an emulsion of liquid crystal droplets in a polymer solution (Fergason, 1984; Drzaic, 1986), or by phase separation. The latter may be induced by thermally initiated polymerization (Doane *et al.*, 1986) or by photopolymerization of a polymer precursor mixed with the liquid crystal (Vaz *et al.*, 1987), by cooling a liquid crystal/polymer mixture into a miscibility gap (Wu *et al.*, 1987), or by evaporation of the solvent from a liquid crystal/polymer solution (West, 1988). (For general reviews, see e.g. Doane *et al.* (1988), Doane (1990) and Kitzerow (1994).)

Beyond the scope of applications, the interest in PDLC systems has considerably stimulated *fundamental research*, concerning the phase separation and polymerization processes (see e.g. Kim and Palffy-Muhoray, 1991, Golemme *et al.*, 1994; Smith, 1994), the optical properties of PDLC systems (Khoo and Wu, 1993; Kelly and Palffy-Muhoray, 1994), and especially the effects which are due to the confinement of liquid crystals to small cavities. The nematic mesophase of rod-like molecules is characterized by a preferred parallel alignment of these molecules. Their average orientation can be described by a unit vector $\mathbf{n}$, the director. The degree of order is characterized by the orientational order parameter $S = \frac{1}{2}\langle 3\cos^2\theta_i - 1\rangle$. The confinement of the liquid crystal to a small droplet and the interactions at the interface cause deformations of the director field $\mathbf{n}(\mathbf{r})$ and even the occurrence of defects. Taking into account the elastic forces and the reorienting influence of external electric fields $\mathbf{E}$ due to the dielectric anisotropy ε_a, the competing influences on $\mathbf{n}$ can be expressed by minimizing the free energy

$$F = \tfrac{1}{2} \int [K_{11}(\text{div } \mathbf{n})^2 + K_{22}(\mathbf{n} \cdot \text{rot } \mathbf{n})^2 + K_{33}(\mathbf{n} \times \text{rot } \mathbf{n})^2$$

$$+ K_{13} \text{ div}(\mathbf{n} \text{ div } \mathbf{n}) - K_{24} \text{ div}(\mathbf{n} \times \text{rot } \mathbf{n} + \mathbf{n} \text{ div } \mathbf{n})] \, dr^3$$

$$- \tfrac{1}{2} \int \varepsilon_0 \varepsilon_a (\mathbf{E} \cdot \mathbf{n})^2 \, dr^3$$

$$+ \tfrac{1}{2} \int W_\phi \sin^2(\phi - \phi_0) \, dr^2 + \tfrac{1}{2} \int W_\Theta \sin^2(\Theta - \Theta_0) \, dr^2 \qquad (8.1)$$

Here the volume integrals describe elastic deformations of $\mathbf{n}(\mathbf{r})$ and the influence of external fields. The surface integrals describe a finite anchoring of $\mathbf{n}$ at the liquid crystal surface, with ϕ and Θ being the azimuthal and the polar angle of the director in a local coordinate system at the surface. W_ϕ and W_Θ are the azimuthal and the polar anchoring energy, respectively. The second volume integral (containing K_{13} and K_{24}) can be converted into a surface integral using Gauss's theorem, and thus its value is negligible for liquid crystals in the bulk, where the ratio between the surface and the volume is small. However, for confined systems, the saddle–splay elastic coefficient K_{24} was determined for the first time and found to be of the same value as $K_{11} \approx K_{33} \approx 5 \times 10^{-12}$ N (Crawford *et al.*, 1991) or even larger (Polak *et al.*, 1994). It is interesting to note that the confinement may induce twist of the director field even for non-chiral materials if the twist coefficient K_{22} is sufficiently small compared with K_{11} and K_{33} (Cladis and Kléman, 1972; de Gennes, 1974). Figure 8.1 shows some examples of nematic droplets with a twisted structure. The finite size of liquid crystal droplets can also affect the order parameter S. Thus, the nematic–isotropic phase transition which is discontinuous in the bulk (Maier and Saupe, 1958) may become continuous for droplet radii below a critical value $R_c \approx 20$ nm (Golemme *et al.*, 1988).

The first studies on PDLC systems focused on non-chiral nematic liquid crystals. However, many interesting optical properties and electro-optical phenomena occur in *chiral* liquid crystals. Twisted nematic structures and cholesteric liquid crystals with very large pitch (Figure 8.2(a)) show a wave-guiding effect which is used in the TN cell (Schadt and Helfrich, 1971). Cholesteric liquid crystals with medium chirality exhibit a high optical activity (de Vries, 1951; Chanishvili *et al.*, 1991). If the pitch is very small, Bragg reflection occurs in the visible wavelength range (de Vries, 1951). The chiral modifications of smectic phases which are characterized by a layer structure can show ferroelectric properties (Meyer *et al.*, 1975), and a field-induced rotation of the optical axis with very short response times (Clark and Lagerwall, 1980; Ostrovski *et al.*, 1980).

Consequently, the development of PDLC systems has also been extended to chiral liquid crystals. Crooker and Yang (1990) initiated their electro-optic application in PDLC films by developing a reflective colour display which utilizes the selective reflection of short-pitch cholesteric substances. More recently, the ferroelectric chiral smectic liquid crystals proved to be useful for PDLC applications (Kitzerow *et al.*, 1992b). A precondition for the latter development is a uniform orientation of the liquid crystal droplets which was known to occur if the PDLC film is stretched after its formation (West *et al.*, 1987), or if the sample is sheared (Wu *et al.*, 1989; Lackner *et al.*, 1990) or exposed to external fields during the phase separation (Margerum *et al.*, 1989). Nazarenko *et al.* (1993) succeeded even in promoting alignment by covering the inner surfaces of a porous film with a photoinduced anisotropic polymer layer before filling the pores with liquid crystal. For ferroelectric PDLC systems, the shearing (Kitzerow *et al.*, 1992b) and the stretching procedure (Zyryanov *et al.*, 1992b; Komitov *et al.*, 1994a) are suitable to obtain uniform alignment. The latter method has also led to the application of the flexoelectric effect in cholesteric PDLC systems (Komitov *et al.*, 1994b).

Anisotropic gels (Hikmet, 1990) or liquid-crystal-dispersed polymers (Stannarius *et al.*, 1991) represent a different system in which the alignment of a low-molecular-mass liquid crystal is promoted or stabilized by a polymer. These systems contain only a small volume fraction of

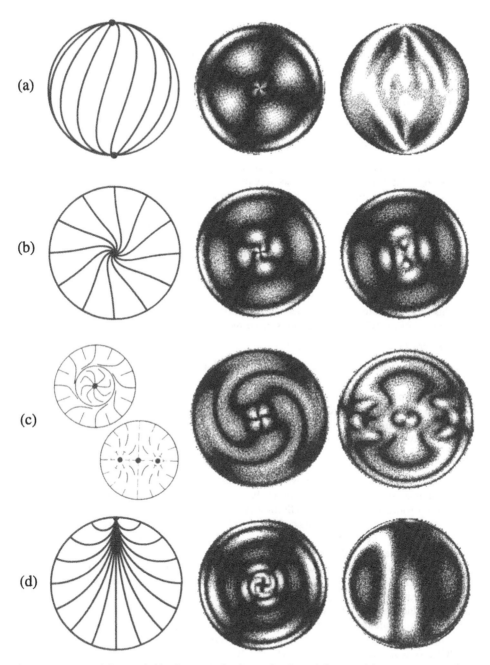

Figure 8.1 Twisted director fields of nematic droplets with spherical shape and their transmission when the sample is placed between crossed polarizers (with polarization planes at 30° and 120° with respect to a vertical line): (a) twisted bipolar (TB); (b) twisted radial (TR); (c) twisted hyperbolic (TH); (d) escaped twisted radial (ER) structure.

a highly cross-linked mesogenic polymer network dispersed in the liquid crystal. Again, these systems are particularly interesting in connection with chiral liquid crystals (Yang *et al.*, 1992; Hikmet and Zwerver, 1992, Pirš *et al.*, 1992). In this chapter, the properties of polymer-dispersed cholesteric (section 8.2.1) and ferroelectric (section 8.3.1) liquid crystals are reviewed. Only some examples are presented for the polymer-stabilized cholesteric (section 8.2.2) and ferroelectric (section 8.3.2) liquid crystals. Detailed presentations on the latter systems are given by other authors in this book.

8.2 Cholesteric Liquid Crystals

8.2.1 *Polymer-dispersed Cholesteric Liquid Crystals*

8.2.1.1 *Selective reflection*

Cholesteric liquid crystals with sufficiently short pitch show selective reflection of circularly polarized light (de Vries, 1951). The discovery of liquid crystals (Planer, 1861; Reinitzer, 1888) and one of their first applications, the use of encapsulated thermochromic cholesteric liquid crystals for thermometers (Jones, 1969, McDonnell, 1987), were due to this phenomenon. More recently, Crooker and Yang (1990) have invented a reflective display which makes use of polymer-dispersed cholesteric liquid crystals (PDCLCs). In the field-off state, the cholesteric droplets show a radial arrangement of the pitch axes and thus no selective reflection (Figure 8.2(b)). However, a uniform orientation of the pitch axis perpendicular to the surface of the PDCLC film can be induced by an alternating voltage (typically with a frequency of 1 kHz) if the dielectric anisotropy ε_a of the cholesteric liquid crystal is negative, $\varepsilon_a < 0$.

In this uniform state, a bright selective reflection appears. According to Bragg's law the wavelength λ of the selective reflection is given by

$$\lambda = np \cos \Theta' = np(1 - \sin^2\Theta/n^2)^{1/2} \tag{8.2}$$

where p is the pitch of the helical structure, $n = (\frac{1}{2}n_\parallel^2 + + \frac{1}{2}n_\perp^2)^{1/2}$ is an average value of the refractive indices, Θ' is the angle between the direction of light incidence and the helix axis within the sample, and Θ is the angle of light incidence with respect to the surface normal outside the sample (which is related to Θ' by Snell's law). In liquid crystal mixtures consisting of a nematic liquid crystal and a chiral additive, the reciprocal value of the pitch p is approximately proportional to the concentration x_{ch} of the chiral dopant,

$$p^{-1} = hx_{ch} \tag{8.3}$$

with h being the helical twisting power. Thus, the colour of the selective reflection can be controlled by adjusting the composition of the liquid crystal mixture (Kitzerow *et al.*, 1992a). The spectral width of the selective reflection peaks is quite small, typically about 40–50 nm, so that the chromaticity coordinates (Figure 8.2(c)) are close to the pure spectral colours (Kitzerow and Crooker, 1991; Kitzerow *et al.*, 1992a). Since PDCLC films are transparent apart from the selective reflection band described by eqn (8.2), these systems can be used to realize multicolour reflective displays by stacking several PDCLC films with different colours (Kato *et al.*, 1993a, 1994a). When a cholesteric liquid crystal is illuminated with unpolarized light, the maximum reflectivity is 50% since only the circularly polarized light exhibiting the same handedness as the cholesteric helix is reflected. However, stacking two PDCLC films with different handedness of the cholesteric structures, e.g. containing the two opposite enantiomers of the chiral dopant, can be used to enhance the reflectivty (Kato *et al.*, 1994b).

PDCLC systems with negative dielectric anisotropy have been prepared by thermally induced phase separation using the polymer poly-(vinylbutyral), PVB,

$$\left(-CH_2 -CH \overset{\displaystyle CH_2}{\underset{\displaystyle O}{\diagup}} \overset{}{\underset{\displaystyle CH}{\diagdown}} CH- \right)_p \tag{I}$$

$$\underset{C_3H_7}{\overset{\displaystyle |}{CH}}$$

and by photopolymerization-induced phase separation using the thiol-ene photopolymer NOA-65 (Norland). The latter polymer precursor consists of trimethylolpropane diallyl ether,

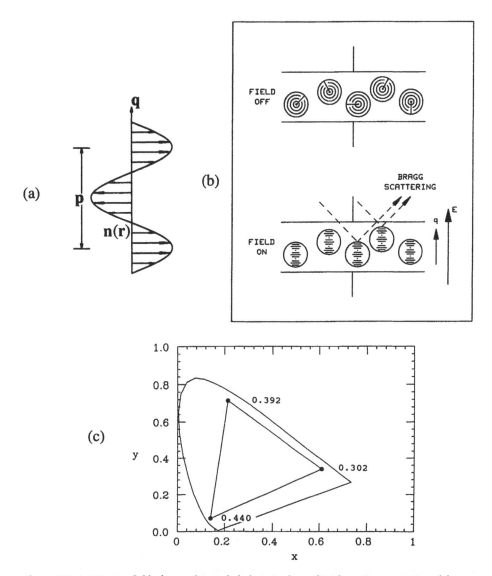

Figure 8.2 (a) Director field of an undistorted cholesteric phase. (b) Schematic presentation of the reflective PDCLC display invented by Crooker and Yang (1990). (c) Chromaticity diagram for three PDCLC samples with different concentrations of the chiral component. Composition: x_{ch} 50% CE2, $(1 - x_{ch})$ 50% ZLI-2806, 50% PVB. The respective values of x_{ch} are given in the diagram. From Kitzerow *et al.* (1992a).

trimethylolpropane tris thiol, isophorone diisocyanate ester, and benzophenone photoinitiator (Smith, 1991). The liquid crystal is usually composed of a wide-temperature-range nematic mixture with $\varepsilon_a < 0$, e.g. ZLI 2806 or ZLI 4788-000 (Merck, Darmstadt, Germany), or EN 18 (Chisso Corporation, Tokyo, Japan), and a chiral additive with high helical twisting power, such as CE2 (Merck Ltd, Poole, UK)

$$H_3C-CH_2-\underset{\underset{CH_3}{|}}{CH}-CH_2-\bigcirc\!\!\!-\bigcirc\!\!\!-COO-\bigcirc\!\!\!-CH_2-\underset{\underset{CH_3}{|}}{CH}-CH_2-CH_3 \quad \text{(II)}$$

or S 811 (Merck, Darmstadt, Germany)

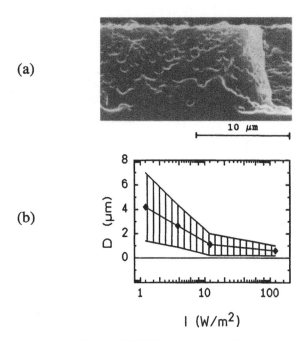

Figure 8.3 (a) Surface of a freeze-fractured PDCLC sample, observed in a scanning electron microscope (9.15% S811, 27.98% ZLI-4788-000, 62.87% NOA-65; $I_{UV} = 3.8$ mW/cm^2; $t_{cure} = 90$ s). (b) Drop diameter against UV intensity I_{UV}, for $I_{UV}t_{cure} = $ constant $= 360$ mJ/cm^2.

$$H_{13}C_6 \longrightarrow O \longrightarrow COO \longrightarrow COO \overset{CH_3}{\underset{|}{-CH}} -C_6H_{13} \qquad \text{(III)}$$

The weight fraction of the liquid crystal in the PDLC film is typically 30–50%. The photopolymerization technique has the advantage that the size and shape of the droplets is temperature independent after completion of the polymerization. Thus, PDCLC films working at room temperature could be easily realized (Kitzerow and Crooker, 1993b). Moreover, faster switching times have been found than in the PDCLC films prepared by thermally induced phase separation. This is probably due to a more complete phase separation in the case of polymerization-induced phase separation.

For the photopolymerization process, the size of the droplets is determined by the UV intensity (Figure 8.3). A slow phase separation (small UV intensity) leads to the growth of large drops, while fast phase separation (high UV intensity) causes the formation of submicrometre-sized droplets. The electro-optical properties of PDCLC films depend essentially on the diameter D of the droplets. The intensity I_{drop} scattered from a single droplet is approximately proportional to D^6, since the field strength of the radiation scattered from a spherical Bragg scattering particle is proportional to its volume (Landau *et al.*, 1984). For a given volume ratio V_{LC}/V of the liquid crystal within the PDLC system, the number density of droplets is $N/V = (V_{LC}/V)/(\pi D^3/6)$, and the total intensity $I = NI_{drop} \propto D^3$ (Kitzerow and Crooker, 1993b). Consequently, small UV intensities used for the curing process result in large droplets and thus high brightness of the selective reflection of the PDCLC film (Figure 8.4).

Figure 8.4(a) shows that large droplets exhibit a lower threshold voltage than small droplets. This behaviour can be explained by the competing effects of dielectric polarization and elastic

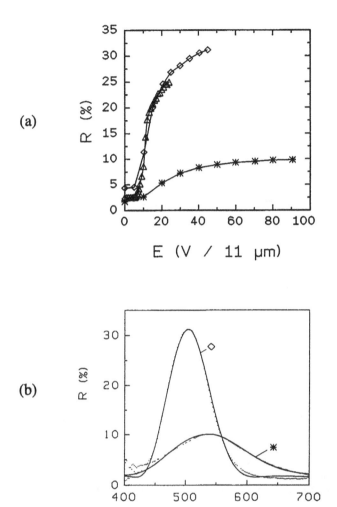

Figure 8.4 (a) Reflectivity against field strength for reflective PDCLC displays, cured with different UV intensities I_{UV} and exposure times t_{cure}: *, $I_{UV}t_{cure}$ = 12 mW/cm^2 30 s; ◇, $I_{UV}t_{cure}$ = 1.2 mW/cm^2 5 min; △, $I_{UV}t_{cure}$ = 0.12 mW/cm^2 50 min. Composition: 9.15% S811, 29.98% ZLI-4788-000, 62.87% NOA-65. (b) Reflection spectra for two of the same samples in the field-on state. From Kitzerow and Crooker (1993b).

deformation on the free energy, and will be discussed in connection with observations on large-pitch cholesteric droplets in the next section. The finite size of selectively reflecting cholesteric droplets affects also the half width of the selective reflection band since the field strength of the radiation reflected from a Bragg scattering particle is proportional to the Fourier transform of its periodic structure (Landau *et al.*, 1984). The full width at half maximum intensity (FWHM) of the selective reflection bands at normal light incidence is related to the droplet diameter D by

$$\Delta\lambda_{FWHM} = 3.62\lambda_0^2/\pi n D \qquad (8.4)$$

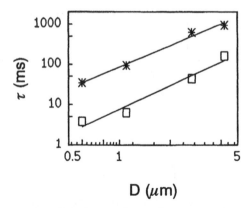

Figure 8.5   Response times versus drop diameter for switching $E = 52$ V/11 μm ($\square$) on and (*) off, respectively. Same sample composition as in Figures 8.3 and 8.4.

where λ_0 is the Bragg wavelength given by eqn (8.2) for $\Theta = 0$. The width of the angular dependence of the scattered light for normal incidence of monochromatic light with $\lambda = \lambda_0$ is given by

$$\Delta\theta'_{\text{FWHM}} = 4 \sin^{-1}(1.81\lambda_0/2\pi nD) \tag{8.5}$$

If the Bragg scattering regions are not spherical, the effective values of the diameter D in eqns (8.4) and (8.5) may deviate slightly from each other. The response times τ increase linearly with increasing drop size (Figure 8.5) and decrease with increasing field strength. According to the respective degree of phase separation, the switching time may be in the millisecond range in the system prepared by photopolymerization, but was found to be several seconds in a system prepared by thermally induced phase separation (Kitzerow *et al.*, 1992d).

In conclusion, PDCLC films containing a cholesteric liquid crystal with very small pitch and negative dielectric anisotropy are suitable for reflective colour displays with a bright selective reflection. In order to optimize these systems, a compromise has to be found, since high brightness and low switching voltages can be achieved by large droplet sizes, while low switching times are favoured by small droplets.

8.2.1.2   *Large droplets as model systems*

Large droplets with large cholesteric pitch can be studied as model systems in order to investigate the structural changes under the influence of an electric field (Yang and Crooker, 1991; Kitzerow and Crooker, 1992, 1993a). If the pitch is larger than 1 μm, fingerprint lines can be seen in the polarizing microscope. The concentric arrangement of these lines in systems with planar anchoring of the director (Figure 8.6) indicates a radial arrangement of the pitch axes $\mathbf{q}(\mathbf{r})$ (Frank and Pryce, 1958). However, the cholesteric structure is not solely characterized by the direction of the pitch axis $\mathbf{q}$ and the size of the pitch $p = 2\pi/|\mathbf{q}|$, but also by the phase Ω of the helicoidal structure. For the radial $\mathbf{q}$ field, different relations of the phase Ω may lead to the occurrence of either a radial disclination line (Frank and Pryce, 1958) or a diametrical disclination line (Bouligand and Livolant, 1984). According to Bezić and Žumer (1992), the director $\mathbf{n}$ at position $\mathbf{r}$ with respect to the drop centre may be described in a local coordinate system with the unit vectors $\mathbf{e}_r \parallel \mathbf{r}$, $\mathbf{e}_\phi \parallel (\mathbf{z} \times \mathbf{r})$ and $\mathbf{e}_\theta \parallel (\mathbf{e}_\phi \times \mathbf{e}_r)$. Then, $\mathbf{n}(\mathbf{r})$ is given by

$$\mathbf{n} = \mathbf{e}_\phi \cos \Omega(\mathbf{r}) + \mathbf{e}_\theta \sin \Omega(\mathbf{r}) \tag{8.6}$$

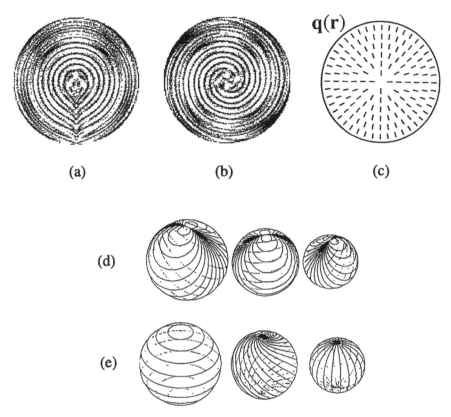

(a)　　　　　　　　(b)　　　　　　　　(c)

(d)

(e)

Figure 8.6 (a), (b) Transmission of cholesteric droplets between crossed polarizers for two different orientations (from Kitzerow and Crooker, 1993a). (c) Field of the pitch axes $\mathbf{q}(\mathbf{r})$ corresponding to the droplets in (a) and (b). (d) Representation of the radial spherical structure (RSS). (e) Representation of the director field for the diametrical spherical structure (DSS). Figures (d) and (e) are from Bezić and Žumer (1992).

where the position-dependent phase Ω is given by

$$\Omega = (s - 1)\phi + \Omega_0 + \mathbf{q} \cdot \mathbf{r} \tag{8.7}$$

The corresponding structures can be visualized by spherical shells with a tangential director field (Figures 8.6(d) and (e)). The closed surface of each shell requires the occurrence of a defect with strength $s = 2$, or the occurrence of several defects with $\Sigma\, s_i = 2$ (Bezić and Žumer, 1992). The helical structure of the droplet is described by a continuous rotation of the director field with varying radius $|r|$ of the shell. Thus, the occurrence of just one defect point on the quasi-planar director field on one of these spherical surfaces corresponds to a radial disclination line (Figure 8.6(d)) as observed by Robinson *et al.* (1958). The occurrence of two defects, e.g. with $s_1 = s_2 = 1$, may lead to a diametrical defect line in the drop (Figure 8.6(e)) as observed by Bouligand and Livolant (1984).

Investigations of the field effects on droplets with a radial $\mathbf{q}$ field and a radial disclination line show the appearance of a region with uniform orientation, $\mathbf{q} \parallel \mathbf{E}$, in the droplet centre for $\varepsilon_a < 0$. This region is surrounded by a disclination ring perpendicular to the electric field. Its radius r increases continuously with increasing field strength (Figure 8.7(a)) until the entire drop is uniformly oriented ($r = R$). The field-induced reorientation of $\mathbf{q}(\mathbf{r})$ for $\varepsilon_a < 0$ does *not*

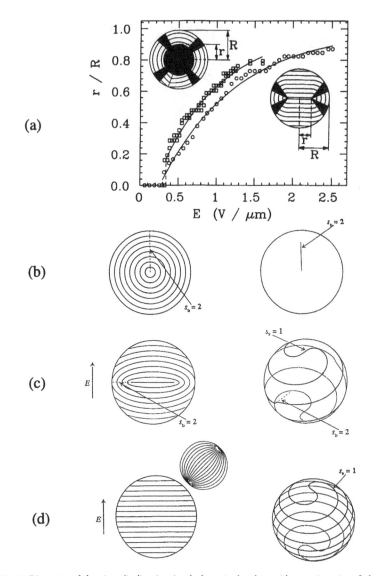

Figure 8.7 (a) Diameter of the ring disclination in cholesteric droplets with $\varepsilon_a < 0$ against field strength: $\square$, observation paralel and $\bigcirc$ perpendicular to the applied electric field. Composition: $\approx 3\%$ CE2 (compound II), 47% ZLI-2806 and 50% PVB (compound (I)). From Kitzerow and Crooker (1993a). (b)–(d) Model for the radial ellipsoidal structure (RES) for (b) zero, (c) intermediate, and (d) very high field strength (bottom). The respective picture to the left shows a cross-section of the planes of director orientation, the figure to the right represents the corresponding disclination lines. From Bajc et al. (1995).

affect the size of the pitch p. Bajc et al. (1995) have proposed radial elliptical structures (RESs) (Figures 8.7(b)–(d)) where the surfaces tangential to the director field become elliptical for $E \neq 0$, instead of spherical ($E = 0$). According to their $\text{RES}_\perp$ model which fits the experimental observations, the disclination ring perpendicular to the field can be classified as an ($s = 1/2$) λ line, i.e. as a non-singular disclination which does not considerably contribute to the elastic energy. (For the classification scheme of defects in cholesterics, see e.g. de Gennes (1974)). However, a uniform orientation of $\mathbf{q}$ leads also to the occurrence of a pair of χ ($s = 1$) defect lines with spiral shape on the surface of the droplet (Quigley and Benton, 1977; Bajc et al., 1995). The curvature of the director field close to a χ line in the bulk or at the surface results in a contribution to the elastic free energy of $f_{d,\,bulk} = \pi K l s^2 \ln(R/r_c)$ or $f_{d,\,surface} = \frac{1}{2}\pi K l s^2$ $\ln(R/r_c)$, respectively. Here, s is the disclination strength, l the length of the line, and $r_c \approx 10^{-8}$ m, the core radius of the disclination. In the model by Bajc et al. (1995), the equilibrium structure is mainly the result of the aligning effect of the electric field $\mathbf{E}$ on $\mathbf{q}$ and the destabilization of this uniform alignment due to the increase of the elastic free energy with increasing length l of the surface disclination. The first of these competing effects is described by a volume integral in the free energy (eqn (8.1)), and thus is proportional to R^3. The number of turns of a χ ($s = 1$) spiral scales with R/p, and thus its absolute length is proportional to R^2/p. Consequently, the model predicts that the characteristic field strength E_c which is necessary to overcome the elastic stability of the spherulite structure decreases with increasing drop radius R. The experimental results shown in Figure 8.4(a) and the observations on large drops (Kitzerow and Crooker, 1992) are in agreement with this expectation. For constant drop radius R, the critical field E_c is expected to decrease with increasing helical pitch since $l \propto p^{-1}$, which is again in agreement with earlier experimental observations (Kitzerow and Crooker, 1992). From a balance equation for a viscous term and the deviation of the free energy with respect to the field-dependent parameter r, it follows that the switching time increases with increasing drop size and decreases with increasing voltage.

Besides the special case of planar anchoring and $\varepsilon_a < 0$, different surface interactions, elastic coefficients and dielectric properties of liquid crystals provide a large variety of director configurations (Figure 8.8). The untwisted director configurations shown for infinite intrinsic pitch $(p_0^{-1} = 0)$ and $\varepsilon_a > 0$ represent the well-investigated field-induced reorientation of a bipolar (B) droplet configuration (Wu et al., 1989) and the transition from a radial (R) to an axial (A) configuration (Erdmann et al., 1990), respectively. These two mechanisms are responsible for the electro-optic effect in nematic PDLC films. In addition to these structures, deformed radial (R^+ and R^-) structures have been observed in electric fields (Kitzerow et al., 1992e). The concentric (C) configuration occurs for a large ratio of K_{11}/K_{33} (Drzaic, 1988). It may be stabilized by electric fields for $\varepsilon_a < 0$. Besides these non-twisted structures, several twisted configurations with $R < p < \infty$ may occur, even for nematic liquid crystals with infinite intrinsic pitch p_0. The twisted-bipolar (TB) structure is transformed to an escaped-bipolar (EB) structure by electric fields (Xu et al., 1994) and the twisted-radial structure (TR) is deformed into a displaced-radial (DR) and finally to an escaped-radial (ER) structure by the field (Xu et al., 1992). The cholesteric structures with $p_0 \ll R$ are for simplicity just represented by their respective $\mathbf{q}$ fields in Figure 8.8. However, it is important to note that each of the $\mathbf{q}$ fields may be realized by several structures for which the phase Ω of the helix and the occurrence of defects are different.

Since the pitch axis $\mathbf{q}$ is perpendicular to the director, there is a formal (but otherwise limited) analogy between the $\mathbf{q}$ fields for parallel anchoring of the director, and the $\mathbf{n}$ fields for perpendicular anchoring, and vice versa. Consequently, a bipolar (B*) arrangement of the pitch axes in cholesteric drops can occur for perpendicular anchoring of the director (Kitzerow and Crooker, 1993a). The analogy may also be extended to the behaviour in electric fields. If the dielectric polarization of a nematic phase parallel and perpendicular to the director is described by the dielectric constants $\varepsilon_\parallel$ and $\varepsilon_\perp$, the respective cholesteric phase exhibits the effective dielectric constants $\varepsilon_\perp$ parallel and $\frac{1}{2}(\varepsilon_\parallel + \varepsilon_\perp)$ perpendicular to the pitch axis $\mathbf{q}$. Thus, an electric field has an aligning effect on the $\mathbf{q}$ field for $\varepsilon_a < 0$ (analogous to the $\mathbf{n}$ field of a nematic phase for $\varepsilon_a > 0$), and a disaligning effect on $\mathbf{q}$ for $\varepsilon_a > 0$ (analogous to $\mathbf{n}$ for a nematic phase with $\varepsilon_a < 0$). The cholesteric droplet configuration described by Frank and Pryce (1958) is simply represented by its radial $\mathbf{q}$ field (R*) in Figure 8.8. For $\varepsilon_a < 0$, the influence of an electric field causes this vector field to transform into an axial (A*) and finally completely parallel (P*) field of the pitch axes. However, for $\varepsilon_a > 0$ the R* structure is deformed by the field to a structure where the pitch axes are oriented in a planar radial (PR*) fashion (Kitzerow and Crooker, 1993a). The same structure has been observed in magnetie fields (Candau et al., 1973) for

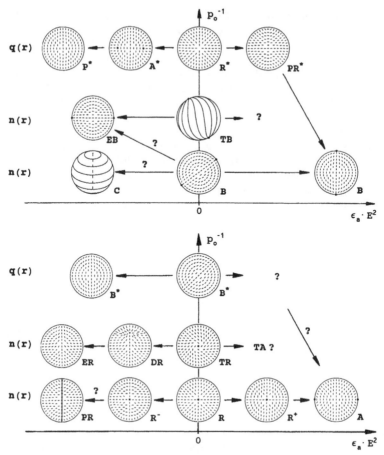

Figure 8.8 Nematic director fields **n(r)** and cholesteric fields **q(r)** in spherical droplets for planar anchoring (top) and perpendicular anchoring (below). B, bipolar; C, concentric; TB, twisted bipolar; EB, escaped bipolar. $|p^{-1}| \gg 0$: R*, chiral radial (Frank and Pryce, 1958); A*, chiral axial; P*, chiral parallel; PR*, chiral planar radial; R, radial; PR, planar radial; A, axial; TR, twisted radial; DR, displaced radial; ER, escaped radial; TA, twisted axial. $|p^{-1}| \gg 0$: B*, chiral bipolar.

positive magnetic anisotropy. For $\varepsilon_a > 0$, the field affects not only the direction of the pitch axes but leads also to helical unwinding (de Gennes, 1968; Meyer, 1968, 1969). Figure 8.8 demonstrates that the recent interest in twisted structures with $\varepsilon_a < 0$ has raised some new questions concerning $\varepsilon_a > 0$. Since twisted-radial (TR) and twisted-bipolar (TB) structures can occur in non-chiral nematic droplets, their behaviour in electric fields for $\varepsilon_a > 0$ is worthwhile studying with respect to the electro-optic characteristics of nematic PDLC displays and PDCLC films with medium chirality as described in the next section. For a more detailed overview on nematic and cholesteric structures in cylinders and spheres, see Crooker and Xu (1996).

8.2.1.3 PDCLC films with medium chirality

The optical properties of cholesteric liquid crystals depend considerably on the size of the helical pitch p. For very short pitch, selective reflection occurs at the Bragg wavelength $\lambda_0 = np$.

However, if the Mauguin condition $\lambda \ll \Delta np$ is fulfilled, i.e. for very large pitch, twisted structures are known to rotate the plane of polarization continuously along the director field (wave guiding). Cholesteric liquid crystals with *medium chirality* show a high optical activity which is (in contrast to the wave-guiding effect) independent of the orientation of the liquid crystal (Chanishvili *et al.*, 1991, 1994). The preconditions for this optical behaviour are

$$\lambda \gtrsim \Delta np/2 \quad \text{and} \quad p > d \tag{8.8}$$

where d is the sample thickness. Chanishvili *et al.* have pointed out that chiral liquid crystals with positive dielectric anisotropy can be used for electro-optic switching between this optically active state and a uniform alignment which shows no optical activity. The same effect can be achieved in PDCLC samples with medium chirality (Behrens *et al.*, 1994). In this case, the quantity d in the condition (8.8) means the drop size. Using a cholesteric liquid crystal with $p \approx 7$ μm and drops with diameters of a few micrometres in NOA-65, switching between a bright field-off state and a dark field-on state is possible if the PDCLC film is placed between crossed polarizers. Rotation of the PDCLC sample between the crossed polarizers does not affect the transmittance. The PDCLC samples with medium chirality operate in a reverse mode compared with the nematic scattering PDLC device by Doane *et al.* (1986) where the powered state is transparent. PDCLC films with medium chirality show an improved angular dependence of the transparent state, but due to the polarizers their brightness is reduced compared with scattering PDLC films.

8.2.1.4 Flexoelectric effect

For liquid crystal molecules of non-cylindrical shape, a splay or bend distortion of the director field $\mathbf{n}(\mathbf{r})$ may cause a flexoelectric polarization (de Gennes, 1974)

$$\mathbf{P}_f = e_s \mathbf{n}(\text{div } \mathbf{n}) - e_b \mathbf{n} \times (\text{rot } \mathbf{n}) \tag{8.9}$$

where e_s and e_b are the flexoelectric coefficients for splay and bend deformation, respectively. In this case, an additional term $-\int \mathbf{P}_f \cdot \mathbf{E} \, dr^3$ has to be considered in the free energy (eqn (8.1)). The coupling between the flexoelectric polarization and an external electric field E can lead to a field-induced tilt of the director $\mathbf{n}$ with respect to the helix axis $\mathbf{q}$ in cholesteric phases (Patel and Meyer, 1987). The director field in Cartesian coordinates with $\mathbf{q} \parallel \mathbf{z}$ and $\mathbf{E} \parallel \mathbf{x}$ (Figure 8.9(a)) is given by $\mathbf{n} = [\cos(qz), \cos \psi(E) \sin(qz), \sin \psi(E) \sin(qz)]$. For $\psi \neq 0$, this director field exhibits a periodic splay–bend pattern for which the terms in eqn (8.9) do not vanish. The tilt angle ψ depends approximately linearly on the electric field strength E. Using the approximations $K_{11} = K_{33}$, $\varepsilon_a = 0$ and $e_s = e_b =: e_f$ (Patel and Meyer, 1987), the free energy is minimized for

$$\tan \psi = e_f E / K_{11} q_0 \tag{8.10}$$

Cholesteric liquid crystals with sufficiently short pitch are optically uniaxial. The rotation of the optical axis in the plane perpendicular to the field can be used for electro-optic effects. If a uniaxial (liquid) crystal is placed between crossed polarizers and oriented with its optical axis perpendicular to the direction of light propagation, the intensity of a transmitted light beam with initial intensity I_0 is given by

$$I(\phi) = \tfrac{1}{2} I_0 \sin^2 2\phi \sin^2 \tfrac{1}{2}\delta \tag{8.11}$$

where ϕ is the angle between the azimuthal orientation of the optical axis of the liquid crystal and the polarization plane of the incident light. If the initial alignment of the optical axis is ϕ_0, electro-optic modulation between $\phi_1 = \phi_0 - \psi$ and $\phi_2 = \phi_0 + \psi$ causes a change of the transmitted intensity given by

$$\Delta I(\phi_0, \psi) = \tfrac{1}{2} I_0 \sin^2 \tfrac{1}{2}\delta \sin 4\phi_0 \sin 4\psi \tag{8.12}$$

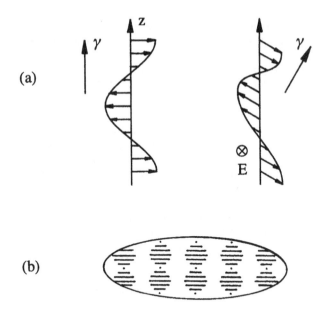

Figure 8.9 (a) Change of the cholesteric director field due to flexoelectric polarization. In the planes perpendicular to the z axis, the director field is uniform. (b) Orientation of the cholesteric structure in the elliptical droplets of stretched PDCLC films, as observed by Komitov *et al.* (1994b).

This change of the intensity is maximized for an initial alignment of $\phi_0 = 22.5°$. In this case, eqn (8.11) becomes

$$I(22.5° + \psi) = \tfrac{1}{4}I_0 \sin^2\tfrac{1}{2}\delta[1 + \sin 4\psi] \tag{8.13}$$

i.e. for small field-induced tilt $\psi(E) \approx \tfrac{1}{4} \sin 4\psi$ the intensity depends approximately linearly on the applied voltage.

The retardation δ in eqns (8.11)–(8.13) is given by $\delta = 2\pi(d/\lambda_e - d/\lambda_o) = 2\pi(n_e - n_o)d/\lambda$, where λ is the wavelength outside of the sample while the indices 'o' and 'e' refer to the ordinary and the extraordinary beams within the liquid crystal, respectively. For short-pitch cholesterics, the ordinary refractive index is given by $n_o \approx \tfrac{1}{2}(n_\parallel^2 + n_\perp^2)^{1/2} \approx \tfrac{1}{2}(n_\parallel + n_\perp)$ and the extraordinary refractive index by $n_e \approx n_\perp$, where $n_\parallel$ and $n_\perp$ are the refractive indices in the nematic phase of the equivalent racemic mixture for polarization of the light parallel and perpendicular to the director, respectively.

Komitov *et al.* (1994a,b) obtained flexoelectric films using the liquid crystal 6415L (Hoffmann-La Roche, Basle, Switzerland) in poly-(vinylbutyral). Their films are formed by phase separation, induced by solvent evaporation. In order to get a uniform alignment of the helix axis in the film plane, Komitov *et al.* (1994) dope the system with lecithin (thereby promoting a perpendicular anchoring of the liquid crystal molecules at the droplet surface) and stretch the PDCLC film, thereby creating an elliptical shape of the droplets. The pitch axis was found to be oriented along the minor axis of the elliptical cavities (Figure 8.9).

The switching times for the flexoelectric effect are given (Patel and Lee, 1989) by

$$\tau = \gamma/Kq_0^2 \tag{8.14}$$

where γ is an effective rotational viscosity. The effect can be much faster than the Freedericks transition if the intrinsic pitch $p_0 = 2\pi/q_0$ is small. Consequently, Komitov *et al.* (1994b) found the switching times as low as $\tau \approx 100$ μs in the PDFLC film, which is the same order of magnitude as observed for a pure liquid crystal (Patel and Lee, 1989).

8.2.2 *Polymer-stabilized Cholesteric Liquid Crystal*

Electro-optic switching between a uniform cholesteric state and a diffuse scattering state as proposed by Crooker and Yang (1990) can also be achieved in gel systems which contain a high amount of liquid crystal and only a few per cent of a highly cross-linked polymer network (Yang and Doane, 1992). The uniform state (Grandjean texture) appears coloured if the wavelength of the selective reflection is in the visible range, or transparent if λ_0 is out of the visible range (Yang *et al.*, 1992). Yang *et al.*, used the diacrylate 4,4'-bis-acryloylbiphenyl which is mesogenic, but does not show a liquid crystalline phase. However, liquid crystalline diacrylates of the type

$$\diagdown\!\!\!-\text{COO}-(\text{CH}_2)_{\overline{n}}\text{O}-\bigcirc-\text{COO}-\bigcirc\overset{R}{-}\text{OOC}-\bigcirc-\text{O}-(\text{CH}_2)_{\overline{n}}\text{OOC}-\diagup \quad (IV)$$

as used by Broer *et al.* (1989) and Hikmet and Zwerver (1992) can provide bistable switching, as well (Behrens and Kitzerow, 1994). Figure 8.10(a) shows the reflectivity against voltage for a gel consisting of 39.4% (by weight) of the chiral compound CB15 (Merck Ltd., Poole, UK), 59.1% of the wide-temperature-range nematic mixture RO-TN 403 (Hoffmann-La Roche, Basle, Switzerland) and 1.5% of the compound (IV) with $n = 6$ and $R = -CH_3$. The Grandjean texture was present during the photopolymerization process so that the selectively reflecting state is stabilized in the field-off state. In contrast to the PDCLC display, the liquid crystal used here shows positive dielectric anisotropy $\varepsilon_a > 0$. Thus, the non-uniform focal-conic texture is induced by the electric field, and above a critical field strength

$$E_c = \pi^2 p^{-1} K_{22}^{1/2} (\varepsilon_0 \varepsilon_a)^{-1/2} \tag{8.15}$$

helical unwinding (de Gennes, 1968; Meyer, 1968, 1969) leads to a homeotropically oriented nematic phase. The Grandjean texture reappears when this high voltage is switched off (Figure 8.10(b)). However, the focal-conic state induced by medium field strength persists when this medium field is switched off. Thus, bistable switching between two stable states can be obtained by pulses with appropriate voltage.

Using the type of liquid crystalline diacrylates with structure (IV), the electro-optic performance was found to depend considerably on the orientation which was present during the polymerization process. For weight fractions x_p higher than about 5% of the polymer this respective texture is solely stable at $E = 0$. For much higher concentration x_p, the structure is completely fixed and not at all affected by external fields. In the range of small concentrations ($x_p < 3.5\%$), the response times for the switching from the Grandjean to the focal-conic texture (at 4 V/μm) and to the nematic state (at 7 V/μm) are approximately 10 ms and 2 ms, respectively (Figure 8.11). These response times depend only slightly on the concentration of the polymer. However, the relaxation to the Grandjean texture is much slower and exhibits a minimum of $\tau \approx 100$ ms at $c_p = 1.5\%$. Like the systems studied by Yang *et al.* (see their chapter in this volume), the bistable system containing liquid crystalline diacrylates displays high brightness of the selective reflection and chromaticity coordinates suitable for colour mixing.

8.3 Ferroelectric Liquid Crystals

8.3.1 *Polymer-dispersed Ferroelectric Liquid Crystals*

The symmetry of a non-chiral tilted smectic SmC phase is described by the point group C_{2h} so that no polar properties can occur on a macroscopic scale. However, for chiral molecules

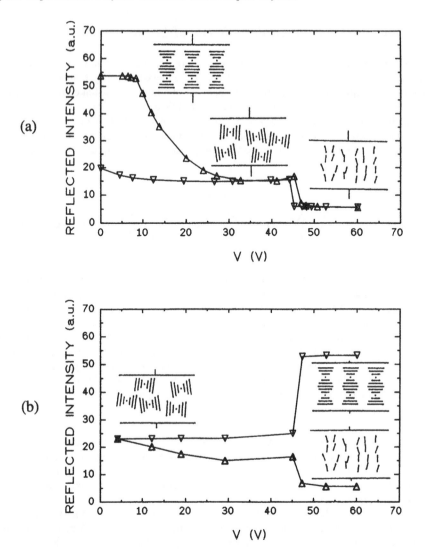

Figure 8.10 (a) Intensity of the reflected light against voltage for a cholesteric gel display, consisting of 39.19% CB 15, 59.27% RO–TN 403 and 1.54% compound (IV) with $n = 6$ and R = –CH$_3$: △, slow increase of the voltage; ▽, slow decrease of the voltage. (b) Intensity of the reflected light for same sample (△) in the field-on state and (▽) in the field-off state which is obtained after sudden removal of the respective voltage. From Behrens and Kitzerow (1994).

the local symmetry is reduced to the point group C$_2$. Since only one rotational axis is left in this case and additional mirror planes are absent in chiral tilted smectic phases, a local spontaneous polarization P$_s$ may occur along the twofold axis which is perpendicular to both the layer normal $\mathbf{q}_l$ and the director $\mathbf{n}$ (Meyer *et al.*, 1975), i.e.

$$\mathbf{P}_s = P_0 |\mathbf{q}_l|^{-1} \mathbf{q}_l \times \mathbf{n} \tag{8.16}$$

In spite of this local polarization, the macroscopic average value of the spontaneous polarization in the SmC* phase is zero due to its helical structure (Figure 8.12(a)). Nevertheless, the coupling of P$_s$ to an external DC field can lead to helical unwinding (Meyer, 1977), or the suppression of the helical structure by strong surface anchoring can be used for bistable switching between

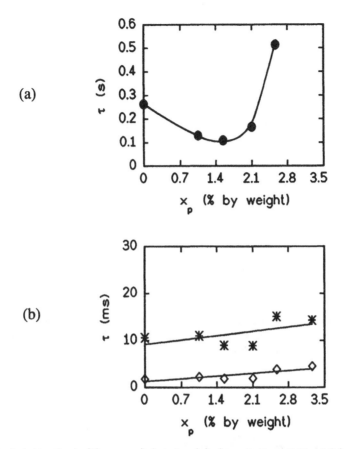

(a)

(b)

Figure 8.11 Switching times of the same cholesteric gel display as in Figure 8.10. (a) Relaxation time from the nematic state to the Grandjean texture when 70 V/10 μm are switched off. (b) Response times for the transitions from the Grandjean texture to ($\Diamond$) the focal conic texture ($E = 40$ V/10 μm) and (*) the homeotropic nematic state ($E = 70$ V/10 μm). From Behrens and Kitzerow (1994).

two unwound states with different polarity (Clark and Lagerwall, 1980). To achieve electro-optic contrast utilizing these and similar effects, the liquid crystal is placed between crossed polarizers and behaves essentially like an optical retarder (wave plate). The azimuthal orientation of its optical axis in the plane perpendicular to the field is controlled by the field strength. This birefringence mode can be used in polymer-dispersed ferroelectric liquid crystals (PDFLCs) with uniform alignment of the liquid crystal droplets (Kitzerow *et al.*, 1992b). However, another elegant method to combine the fast switching times of ferroelectric liquid crystals with the unique features of PDLC systems is to use scattering effects in PDFLC (Zyryanov *et al.*, 1992b). A review of these different approaches is given in the following section.

8.3.1.1 *The liquid crystal as a wave plate*

Electro-optic effects based on the field-induced *helical unwinding* of the SmC* structure can operate in PDFLC systems at room temperature with short switching times. For this purpose liquid crystal mixtures have been employed, e.g. a mixture of 20% (by weight) of the SmC* compound

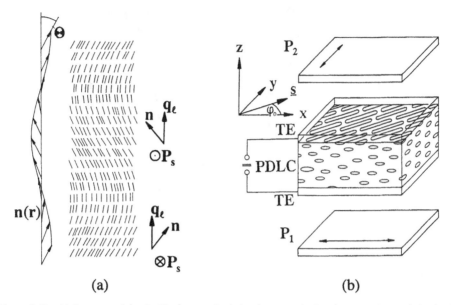

Figure 8.12 (a) Structure of the SmC* phase and relation between the local spontaneous polarization P_s and the tilt of the director **n** with respect to the layer normal $q_{\parallel}$. (b) Schematic representation of a PDFLC display. P_1, P_2, crossed polarizers; TE, transparent electrodes, s, shear direction. From Kitzerow et al. (1992b).

$$C_7H_{15}O-\!\!\!\bigcirc\!\!\!-\!\!\!\bigcirc\!\!\!-OOC\overset{*}{C}HCH(CH_3)_2 \qquad\qquad (V)$$
$$\underset{\displaystyle \quad\;\; Cl}{}$$

with 80% of the non-chiral wide-temperature-range SmC liquid crystal

$$C_8H_{17}-\!\!\!\bigcirc\!\!\!-\!\!\!\bigcirc\!\!\!-OC_8H_{17} \qquad\qquad (VI)$$

(Kitzerow et al., 1992b,c), or ready-for-use ferroelectric liquid crystal mixtures, such as 6430 from Hoffmann-La Roche (Komitov et al., 1994a), CS 1024 or CS 2004 from Chisso Corporation (Lee et al., 1994a,b). The films are produced by photopolymerization of commercial optical adhesives (NOA-65 and NOA-61 from Norland, North Brunswick, NJ, USA), or by phase separation due to solvent evaporation. Uniform alignment of the liquid crystal droplets may be achieved by shearing of the PDFLC film during the curing process (Kitzerow et al., 1992b), by stretching (Komitov et al., 1994a), or by a surface-induced alignment as in conventional displays (Lee et al., 1994a,b). The last method is only suitable for a very large droplet size which is in the range of the sample thickness (several micrometres). If a PDFLC film with uniform orientation of the layer normal is placed between crossed polarizers (Figure 8.12(b)), the helical unwinding can be used for a modulation of the transmitted light which is linear to the applied field for small field strength (Figure 8.13). In SmC* phases with short pitch, the optical axis is parallel to the layer normal in the field-off state, and parallel to the director in the unwound state. To describe the rotation of the optical axis in the film plane, the equations

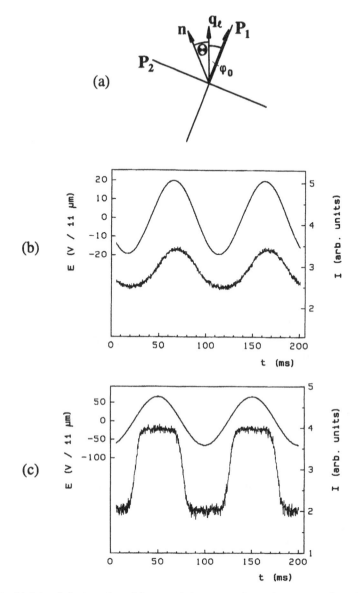

Figure 8.13 (a) Azimuthal orientation of the smectic layer normal $\mathbf{q}_l$ with respect to the polarizers $\mathbf{P}_1$, $\mathbf{P}_2$ for maximum contrast in the birefringence mode. (b) Linear dependence of the transmitted intensity on the applied field strength for a PDFLC sample showing helical unwinding. (c) Saturation for high field strength (from Kitzerow *et al.*, 1992c).

(8.11)–(8.13) can be applied as discussed in connection with the flexoelectric effect. However, above a threshold voltage E_u, the modulation of the intensity becomes saturated owing to complete unwinding (Figure 8.13(c)). For the pure liquid crystal, E_u is given by

$$E_u = \pi^2 K q_0^2 / 16 P_s \tag{8.17}$$

where K is an effective elastic coefficient which scales as $\sin^2 \Theta$ (Ostrovski *et al.*, 1980; Beresnev *et al.*, 1989). For PDFLC films containing the mixture of compounds (V) (20%) and (VI) (80%),

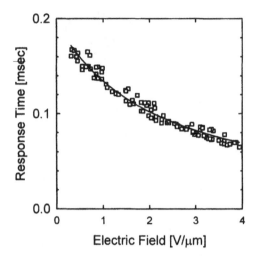

Figure 8.14   Response time τ_{on} against voltage in a surface-aligned PDFLC film (CS 2004, Chisso, in NOA 65, Norland). From Lee *et al.* (1994a).

the threshold field strength is $E_u \approx 1\text{--}2$ V/μm, which is about five times larger than E_u for the pure liquid crystal. This deviation is due to the different values of the external field E and the field strength E' effective on the droplets. For spherical drops, the relation between E' and E is given by

$$E' \approx 3E/(\varepsilon_{1c}/\varepsilon_p + 2) \approx 3E/(\rho_p/\rho_{1c} + 2) \tag{8.18}$$

(Wu *et al.*, 1989), where ρ_p and ρ_{1c} are the resistivities of the polymer and the liquid crystal.

The switching times for the helical unwinding in pure liquid crystals are described by $\tau = \gamma_\phi/Kq^2$ for $E \ll E_u$, $\tau \propto (E - E_u)^{-1}$ for $E > 0$, and diverge at $E = E_u$ (Ostrovski *et al.*, 1980). Qualitatively the same behaviour is found for the PDFLC samples described above (Molsen, 1994). The response time τ is a few milliseconds for $E \ll E_u$, diverges at $E \approx E_u$, and decreases with increasing field strength for $E \gg E_u$, reaching values as small as 30 μs (at $E = 15$ V/μm). However, the dependence of τ^{-1} on E is not described by a linear relation, which indicates that the mechanism in small separated droplets is more complicated than the behaviour of the liquid crystal in the bulk. In contrast to these samples, Lee *et al.* (1994a,b) have studied very large elliptical drops which are attached to the surfaces of an aligning substrate. For this system, the relation

$$\tau_{on}^{-1} = PE'/\gamma_\phi + \tau_{off}^{-1} \tag{8.19}$$

was confirmed (Figure 8.14) which is expected from a balance equation in which the torque $PE' \sin \phi$ due to effect of the electric field E', a torque due to the elastic deformation and a viscous term $\gamma_\phi \, d\phi/dt$ are considered.

Linear electro-optic response can be achieved even in non-tilted chiral smectic phases, e.g. the SmA* phase, owing to the *electroclinic effect* (Blinc, 1976; Garoff and Meyer, 1977). The coupling between tilt angle and polarization leads to a field-induced tilt. The tilt angle Θ_T, and thus the azimuthal orientation of the optical axis, depend linearly on the field strength, $\Theta_T = e_c E$. In sheared samples of the pure compound (V) dispersed in NOA-65 (Norland), this effect proved to be useful for PDFLC applications (Kitzerow *et al.*, 1992b). In the polymer-dispersed sample, the electroclinic coefficient e_c is between 0.5 and 2.1×10^{-3} rad μm/V, while the value for the pure compound is between 7.9 and 20.9×10^{-3} rad μm/V (Bahr and Heppke, 1987). Komitov *et al.* (1994a) have studied the electroclinic effect in PDFLC films containing the liquid crystal W317

$$H_{21}C_{10}-O-\bigcirc-COO-\bigcirc-\bigcirc-O-\underset{\underset{NO_2}{|}}{\overset{\overset{CH_3}{|}}{CH}}-C_6H_{13} \qquad\qquad \text{(VII)}$$

which exhibits a similarly high electroclinic coefficient, $e_c \approx 0.9$–2.8×10^{-2} rad μm/V (Williams *et al.*, 1991), and have found $e_c \approx 8.7 \times 10^{-3}$ rad μm/V for the polymer-dispersed liquid crystal. The PDFLC films were made of poly-vinylbutyral, compound (I), and prepared using a phase separation induced by solvent evaporation. The uniform alignment was achieved by pressing and stretching the film. A comparison of the different linear electro-optic effects (Figure 8.15) shows that the field-induced rotation of the optical axis is rather small for the flexoelectric effect, intermediate for the electroclinic effect and relatively high for the helical unwinding. However, the dynamic behaviour of the three effects is quite different.

In agreement with a simple Landau theory approach (see e.g. Goodby *et al.*, 1991) which predicts

$$\Theta_T = e_c E = s\alpha^{-1}(T - T_c)^{-1}E \qquad\qquad (8.20)$$

the electroclinic coefficient e_c decreases with increasing temperature (Heppke *et al.*, 1993). For temperatures close to the SmC*–SmA transition temperature T_c, pretransitional effects may lead to a non-linear dependence of the tilt angle on the field strength. In contrast to many other effects, the switching time τ does not depend on the field strength. It decreases with increasing temperature. This behaviour is in agreement with the relation

$$\tau = \gamma_\Theta \alpha^{-1}(T - T_c)^{-1} \qquad\qquad (8.21)$$

which is also obtained from Landau theory. Values of the switching time between 100 and 170 μs have been found for the PDLC films consisting of compound (V) and NOA-65. Response times smaller than 20 μs are reported by Komitov *et al.* (1994a) for compound (VII).

Bistable switching instead of a linear electro-optic effect occurs in PDFLC films when the droplet size is sufficiently small (Molsen and Kitzerow, 1994). Figures 8.16(a) and (b) show the optical transmission as a function of time for different samples addressed with short pulses.

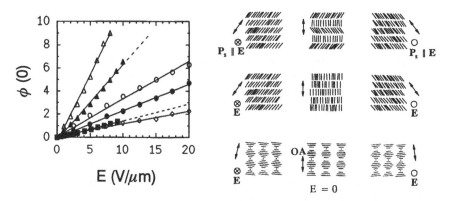

Figure 8.15 Field-induced tilt of the optical axis versus field strength for (■) the flexoelectric effect (Hoffmann-La Roche mixture 6415 L in PVB, data from Komitov *et al.* (1994b)), (◇) the electroclinic effect (A7 in NOA65 at 70 °C, data from Kitzerow *et al.* (1992b)), (○, ●) the electroclinic effect (W317 in PVB at (○) 38 °C and (●) 47 °C, data from Komitov *et al.* (1994a)), and (▲, △) the helical unwinding (Hoffmann–La Roche mixture 6430 in PVB at (▲) 22.4 °C and (△) 41.6 °C, data from Komitov *et al.* (1994a)).

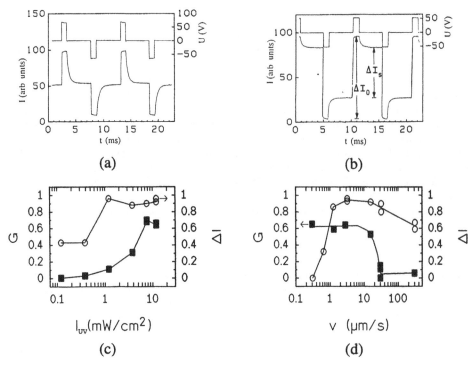

Figure 8.16 Bistability of PDFLC light valves. (a) Applied voltage pulses (above) and optical response (below) against time for a sample without memory effect (6.6% A7, 26.4% Pyp 808, 67% NOA 65, I_{UV} = 12 mW/cm^2, v = 31.3 μm/s). (b) Electro-optic response of a sample with bistable switching behaviour (same composition as in (a), different shear velocity, v = 3.1 μm/s). (c) Dependence of (■) the bistability ratio and (○) the contrast on the UV intensity used for the curing process (v = 3.1 μm/s = constant). (d) Dependence of (■) the bistability ratio and (○) the contrast on the shear velocity applied during the curing process (I_{UV} = 12 mW/cm^2 = constant). Data from Molsen and Kitzerow (1994).

Depending on the conditions of the preparation, the liquid crystal may either relax to its initial state after applying the pulse voltage (Figure 8.16(a)) or remain in a memory state close to the field-on state (Figure 8.16(b)). Systematic studies on the influence of the curing process (Molsen, 1994) have shown that the degree of bistability is affected by both the UV intensity I_{UV} which controls the droplet size, and the shearing velocity v which controls the flow and the final shape of the drops. The quality of the memory effect may be described by a bistability ratio

$$G = \Delta I_s/\Delta I_0 = (I_{s,1} - I_{s,2})/(I_1 - I_2) \tag{8.22}$$

with $0 \leqslant G \leqslant 1$, where I_1 and I_2 are the transmitted intensities in the two field-on states with opposite polarity of the applied voltage, and $I_{s,1}$ and $I_{s,2}$ are the respective intensities in the subsequent stored state. As expected, the bistability is enhanced by high UV intensities, i.e. small droplets (Figure 8.16(c)). This is in qualitative agreement with observations on the surface-stabilized ferroelectric liquid crystal display (Clark and Lagerwall, 1980) which requires a sample thickness smaller than the helical pitch of the SmC* structure. However, the bistability of the PDFLC films vanishes if the shear velocity exceeds a critical value (≈ 30 μm/s). Of course, the shear velocity v is essential for the quality of the alignment of the liquid crystal in the PDFLC film. This quality may be described by

$$\Delta I := (I_{max} - I_{min})/(I_{max} + I_{min}) \tag{8.23}$$

with $0 \leqslant \Delta I \leqslant 1$, where $I_{\min}$ and $I_{\max}$ are the minimum and the maximum values of the transmitted intensity when rotating the film between crossed polarizers. Figure 8.16(d) shows that a medium shear rate between 1 and 20 μm/s provides both good contrast and bistability. Compared with the response times for samples with helical unwinding, the bistable samples show larger switching times, between 2 ms at 6 V/μm and 70 μs at 15 V/μm. This may be due to a higher content of polymer or polymer precursor in the liquid crystal droplets for the samples which are rapidly cured at high UV intensity. Similar observations about the influence of UV intensity and shear velocity on the occurrence of bistability in PDFLC systems have been made by Leader *et al.* (1994). Patel *et al.* (1994) have observed bistable switching in PDFLC films using the commercial mixtures ZLI-3654 (E. Merck, Germany) and SCE-13 (Merck Ltd, UK) which were dispersed either in a UV-cured polymer or in poly-(methyl-methacrylate). In the latter case, the samples were sheared during thermally induced phase separation in order to achieve the uniform alignment of the liquid crystal.

The preparation of very small, uniformly aligned droplets can also be used to achieve *multistable switching* by means of *antiferroelectric liquid crystals* (Molsen *et al.*, 1992). The liquid crystal MHPOBC

$$C_8H_{17}O\!-\!\langle\bigcirc\rangle\!-\!\langle\bigcirc\rangle\!-\!COO\!-\!\langle\bigcirc\rangle\!-\!COOC^*H(CH_3)C_6H_{13} \qquad\text{(VIII)}$$

is well known to exhibit an antiferroelectric SmC$_A^*$ phase, characterized by an alternating tilt angle, in the temperature range between 84°C (on cooling 65°C) and 118.4°C (Chandani *et al.*, 1989; Hiraoka *et al.*, 1990). This liquid crystal was embedded in a sheared sample formed by NOA-65. The electro-optic characteristics as well as the electric polarization as a function of applied voltage show a double hysteresis which can be attributed to discontinuous transitions from the antiferroelectric state (with alternating tilt) to a ferroelectric state with uniform tilt. Switching times down to 150 μs (at 100°C, $E = 10$ V/μm) were found for the latter system.

8.3.1.2 Scattering effects

In the PDFLC applications mentioned so far, the liquid crystal is used as a variable wave plate between crossed polarizers in order to modulate the light intensity. However, a suspension of uniformly oriented birefringent particles can also be used as a *scattering polarizer*, as demonstrated by Land (1938) by embedding urea crystals with uniform orientation in a polymer film. Analogously, a stretched PDLC film incorporating uniformly oriented elliptical nematic droplets may serve as a scattering polarizer (West *et al.*, 1987; Zyryanov *et al.*, 1992a). If the effective refractive index $n_\perp$ perpendicular to the stretching direction (i.e. the ordinary refractive index in the case of a nematic liquid crystal) is equal to the refractive index n_p of the polymer, the PDLC film is transparent for light polarized perpendicular to the direction of elongation. However, it is scattering for light polarized along the direction of elongation owing to the mismatching of the refractive indices $n_\parallel$ and n_p. In a single scattering approximation the respective light intensity transmitted through the sample is described by

$$I_{\parallel,\perp} = I_0 \exp(-\sigma_{\parallel,\perp} dN/V) \qquad\text{(8.24)}$$

where $\sigma_\parallel$ and $\sigma_\perp$ are the average scattering cross-sections, d the thickness of the film, and N/V the number density of the droplets. Depending on the ratio between the drop radius R and the wavelength λ, the Rayleigh–Gans approximation may be applied (Žumer and Doane, 1986), or for sufficiently large drop sizes ($2\pi R/\lambda \gg 1$) the anomalous diffraction approach (Žumer, 1988). In the latter case, the total cross-section of a spherical drop with uniform director orientation is approximately given by

$$\sigma \approx 2\sigma_0(kR)^2[(n_\parallel/n_p - 1)^2 \cos^2\phi + (n_\perp/n_p - 1)^2 \sin^2\phi] \qquad\text{(8.25)}$$

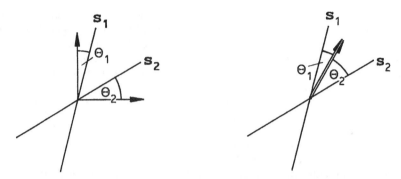

Figure 8.17 Setup for a scattering display consisting of two stacked PDFLC films, as proposed by Zyryanov et al. (1994a). s_1, s_2: azimuthal orientation of the stretching directions of film 1 and film 2. Θ_1, Θ_2: tilt angles of the SmC* phase in film 1 and film 2, respectively. The optimum dark (left) and bright state (right) are obtained, if $\Theta_1 + \Theta_2 = 45°$. In a simpler setup (Zyryanov et al., 1992b), film 2 is replaced by a fixed polarizer and the tilt angle has to be $\Theta_1 = 45°$ to achieve good contrast.

where σ_0 is the geometrical cross-section and ϕ is the angle between the director and the polarization plane of the light. Thus, for completely uniform orientation $\phi_\parallel \approx 2\sigma_0(kR)^2 \cdot (n_\parallel/n_p - 1)^2$ and $\sigma_\perp \approx 0$. The ratio of the respective intensities is given by

$$I_\parallel/I_\perp = \exp[-(\sigma_\parallel - \sigma_\perp)dN/V] \tag{8.26}$$

Zyryanov et al. (1992b, 1993) have shown that stretched PDFLC films consisting of the ferroelectric liquid crystal **DOBAMBC**

$$H_{21}C_{10}\!-\!O\!-\!\bigcirc\!-\!CH\!=\!N\!-\!\bigcirc\!-\!CH\!=\!CH\!-\!COO\!-\!CH_2\!-\!\overset{CH_3}{\underset{|}{CH}}\!-\!C_2H_5 \qquad \text{(IX)}$$

dispersed in poly-(vinyl-acetate) may serve as scattering polarizers to achieve electro-optic modulation. The field causes a rotation of the optical axis of the liquid crystal, and thus a rotation of the plane with maximum transmittance for linearly polarized light. This may be used for the modulation of light intensity, if just *one* polarizer is used. The azimuthal angle between the smectic layer normal q_l and the polarization plane of the incident light is adjusted at $\phi_0 = 45°$. Optimum contrast is obtained if the tilt angle Θ is also $45°$, so that rotation of the optical axis between $\phi_0 - \Theta$ and $\phi_0 + \Theta$ corresponds to a rotation of the scattering polarizer by $90°$. Unfortunately, chiral smectic liquid crystals with a tilt angle as high as $45°$ are hardly available. To overcome this problem, Zyryanov et al. (1994a) have proposed a setup which consists of *two PDFLC films* with tilt angles $\Theta_1 + \Theta_2 = 45°$ and *no polarizer* (Figure 8.17). In the field-off state, the polarization planes are adjusted at azimuthal angles ϕ_1 and ϕ_2, with $\phi_1 - \phi_2 = 45°$. The field-induced rotation of the optical axis is opposite for the two films so that the planes of maximum transmittance of the two films are crossed for one field-on state, $\phi_1 + \Theta_1 - (\phi_2 - \Theta_2) = 90°$, and coincide for the opposite field-on state, $\phi_1 - \Theta_1 - (\phi_2 + \Theta_2) = 0°$. The tilt of the optical axes to opposite directions can be realized by addressing the films separately, or – more easy – by using liquid crystals with opposite sign of the spontaneous polarization P_s, e.g. the two different enantiomers of the same ferroelectric liquid crystal.

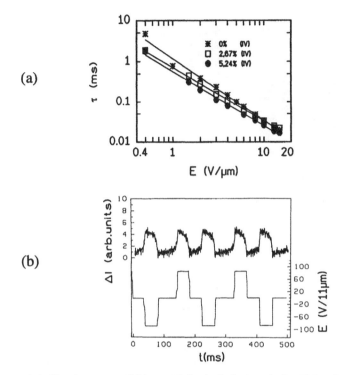

Figure 8.18 (a) Switching times versus field strength for the helical unwinding (*) in a ferroelectric liquid crystal oriented by surface effects and (□), (●) in a ferroelectric gel display, containing a liquid crystal mixture consisting of 20% compound (V) and 80% compound (VI), and a polymer network formed by compound (IV) (with $n = 6$ and $R = -H$. (b) Applied voltage (below) and electro-optic response of a ferroelectric gel with homeotropic prealignment. (Behrens and Kitzerow, unpublished).

8.3.2 *Polymer-network-stabilized FLC*

In volume-stabilized ferroelectric liquid crystals (Pirš *et al.*, 1992) (see Blinc *et al.* in this volume), a small amount of polymer dispersed in a ferroelectric liquid crystal is used to reduce the shock sensitivity of the display due to the stabilizing effect of the polymer on the liquid crystal orientation. Pirš *et al.* have found that addition of up to 3% of the acrylate Desolite D044 does not deteriorate the electro-optic performance of a surface-stabilized ferroelectric cell. For the same purpose, we have investigated ferroelectric samples stabilized by a polymer network formed by the liquid crystalline diacrylate (IV) with $n = 6$ and $R = H$ (Behrens and Kitzerow, to be published). The liquid crystal was a mixture consisting of 20% of compound(V) and 80% of compound(VI), as described in the previous section. Mixtures containing up to 6% of the diacrylate and very small amounts of photoinitiator were poured into 4 μm thick cells coated with polyimide and rubbed in order to promote parallel alignment. In the aligned state, the samples were polymerized by UV radiation with $I_{UV} \approx 12$ mW/cm². Unfortunately, the formation of the polymer network led to disclinations which diminish the contrast. The contrast ratio I_{max}/I_{min} for switching at room temperature was found to decrease with increasing weight fraction x_p of the polymer from $I_{max}/I_{min} \approx 25$ for $x_p = 0$, to $I_{max}/I_{min} \approx 5$ for $x_p = 2.5\%$ and $I_{max}/I_{min} \approx 3$ for $x_p = 5\%$. Better contrast ratios have been recently obtained by Hikmet *et al.* (1995), using polymer concentrations of 10% and higher. It is interesting to note that the switching times of the samples decrease slightly with increasing polymer concentration owing

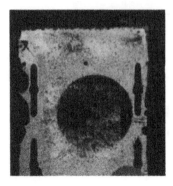 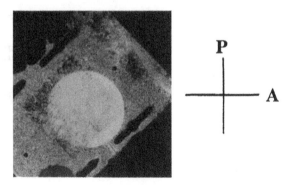

Figure 8.19 Photograph of an SmA liquid crystal (CCN-47), aligned by photopolymerization of PVMC with linearly polarized light in the bulk mixture. The orientation of the sample differs by 45° in the two photographs. 'P' and 'A' indicate the aximuthal orientation of the polarizer and analyzer. The round area in the centre was exposed to the linearly polarized UV radiation. The area outside of this spot is not aligned. (For details, see Jain and Kitzerow (1994a,b).)

to the anchoring of the liquid crystal at the polymer network (Figure 8.18(a)). At temperatures above the clearing point, a network of parallel birefringent lines could be seen in the polarizing microscope. It seems that not only the small amount of the polyacrylate but also monomer liquid crystal molecules anchored at the network contribute to this pattern. This is in agreement with observations in cholesteric systems by Hikmet and Zwerver (1992) who have found two different pitch values for free and anchored liquid crystal fractions, respectively, and with Crawford *et al.* (1994) who conserved even complicated defect structures above the clearing point by *in situ* photopolymerization of mesogenic diacrylates in the liquid crystalline state.

The polymer network created by photopolymerization of the diacrylate (IV) can also be used to achieve a homeotropic prealignment of the smectic layer normal. In this case, the sample appears nearly dark between crossed polarizers. Application of an electric field leads to a planar alignment of the director and thus to an increase of the intensity, independent of the sign of the applied voltage (Figure 8.18(b)). However, the contrast for this effect is even much smaller than for an initial planar alignment.

A very useful method for using the aligning effect of an anisotropic polymer is the formation of anisotropic networks by photopolymerization of poly-(vinyl-4-methoxycinnamate) (PVMC) with linearly polarized light (LPP effect, Schadt *et al.* (1992)). The procedure which has been developed by Schadt *et al.* leads to the creation of uniformly aligned mesogenic units and proved to be very suitable for the formation of surface alignment layers on solid substrates. The azimuthal anchoring strength of these alignment layers is about $W_\phi = 2.3 \times 10^{-6}$ J/m^2 (Vorflusev *et al.*, 1995). Nazarenko *et al.* (1993) have used PVMC to treat the pore surfaces in PDLC samples in order to get a uniform alignment. Instead of using surface treatment, it is also possible to disperse a small amount ($\leqslant 1\%$) of the polymer precursor PVMC in the liquid crystal and to photopolymerize the bulk mixture above its clearing point with linearly polarized light in order to promote alignment (Jain and Kitzerow, 1994a,b). A uniform alignment is obtained after slow cooling of the sample. This method leads to good results for non-chiral nematic and smectic A phases (Figure 8.19).

However, it is more difficult to align chiral materials. Nevertheless, contrast ratios of 4–5 could be achieved even for the SmC* phase in the commercial mixtures ZLI 4237-000 and ZLI 3654 (Merck, Germany). The spontaneous polarization is only little affected by the addition of the polymer network (Figure 8.20(a)) which indicates that small concentrations of the polymer do not restrict the rotational mobility of the liquid crystal molecules (Behrens and Kitzerow, to be published). The switching times (Figure 8.20(b)) of the composite system are slightly larger than the switching times of the pure liquid crystal, but again the influence of the polymer

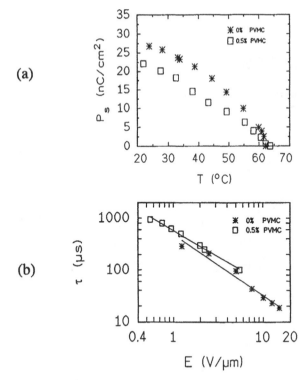

Figure 8.20 (a) Spontaneous polarization versus temperature and (b) switching times versus field strength for (*) ZLI 3654 with conventional surface alignment and (□) photoinduced alignment of the ferroelectric mixture ZLI 3654 (Merck) with PVMC dispersed in the bulk liquid crystal. (Behrens and Kitzerow, to be published.)

network is not dramatic. If the contrast ratios can be enhanced, the photopolymerization of PVMC in bulk liquid crystal mixtures may provide an easy way to align liquid crystals without surface layers, or at least to stabilize their orientation.

8.4 Conclusions and Prospect

The extension of polymer-dispersed and polymer-stabilized liquid crystal systems to chiral liquid crystals has opened up a large variety of new electro-optic effects. Polymer-dispersed cholesteric liquid crystals and cholesteric gels have proved to be useful for *colour effects*. Like the scattering nematic PDLC films, the use of polarizers is not necessary for electro-optic application of these systems so that high brightness can be obtained. The use of the *optical activity* in PDCLC samples with medium chirality (Behrens *et al.*, 1994) is an interesting new electro-optic effect. However, with respect to applications for intensity modulation it has few advantages over nematic PDLC modulators. Its angular dependence is improved with respect to nematic PDLC but the necessity of polarizers prevents high brightness.

The possibility of aligning *flexoelectric* and *ferroelectric liquid crystals* in PDLC films by shearing or stretching provides the application of linear, bistable and multistable effects with switching times down to 20 μs. The response times can probably be diminished by optimizing the phase separation process. A promising direction to make the fabrication of PDFLC films

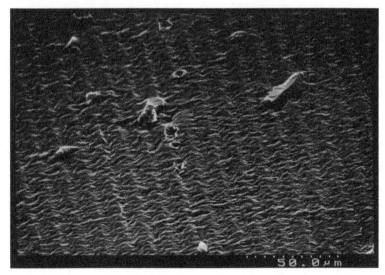

Figure 8.21 Scanning electron micrograph of a freeze-fractured PDLC film with thermally induced holographic grating.

even easier, is their use in a *scattering mode* as proposed by Zyryanov *et al.* (1992b). To operate effectively in this mode, suitable refractive-index-matching PDFLC systems have to be developed. However, the setup proposed by Zyryanov *et al.* does not provide as high a brightness as a nematic PDLC display, since even in the transparent state linearly polarized light with a certain polarization plane is scattered. To understand the complicated dynamics of PDFLC systems, basic investigations on the precise structure of the director field in small chiral smectic droplets are necessary. In addition to defects of the director field, discontinuities in the direction of the smectic layers (as in the chevron structure) and dislocations of the smectic layers are of importance in these systems.

Apart from electro-optic effects, PDLC systems are also suitable for non-linear optical effects (Li *et al.*, 1991; Simoni *et al.*, 1992) and storage effects (Yamaguchi and Sato, 1991; Kato *et al.*, 1993b). As an example of optically induced storage, Figure 8.21 shows a grating structure in a PDLC film which was obtained from the thermal effect of a holographic amplitude grating focused on the sample (Kitzerow, to be published). The temperature dependence of the miscibility of the liquid crystal and the polymer leads to a spatial variation of the droplet size, which in turn affects the scattering properties and the switching behaviour of the film, thereby revealing the possibility of switchable grating structures. In a similar way, thermoelectro-optical effects in PDCLC (Zyryanov *et al.*, 1994b) or stored patterns in PDFLC may be utilized.

The application of anisotropic polymer networks in gel systems may help to stabilize the orientation of the liquid crystal in conventional displays, or even lead to new effects. The stabilizing effect of the network should be especially useful to enhance the shock resistivity of ferroelectric liquid crystal displays (see also Blinc *et al.* and Hikmet and Broer in this volume).

Acknowledgements

The developments presented in this chapter would not have been possible without the pioneering work of Professor J. W. Doane and his coworkers at the Liquid Crystal Institute in Kent State University. In addition to these scientists, U. Behrens, G. Heppke and H. Molsen from the Technical University Berlin, P. P. Crooker and F. Xu from the University of Hawaii, G. Chilaya from the Academy of Sciences in Tbilisi (Georgia), S. C. Jain from the National Physical Laboratory, New Delhi (India), as well as V. G. Chigrinov and V. Vorflusev from the Organic Intermediates and Dyes Institute, Moscow (Russia), deserve thanks and acknowledgement from the author for fruitful collaborations. Financial support by the Deutsche Forschungsgemeinschaft (Sfb 335) and the Volkswagenstiftung (I/70362) is acknowledged.

References

BAHR, Ch. and HEPPKE, G. (1987) Optical and dielectric investigations on the electroclinic effect exhibited by a ferroelectric liquid crystal with high spontaneous polarization, *Liq. Cryst.* **2**, 825–831.

BAJC, J., BEZIĆ, J. and ŽUMER, S. (1995) Chiral nematic droplets with tangential anchoring and negative dielectric anisotropy in electric field, *Phys. Rev. E* **51**, 2176–2189.

BEHRENS, U. and KITZEROW, H.-S. (1994) Utility of liquid crystalline diacrylates for bistable cholesteric gel displays, *Polym. Adv. Technol.* **5**, 433–437.

BEHRENS, U., KITZEROW, H.-S. and CHILAYA, G. (1994) Electrooptic effects in polymer-dispersed cholesteric liquid crystals with medium chirality, *Liq. Cryst.* **17**, 597–603.

BERESNEV, L. A., CHIGRINOV, V. G., DERGACHEV, D. I., POSHIDAEV, E. P., FÜNFSCHILLING, J. and SCHADT, M. (1989) Deformed helix ferroelectric liquid crystal display: A new electrooptic mode in ferroelectric chiral smectic C liquid crystals, *Liq. Cryst.* **5**, 1171–1177.

BEZIĆ, J. and ŽUMER, S. (1992) Structures of cholesteric liquid crystal droplets with parallel surface anchoring, *Liq. Cryst.* **11**, 593–619.

BLINC, R. (1976) Soft mode dynamics in ferroelectric liquid crystals, *Ferroelectrics* **14**, 603–606.

BOULIGAND, Y. and LIVOLANT, F. (1984) The organization of cholesteric spherulites, *J. Phys.* **45**, 1899–1923.

BROER, D. J., HIKMET, R. A. M. and CHALLA, G. (1989) *In-situ* photopolymerization of oriented liquid-crystalline acrylates: Influence of a lateral methyl substituent on monomer and oriented polymer network properties of a mesogenic diacrylate, *Makromol. Chem.* **190**, 3201–3215.

CANDAU, S., LE ROY, P. and DEBEAUVAIS, F. (1973) Magnetic field effects in nematic and cholesteric droplets suspended in an isotropic liquid, *Mol. Cryst. Liq. Cryst.* **23**, 283–297.

CHANDANI, A. D. L., GORECKA, E., OUCHI, Y., TAKEZOE, H. and FUKUDA, A. (1989) Novel phases exhibiting tristable switching, *Jpn. J. Appl. Phys.* **28**, L1261–L1264.

CHANISHVILI, A., CHILAYA, G. and SIKHARUDLIDZE, D. (1991) Light modulator based on optically active nematic-chiral liquid crystal structure, *Mol. Cryst. Liq. Cryst.* **207**, 53–57.

CHANISHVILI, A., CHILAYA, G. and SIKHARUDLIDZE, D. (1994) Electro-optic effect in an optically active nematic chiral liquid-crystal structure, *Appl. Opt.* **33**, 3482–3485.

CLADIS, P. E. and KLÉMAN, M. (1972) Non-singular disclinations of strength $s = +1$ in nematics, *J. Phys.* **33**, 591–598.

CLARK, N. A. and LAGERWALL, S. T. (1980) Submicrosecond bistable electro-optic switching in liquid crystals, *Appl. Phys. Lett.* **36**, 899–901.

CRAWFORD, G. P., ALLENDER, D. W., DOANE, J. W., VILFAN, M. and VILFAN, I. (1991) Finite molecular anchoring in the escaped-radial nematic configurations: a ^{2}H-NMR study, *Phys. Rev. A* **44**, 2570–2577.

CRAWFORD, G. P., POLAK, R. D., SCHARKOWSKI, A., CHIEN, L.-C., DOANE, J. W. and ŽUMER, S. (1994) Nematic director-fields captured in polymer networks confined in spherical droplets, *J. Appl. Phys.* **75**, 1968–1971.

CROOKER, P. P. and XU, F. D. (1996) Cholesteric liquid crystals confined to cylinders and spheres, *Chiral Liquid Crystals*, ed. L. Komitov *et al.*, Singapore: World Scientific.

CROOKER, P. P. and YANG, D. K. (1990) Polymer-dispersed chiral liquid crystal color display, *Appl. Phys. Lett.* **57**, 2529–2531.

DE GENNES, P. G. (1968) Calcul de la distorsion d'une structure cholesterique par un champ magnetique, *Solid State Commun.* **6**, 163–165.

DE GENNES, P. G. (1974) *The Physics of Liquid Crystals*, Oxford: Clarendon Press.

DE VRIES, H. (1951) Rotary power and other optical properties of certain liquid crystals, *Acta Crystallogr.* **4**, 219–226.

DOANE, J. W. (1990) Polymer dispersed liquid crystal displays, in B. Bahadur (ed.) *Liquid Crystals – Applications and Uses*, Vol. 1, Ch. 14, Singapore: World Scientific, pp. 361–395.

DOANE, J. W., VAZ, N. A., WU, B.-G. and ŽUMER, S. (1986) Field controlled light scattering from nematic microdroplets, *Appl. Phys. Lett.* **48**, 269–271.

DOANE, J. W., CHIDISHIMO, G. and VAZ, N. A. (1987) Light modulating material comprising a liquid crystal dispersion in a plastic matrix, US Patent 4,688,900.

DOANE, J. W., GOLEMME, A., WEST, J. L., WHITEHEAD, J. B. and WU, B.-G. (1988) Polymer dispersed liquid crystals for display application, *Mol. Cryst. Liq. Cryst.* **165**, 511–532.

DRZAIC, P. (1986) Polymer dispersed nematic liquid crystal for large area displays and light valves, *J. Appl. Phys.* **60**, 2142–2148.

DRZAIC, P. (1988) A new director alignment for droplets of nematic liquid crystal with low bend-to-splay ratio, *Mol. Cryst. Liq. Cryst.* **154**, 289–306.

ERDMANN, J. H., ŽUMER, S. and DOANE, J. W. (1990) Configuration transition in a nematic liquid crystal confined to a small spherical cavity, *Phys. Rev. Lett.* **64**, 1907–1910.

FERGASON, J. L. (1984) Encapsulated liquid crystal and method, US Patent 4,435,047.

FRANK, F. C. and PRYCE, M. H. L. (1958) Quoted on page 41 in Robinson, C., Ward, J. C. and Beevers, R. B. (1958) Liquid crystalline structure in polypeptide solutions, *Discuss. Faraday Soc.* **25**, 29–42.

GAROFF, S. and MEYER, R. B. (1977) Electroclinic effect at the A–C phase change in a chiral smectic liquid crystal, *Phys. Rev. Lett.* **38**, 848–851.

GOLEMME, A., ŽUMER, S., ALLENDER, D. W. and DOANE, J. W. (1988) Continuous nematic-isotropic transition in submicron-size liquid-crystal droplets, *Phys. Rev. Lett.* **61**, 2937–2940.

GOLEMME, A., ARABIA, G. and CHIDICHIMO, G. (1994) Influence of the phase separation mechanism on polymer dispersed liquid crystalline materials, *Mol. Cryst. Liq. Cryst.* **243**, 185–199.

GOODBY, J. W., BLINC, R., CLARK, N. A., LAGERWALL, S. T., OSIPOV, M. A., PIKIN, S. A., SAKURAI, T., YOSHINO, K. and ŽEKŠ, B. (1991) *Ferroelectric Liquid Crystals: Principles, Properties and Applications*, Philadelphia: Gordon & Breach.

HEPPKE, G., KITZEROW, H.-S. and MOLSEN, H. (1993) Electroclinic effect in polymer dispersed ferroelectric liquid crystals, *Mol. Cryst. Liq. Cryst.* **237**, 471–476.

HIKMET, R. A. M. (1990) Electrically induced light scattering from anisotropic gels, *J. Appl. Phys.* **68**, 4406–4412.

HIKMET, R. A. M., BOOTS, H. M. J., and MICHIELSEA, M. (1995) Ferroelectric liquid crystal gels—Network stabilized ferroelectric displays, *Liq. Cryst.* **19**, 65–76.

HIKMET, R. A. M. and ZWERVER, B. H. (1992) Structure of cholesteric gels and their electrically induced light scattering and colour changes, *Liq. Cryst.* **12**, 319–336.

HIRAOKA, K., CHANDANI, A. D. L., GORECKA, E., OUCHI, Y., TAKEZOE, H. and FUKUDA, A. (1990) Electric-field-induced transitions among antiferroelectric, ferrielectric and ferroelectric phases in a chiral smectic MHPOBC, *Jpn. J. Appl. Phys.* **29**, L1473–L1476.

JAIN, S. C. and KITZEROW, H.-S. (1994a) Bulk-induced alignment of nematic liquid crystals by photopolymerization, *Appl. Phys. Lett.* **64**, 2946–2948.

JAIN, S. C. and KITZEROW, H.-S. (1994b) A new method to align smectic liquid crystals by photo-polymerization, *Jpn. J. Appl. Phys.* **33**, Part 2, L656–L659.

JONES, W. J. (1969) Thermometer, US Patent 3,440,882.

KATO, K., TANAKA, K., TSURU, S. and SAKAI, S. (1993a) Color image formation using polymer-dispersed cholesteric liquid crystal, *Jpn. J. Appl. Phys.* **32**, Part 1, 4600–4604.

KATO, K., TANAKA, K., TSURU, S. and SAKAI, S. (1993b) Multipage display using stacked polymer-dispersed liquid crystal films, *Jpn. J. Appl. Phys.* **32**, Part 1, 4594–4599.

KATO, K., TANAKA, K., TSURU, S. and SAKAI, S. (1994a) Reflective color display using polymer-dispersed cholesteric liquid crystal, *Jpn. J. Appl. Phys.* **33**, Part 1, 2635–2640.

KATO, K., TANAKA, K., TSURU, S. and SAKAI, S. (1994b) Characteristics of right- and left-handed polymer-dispersed cholesteric liquid crystals, *Jpn. J. Appl. Phys.* **33**, Part 1, 4946–4949.

KELLY, J. R. and PALFFY-MUHORAY, P. (1994) The optical response of polymer dispersed liquid crystasls, *Mol. Cryst. Liq. Cryst.* **243**, 11–29.

KHOO, I.-C. and WU, S.-T. (1993) *Optics and Nonlinear Optics of Liquid Crystals*, Ch. 2, Singapore: World Scientific, p. 212.

KIM, J. Y. and PALFFY-MUHORAY, P. (1991) Phase separation kinetics of a liquid crystal-polymer mixture, *Mol. Cryst. Liq. Cryst.* **203**, 93–100.

KITZEROW, H.-S. (1994) Polymer-dispersed liquid crystals – From the nematic curvilinear aligned phase to ferroelectric films, *Liq. Cryst.* **16**, 1–31.

KITZEROW, H.-S. and CROOKER, P. P. (1991) Polymer-dispersed cholesteric liquid crystals – Challenge for research and applications, *Ferroelectrics* **122**, 183–196.

KITZEROW, H.-S. and CROOKER, P. P. (1992) Behavior of polymer-dispersed cholesteric droplets with negative dielectric anisotropy in electric fields, *Liq. Cryst.* **11**, 561–568.

KITZEROW, H.-S. and CROOKER, P. P. (1993a) Electric field effects on the droplet structure in polymer-dispersed cholesteric liquid crystals, *Liq. Cryst.* **13**, 31–43.

KITZEROW, H.-S. and CROOKER, P. P. (1993b) UV-cured cholesteric polymer-dispersed liquid crystal display, *J. Phys. II* **3**, 719–726.

KITZEROW, H.-S., CROOKER, P. P. and HEPPKE, G. (1992a) Chromaticity of polymer-dispersed cholesteric liquid crystals, *Liq. Cryst.* **12**, 49–58.

KITZEROW, H.-S., MOLSEN, H. and HEPPKE, G. (1992b) Linear electrooptic effects in polymer-dispersed ferroelectric liquid crystals, *Appl. Phys. Lett.* **60**, 3093–3095.

KITZEROW, H.-S., MOLSEN, H. and HEPPKE, G. (1992c) Helical unwinding in polymer-dispersed ferroelectric liquid crystals, *Polym. Adv. Technol.* **3**, 231–235.

KITZEROW, H.-S., RAND, J. and CROOKER, P. P. (1992d) Dynamics of polymer-dispersed cholesteric liquid crystals, *J. Phys. II* **2**, 227–234.

KITZEROW, H.-S., XU, F. and CROOKER, P. P. (1992e) Microscopic observations of axial and radial nematic droplets for a dual-frequency addressable liquid crystal, *Liq. Cryst.* **12**, 1019–1024.

KOMITOV, L., LAGERWALL, S. T. and CHIDICHIMO, G. (1994a) Linear light modulation by polymer dispersed chiral liquid crystals, *Liquid Crystal Materials, Devices, and Applications III*, ed. R. Shashidar, *SPIE Proc.* **2175**, 160–172.

KOMITOV, L., RUDQUIST, P., ALOE, R. and CHIDICHIMO, G. (1994b) Linear light modulation by a polymer dispersed chiral nematic, *Mol. Cryst. Liq. Cryst.* **251**, 317–327.

LACKNER, A. M., MARGERUM, D., RAMOS, E., SMITH, H. and LIM, C. (1990) Film with partial pre-alignment of polymer-dispersed liquid crystals for electro-optical devices, and method of forming the same, US Patent 4,944,576.

LAND, E. H. (1938) Light polarizing material, US Patent 2,123,902.

LANDAU, L. D., LIFSHITZ, E. M. and PITAEVSKII, L. P. (1984) *Electrodynamics of Continuous Media*, 2nd edn, Section 124, New York: Pergamon.

LEADER, C., ZHENG, W., TIPPING, J. and COLES, H. J. (1994) Electro-optics of polymer-dispersed ferroelectric liquid crystal devices, Poster presentation at the *15th Int. Liquid Crystal Conf., Budapest, July*.

LEE, K., SUH, S.-W. and LEE, S.-D. (1994a) Fast linear electro-optical switching properties of polymer-dispersed ferroelectric liquid crystals, *Appl. Phys. Lett.* **64**, 718–720.

LEE, K., SUH, S.-W., LEE, S.-D. and KIM, C. Y. (1994b) Ferroelectric response of polymer-dispersed chiral smectic C* liquid crystal composites, *J. Korean Phys. Soc.* **27**, 86–90.

LI, L., YUAN, H. J. and PALFFY-MUHORAY, P. (1991) Second harmonic generation by polymer dispersed liquid crystal films, *Mol. Cryst. Liq. Cryst.* **198**, 239–246.

MAIER, W. and SAUPE, A. (1958) Eine einfache molekulare Theorie des nematischen flüssigkristallinen Zustandes, *Z. Naturforsch.* **13a**, 564–566.

MARGERUM, J. D., LACKNER, A. M., RAMOS, E., LIM, K.-C. and SMITH, W. H. (1989) Effects of off-stage alignment in polymer-dispersed liquid crystals, *Liq. Cryst.* **5**, 1477–1487.

MCDONNELL, D. G. (1987) Thermochromic cholesteric liquid crystals, in G. W. Gray (ed.) *Thermotropic Liquid Crystals, Critical Reports on Applied Chemistry*, Vol. 22, Ch. 5, Chichester: John Wiley, pp. 120–144.

MEYER, R. B. (1968) Effects of electric and magnetic fields on the structure of cholesteric liquid crystals, *Appl. Phys. Lett.* **12**, 281–282.

MEYER, R. B. (1969) Magnetically induced cholesteric-to-nematic phase transition in liquid crystals, *Phys. Rev. Lett.* **22**, 227–228.

MEYER, R. B. (1977) Ferroelectric liquid crystals: A review, *Mol. Cryst. Liq. Cryst.* **40**, 33–48.

MEYER, R. B., LIÉBERT, L., STRZELECKI, L. and KELLER, P. (1975) Ferroelectric liquid crystals, *J. Phys. Lett.* **36**, L69–L71.

MOLSEN, H. (1994) Über polymer-dispergierte chiral-smektische Flüssigkristalle, PhD Thesis, Technical University Berlin.

MOLSEN, H. and KITZEROW, H.-S. (1994) Bistability in polymer-dispersed ferroelectric liquid crystals, *J. Appl. Phys.* **75**, 710–716.

MOLSEN, H., KITZEROW, H.-S. and HEPPKE, G. (1992) Antiferroelectric switching in polymer-dispersed ferroelectric liquid crystals, *Jpn. J. Appl. Phys.* **31**, Part 2, L1083–L1085.

NAZARENKO, V. G., REZNIKOV, Yu. A., RESHETNYAK, V. Yu., SERGAN, V. V. and ZYRYANOV, V. Ya. (1993) Oriented dispersion of liquid crystal droplets in a polymer matrix with light induced anisotropy, *Mol. Mater.* **2**, 295–299.

OSTROVSKI, B. I., RABINOVICH, A. Z. and CHIGRINOV, V. G. (1980) Behaviour of ferroelectric smectic liquid crystals in electric field, *Advances in Liquid Crystal Research and Applications*, ed. L. Bates, Oxford: Pergamon Press, pp. 469–482.

PATEL, J. S. and LEE, S.-D. (1989) Fast linear electro-optic effect based on cholesteric liquid crystals, *J. Appl. Phys.* **66**, 1879–1881.

PATEL, J. S. and MEYER, R. B. (1987) Flexoelectric electro-optics of a cholesteric liquid crystal, *Phys. Rev. Lett.* **58**, 1538–1540.

PATEL, P., CHU, D., WEST, J. L. and KUMAR, S. (1994) Bistable switching in polymer-dispersed ferroelectric smectic-C* displays, *SID 94 Dig.* pp. 845–847.

PIRŠ, J., BLINC, R., MARIN, B., MUŠEVIČ, I., PIRŠ, S., ŽUMER, S. and DOANE, J. W. (1992) Polymer network volume stabilized ferroelectric liquid crystal displays, Poster C-P89 at the *14th Int. Liquid Crystal Conf., Pisa (Italy)*, June.

PLANER, (1861) Notiz über das Cholestearin, *Liebigs Ann. Chem.* **118**, 25–27.

POLAK, R. D., CRAWFORD, G. P., KOSTIVAL, B. C., DOANE, J. W. and ŽUMER, S. (1994) Optical determination of the saddle-splay elastic constant K_{24} in nematic liquid crystals, *Phys. Rev. E* **49**, R978–R981.

QUIGLEY, J. R. and BENTON, W. J. (1977) Some optical properties of dispersed cholesteric liquid crystals, *Mol. Cryst. Liq. Cryst.* **42**, 43–51.

REINITZER, F. (1888) Beiträge zur Kenntniss des Cholesterins, *Monatsh. Chem.* **9**, 421–441.

ROBINSON, C., WARD, J. C. and BEEVERS, R. B. (1958) Liquid crystalline structure in polypeptide solutions, *Discuss. Faraday Soc.* **25**, 29–42.

SCHADT, M. and HELFRICH, W. (1971) Voltage dependent optical activity of a twisted nematic liquid crystal, *Appl. Phys. Lett.* **18**, 127–128.

SCHADT, M., SCHMITT, K., KOZINKOV, V. and CHIGRINOV, V. (1992) Surface-induced parallel alignment of liquid crystals by linearly polymerized photopolymers, *Jpn. J. Appl. Phys.* **31**, 2155–2164.

SIMONI, F., BLOISI, F. and VICARI, L. (1992) Transient amplitude gratings in polymer-dispersed liquid crystals, *Mol. Cryst. Liq. Cryst.* **223**, 169–179.

SMITH, G. W. (1991) Cure parameters and phase behavior of an ultraviolet-cured polymer-dispersed liquid crystal, *Mol. Cryst. Liq. Cryst.* **196**, 89–102.

SMITH, G. W. (1994) A calorimetric study of phase separation in liquid crystal/matrix systems: Determination of the excess specific heat of mixing, *Mol. Cryst. Liq. Cryst.* **239**, 63–85.

STANNARIUS, R., CRAWFORD, G. P., CHIEN, L.-C. and DOANE, J. W. (1991) Nematic director orientation in a liquid-crystal-dispersed polymer: A deuterium nuclear-magnetic-resonance approach, *J. Appl. Phys.* **70**, 135–143.

VAZ, N. A., SMITH, G. W. and MONTGOMERY, G. P. (1987) A light control film composed of liquid crystal droplets dispersed in a UV-curable polymer, *Mol. Cryst. Liq. Cryst.* **146**, 1–15.

VORFLUSEV, V. P., KITZEROW, H.-S. and CHIGRINOV, V. G. (1995) Azimuthal anchoring energy in photo-induced anisotropic films, *Jpn. J. Appl. Phys.* **34** (9A), L1137–L1140.

WEST, J. L. (1988) Phase separation of liquid crystals in polymers, *Mol. Cryst. Liq. Cryst.* **157**, 427–441.

WEST, J. L., DOANE, J. W. and ŽUMER, S. (1987) Liquid crystal display material comprising a liquid crystal dispersion in a thermoplastic resin, US Patent 4,685,771.

WILLIAMS, P. A., CLARK, N. A., ROS, M. B., WALBA, D. M. and WAND, M. D. (1991) Large electroclinic effect in new liquid crystal material, *Ferroelectrics* **121**, 143–146.

WU, B.-G., WEST, J. L. and DOANE, J. W. (1987) Angular discrimination of light transmission through polymer-dispersed liquid crystal films, *J. Appl. Phys.* **62**, 3925–3931.

WU, B.-G., ERDMANN, J. H. und DOANE, J. W. (1989) Response times and voltages for PDLC light shutters, *Liq. Cryst.* **5**, 1453–1465.

XU, F., KITZEROW, H.-S. and CROOKER, P. P. (1992) Electric-field effects on nematic droplets with negative dielectric anisotropy, *Phys. Rev. A* **46**, 6535–6540.

XU, F., KITZEROW, H.-S. and CROOKER, P. P. (1994) The director fields of nematic droplets with negative dielectric anisotropy and parallel anchoring, *Phys. Rev. E* **49**, 3061–68.

YAMAGUCHI, R. and SATO, S. (1991) Memory effects of light transmission properties in polymer-dispersed liquid crystal films, *Jpn. J. Appl. Phys.* **30**, L616–L618.

YANG, D.-K. and CROOKER, P. P. (1991) Field-induced textures of polymer-dispersed chiral liquid crystal microdroplets, *Liq. Cryst.* **9**, 245–251.

YANG, D.-K. and DOANE, J. W. (1992) Cholesteric liquid crystal/polymer gel dispersion: Reflective display application, *SID 92 Dig.*, pp. 759–761.

YANG, D.-K., CHIEN, L.-C. and DOANE, J. W. (1992) Cholesteric liquid crystal/polymer dispersion for haze-free light shutters, *Appl. Phys. Lett.* **60**, 3102–3104.

ŽUMER, S. (1988) Light scattering from nematic droplets: Anomalous-diffraction approach, *Phys. Rev. A* **37**, 4006–4015.

ŽUMER, S. and DOANE, J. W. (1986) Light scattering from a small nematic droplet, *Phys. Rev. A* **34**, 3373–3386.

ZYRYANOV, V. YA., SMORGON, S. L. and SHABANOV, V. F. (1992a) Elongated films of polymer-dispersed liquid crystals as scattering polarizers, *Mol. Eng.* **1**, 305–310.

ZYRYANOV, V. YA., SMORGON, S. L. and SHABANOV, V. F. (1992b) Polymer-dispersed ferroelectric liquid crystals as display materials, *SID 92 Dig.* pp. 776–777.

ZYRYANOV, V. YA., SMORGON, S. L. and SHABANOV, V. F. (1993) Light modulation by a planar-oriented film of a polymer-encapsulated ferroelectric liquid crystal, *JETP Lett.* **57**, 15–18.

ZYRYANOV, V. YA., SHABANOV, V. F., SMORGON, S. L. and POZHIDAEV, E. P. (1994a) Polymer-dispersed ferroelectric liquid crystal light valves, *SID 94 Dig.* pp. 605–607.

ZYRYANOV, V. YA., SMORGON, S. L., ZHULKOV, V. A. and SHABANOV, V. F. (1994b) Memory effects in polymer-encapsulated cholesteric liquid crystals, *JETP Lett.* **59**, 547–550.

9

Confined and Polymer-stabilized Ferroelectric Liquid Crystals

R. BLINC, I. MUŠEVIČ, J. PIRŠ, M. ŠKARABOT AND B. ŽEKŠ

9.1 Introduction

In contrast to 3D periodic solids, many liquid crystalline phases exhibit a continuous rotational symmetry which can be broken either spontaneously or by the presence of external fields or restricted geometry. Whereas the effects of the symmetry breaking by the external electric or magnetic fields seem to be well understood, our understanding of the effects of confinement on the static and dynamic properties of liquid crystals near phase transitions is still lacking a coherent framework. This is in particular true for the case of smectic phases of lower symmetry, confined to complex media like random porous matrices, which are the subject of great current interest.

Phase transitions in ferroelectric and antiferroelectric liquid crystals seem to be excellent candidates for these studies, because there has been tremendous progress in our understanding of these phases in recent years. This progress has been initiated not only by the discovery of the technological importance of these materials (Clark and Lagerwall, 1980, 1984; Handschy and Clark, 1984) but also by the variety of outstanding thermodynamical analogies to other systems like solid ferroelectrics and antiferroelectrics, incommensurate systems, etc. In addition, the symmetry and the linear coupling to an external electric field offers a unique possibility of studying phase transitions and collective behaviour in these systems with dielectric (Levstik et al., 1987; Vallerien et al., 1989; Biradar et al., 1989) and electro-optic spectroscopy (Garoff and Meyer 1977, 1979), as well as quasi-elastic light scattering spectroscopy (Muševič et al., 1988; Drevenšek et al., 1990).

The observations of the behaviour of liquid crystals in restricted geometries clearly show that complex behaviour in liquid crystals emerges as a consequence of complex geometry. However, in most cases the degree of complexity of the constraining media (i.e. aerogels or other porous media) is so high that it does not enable one to identify various sources of the complex response of confined liquid crystals. This is the reason why it seems reasonable to start from simple but extreme geometry like the two flat, solid surfaces separated by a very small distance. For the nematic liquid crystals, phase transitions in this geometry have been extensively studied both from the experimental (Miyano, 1979; Tarczon and Miyano, 1980; Van Sprang, 1983; Hsiung et al., 1986; Yokohama, 1988) and theoretical point of view (Sheng, 1976; Allender et al., 1981; Mauger et al., 1984, Sluckin and Poniewierski, 1985, 1990). They have also been the subject of wide experimental and theoretical efforts in the field of ferroelectric liquid crystals confined to plane parallel plates separated by micrometre distances (for a review, see Clark and Lagerwall (1984)). To our knowledge, however, very little is known about the behaviour and phase transitions of ferroelectric and antiferroelectric liquid crystals in extremely thin, submicrometre layers, confined between the two plane parallel surfaces. The exceptions here are the experimental studies of the helix unwinding in thin, wedge-type cells (Kondo *et al.*, 1982, 1983; Kai *et al.*, 1983) and recent experiments on the dielectric response of surface-stabilized ferroelectric cells (Yang *et al.*, 1991; Panarin *et al.*, 1994).

The aim of this contribution is to show that the onset of complex behaviour of chiral ferroelectric liquid crystals can be observed in the vicinity of the phase transition of ferroelectric liquid crystals in a simple geometry. The geometry of the problem is very simple, i.e. two plane parallel glass plates, separated by a small distance d. We shall start with a discussion of the phase transitions in the (d, T) phase diagram for non-polar boundary conditions, as predicted within the Landau theory, and present the experimental results. We continue with a discussion of our recent observations (Škarabot *et al.*, 1995) of the order parameter dynamics of a ferroelectric liquid crystal in the vicinity of the SmA $\rightarrow$ SmC* phase transition in submicrometre cells that clearly show the onset of complex behaviour. In the last section we discuss the concept of 'volume-stabilized' ferroelectric liquid crystals (Pirš *et al.*, 1992, 1995a,b) and show experimental evidence that this system generates by itself the randomness at the confining surfaces, which is a result of polymer phase separation at the surface. It is therefore a good candidate for the experimental study of ferroelectric phase transitions in the presence of a random field that can be externally controlled to a certain extent.

9.2 Phase Diagram of a Ferroelectric Liquid Crystal in a Restricted Geometry

We shall discuss the phase diagram of a ferroelectric liquid crystal in the most simple restricted geometry where it is confined in a bookshelf geometry in between two flat, plane parallel plates, as shown in the inset to Figure 9.1(a). Although it seems at first sight a geometry of marginal interest for the broad field of complex geometries discussed here, a close look will reveal some very interesting features. Apart from the fact that this is a basic geometry for all liquid-crystalline-based devices, one may expect that by decreasing the distance d between the two plates, the surface effects will become more and more pronounced in the thermodynamic properties of such a system. This offers a unique possibility for experimentally controllable probing of the thermodynamic properties of ferroelectric liquid crystals close to the surface.

Here we shall assume homogeneous and non-polar Rapini–Papoular surface coupling (Rapini and Papoular, 1969). The ferroelectric liquid crystal in the so-called bookshelf geometry is confined between the two parallel plates, separated by a distance $d = 2L$. The surface anchoring energy density is taken in a simple quadrupolar form

$$g_S = [\delta(x + L) + \delta(x - L)]\tfrac{1}{2}C_S\xi_x^2 \tag{9.1}$$

Here $\xi = (\xi_x, \xi_y)$ is the projection of the director η on the smectic layers, ξ_y is its out-of-plane component and C_S is assumed to be constant through the interface. For a positive value of the coupling constant $C_S > 0$, parallel alignment of ξ with respect to the interface is favoured. The homogeneous form of the surface anchoring energy in the z–y plane, which favours homogeneous ordering, is competing with the 'bulk' elastic energy, which favours helicoidal ordering far from the surfaces. For a thin enough sample we may thus expect that the helical structure would be unwound by the surface. In a certain way the role of the surface term is thus similar to the role of homogeneous magnetic or AC electric fields, which tend to unwind the helical SmC* structure. We can thus expect that the (d, T) phase diagram of a ferroelectric liquid crystal would resemble the (H, T) phase diagram because of the similar couplings (Michelson, 1977).

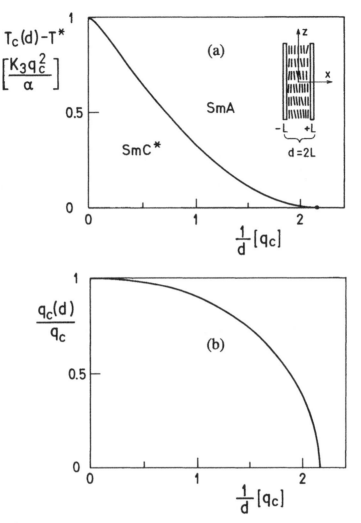

Figure 9.1 (a) Shift of the SmA → SmC* phase transition temperature as a function of the inverse cell thickness. The inset shows the cell geometry. (b) Inverse thickness dependence of the modulation wavevector along the SmA → SmC* transition line.

The phase boundaries in the (d, T) phase diagram of a ferroelectric liquid crystal can be determined by stability analysis of the SmA phase using the free-energy density (Povše *et al.*, 1993)

$$g(x,z) = \tfrac{1}{2}a(T)(\xi_x^2 + \xi_y^2) - \Lambda\left(\xi_x\frac{\mathrm{d}\xi_y}{\mathrm{d}z} - \xi_y\frac{\mathrm{d}\xi_x}{\mathrm{d}z}\right) + \tfrac{1}{2}K\left[\left(\frac{\mathrm{d}\xi_x}{\mathrm{d}z}\right)^2 + \left(\frac{\mathrm{d}\xi_y}{\mathrm{d}z}\right)^2\right]$$

$$+ \tfrac{1}{2}K\left[\left(\frac{\mathrm{d}\xi_x}{\mathrm{d}x}\right)^2 + \left(\frac{\mathrm{d}\xi_y}{\mathrm{d}x}\right)^2\right] + [\delta(x + L) + \delta(x - L)]\tfrac{1}{2}C_s\xi_x^2 \tag{9.2}$$

We have used here the one constant approximation for the elastic distortion in the x and z directions, respectively. The SmA → SmC* transition for an unconfined bulk sample is $T_c = T^* + (1/\alpha)Kq_c^2$, where $q_c = \Lambda/K$ is the wavevector of the undeformed helix, and T^* is the

223

phase transition temperature for the racemic mixture. From the free-energy expansion, the stability analysis leads to a system of Euler–Lagrange equations for the bulk and the surface, which can be solved by

$$\xi_x(x, z) = f(x) \cos(qz) \tag{9.3a}$$

$$\xi_y(x, z) = g(z) \sin(qz) \tag{9.3b}$$

As a result (Povše *et al.*, 1993), one obtains a thickness-dependent stability limit of the SmA phase $T_c(d)$, which is shown in Figure 9.1(a), together with a thickness dependence of the critical wavevector q_c along this phase transition boundary, shown in Figure 9.1(b). Because the boundary conditions interfere with the helicoidal ordering, characterized by the wavevector q_c in the SmC* phase, for finite d the phase transition SmA → SmC* will take place at a lower temperature than in the bulk. There will, however, be some limiting thickness d_{LP}, below which the phase transition SmA → SmC̄ into the unwound SmC̄ phase will take place. Since the boundary conditions for $d < d_{LP}$ do not interfere with homogeneous ordering of the surface-unwound SmC̄ phase, the phase transition boundary for $d < d_{LP}$ will be thickness independent:

$$d > d_{LP} : T_c(d) < T_c(d \to \infty) \tag{9.4a}$$

$$d < d_{LP} : T_c(d) = T^* = \text{constant} \tag{9.4b}$$

The predicted (d, T) phase diagram is shown schematically in the inset to Figure 9.2 and is very similar to the (H, T) phase diagram of a ferroelectric liquid crystal with negative diamagnetic anisotropy. If one calculates the behaviour of the critical wavevector $q_c(d)$ along the phase transition line $T_c(d)$, one can observe that this analogy is even deeper. The wavevector $q_c(d)$ goes continuously to zero at d_{LP}, which is thus a Lifshitz thickness, where the SmA, distorted SmC* and the unwound SmC̄ phases coexist. The value of the Lifshitz thickness depends on the surface coupling constant and liquid crystalline material constants, and can be evaluated in the limit of weak and strong surface anchoring, respectively:

$$C_s \to 0 \qquad d_{LP} \to 0 \tag{9.5a}$$

$$C_s \to \infty \qquad d_{LP} \to \frac{\sqrt{3}}{2\pi} p_0 = 0.276 p_0 \tag{9.5b}$$

Here, p_0 is the period of the helix in bulk liquid crystal. The phase boundary between the distorted SmC* phase and the surface unwound SmC̄ phase can be treated analogously to the cholesteric–nematic transition (Luban *et al.*, 1974). In the constant-amplitude approximation (CAA) the SmC* → SmC̄ transition is of second order at a critical thickness d_c. The analysis gives the variation of the critical cell thickness d_c and the Lifshitz thickness d_{LP} with anchoring strength C_s and shows that the Lifshitz thickness is always smaller than the critical thickness. This means that for cell spacing larger than the Lifshitz spacing there is a re-entrant, distorted SmC* phase present just below the SmA phase. This may have important implications for the technology of FLC devices, where the cell spacing is of the order of the helical pitch in bulk material. When cooling such a cell from the SmA phase, one may cross the re-entrant, distorted SmC* phase and generate disclination lines that are always present in partially unwound ferroelectric phases (Bourdon *et al.*, 1982; Glogarova *et al.*, 1983; Glogarova and Pavel, 1984a,b; Lejčeh, 1984).

A careful analysis of the results of the CAA approximation shows that this approximation breaks down for large coupling strengths. Experimental observations of the SmC* structures in thin samples indeed indicate (Bourdon *et al.*, 1982: Glogarova *et al.*, 1983) that instead of a continuous director field, a system of disclination lines mediates the transition between the homogeneous orientation dictated by the surface and the helical structure in the interior of the liquid crystal. This means that we are always in the strong coupling regime and, in this case, the critical thickness is (Povše *et al.*, 1993)

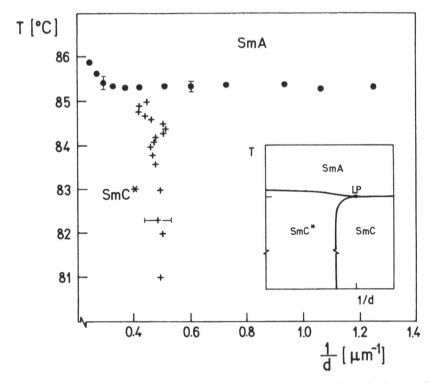

Figure 9.2 Phase diagram on CE-8 on a polyimide-coated surface. The inset shows the theoretically predicted phase diagram.

$$d_c = \frac{p_0}{2} \sqrt{\frac{K}{K_3}} \qquad (9.6)$$

Here K is the bend elastic constant, which is assumed to be equal to the splay elastic constant, whereas K_3 is the twist elastic constant of the SmC* phase.

The (d, T) phase diagram has been determined in wedge-type cells of liquid crystal 4-(2'-methylbutyl)phenyl 4'-n-octylbiphenyl-4-carboxylate (CE-8), using untreated polyimide-coated surfaces. The observed phase diagram is shown in Figure 9.2 and is in qualitative agreement with theory. At $T_c - T = 5$ K the critical thickness is $d_c = 2$ μm and increases slightly on approaching the SmA phase. The experiment allows for the estimation of the coupling constant, $C_s > 3 \times 10^{-3}$ J m^{-2} and indicates that the system is indeed in the strong surface coupling regime. Because of the experimental limitations, the critical thickness very near the SmA phase could not be determined unambiguously, so the question of the existence of a Liftshitz point in such a system is still open.

Recently, the study of the thickness dependence of the SmA $\rightarrow$ SmC phase transition temperature has been extended to CE-8 and SCE-9 ferroelectric liquid crystals confined in submicrometre-thick cells. The surface was coated with nylon and unidirectionally rubbed. The minimum spacing of the wedge-type cells was as low as 0.4 μm. The results for SCE-9 are shown in Figure 9.3 and reveal a surprising result. Contrary to the theoretical predictions (see eqn (9.4b)), the phase transition boundary between the SmA phase and the unwound Sm$\bar{C}$ phase strongly decreases for very small thickness. The shift in the phase transition temperature is proportional to $T_c - T_\infty \propto -1/d^2$ and is as large as 4 K for 0.4 μm thickness. The precise reason for such a large depression of the phase transition temperature is not yet identified but

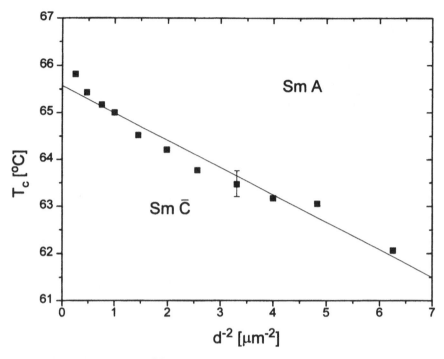

Figure 9.3 Thickness dependence of the SmA → SmC* phase transition temperature for ultrathin homogeneous samples of SCE-9 on a nylon-coated surface.

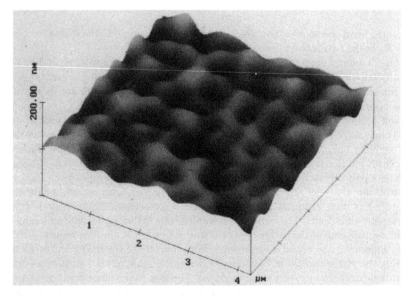

Figure 9.4 AFM image of a nylon-coated glass surface.

it is clearly due to the boundary conditions which are not as simple as the ones assumed in eqn (9.1). The effect can probably be related to the well-known phenomenon of the depression of the I → N phase transition that has its origin in the disordering effects of the surface. Indeed, if one looks closer to the surface profiles of the aligning layers that were used in the experiments, one observes huge irregularities on the molecular length scale that are certainly generators of disorder at the surface. An example of a surface profile of a spin-coated nylon alignment layer as observed by an atomic force microscope is shown in Figure 9.4. It should be stressed that the same order of magnitude for the depression of the SmA → SmC̄ phase transition temperature was observed in recent light scattering experiments in wedge-type cells (Rastegar *et al.*, 1995) or porous matrices (Aliev, this volume).

9.3 Order Parameter Dynamics of a Ferroelectric Liquid Crystal in a Restricted Geometry

Order parameter dynamics in the vicinity of the SmA → SmC* transition has been a topic of significant interest since the discovery of ferroelectric liquid crystals (Meyer *et al.*, 1975). The basic features of this phase transition seem to be qualitatively well understood both from experimental and theoretical points of view and can be described within the framework of Landau theory (Blinc and Žekš, 1978; Carlsson *et al.*, 1988; Dumrongrattana *et al.*, 1986).

The primary order parameter of the transition is the tilt angle that has its magnitude and the phase, so that the SmA → SmC* transition belongs to the 3D XY universality class. The collective excitations (or the order parameter fluctuations) can be decomposed into the polarization and the director branch of excitations. Whereas the polarization excitations do now show interesting behaviour, the director branch of excitations exhibits a critical slowing down in the vicinity of the transition in the bulk. In the SmA phase there is only one, doubly degenerate branch of soft collective excitations of the director field (Mušević *et al.*, 1988) that has a minimum at $\mathbf{q} = \mathbf{q}_c$ in the reciprocal space

$$\tau_s^{-1} = \frac{\alpha}{\gamma} (T - T_c) + \frac{K_3}{\gamma} (q_z - q_c)^2 \tag{9.7}$$

Here $|\mathbf{q}| = 2\pi/\mathbf{p}_0$, $\mathbf{p}_0$ is the period of the helix in the SmC* phase, and γ is the viscosity. The relaxation rate of the soft mode excitation goes to zero at T_c thus leading to the condensation of a helical wave at the phase transition point. For general q_z, these modes are helical-like excitations of the director field and are thus globally non-polar fluctuations. Only in the limit $\mathbf{q} \to \mathbf{0}$ do we find linearly polarized polar modes that can couple to a probing electric field in, for example, a dielectric experiment.

In the unperturbed SmC* phase this doubly degenerate, soft mode branch splits into an amplitudon and a phason branch with dispersion relations

$$\tau_A^{-1} = \frac{2\alpha}{\gamma} (T_c - T) + \frac{K_3}{\gamma} (q_z - q_c)^2 \tag{9.8}$$

$$\tau_{Ph}^{-1} = \frac{K_3}{\gamma} (q_z - q_c)^2 \tag{9.9}$$

Here we have considered only the excitations with a wavevector along the helical axis. Dispersion relations for a wavevector oblique to the helical axis have been treated by Drevenšek *et al.* (1990). The behaviour of collective excitations of the director field in unperturbed bulk samples has been extensively studied in the past, both with a quasi-elastic light scattering and other optical techniques (Mušević *et al.*, 1988; Drevenšek *et al.*, 1990; O'Brien *et al.*, 1993) or dielectric spectroscopy (Levstik *et al.*, 1987; Biradar *et al.*, 1989; Vallerien *et al.*, 1989). On the other hand, little is known about the SmA → SmC* transition in restricted geometry, except for some recent experimental (Cava *et al.*, 1987; Yang *et al.*, 1991; Panarin *et al.*, 1994; Rastegar *et al.*, 1995, Škarabot *et al.*, 1995) and theoretical studies (Limat and Prost, 1993; Kralj and Sluckin, 1994).

The effects of confinement on the dynamics near the SmA → SmC* and SmA → SmC̄ transitions in submicrometre layers can be studied by the quasi-elastic (Rasteger *et al.*, 1995) or linear electro-optic response spectroscopy (Škarabot *et al.*, 1995), which is an optical analogue of dielectric spectroscopy. Here, a polarized laser light is focused to a small spot (10 μm) on the liquid crystal sample and a crossed analyzer is placed in front of the detector. When a small measuring electric field $\mathbf{E} = (E_x, 0, 0)\exp(i\omega t)$ is applied in the direction of the smectic layers of an SmC* ferroelectric liquid crystal, it will couple to those collective eigenmodes that have a space-averaged non-zero net electric polarization $\langle\delta\mathbf{P}(\mathbf{r}, t)\rangle = (\delta P_x(t), 0, 0)$. Because of the interconnection between the polarization and director field this will result in the non-zero value of the space-averaged director field $\langle\mathbf{n}(\mathbf{r}, t)\rangle = (0, \delta n_y(t), 0)$. The dielectric tensor for the optical frequencies $\varepsilon_{ij}(\mathbf{r})$ has the same symmetry properties as the tensor $\mathbf{n} \otimes \mathbf{n}$ so that any change $\delta\mathbf{n}(\mathbf{r}, t)$ causes a corresponding change of the dielectric tensor field $\delta\varepsilon(\mathbf{r}, t)$. To within first order in $\delta\mathbf{n}(\mathbf{r}, t)$ we have

$$\delta\varepsilon_{ij} \approx \delta(n_i n_j) \approx n_i^o \delta n_j + n_j^o \delta n_i \qquad (9.10)$$

It was shown both theoretically and experimentally (Muševič *et al.*, 1993) that the optical properties of modulated birefringent phases like the SmC* phase can be very well described within the first-order perturbation approach to the wave equation. Within this approximation, the optical properties are determined by the space-averaged value of the dielectric tensor of the sample, which can be expressed in terms of $\langle\mathbf{n}(\mathbf{r}, t)\rangle$. In particular, a small electric field in the y direction will induce the following dominant change in the dielectric tensor of the SmC* phase:

$$\delta\varepsilon(\mathbf{r}, t) \propto \begin{bmatrix} 0 & 0 & \delta n_x \\ 0 & 0 & \delta n_y \\ \delta n_x & \delta n_y & 0 \end{bmatrix} \propto \begin{bmatrix} 0 & 0 & \delta P_y \\ 0 & 0 & -\delta P_x \\ \delta P_y & -\delta P_x & 0 \end{bmatrix} \qquad (9.11)$$

Here c_1, $c_2 \ll 1$ describes a coupling to the measuring electric field. One can see that the space-averaged perturbation of the dielectric tensor is directly related to the space-averaged induced polarization. In particular, we have a contribution of the form

$$\langle\delta\varepsilon\rangle \propto \begin{bmatrix} 0 & 0 & 0 \\ 0 & 0 & \langle\delta P_x\rangle \\ 0 & \langle\delta P_x\rangle & 0 \end{bmatrix} \qquad (9.12)$$

which represents a uniform rotation of the unperturbed dielectric tensor as a whole around the direction of the external field. This rotation of the axis of the dielectric tensor in a plane perpendicular to the direction of light propagation can be detected with a very high accuracy, using a suitable optical setup and lock-in technique (Kuczynski *et al.*, 1990). One thus measures the frequency dependence of the response of the reorientations of the dielectric tensor and obtains information on the spectrum of the eigenmodes that couple linearly to the small measuring field. In other words, we are able to detect a linear dielectric response of the sample with an optical technique. The advantage of this technique is that it is a local technique, capable of probing dynamical properties of the sample over the illuminated region, which is of the order of 10 μm^2. It should be mentioned that this technique was originally used to detect the soft mode response in the SmA phase (Garoff and Meyer, 1977), but it is obvious from the above considerations that it can be applied to optically inhomogeneous media, such a distorted SmC* phase, as well. The setup for the optically detected linear response of a wedge-type cells is shown in Figure 9.5 and the inset shows typical real and imaginary parts of the detected light intensity on the photodiode. With the EG&G 5302 lock-in amplifier and an ultrafast photodiode the response of the sample can be measured in the range from 1 mHz to 1 MHz, thus covering nine orders of the time scale.

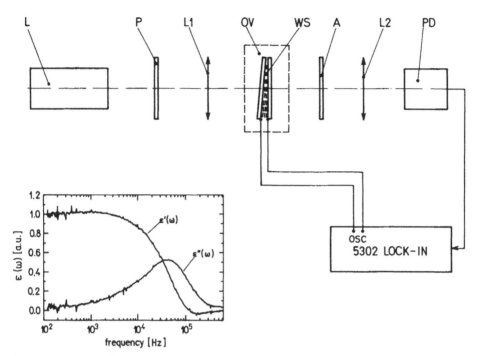

Figure 9.5 Setup for the measurements of the spectrum of the linear electro-optic response. Definitions: L, laser; P, polarizer; L1, L2, lenses; OV, temperature-stabilized oven; WS, wedge-type sample; A, analyser; PD, photodiode. The inset shows the real and imaginary parts of the photodiode current.

Using the above technique we were able to analyze the order parameter dynamics as a function of the cell thickness of a ferroelectric liquid crystal in bookshelf geometry, using a single wedge-type cell with thickness in the range from 4 μm to 400 nm. The magnitude of the measuring voltage was typically of the order of 50 mV, depending on the thickness and the spontaneous polarization of the sample, and was adjusted below the value where the non-linearities in the electro-optical response were observed.

Figure 9.6 shows the observed slowing down of the order parameter dynamics in the ferroelectric liquid crystal CE-8 for different thicknesses, ranging from 4.4 to 0.4 μm. One can clearly observe that the slope of the soft mode decreases with decreasing thickness and is accompanied by a smearing of the phase transition. Similar observations were reported in the experiments of Yang *et al.* (1991). The smearing of the phase transition seems to be related to the disordering effects of the surface, which acts as a random field.

Furthermore, one can observe that the minimum frequency of the $q = 0$ soft mode at the phase transition point increases with decreasing layer thickness. This can be explained by a simple analysis of the soft mode dynamics with fixed boundary conditions $\xi_x = \xi_y = 0$ at the surface of the cells. For $d < d_c$ this gives us the thickness dependence of the $q = 0$ soft mode relaxation rate,

$$\tau_s^{-1}(T, d, q = 0) = \frac{\alpha}{\gamma}(T - T_c) + \frac{K'}{\gamma}\frac{\pi^2}{d^2} \tag{9.13}$$

Here, K' is the splay elastic constant and γ is the corresponding viscosity. The effect of the confinement is thus reflected in an induced frequency gap at $q = 0$ in the soft mode dispersion at the phase transition point,

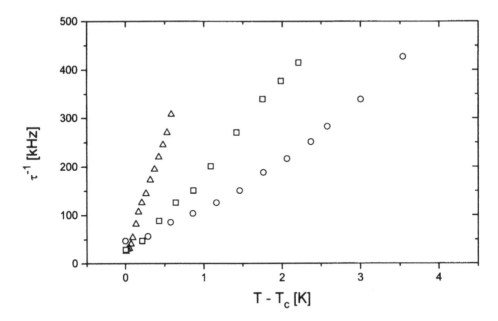

Figure 9.5 Setup for the measurements of the spectrum of the linear electro-optic response. Definitions: L, laser; P, polarizer; L1, L2, lenses; OV, temperature-stabilized oven; WS, wedge-type sample; A, analyser; PD, photodiode. The inset shows the real and imaginary parts of the photodiode current.

$$T = T_c, \quad q = 0: \quad G(d) = \frac{K \pi^2}{\gamma d^2} \tag{9.14}$$

It should be stressed that $q = 0$ is not the critical wavevector in the confined geometry of this experiment. As a result, the $q = 0$ soft mode anomaly at T_c is expected to be less and less pronounced as we move to smaller thickness. Such a behaviour was indeed observed in the experiments and is shown in the inset to Figure 9.7. Here, the soft mode gap at $q = 0$ and $T = T_c$ is shown as a function of the inverse-square thickness for a ferroelectric liquid crystal CE-8. One can observe good agreement with theory up to some minimum thickness. Similar behaviour was observed in the ferroelectric liquid crystal SCE-9 and will be published elsewhere (Škarabot *et al.*, 1995).

The effect of the confinement on the SmA → SmC̄ phase transition is thus reflected in the reduced soft mode slope and in an increase of the soft mode gap for $q = 0$ at the phase transition temperature. In addition, the SmA → SmC̄ phase transition is broadened and smeared at very small thicknesses.

The analysis of the recorded spectra of the linear electro-optic response reveals another surprising result: the spectral lineshape is more and more asymmetric as we move to smaller thicknesses, as is shown in Figure 9.8. The imaginary part of the linear response, which is symmetric on a log scale for a single Debye relaxation, is broadened on the low-frequency side. The broadening is more pronounced at smaller thicknesses and close to the transition point. This indicates that low-frequency excitations are present in the system for strong confinement, which is surprisingly similar to the 'slow' dynamics, observed in nematics, confined in porous media (Wu *et al.*, 1992; Bellini *et al.*, 1995). It should be stressed that this broadening

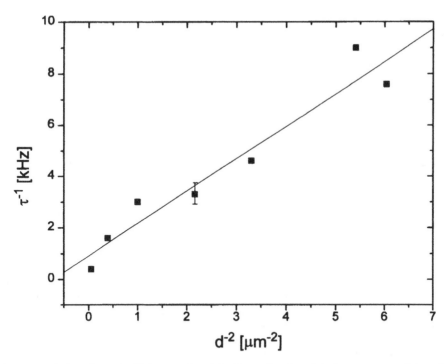

Figure 9.7 Soft mode gap in CE-8 at $q = 0$ and $T = T_c$ as a function of the inverse square of the cell thickness.

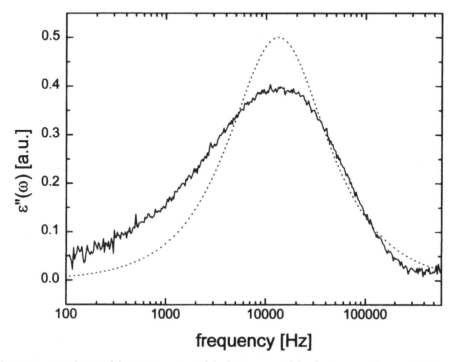

Figure 9.8 Broadening of the imaginary part of the linear susceptibility for strong confinement. CE-8, thickness $d = 0.5 \, \mu$m. The dashed line represents the single relaxation Debye response.

of the linear response on the low-frequency side is also in qualitative agreement with the reported observations of the so-called 'tail' of the order parameter time autocorrelation function in quasi-elastic light scattering experiments near the SmA → SmC* transition (Rastegar *et al.*, 1995). The origin of the low-frequency broadening of the imaginary part of the linear response is not clear, but is somewhat reminiscent of glassy dynamics and very probably also originates from the disordering effects of the confining surfaces.

9.4 Polymer Stabilization of Ferroelectric Liquid Crystals in Thin Cells

The concept of 'volume stabilization' of the molecular ordering in ferroelectric liquid crystals is based on the polymerization of a small amount of polymer precursor that is added to the liquid crystal. The technique was first used by Mariani *et al.* (1986) for smectic phases and more recently by Hikmet and coworkers for nematic liquid crystals (Hikmet 1990, 1991; Hikmet and de Witz, 1991). The amount of added polymer precursor is small enough so that it does not affect significantly the degree of molecular ordering or alignment of the liquid crystal. The polymerization of the polymer precursor is initiated when the liquid crystalline film acquires the desired orientation, which can be determined either by the boundary conditions or by the interaction with external electric or magnetic fields. Under these conditions, the polymerization leads to a highly anisotropic texture. It was observed (Pirš *et al.*, 1992, 1995a) that the polymer forms a network through the entire liquid crystal and 'volume-stabilizes' the structure. It was also observed that a part of the polymer phase separates upon polymerization and is expelled from the bulk of the liquid crystal to the bounding surfaces where it forms an anisotropic texture. In some parts of the cell the phase separation leads to a spontaneous nucleation and formation of droplets that grow from the surface into the bulk and eventually join the other surface.

As a result of polymerization and phase separation, one thus obtains a polymer network that is built into the liquid crystalline structure and eventually binds the two confining surfaces. This has very important technological implications, as the polymer structure significantly hinders the flow of liquid crystal under external mechanical stress and thus preserves the director configuration of smectic liquid crystalline phases that are well known for their high sensitivity to external forces and mechanical shock.

The volume-stabilized ferroelectric and antiferroelectric liquid crystalline cells were prepared essentially in the same way as conventional surface-stabilized devices. ITO-covered glass plates were spin or dip coated with a nylon orienting layer that was uniformly buffed in one direction. The thickness of the cells was 3 μm for the ferroelectric and 1.9 μm for the antiferroelectric liquid crystal. Cells were sealed with UV curable epoxy and vacuum filled either with ZLI 4237-100 (E. Merck) ferroelectric or MHPOBC ((R)-4-(1-methylheptoxyloxycarbonyl)-phenyl-4'-octyloxybiphenyl-4-carboxylate) antiferroelectric liquid crystal. A small amount (0.5 to 3 wt.%) of monomeric acrylate Desolite D044 was added in the isotropic phase of the liquid crystals. The cells were filled with liquid crystal–monomer solution and then slowly cooled down to the SmC* or SmC$_A^*$ phases, respectively.

The helical period of the ferroelectric liquid crystal ZLI 4237-100 is around 10 μm, so the cell thickness of 3 μm is sufficient to surface-unwind the helix. Together with homogeneous boundary conditions this results in a homogeneous chevron molecular ordering in the ferroelectric phase. Using a strong (80 V), slowly alternating ($\approx$ 3 Hz) electric field pulse, a well-known electric-field-induced stripe texture (Clark and Rieker, 1988; Shao *et al.*, 1991) was obtained in pure as well as polymer-doped samples. The angle between the directions of extinction of the two neighbouring stripe-like domains was of the order of a few degrees. This resulted in high contrast, very good bistability and very high angle between the optical axes

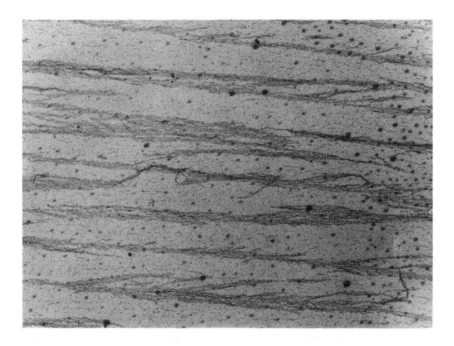

Figure 9.9 TEM micrograph of the ordered polymer texture formed during the UV-activated polymerization process in the polymer–ferroelectric liquid crystal mixture.

of the two bistable states. As the liquid crystal–monomer solution was filled in the isotropic phase of the liquid crystal, part of the polymer precursor phase separated when the solution was cooled down to the smectic phase. This phase separation was optically observed as the appearance of small ($< 1\ \mu m$) microdroplets that condensed on the two surfaces.

The polymerization of the polymer precursor was induced in the ferroelectric phase using the UV light source for a few minutes. As a result of polymerization, the polymer phase separated mainly in the vicinity of the domain walls, separating the two stripe-like domains. Defects in the structure thus act as polymerization centres, nucleating the growth of phase-separated polymer. This observation was proved by heating the ferroelectric liquid crystal into the isotropic phase and then cooling to the ferroelectric phase. The structure that reappeared in the SmC* phase was identical to the one observed before heating the cell. This indicates that the polymer structure is indeed formed in the disordered regions close to domain walls.

The surface of the confining glass plates was analysed using electron microscopy and atomic force microscopy (AFM). The cells were disassembled and flushed with solvent, thus removing the liquid crystal and hopefully preserving the polymer network. The quasi-ordered polymer network was indeed observed by electron microscopy as shown in Figure 9.9. One can clearly see that the preferential direction of the ordering of the network globally matches the direction of the smectic layer normal and thus also the direction of domain walls. The separation between the massive polymer structure is of the order of a few micrometres which is consistent with the observed width of stripe-like ferroelectric domains. One can also see in Figure 9.9 the presence of microdroplets and a very fine structure between them. This fine structure was

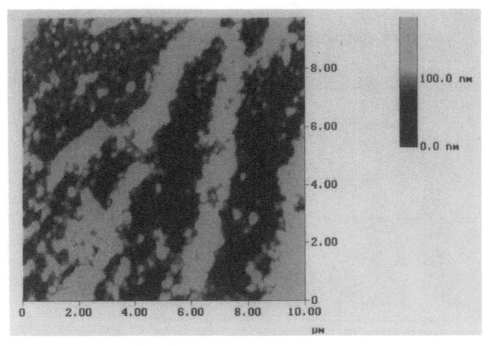

Figure 9.10   AFM image of the polymer texture formed on the surface of the confining glass plates.

further analysed with atomic force microscopy to reveal the structural details on the nanometre scale. The samples were analysed with Nanoscope III AFM in the so-called 'tapping mode'. Here, the surface repulsive forces are detected by a very sharp tip, mounted on a vibrating cantilever. In the tapping mode of operation the order of magnitude of the detected surface repulsive forces is smaller than in the contact mode of operation and is thus suitable for observing soft-matter surfaces. We have indeed observed that scanning in the contact mode of operation significantly changes the polymer-coated surfaces even on large height scales, because the observed profile changes with time. On the contrary, the polymer surface profile as observed in the tapping mode does not change significantly on a height scale of 100 nm. Still, when observing the soft polymer surface on the nanometre scale, one can notice that the surface profile slightly changes with time. This is an indication that, on the nanometre scale, soft polymer surfaces are significantly affected by the perturbation of the scanning tip.

The results of the AFM study of surfaces of disassembled volume-stabilized liquid crystalline cells are shown for different length scales in Figure 9.10, 9.11, and 9.12. Figure 9.10 shows a 10×10 μm^2 region that clearly reveals the local preferential direction of a polymer network. The width of the 'polymer stripes' is of the order of a micrometre and the height is of the order of 100 nm. This indicates that the observed polymer stripes are the remnants of polymer that was phase separated in the region of the domain walls. Figure 9.11 shows an enlarged part of the surface in Figure 9.10. One can see a fine-grained structure that is present in between the polymer strips. The width of the grains is of the order of 100 nm, whereas the height is of the order of 20 nm. The grains are elongated in the direction of the smectic layer normal. Figure 9.12 shows an enlarged part of the surface shown in Figure 9.11. Here, one can see more clearly the anisotropic shape of the 'polymer hills' and the dimensions are in agreement with the objects shown in the previous figure. We have to stress, however, that on this length and height scale, the AFM image changes with time of scanning, indicating that the surface is perturbed on the nanometre scale even in the low-force regime. This indicates that the polymer surface is a rather soft and not very compact surface.

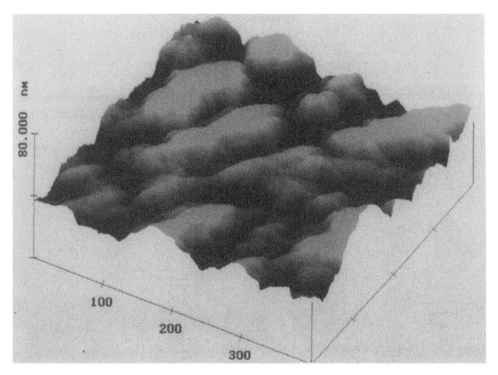

Figure 9.11 Enlarged AFM image of the polymer texture shown in Figure 9.10.

From the above measurements we can estimate the distribution and the amount of polymer material that is phase separated from the ferroelectric phase. A rough calculation shows that almost all of the polymer is expelled from the bulk of the liquid crystal into 'polymer stripes', microdroplets or surface-condensed 'polymer hills'. Approximately 30 % of the material is condensed at the surface, 30 % in the disordered regions in between the stripe-like domains, and the rest is phase separated in the form of microdroplets.

Qualitatively different behaviour was observed for phase separation in the antiferroelectric phase. Here, phase separation occurs predominantly via the formation of microdroplets growing from one surface to the other, finally binding the two surfaces. The reason for such different behaviour is not known.

In conclusion, we have shown that polymer phase separation induced in the smectic phases leads to the formation of three different structures on the nanometre scale. Owing to the presence of wall defects and disordered regions in smectic phases, a stripe-like polymer network is phase separated, extending in between the two confining surfaces. In addition, a large amount of the polymer 'condenses' on the confining surfaces, where it forms 'hill-like' structures on the 10 nm scale. In some parts of the sample, spontaneous growth of phase-separated microdroplets eventually leads to the formation of large droplets, binding the two surfaces firmly together. Whereas the appearance of these large microdroplets is important from the technological point of view (Pirš *et al.*, 1992, 1995a,b), the appearance of surface-condensed polymer may be important from the fundamental point of view, bridging the gap between the smooth, ordered polymer surfaces and disordered surfaces. Here, the extent of the randomness of the confining surfaces could be externally controlled by the amount of polymer that phase separates at the surface.

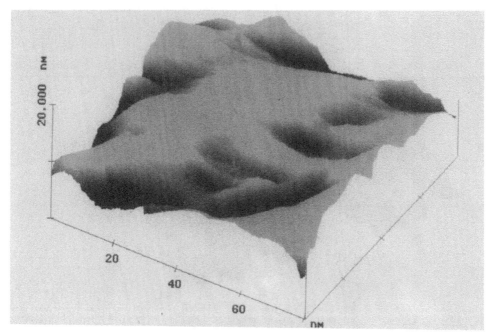

Figure 9.12   Enlarged AFM image of the polymer texture shown in Figure 9.11.

9.5   Conclusions

We have presented experimental evidence that clearly shows the onset of complex behaviour in the vicinity of the SmA → SmC* and SmA → SmC̄ phase transitions in ultrathin cells of submicrometre thickness. The complex behaviour is reflected both in the confinement-induced gap at $q = 0$ for the order parameter excitations and in the unusual and asymmetric form of the complex part of the linear response spectrum. These observations are analogous to the observed dynamics of the order parameter in complex geometries such as porous matrices and thus must have the same physical framework and origin. We have also shown that volume stabilization of smectic phases leads to enhanced randomness of the confining surfaces, which is a result of the polymer phase separation. This system is thus a promising candidate in which the magnitude of surface randomness could be externally varied and this should be reflected in the thermodynamic properties of ultrathin ferroelectric layers.

References

ALLENDER, D. W., HENDERSON, G. L. and JOHNSON, D. L. (1981) Landau theory of wall-induced phase nucleation and pretransitional birefringence at the isotropic–nematic transition, *Phys. Rev. A* **24**, 1086–1089.

BELLINI, T., CLARK, N. A. and SCHAEFER, D. W. (1995) Dynamic light scattering study of nematic and smectic-A liquid crystal ordering in silica aerogel, *Phys. Rev. Lett.* **74**, 2740–2743.

BIRADAR, A. M., WROBEL, S. and HAASE, W. (1989) Dielectric relaxation in the smectic-A and smectic-C* phases of a ferroelectric liquid crystal, *Phys. Rev. A* **39**, 2693–2702.

BLINC, R. and ŽEKŠ, B. (1978) Dynamics of helicoidal ferroelectric smectic-C* liquid crystal, *Phys. Rev. A* **18**, 740–745.

BOURDON, L., SOMMERIA, J. and KLÉMAN, M. (1982) Sur l'existence de lignes singulières dans les domaines focaux en phases SmC et SmC*, *J. Phys.* **43**, 77–90.

CARLSSON, T., ŽEKŠ, B., FILIPIČ, C., LEVSTIK, A. and BLINC, R. (1988) Thermodynamic model of ferroelectric chiral smectic C* liquid crystal, *Mol. Cryst. Liq. Cryst.* **163**, 11–71.

CAVA, R. J., PATEL, J. S., COLLEN, K. R., GOODBY, J. W. and RIETMAN, E. A. (1987) Thin cell dielectric response of a ferroelectric liquid crystal, *Phys. Rev. A* **35**, 4378–4388.

CLARK, N. A. and LAGERWALL, S. T. (1980) Submicrosecond bistable electro-optic switching in liquid crystals, *Appl. Phys. Lett.* **36**, 899–901.

CLARK, N. A. and LAGERWALL, S. T. (1984) Surface-stabilized ferroelectric liquid crystal electro-optics: New multistate structures and devices, *Ferroelectrics* **59**, 25–67.

CLARK, N. A. and RIEKER, T. P. (1988) Smectic-C 'chevron', a planar liquid-crystalline defect: Implications for the surface stabilized ferroelectric liquid crystal geometry, *Phys. Rev. A* **37**, 1053–1057.

DREVENŠEK, I., MUŠEVIČ, I. and ČOPIČ, M. (1990) Dispersion of the SmC* order parameter fluctuations in the SmA and SmC* phases of 4-(2′-methylbutyl)phenyl 4′-n-octylbiphenyl-4-carboxylate, *Phys. Rev. E* **41**, 923–928.

DUMRONGRATTANA, S., HUANG, C. C., NOUNESIS, G., LIEN, S. C. and VINNER, J. M. (1986) Tilt angle, polarization, and heat-capacity measurements near the smectic-A–chiral-C phase transition of p-(n-decyloxybenzylidene)-p-amino-(2-methylbutyl) cinnamate (DOBAMBC), *Phys. Rev. A* **34**, 5010–5019.

GAROFF, S. and MEYER, R. B. (1977) Electroclinic effect at the A–C* phase change in a chiral smectic liquid crystal, *Phys. Rev. Lett.* **38**, 848–851.

GAROFF, S. and MEYER, R. B. (1979) Electroclinic effect at the A–C* phase change in a chiral smectic liquid crystal, *Phys. Rev. A* **19**, 338–347.

GLOGAROVA, M. and PAVEL, J. (1984a) The structure of chiral SmC* liquid crystal in planar samples and its change in an electric field, *J. Phys.* **45**, 143–149.

GLOGAROVA, M. and PAVEL, J. (1984b) The behaviour of thin samples of ferroelectric liquid crystal, *Mol. Cryst. Liq. Cryst.* **114**, 249–257.

GLOGAROVA, M., LEJČEK, L., PAVEL, J., JANOVEC, V. and FOUSEK, J. (1983) The influence of an external electric field on the structure of chiral SmC* liquid crystal, *Mol. Cryst. Liq. Cryst.* **91**, 309–325.

HANDSCHY, M. A. and CLARK, N. A. (1984) Structures and responses of ferroelectric liquid crystals in the surface stabilized geometry, *Ferroelectrics* **59**, 69–116.

HIKMET, R. A. M. (1990) Electrically induced light scattering from anisotropic gels, *J. Appl. Phys.* **68**, 4406–4412.

HIKMET, R. A. M. (1991) Anisotropic gels and plasticized networks by liquid crystalline molecules, *Liq. Cryst.* **9**, 405–416.

HIKMET, R. A. M. and DE WITZ, C. (1991) Gel layers for inducing adjustable pretilt angles in liquid crystalline system, *J. Appl. Phys.* **70**, 1265–1269.

HSIUNG, H., RASING, Th. and SHEN, Y. R. (1986) Wall-induced orientational order of a liquid crystal in the isotropic phase — An evanescent-wave-ellipsometry study, *Phys. Rev. Lett.* **57**, 3065–3068.

KAI, S., NAKAGAWA, M., NARUSHIGE, Y. and IMASAKI, M. (1983) Anomalous behaviour of helicoidal pitch in surface induced transition from winding to unwinding chiral smectic C, *Jpn. J. Appl. Phys.* **22**, L488–L490.

KONDO, K., TAKEZOE, H., FUKUDA, A. and KUZE, E. (1982) Temperature sensitive helical pitches and wall anchoring effects in homogeneous monodomains of ferroelectric SmC* liquid crystal nOBAMBC (n = 6–10), *Jpn. J. Appl. Phys.* **21**, 224–229.

KONDO, K., TAKEZOE, H., FUKUDA, A., KUZE, E., FLATISCHER, K. and SKARP, K. (1983) Surface induced helix-unwinding process in thin homogeneous ferroelectric smectic cells of DOBAMBC, *Jpn. J. Appl. Phys.* **22**, L294–L296.

KRALJ, S. and SLUCKIN, T. J. (1994) Landau-de Gennes theory of the chevron structure in a smectic-A liquid crystal, *Phys. Rev. E* **50**, 2940–2951.

KUCZYNSKI, W., STRYLA, B., HOFFMAN, J. and MALECKI, J. (1990) Investigation of the smectic C*-smectic A phase transition in electric field, *Mol. Cryst. Liq. Cryst.* **192**, 301–305.

LEJČEK, L. (1984) The disclination model of field induced transition between twisted and uniform structure in thin planar SmC* samples, *Ferroelectrics* **58**, 139–148.

LEVSTIK, A., CARLSSON, T., FILIPIČ, C., LEVSTIK, I. and ŽEKŠ, B. (1987) Goldstone mode and soft mode at the smectic-A–smectic-C* phase transition studied by dielectric relaxation, *Phys. Rev. A* **35**, 3527–3534..

LIMAT, L. and PROST, J. (1993) A model for the chevron structure obtained by cooling a smectic-A liquid crystal in a cell of finite thickness, *Liq. Cryst.* **13**, 101–113.

LUBAN, M., MUKAMEL, D. and SHTRIKMAN, S. (1974) Transition from the cholesteric storage mode to the nematic phase in critical restricted geometries, *Phys. Rev. A* **10**, 360–367.

MARIANI, P., SAMORI, B., ANGELONI, A. S. and FERRUTI, P. (1986) Polymerization of bisacrylic monomers within a liquid crystalline smectic-B solvent, *Liq. Cryst.* **1**, 327–336.

MAUGER, A., ZRIBI, G., MILLS, D. L. and TONER, J. (1984) Substrate-induced orientational order in the isotropic phase of liquid crystals, *Phys. Rev. Lett.* **53**, 2485–2488.

MEYER, R. B., LIEBERT, L., STRZELECKI, L. and KELLER, P. (1975) Ferroelectric liquid crystals, *J. Phys.* **36**, L69–L71.

MICHELSON, A. (1977) Physical realization of a Lifshitz point in liquid crystal, *Phys. Rev. Lett.* **39**, 464–467.

MIYANO, K. (1979) Wall-induced pretransitional birefringence: A new tool to study boundary aligning forces in liquid crystals, *Phys. Rev. Lett.* **43**, 51–54.

MUŠEVIČ, I., BLINC, R., ŽEKŠ, B., FILIPIČ, C., ČOPIČ, M., SEPPEN, A., WYDER, P. and LEVANYUK, A. P. (1988) Observation of phason dispersion in a ferroelectric liquid crystal by light scattering, *Phys. Rev. Lett.* **60**, 1530–1533.

MUŠEVIČ, I., ŽEKŠ, B., BLINC, R. and RASING, Th. (1993) Magnetic field induced biaxiality in an antiferroelectric liquid crystal, *Phys. Rev. E* **47**, 1094–1100.

O'BRIEN, J. P., MOSES, T., CHEN, W., FREYSZ, E., OUCHI, Y. and SHEN, Y. R. (1993) Reorientation dynamics of ferroelectric liquid crystal molecules near the smectic-A–smectic-C* transition, *Phys. Rev. E* **47**, R2269–R2272.

PANARIN, Y. P., KALMYKOV, Y. P., MacLUGHADA, S. T., XU, H. and VIJ, J. K. (1994) Dielectric response of surface stabilized ferroelectric liquid crystal cells, *Phys. Rev. E* **50**, 4763–4772.

PIRŠ, J., BLINC, R., MARIN, B., PIRŠ, S., and DOANE, J. W. (1995a) Polymer network volume stabilized ferroelectric liquid crystal display, *Mol. Cryst. Liq. Cryst.* **264**, 155–163.

PIRŠ, J., BLINC, R., MARIN, B., PIRŠ, S., MUŠEVIČ, I., ŽUMER, S. and DOANE, J. W. (1992) Poster presented at *14th ILCC Conf. Pisa, Italy*.

PIRŠ, J., BLINC, R., ŽUMER, S., MUŠEVIČ, I., MARIN, B., PIRŠ, S., and DOANE, S. W. (1995b) Ferroelectric liquid crystal cell, a method of making it, and its use, US Patent 5,434,685.

POVŠE, T., MUŠEVIČ, I., ŽEKŠ, B. and BLINC, R. (1993) Phase transitions in ferroelectric liquid crystals in a restricted geometry, *Liq. Cryst.* **14**, 1587–1598.

RAPINI, A. and PAPOULAR, M. (1969) Distortion d'une lamelle nématique sous champ magnétique conditions d'ancrage aux parois, *J. Phys. Coll.* **30**, C4-54-56.

RASTEGAR, A., MUŠEVIČ, I. and ČOPIČ, M. (1995) Dynamic light scattering in the vicinity of the SmA–SmC* phase transition in a ferroelectric liquid crystal in thin cells, Poster SC-23 presented at the *Eur. Conf. on Liquid Crystals, 5–10 March 1995, Bovec* (to be published).

SHAO, R. F., WILLIS, P. C. and CLARK, N. A. (1991) The field induced stripe texture in surface stabilized ferroelectric liquid crystalline cells, *Ferroelectrics* **121**, 127–136.

SHENG, P. (1976) Phase transition in surface-aligned nematic films, *Phys. Rev. Lett.* **37**, 1059–1062.

ŠKARABOT, M., MUŠEVIČ, I. and BLINC, R. (1995) Dynamics of ferroelectric liquid crystals in submicron layers, Poster SC-24 presented at the *Eur. Conf. on Liquid Crystals, 5–10 March 1995, Bovec* (to be published).

SLUCKIN, T. J. and PONIEWIERSKI, A. (1990) Wetting and capillary condensation in liquid crystal system, *Mol. Cryst. Liq. Cryst.* **179**, 349–364.

SLUCKIN, T. J. and PONIEWIERSKI, A. (1985) Novel surface phase transition in nematic liquid crystals: Wetting and the Kosterlitz–Thouless transition, *Phys. Rev. Lett.* **55**, 2907–2910.

TARCZON, J. C. and MIYANO, K. (1980) Surface-induced ordering of a liquid crystal in the isotropic phase, *J. Chem. Phys.* **73**, 1994–1998.

VALLERIEN, S. U., KREMER, F., KAPITZA, H., ZENTEL, R. and FRANK, W. (1989) Field dependent soft and Goldstone mode in a ferroelectric liquid crystal as studied by dielectric spectroscopy, *Phys. Lett. A* **138**, 219–222.

VAN SPRANG, H. A. ((1983) Surface-induced order in a layer of nematic crystal, *J. Phys.* **44**, 421–426.

WU, X. I., GOLDBURG, W. I., LIU, M. X. and XUE, J. Z. (1992) Slow dynamics of isotropic–nematic phase transition in silica gels, *Phys. Rev. Lett.* **69**, 470–473.

YANG, Y. B., BANG, T., MOCHIZUKI, A. and KOBAYASHI, S. (1991) The surface anchoring dependence of the dielectric constant of an FLC material showing electroclinic effect, *Ferroelectrics* **121**, 113–125.

YOKOHAMA, H. (1988) Nematic–isotropic transition in bounded thin films, *J. Chem. Soc., Faraday Trans.* **84**, 1023–1040.

10

Liquid Crystalline Networks Formed by Photoinitiated Chain Cross-linking

D. J. BROER

10.1 Introduction

Polymer and gel systems with three-dimensional control over the molecular order are of great theoretical and practical interest because of their unusual, but very accurately adjustable and addressable, properties. The methods to produce these structures are limited in number and no general procedure can be given which enables complete and unrestricted control over the position of the molecules in all dimensions. However, a method which comes close to this is the photoinitiated polymerization of liquid crystalline (LC) monomers (Broer et al., 1989b, c, 1991; Broer, 1993a). The different LC states of low-molar-mass reactive mesogens provide a large variety of molecular order, all being accessible to be fixed by the polymerization process (Broer and Heynderickx, 1990, Broer, 1995; Kitzerow et al., 1993; Lub et al., 1995). The various known techniques to establish monolithic molecular order in liquid crystals can be applied or even combined with each other to create films of complicated molecular architectures that are optically free of defects (Broer, 1995). The monomers allow a large degree of freedom in their molecular structure to tailor the mechanical and optical properties of the films. Moreover, blends of monomers can be used to adjust the properties in the monomeric state, e.g. the LC transition temperatures and the flow viscosity, and in the polymeric state, e.g. the elastic modulus, the glass transition temperature and the refractive indices. When LC monomers, or mixtures thereof, with a higher functionality are polymerized in the bulk, the so-called LC networks are produced which exhibit a stable molecular organization up to the degradation temperature of the polymers. When the LC monomers are polymerized in the presence of non-reactive liquid crystals the so-called plasticized LC networks and LC gels are formed depending on the ratio between the reactive and non-reactive LC component (Hikmet 1991a,b; Hikmet and Yang et al., this volume). In the case of plasticized LC networks the physical properties of the polymeric films can be modified and tailored by the addition of relatively small amounts of liquid crystals, which are nowadays available in a large variety. The attractiveness of the LC gels is their ability to alter the molecular organization of the mobile LC molecules by electrical addressing, whereas the polymeric part maintains its initial order, as such providing a memory function for the molecular architecture as soon as the electrical field is switched off.

The history of LC networks goes back to the end of the 1960s where different authors suggested polymerizing and cross-linking liquid crystals in their mesophase aiming for highly ordered polymers (Wendorff, 1967; de Gennes, 1969). Thermally initiated bulk polymerization (thermosetting) of LC diacrylates was reported to yield three-dimensionally cross-linked polymer networks with a strong optical anisotropy and the molecular order of the frozen-in monomeric phase was retained to the decomposition temperature (Strzelecki and Liebert, 1973; Bouligand *et al.*, 1974; Clough *et al.*, 1976; Arslanov and Nikolajeva, 1984). However, the use of high temperatures to initiate the polymerization often conflicts with the temperature ranges of the LC phases of many of the reactive liquid crystals. During the heating to, and the processing in, the LC phase, polymerization and cross-linking already starts before the desired phase and orientation are achieved and defects become permanently frozen in. It is for this reason that photoinitiation is highly preferable for bulk polymerization and network formation. In the presence of small amounts of polymerization inhibitors the monomers can be processed in their LC state until the desired order has been obtained. From this moment on the molecular arrangement can be fixed very rapidly by exciting a dissolved photoinitiator with actinic light of an apropriate wavelength, mostly around 360 nm. The first reports on the bulk photo-polymerization of reactive liquid crystals were on monoacrylates forming linear LC side-chain polymers (Shannon, 1984; Broer *et al.*, 1988, 1989a; Hoyle *et al.*, 1988, 1989). Real fixation of the molecular order in this case very often does not occur and the formed polymer still exhibits various mesophases, though the temperatures differ from those of the initial monomer. As a result phase transitions from one LC phase into another or from the LC phase to the crystal phase take place during the polymerization process, and consequently the molecular order changes somewhere during the polymerization process. Also phase separation might take place if the LC state of the polymer formed is not mixable with that of the initial monomer, which results in many optical defects in the aimed, and initially established, monolithic order. It is for this particular reason that the bulk photopolymerization of polyfunctional LC monomers (photosetting liquid crystals) became so important. For instance, the free-radical-based photopolymerization of monolithically ordered nematic LC diacrylates resulted in a stable polymer with the same texture and almost the same degree of molecular orientation (Broer *et al.*, 1989b, c, 1991). But also the use of LC diepoxides (Broer *et al.*, 1993; Jahromi *et al.*, 1994) and LC divinylethers (Johnson *et al.*, 1991; Andersson *et al.*, 1992; Hikmet *et al.*, 1993) polymerized by photocationic mechanisms essentially lead to the same results. The photosetting LC gels based on conventional low-molar-mass LCs stabilized by anisotropic LC diacrylate networks were first reported for nematic blends (Hikmet 1990, 1991a, b). Their interesting electro-optical properties lead to a broad interest in these types of structures including cholesterics (Hikmet and Zwerver 1991, 1992, 1993b; Yang *et al.*, 1994; see also the chapters in this volume by Yang *et al.* and Yuan), smectic-A (Hikmet and Howard, 1993) and chiral smectic-C (ferroelectric) phases (Hikmet and Michielsen, 1995, Blinc *et al.*, this volume).

The optical properties of thin films of the monolithically ordered LC networks are quite similar to those of the low-molar-mass liquid crystals. They are transparent for light in the visible wavelength region, exhibit a high birefringence up to 0.25 at 589 nm (Broer *et al.*, 1991), are able to show selective reflection in the case of a cholesteric order (Broer, 1995) and exhibit a half-wave optical retardation in the case of a twisted-nematic molecular arrangement (Broer and Heynderickx, 1990; Heynderickx and Broer, 1991). These properties make the materials ultimately suitable for the creation of various optical devices as polarizing beam splitters, organic Wollaston prisms, retardation films, colour and infrared filters, etc. All properties are temperature insensitive and stable against environmental (light, humidity, temperature) ageing. The mechanical properties are, apart from their anisotropic nature, of the same class as those of the isotropic acrylate and epoxide networks, meaning that the modulus and strength are of the same order and depend strongly on the molecular parameters like cross-link density and the ratio between stiff and flexible units (Hikmet and Broer, 1991). An interesting mechanical feature is that, owing to a unidirectional or planar molecular order, the thermal expansion can be brought back to essentially zero in one or two directions. This implies that the build-up of thermal stresses is avoided when the networks are applied as coatings or encapsulants on an inorganic substrate (Broer and Mol, 1991). The electro-optic behaviour of the LC gels, e.g.

electrically addressed light scattering or suppression of the cholesteric reflection band, make them ultimately suitable for the production of active optical components such as non-mechanical light valves, shutters and adjustable colour filters.

It is the aim of this chapter to discuss the formation and some of the properties of the LC networks. The structure and behaviour of the LC gels will be discussed elsewhere in this book by Hikmet.

10.2 Liquid Crystalline Monomers

Most of the publications on difunctional reactive liquid crystals concern mesogens with central groups containing three aromatic rings. The three-ring mesogens without reactive end groups have been known for a long time (Dewar and Schroeder, 1965). They form stable mesophases with broad processing windows. Table 10.1 shows some examples of LC monomers based on the same mesogens and compares the influence of the different end groups. It illustrates that the various reactive moieties studied have some, but not a tremendous, effect on the melting temperatures. The somewhat bulky acrylate and methacrylate groups destabilize the mesophases (C6H, C5H and C5H-M), i.e. lower the smectic and nematic transition temperatures. The transition temperatures of the vinylether (C6H-V) and the epoxide (C4H-E) functionalized liquid crystals are very similar to their analogues without reactive end groups. As a result these materials in general have broad LC phases with higher order parameters as polymerization at temperatures far from the isotropic transition is possible.

Table 10.1 Influence of functional moiety on mesomorphism of reactive liquid crystals of the type

$$\text{R--O--}\underset{}{\bigcirc}\text{--}\overset{O}{\underset{\parallel}{C}}\text{--O--}\bigcirc\text{--O--}\overset{C}{\underset{\parallel}{}}\text{--}\bigcirc\text{--O--R}$$

Sidegroup R	Abbreviation	Mesomorphism[a]		Ref.
$CH_3-(CH_2)_5-$	–	Cr 111(S_x 120) N 205 I		Dewar and Schroeder (1965)
$CH_2=CH-\overset{O}{\underset{\parallel}{C}}-O-(CH_2)_6-$	C6H	Cr 108 (S_C 88) N 155 I		Broer et al. (1984)
$CH_2=CH-\overset{O}{\underset{\parallel}{C}}-O-(CH_2)_3-\overset{CH_3}{\underset{*}{CH}}-(CH_2)_2-$	C6*H	Cr 69	Ch 97 I	Lub et al. (1995)
$CH_2=CH-\overset{O}{\underset{\parallel}{C}}-O-(CH_2)_5-$	CH5	Cr 92	N 170 I	Broer et al. (1991), Broer (1993)
$CH_2=\overset{CH_3}{\underset{}{C}}-\overset{O}{\underset{\parallel}{C}}-O-(CH_2)_5-$	C5H-M	Cr 74	N 158 I	–
$CH_2=CH-O-(CH_2)_6-$	C6H-V	Cr 110	N 194 I	Hikmet et al. (1992)
$\overset{O}{\underset{}{CH_2-CH}}-(CH_2)_4-$	C4H-E	Cr 119	N 213 I	Boer (1993)

[a] Cr, S_X, S_C, N, I are the crystalline, an unknown smectic, the smectic C, the nematic and the isotropic phase, respectively. The intermediate numbers denote the transition temperature in °C between the adjacent phases.

The general trend is that in the materials studied the mesophases occur at elevated temperatures making high-temperature processing and curing necessary. Substitution of the central ring with methyl or chlorine breaks the symmetry of the reactive liquid crystals and, therefore, lowers the melting temperatures and suppresses smectic phase formation (Broer, 1993). This makes these materials ultimately suitable for applications based on thin films with a nematic molecular arrangement. Bringing a chiral methyl substituent in the side group destabilizes the mesophase even further (C6*H) but introduces the cholesteric molecular order which is interesting for many applications.

10.3 Photo-initiated Polymerization and Cross-linking

10.3.1 *Kinetic Aspects*

In order to enhance the sensitivity to UV light, small amounts of photoinitiator are added to the reactive LC formulations. A typical free-radical initiator for polymerization of the LC diacrylates is α,α-dimethoxydeoxybenzoin (Irgacure 651, CIBA Geigy). For the cationic polymerization of the LC vinylethers and epoxides diphenyliodonium hexafluoroarsenate (Lohse and Zweifel, 1986) is used, optionally sensitized by α,α-dimethoxydeoxybenzoin. The concentrations used are typically of the order of 1 wt.% which does not affect the monomeric transition temperatures too much. Such sensitized LC formulations can be photopolymerized by exposure to UV light in the region between 300 and 400 nm. Very convenient UV sources are the fluorescent lamps emitting at 340 nm (Philips TL08 or TL09) or at 365 nm (Philips PL10). With these lamps polymerization proceeds on a time scale of seconds to minutes, which is fast enough for practical applications and slow enough for kinetic experiments. The use of moderate-intensity lamps also avoids heating of the samples and thus unwanted phase transitions. In the case of the free-radical polymerization of the LC diacrylates the presence of oxygen should be avoided as much as possible as it inhibits chain propagation by the formation of peroxy radicals. It is, however, sufficient to cover the films with a transparent substrate such as glass or polyester. The dissolved oxygen will be consumed at the onset of the reaction and not supplied fast enough during polymerization. In the case of the cationic mechanisms oxygen plays a less dominant role but it is the presence of moisture which might lead to irreproducible results.

The kinetics of photoinitated chain polymerization is usually described in terms of initiation, propagation and termination and under steady-state conditions of diluted systems the following generally accepted equation is valid (Odian, 1981):

$$R_p = k_p [M] \left(\phi I_0 \frac{1 - \exp(-\varepsilon[P]d)}{k_t} \right)^{1/2}$$

in which R_p is the rate of polymerization, k_p and k_t are the rate constants of propagation and termination, $[M]$ is the monomer concentration, Φ is the overall quantum efficiency for dissociation and initiation, ε and $[P]$ are the molar extinction coefficient and the concentration of the photoinitiator, respectively, I_0 is the intensity of the incident light and d is the thickness of the sample. The equation shows that the polymerization rate is linearly related to the monomer concentration, meaning that a gradual and linearly decrease should be expected for the polymerization rate as a function of conversion assuming the photoinitiator concentration, light absorption and the like are not changing during polymerization. For the chain cross-linking polymerization of polyfunctional LC monomers this equation has only a qualitative value as many sources of deviations are present. Figure 10.1 shows the course of the polymerization rate of an LC diacrylate as a function of its conversion into cross-linked polymer. It becomes clear that no real steady state is established as already at a low conversion the system forms a gel in which the hindered molecular mobility of the reactive chain ends causes autoacceleration (Trommsdorff effect). Consequently, the rate reaches high values and the polymer formed at

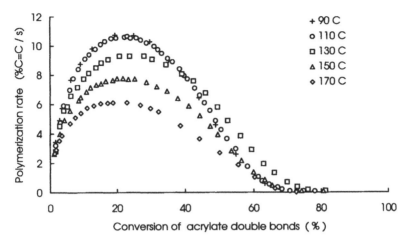

Figure 10.1 Conversion rate of C5H LC diacrylate (cf. Table 10.1) during photoinitiated isothermal chain cross-linking as a function of conversion measured at the indicated polymerization temperatures.

this stage of the polymerization has a large kinetic chain length, estimated to lie between 10^3 and 10^6. The rate reaches its maximum between 20% and 40% conversion. From this stage on a number of factors cause the decrease in polymerization rate:

1   the concentration of the monomer decreases which should lead to a decrease in rate according to the above equation,

2   the monomer becomes hindered in its diffusion through the solidifying medium, and

3   the pendant reactive groups of those monomers which have already reacted at one side are drastically immobilized.

Other factors affecting the polymerization at a higher conversion may be decreasing dissociation efficiency of the photoinitiator in a solidifying medium (cage effect) and the highly hindered monomer mobility in a vitrifying system in which time- and thus rate-dependent free-volume effects may play a role. In this respect the reactive liquid crystals behave exactly the same as isotropic diacrylates. As a result of all these more or less simultaneously occurring phenomena the rate drops to almost zero around a conversion of 80% and is no longer detectable by, for instance, photo-DSC. Infrared studies, however, revealed that the polymerization continues very slowly until a conversion higher than 90% is reached as can be concluded from the complete disappearance of the vinyl absorption bands at 840 and 1636 cm^{-1}.

As can be seen from Figure 10.1 the maximum polymerization rate decreases in the temperature range from 90°C towards higher temperatures. This effect is displayed in Figure 10.2 for different monomers. Below 90°C there is a tendency for the maximum polymerization rate to increase with temperature due to an increased mobility of the monomeric units in the reacting medium (viscosity effect). This effect is very clearly visible for the C6M monomer which has a supercooled nematic phase down to room temperature. The decrease at higher temperatures is a phenomenon also seen for isotropic diacrylates and is caused by the thermodynamic polymerization parameters (ceiling temperature) causing a change in the propagation/depropagation equilibrium. Extrapolating the curves in Figure 10.2 to a zero rate reveals a temperature of around 220°C where monomer and polymer should be in equilibrium with each other.

An interesting question of course is whether the molecular order in the mesophases has any effect on the polymerization kinetics. Although this might very well be possible (there are

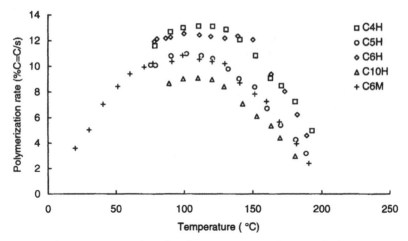

Figure 10.2 Peak conversion rate of LC diacrylate monomers as a function of polymerization temperature. The number in the monomer notation, shown in the key, represents the length of the flexible alkylene spacer between the acrylate ester group and the other ether linkage to the central core. C6M corresponds to C6H (Table 10.1) with a methyl group substituted in the central phenylene group.

indications that the termination reaction as well as free-radical transfer reactions are affected by the molecular order) the temperature-related thermodynamic and diffusion parameters of depropagation and viscosity, however, dominate eventual effects associated with the specific molecular organization in the N or S_C phase. In the polymerization rate curves of Figure 10.2 no discontinuities are to be seen around the phase transitions indicating that the effects, if any, must be relatively small.

The effects observed for the free-radical polymerization of diacrylates are also found for the cationic polymerization of divinylethers and the diepoxides, although the location on both the temperature and the conversion axes might be somewhat different. Divinylethers especially behave very much the same as diacrylates up to a temperature of around 100°C. Above that temperature the reaction becomes unstable because of instabilities in the initiator system. The polymerization of the LC diepoxides is far more complicated than that of the diacrylates because more side reactions occur. The degree of conversion of the epoxy groups remains lower for these monomers and often thermal post-treatments are needed to achieve dimensionally stable products.

10.3.2 Fixation of Order

For the LC diacrylates, the LC diepoxides and the LC divinylethers it has been demonstrated that during the photoiniated bulk polymerization the initial texture of the monomer is preserved. For instance, polymerization in a monodomain N phase yields a network with a monodomain nematic type of molecular organization which is transparent and has a birefringence comparable with its initial monomer. The X-ray diffraction pattern is typical for a system with nematic order. However, when polymerized near an S transition temperature the SAXS diffraction pattern indicates that either fluctuations already present in the monomer are preserved or that some local smectic (cybotactic) ordering is formed upon polymerization (Hikmet and Broer, 1989). Polymerization in the S_C state of the monomer yields a network with the typical smectic-C X-ray pattern. The Bragg spacing derived from the latter is consistent with a stretched conformation of the molecules between the acrylate chains with an oblique angle of the

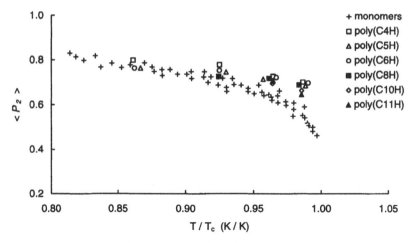

Figure 10.3 Order parameter of LC diacrylates at the polymerization temperature expressed as the reduced temperature T/T_c (K/K) before and after polymerization. Data were derived from UV–VIS dichroic measurements. For the monomer $\langle P_2 \rangle$, the data of the various LC diacrylate monomers all fall on one master curve and are not further indicated.

molecular axis with the smectic layers. During the polymerization the intermolecular distance perpendicular to the long axes tends to decrease enforced by the covalent linkages formed during the growth of the polymer mains (Hikmet *et al.*, 1992). This leads to a somewhat higher packing of the molecular rods with typical spacings between 0.45 and 0.50 nm, actually smaller than the width of a benzene ring and consistent with the high packing of the central units.

The degree of orientation of the polymer networks $\langle P_2 \rangle$ has been quantified and related to that of the corresponding monomers with a variety of techniques and summarized for the LC diacrylates in Broer (1993). In general $\langle P_2 \rangle$ of the network is close to that of the monomer with a tendency to increase during polymerization at low initial order of the monomer at relatively high temperatures close to the clearing point. At high initial order of the monomer the order parameter remains the same or tends to decrease somewhat, especially when groups are laterally substituted in the monomer central unit. This is illustrated in Figure 10.3 showing the molecular order parameter of some liqiud crystals at the polymerization temperature before and after polymerization as determined by UV–VIS dichroism. In general the order parameter of the LC acrylate networks lies between 0.6 and 0.75 depending on the molecular structure, the polymerization conditions and the method of measurement. Looking in somewhat more detail at local order in the molecule using dichroic measurements at localized IR bands one finds, not surprisingly, a somewhat lower $\langle P_2 \rangle$ for the flexible spacer units than for the central core moieties.

At this stage it is also of interest to compare the LC networks formed by the polymerization of different reactive end groups discussed in section 10.2. Figure 10.4 shows some typical refractive index curves as a function of temperature for LC diacrylates and LC diepoxides before and after polymerization. The general behaviour for the two different networks is the same. In the monomeric state, as currently found with non-reactive liquid crystals, when approaching the first-order transition from the nematic to the isotropic state there is a strong temperature dependence of the ordinary (n_o) and extraordinary (n_e) refractive indices. Especially, n_e is very sensitive to the decrease of the molecular order close to the transition to isotropic. During polymerization both n_o and n_e increase mainly owing to an increasing density (polymerization shrinkage). The small decrease of the refractive indices with temperature is mainly determined by the volume expansion of the networks. The order parameter is only slightly dependent on temperature in this region.

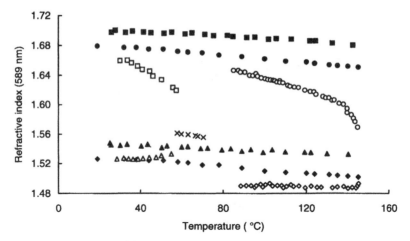

Figure 10.4 Refractive indices of the monomeric LC diacrylates ($\bigcirc = n_e$, $\Diamond = n_o$) and LC diepoxides ($\square = n_e$, $\triangle = n_o$, $\times = n_i$), as well as of the corresponding polymer networks ($\bullet = n_e$ and $\blacklozenge = n_o$ for the poly(diacrylate), $\blacksquare = n_e$ and $\blacktriangle = n_o$ for the poly(diepoxides)) as a function of temperature.

Using the Haller method (Haller *et al.*, 1973) to calculate the order parameter from the refractive indices it was found that the order parameter of the diepoxide networks, although of the same magnitude in the monomeric state, is substantially higher than that of the diacrylate networks. This is explained by a reduced steric hindrance between the aromatic cores as the main chain repetition in epoxides is three instead of two as for the acrylates. The behaviour of the divinylethers comes close to that of the diacrylates, i.e. comparable refractive indices and order parameters.

Of some practical interest is the observation that if the LC monomers are polymerized in their isotropic state, a stable isotropic network is also formed. Only at a polymerization temperature very close to the transition to liquid crystallinity may some local order be observed as, during the chain cross-linking reaction, the LC transition is passed in a moment if there is still sufficient mobility in the vitrifying system. The practical interest for the isotropic phase stems from the fact that one is now able to create within one single film anisotropic and isotropic areas next to each other by local polymerization at different temperatures using masking techniques to cover local areas during the UV illumination step.

10.4 Liquid Crystalline Networks

10.4.1 Nematic Networks

Nematic networks are characterized by a nematic type of molecular organization in a network-restricted environment. The sample texture as observed in a polarizing microscope is exactly the same as that of the low-molar-mass liquid crystals. For example, in the macroscopically oriented state the thin plastic films are nicely transparent just as monodomain nematics. In general they are produced from low-molar-mass nematics which are oriented by conventional techniques such as buffed substrates, surfactant-treated substrates, E and H field alignment, preferably in a closed-cell configuration, and subsequently polymerized by irradiation with a low-to-medium-intensity UV lamp. As shown in Figure 10.4 the refractive indices of such films are highly anisotropic and birefringence can be as high as 0.2.

The mechanical behaviour of the nematic networks compares well with that of the isotropic acrylate and epoxide networks, i.e. in general they form tough films which, however, tend to become brittle at too high a cross-link density. The overall mechanical properties can be adjusted by the molecular structure, e.g. the length of the alkylene spacer, by blending difunctional with monofunctional monomers or by adding a (LC) plasticizer. Owing to the molecular orientation the modulus and strength are anisotropic too. The differences measured parallel and perpendicular to the orientation direction, however, are by far not as spectacular as those found for main-chain LCPs. Table 10.2 shows some orientation-dependent dynamic modulus values for an LC diacrylate compared with the isotropic modulus of the same sample. The reason for this moderate anisotropy is thought to be caused by the presence of an area of low order (polymer main chains) and moderate order (flexible spacers) which can be stretched more readily than the oriented rod-like parts of the network. In addition, the integral order parameter of the networks is of course much lower than that of the main-chain polymers, i.e. around 0.7 instead of close to one.

Of interest for specific applications is the anisotropic thermal expansion of the nematic networks. Table 10.3 summarizes the thermal expansion coefficient along the three main axes of a uniaxially oriented film. Perpendicular to the orientation direction (director) the two thermal expansions measured are about equal at all temperatures. Along the director a low positive thermal expansion is measured at low temperatures and even negative values are obtained at room temperature or higher. Entropy effects, also related to the small reversible change of the order parameter at elevated temperatures, are considered as the main cause for this particular behaviour and the effect can even be enchanced by increasing the length of the alkylene spacer units.

Besides the uniaxial nematic order which can be easily realized in thin films of LC networks one has the possibility to manipulate the local nematic order by various other means. An obvious, but still spectacular to visualize, type of nematic molecular arrangement which can be permanently fixed in the twisted and tumbling texture in which the molecules rotate either in the plane of the film or in the plane of its cross-section (Heynderickx *et al.*, 1992). Figure 10.5 shows two SEM photographs of freeze-fractured surfaces of thin films of poly(C6H) which have a 90° molecular rotation in the plane of the film (twist) and in the plane with fractured surface (tumbling), respectively. Local alignment is made visible by SEM by the difference in texture when the fracture is parallel or perpendicular to the orientational director. The twist can be obtained by the use of two perpendicularly buffed substrates between which the molten monomer is applied before polymerization. It may be helpful to add a chiral dopant that addresses the rotation direction such that the monolithic behaviour is maintained. The tumbling can be obtained by combining a buffed substrate and a surfactant treated substrate. In fact

Table 10.2 Dynamic (1 Hz) tensile modulus of poly(C6H) at 20 °C

	Isotropic	Anisotropic parallel	Anisotropic perpendicular
E (GPa)	1.35	2.10	1.08

Table 10.3 Thermal expansion coefficients of a film of poly(C6H) diacrylate along the three principal axes

Temperature (°C)	In plane; parallel to director	In plane; perpendicular to director	Perpendicular to plane; perpendicular to director
−50	0.27×10^{-4}	1.14×10^{-4}	0.95×10^{-4}
+25	-0.24×10^{-4}	1.69×10^{-4}	1.54×10^{-4}
100	-1.10×10^{-4}	2.91×10^{-4}	2.72×10^{-4}

(a)

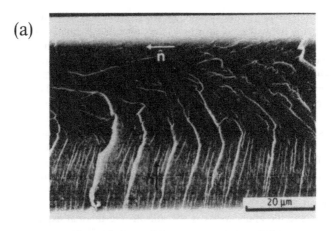

(b)

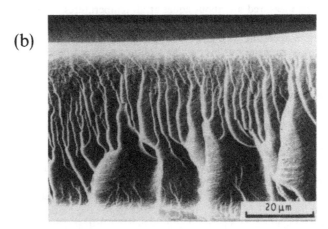

Figure 10.5 Scanning electron microscopy photographs of a C6H network with a tumbled (a) or a twisted (b) nematic order. (From Heynderickx (1993), reproduced by permission of Chapman and Hall.)

these examples illustrate how easy it is to make optical components and films by this technique as the twisted film acts as an achromic half-wave plate, while the tumbled film may be used as an angle-invariant retardation plate.

10.4.2 Cholesteric Networks

Cholesteric networks can be produced either by adding chiral dopants to a reactive mesogen or by modifying the monomer itself by bringing chirality into the side groups. An example of a chiral dopant which can be well dissolved in the reactive liquid crystals is 1-4(4-hexyloxyben-zoyloxy)-benzoic acid-2-octyl ester (S811-Merck Ltd):

$$C_6H_{13}O-\bigcirc-COO-\bigcirc-COO-\overset{\overset{\displaystyle CH_3}{|}}{\underset{*}{CH}}-C_6H_{13}$$

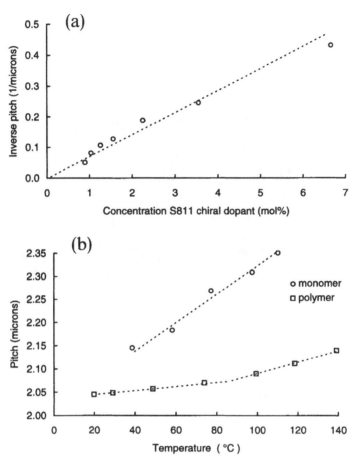

Figure 10.6 Influence of the S811 dopant concentration (a) and the temperature (b) on the pitch of an LC monomer blend of C6H and C6M and on the pitch after polymerization (b).

This left-handed chiral material has a melting point of 48 °C and forms no stable LC phases itself. Even for the highest concentration studied, i.e. 7 mol%, the melting temperature of the LC monomers hardly changes, while T_c decreases from 155 to 144 °C for C6H and from 120 to 114 °C for the eutectic mixture of C6H and C6M. When applied between unidirectionally rubbed substrates the total angle of rotation of the networks obtained after polymerization is accurately set by an appropriate adjustment of the dopant concentration, the polymerization temperature, the film thickness and the angle between the two rubbing directions at the substrates. Figres 10.6(a) and (b) show the influence of the chiral dopant concentration and the temprature on an LC monomer blend of C6H and C6M. Figure 10.6(b) also shows the temperature dependence of the polymeric pitch after the network has been formed, which is much flatter. The change of the pitch during polymerization owing to polymerization shrinkage can be derived from the difference between the two curves at the polymerization temperature being 100 °C for this sample. Taking all these parameters into account one can easily adjust the twist angle to, for instance, 240 °, a value which is relevant for compensation foils for STN displays. The accuracy with which the cholesteric order is frozen into an LC network is demonstrated in Figure 10.7 showing an SEM photograph of a cholesteric film at the location of a Grandjean line (Heynderickx *et al.*, 1993). One can easily see the stepwise addition of one π molecular rotation as the sample thickness gradually increases yielding information on the

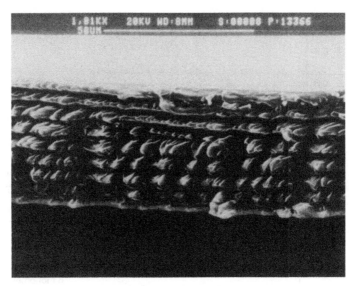

Figure 10.7 Scanning electron microscopy photograph of a chiral nematic C6H, S811-doped, network fractured at the location of a Grandjean step. (From Heynderickx (1993), reproduced by permission of Taylor & Francis Ltd.)

morphology of these particular disclinations.

The chiral dopants are usually used to produce large-pitch chiral nematic networks. In contrast, the cholesteric networks based on chiral reactive liquid crystals are selected for applications where tighter pitches are desired such as for instance in colour filters and TN compensation layers. The pitch of a network consisting of pure C6*H can be as small as 180 nm, as can be estimated from Figure 10.8 showing the SEM photograph of a fractured

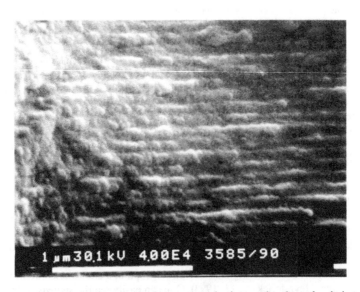

Figure 10.8 Scanning electron microscopy photograph of the fractured surface of a cholesteric network of C6*H.

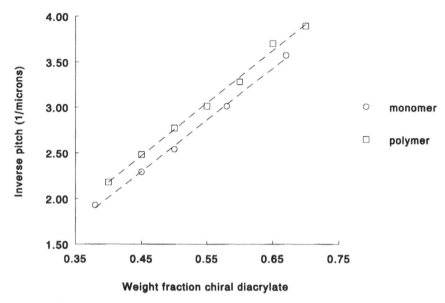

Figure 10.9 Cholesteric pitch as a function of composition of a blend of the cholesteric monomer C6*H and a nematic monomer mix of C6H and C6M.

film of this material. A standard procedure is to blend the chiral reactive liquid crystal with a comparable nematic monomer to adjust the pitch, and thus the reflection wavelength, as shown in Figure 10.9. As was the case for the doped chiral nematic networks the polymerization shrinkage of the cholesteric reactive liquid crystal gives a small decrease in the pitch during polymerization. After polymerization the position of the reflection band becomes temperature insensitive and changes only with the lateral thermal expansion of the thin film. The latter is illustrated in Figure 10.10 showing the temperature dependence of the reflection wavelength of a monomer composition containing 50 wt. % of the chiral LC diacrylate. After polymerization the physical properties of the networks change only marginally, which for instance leads to an anisotropic thermal expansion behaviour with an almost zero thermal expansion in the plane of the film and a higher one along the helical axis. This property might become of importance for inorganic devices to be protected by organic encapsulants to avoid thermal stress in the plane of the interface.

10.5 Conclusions

It has been demonstrated in this chapter that the photoinitiated chain cross-linking of reactive liquid crystals leads to an interesting class of LC networks with high control over the molecular order into three dimensions. Orientational control can be achieved by the molecular design of the materials, e.g. type of reactive groups, spacer lengths, chiral substituents, all affecting the mesomorphic behaviour, but also by external parameters such as the orientation technique or a local polymerization by the use of masks. The properties of the networks thus formed lead to applications especially in the field of optics and electro-optics where control over the state of polarization of light is required. The polymerization of the difunctional monomers in their mesophase proceeds quickly and they are in that sense completely comparable with well-known isotropic cross-linking monomers based on acrylates, methacrylates, vinyl ethers and epoxides.

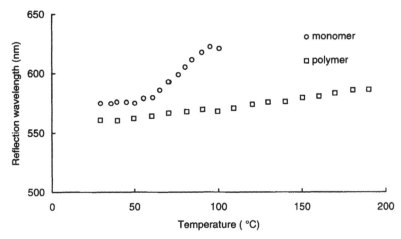

Figure 10.10   Reflection wavelength as a function of temperature of a cholesteric blend of C6*H (50 wt.%) and a nematic monomer mix of C6H and C6M (8 : 2, total 50%) before and after polymerization.

Acknowledgements

Much of the work presented in this chapter was done in colaboration with Dr I. Heynderickx, Dr R. A. M. Hikmet, Dr J. Lub, Mrs G. N. Mol and Mrs C. M. R. de Witz. The author wishes to thank them for valuable discussions, advice and their many experimental contributions.

References

ANDERSSON, H., GEDDE, U. W. and HULT, A. (1992) Preparation of ordered, crosslinked and thermally stable liquid crystalline poly(vinyl ether) films, *Polymer* **33**, 4014.

ARSLANOV, V. V. and NIKOLAJEVA, V. I. (1984) Fixation of the structure of a liquid-crystalline monomer based on aromethine by polymerisation in the mesophase, *Vysokomol. Soedin, Ser. B* **26**, 208.

BOULIGAND, Y., CLADIS, P., LIEBERT, L. and STRZELECKI, L. (1974) Study of sections of polymerized liquid crystals, *Mol. Cryst. Liq. Cryst.* **25**, 233.

BROER, D. J. (1993) Photoinitiated polymerization and crosslinking of liquid crystalline systems, *Radiation Curing in Polymer Science and Technology* – Vol III, *Polymerization mechanisms*, ed. J. P. Foussier and J. F. Rebek, Ch. 12, London and New York: Elsevier.

BROER, D. J. (1995) Creation of supramolecular thin film architectures with liquid-crystalline networks, *Mol. Cryst. Liq. Cryst.* **261**, 513–521.

BROER, D. J. and HEYNDERICKX, I. (1990) Three-dimensionally-ordered polymer networks with a helicoidal structure, *Macromolecules* **23**, 2474.

BROER, D. J. and MOL, G. N. (1991) Anisotropic thermal expansion of densely crosslinked oriented polymer networks, *Polym. Eng. Sci.* **31**, 625.

BROER, D. J., FINKELMANN, H. and KONDO, K. (1988) *In-situ* photopolymerization of oriented-liquid crystalline acrylates, 1. – Preservation of molecular order during photopolymerization, *Macromol. Chem.* **189**, 185.

BROER, D. J., MOL, G. N. and CHALLA, G. (1989a) *In-situ* photopolymerization of oriented liquid-crystalline acrylates, 2. – Kinetic aspects of photopolymerization in the mesophase, *Makromol. Chem.* **190**, 19.

BROER, D. J., BOVEN, J., MOL, G. N. and CHALLA, G. (1989b) In-situ photopolymerization of oriented liquid-crystalline acrylates, 3. – Oriented polymer networks from a mesogeneic diacrylate, *Makromol. Chem.* **190**, 2255.

BROER, D. J., HIKMET, R. A. M. and CHALLA, G. (1989c) *In-situ* photopolymerization of oriented liquid-crystalline acrylates, 4. – Influence of a lateral methyl substituent on monomer and oriented network properties of a mesogenic diacrylate, *Makromol. Chem.* **190**, 3201.

BROER, D. J., MOL, G. N. and CHALLA, G. (1991) *In-situ* photopolymerization of oriented liquid-crystalline acrylates, 5. – Influence of the alkylene spacer on the properties of the mesogenic monomers and the formation and properties of oriented polymer networks, *Makromol. Chem.* **192**, 59.

BROER, D. J., LUB, J. and MOL, G. N. (1993) Synthesis and photopolymerization of a liquid-crystalline diepoxide, *Macromolecules* **26**, 1244.

CLOUGH, S. B., BLUMSTEIN, A. and HSU, E. C. (1976) Structure and thermal expansion of some polymers with mesomorphic ordering, *Macromolecules* **9**, 123.

DE GENNES, P. G. (1969) Possibilités offertes par la réticulation de polymères en presence d'un cristal liquide, *Phys. Lett.* **28A**, 725.

DEWAR, M. J. S. and SCHROEDER, J. P. (1965) *p*-Alkoxy- and *p*-Carbalkoxybenzoates of diphenols. A new series of liquid crystalline compounds, *J. Org. Chem.* **30**, 2296.

HALLER, I., HUGGINS, H. A., LILIENTHAL, H. R. and MCGUIRE, T. R. (1973) Order related properties of some nematic liquid crystals, *J. Phys. Chem.* **77**, 950.

HEYNDERICKX, I. and BROER, D. J. (1991) The use of cholesterically-ordered polymer networks in practical applications, *Mol. Cryst. Liq. Cryst.* **203**, 113.

HEYNDERICKX, I., BROER, D. J. and TERVOORT-ENGELEN, T. (1992) Molecular ordering in a liquid crystalline material visualized by scanning electron microscopy, *J. Mater. Sci.* **27**, 4107.

HEYNDERICKX, I., BROER, D. J. and TERVOORT-ENGELEN, Y. (1993) Visualization of the cholesteric texture near a Grandjean line, *Liq. Cryst.* **15**, 745.

HIKMET, R. A. M. (1990) Electrically induced light scattering from anisotropic gels, *J. Appl. Phys.* **68**, 4406.

HIKMET, R. A. M. (1991a) Anisotropic gels and plasticized networks formed by liquid crystal molecules, *Liq. Cryst.* **9**, 405.

HIKMET, R. A. M. (1991b) From liquid crystalline molecules to anisotropic gels, *Mol. Cryst. Liq. Cryst.* **198**, 357.

HIKMET, R. A. M. and BROER, D. J. (1989) X-ray diffraction study of networks formed by liquid crystalline diacrylates, *Integration of Fundamental Polymer Science and Technology*, ed. L. A. Kleintjes and P. J. Lemstra, Amsterdam and London: Elsevier.

HIKMET, R. A. M. and BROER, D. J. (1991) Dynamic mechanical properties of anisotropic networks formed by liquid crystalline acrylates, *Polymer* **32**, 1627.

HIKMET, R. A. M. and HOWARD, R. (1993) Structure and properties of anisotropic gel and plasticized networks containing molecules with a smectic-A phase, *Phys. Rev. E* **48**, 2752.

HIKMET, R. A. M. and MICHIELSEN, M. C. B. (1995) Anisotropic network stabilized ferroelectric gels, *Adv. Mat.* **7**, 300–307.

HIKMET, R. A. M. and ZWERVER, B. H. (1991) Cholesteric networks containing free molecules, *Mol. Cryst. Liq. Cryst.* **200**, 197.

HIKMET, R. A. M. and ZWERVER, B. H. (1992) Structure of cholesteric gels and their electrically induced light scattering and colour changes, *Liq. Cryst.* **12**, 319.

HIKMET, R. A. M. and ZWERVER, B. H. (1993) Cholesteric gels formed by LC molecules and their use in optical storage, *Liq. Cryst.* **13**, 561.

HIKMET, R. A. M., ZWERVER, B. H. and BROER, D. J. (1992) Anisotropic polymerization shrinkage behaviour of liquid-crystalline diacrylates, *Polymer* **33**, 89.

HIKMET, R. A. M., LUB, J. and HIGGINS, J. A. (1993) Anisotropic networks obtained by *in situ* cationic polymerization of liquid-crystalline divinyl ethers, *Polymer*, **34**, 1736.

HOYLE, C. E., CHAWLA, C. P. and GRIFFIN, A. C. (1988) Photopolymerization of a liquid crystalline monomer, *Mol. Cryst. Liq. Cryst.* **157**, 639.

HOYLE, C. E., CHAWLA, C. P. and GRIFFIN, A. C. (1989) Photoinitiated polymerization of a liquid crystalline monomer: Order and mobility effects, *Polymer* **30**, 1909.

JAHROMI, S., LUB, J. and MOL, G. N. (1994) Synthesis and photoinitiated polymerization of liquid crystalline diepoxides, *Polymer* **35**, 621.

JOHNSON, H., ANDERSON, H., SUNDELL, P. E., GUDDE, U. W. and HULT, A. (1991) Photoinitiated atomic bulk polymerization of mesogenic vinyl ethers, *Polym. Bull.* **25**, 641.

KITZEROW, H.-S., SCHMID, H., RANFT, A., HEPPKE, G., HIKMET, R. A. M. and LUB, J. (1993) Observation of blue phases in chiral networks, *Liq. Cryst.* **14**, 911.

LOHSE, F. and ZWEIFEL, H. (1986) Photocrosslinking of expoxy resin, *Adv. Polym. Sci.* **78**, 61.

LUB, J., BROER, D. J., HIKMET, R. A. M. and NIEROP, K. G. J. (1995) Synthesis and photopolymerization of cholesteric liquid crystalline diacrylates, *Liq. Cryst.* **18**, 319.

ODIAN, G. (1991) *Principles of Polymerization*, 2nd edn, New York: John Wiley, p. 210.

SHANNON, P. J. (1984) Photopolymerization in cholesteric mesophases, *Macromolecules*, **17**, 1873.

STRZELECKI, L. and LIEBERT, L. (1973) Sur la synthèse de quelques nouveaux monomères mésomorphes. Polymérisation de la *p*-acryloyloxybenzylidène *p*-carbocxyaniline, *Bull. Soc. Chim. Fr.* **2**, 597 and 605.

WENDORFF, J. H. (1967) *Liquid Crystalline Order in Polymers*, ed. A. Blumstein, New York: Academic Press.

YANG, D. K., WEST, J. L., CHIEN, L.-C. and DOANE, J. W. (1994) Control of reflectivity and bistability in displays using cholesteric liquid crystals, *J. Appl. Phys.* **761**, 1331.

11

The Challenge of New Applications to Liquid Crystal Displays

J. L. WEST

11.1 Introduction and Overview

The properties of bistable, reflective cholesteric displays are reviewed in other chapters of this book (Yuan and Yang *et al.*). These displays meet the challenges of a variety of new portable electronic applications by providing low power consumption, bright high-resolution displays. In this chapter, I review the status of related liquid crystal technologies and present the development of high-polymer-content bistable, reflective cholesteric formulations and their use for black on white displays with plastic substrates.

Liquid crystals dominate the flat-panel display market. This dominance results from determined research, mainly in Japan, since the early 1980s. This research brought active matrix addressing technology, transforming the twisted-nematic (TN) effect into high-resolution, high-contrast colour images and the super-twisted-nematic (STN) effect which provides high-resolution colour images with relatively simple matrix addressing schemes. These two liquid crystal technologies provide displays for the exploding laptop computer market, as well as hand-held televisions and video recorders. The market has grown to $5 billion world-wide in 1994 and is forecasted to exceed $20 billion by the year 2000.

TN devices have a broad voltage threshold for switching between the on- and off-states, limiting the resolution achievable using simple multiplex addressing techniques. The greater the number of addressed lines, the lower the voltage difference between addressed and unaddressed pixels, effectively limiting the number of addressable lines to less than 100 (Alt and Pleshko, 1974). Active matrix addressing techniques overcome this limitation by electrically isolating each pixel. The STN effect produces a much sharper voltage threshold, greatly increasing the resolution achievable using simple multiplex addressing.

STN liquid crystal displays were first used for laptop computers (Scheffer and Nehring, 1984). These early displays suffered from poor optical characteristics. Rather than a black-on-white image, the STN effect produces a coloured image on a coloured background. The colours and contrast depend on the degree of twist, the orientation of the polarizers, and the optical path difference, Δnd (Scheffer and Nehring, 1985). Early STN displays produced blue images on a yellow background. These early STN cells required tight control of the cell gap and pretilt angle. Improved ST designs relaxed the manufacturing tolerances and led to broad application of STN displays in laptop computers. (Leenhouts and Schadt, 1986). The STN displays also have relatively slow response times that preclude their use for video applications and often produce ghosting in laptop computer screens (Asano *et al.*, 1986).

Many improvements were made in STN displays in the late 1980s and early 1990s. The addition of retardation films (Matsumoto *et al.*, 1984; Akatsuka *et al.*, 1989) produces black-on-white images and increases the viewing angle. The retardation films, typically thin polymer films such as polycarbonate or polyvinylalcohol, compensate for the phase difference induced by the super-twisted structure. Good compensation, however, is only achieved over a limited temperature range because of the different temperature dependence on Δn for the liquid crystal and the polymer films. Double-layer STN cells were developed to eliminate the temperature dependence of compensation. A second cell of the same thickness, with no electrodes and twisted in the opposite direction, compensates for the optical aberrations produced by the driven STN cell (Katoh *et al.*, 1987). The double-layer STN displays weigh considerably more than single-layer cells, an important consideration for laptop computers. Active addressing of STN displays can improve the response time and offers the potential of achieving video rates.

Improvements in STN technology have produced full-colour VGA displays with relatively rapid response times. However, the performance of the best STN displays still lags behind the directly driven TN devices and cathode ray tubes. Manufacturers have therefore developed active matrix addressing schemes. Thin-film transistors or diodes are used as non-linear elements. Most active matrices on the market today use thin-film transistors. For large-area displays, suitable for laptop computers, manufacturers typically use hydrogenated amorphous silicon because of the relatively low processing temperatures ($\approx 300\,^\circ$C) (Firester, 1988).

Each pixel in the active matrix addressing scheme is connected to a non-linear electrical element. The drive voltage is applied across the pixel during the entire frame time, eliminating the problems of cross-talk found in multiplexed TN displays. Production of large-area, active matrix displays introduces many technical problems. While the resolution of the photolithographic steps required to produce active matrix substrates is at least an order of magnitude less than that employed in the semiconductor industry, this resolution must be maintained over a much larger area. For example, 10.4 inch (26 cm) active matrix displays are now the norm for laptop computers. Maintaining resolution over these relatively large areas required entirely new process equipment. The complete process must be carefully optimized to achieve high production yields. The sophisticated equipment pushes the price of an active matrix production facility to over \$300 million.

The increased costs and complexity of active matrix manufacture are offset by the improved performance of active matrix displays. Incorporation of retardation films produces high-contrast, wide-viewing-angle colour displays with video capability. The consumer finds this improved performance well worth the cost.

Both the active matrix and STN displays utilize polarizers, severely limiting their brightness in reflection. For this reason, virtually all laptop computers require a power-hungry backlight to produce the bright images consumers demand. Recent advances using polarized beam splitters rotate and utilize the unwanted light polarizaztion for projection LCDs. There are, however, no prospects foreseen for substantially increasing the efficiency of the sheet polarizers used for direct-view displays. Therefore direct-view active matrix and STN displays cannot achieve or even approach reflectivities of 50%. The addition of colour filters required for colour displays reduces the light efficiency of these devices to well below 10%. Power-hungry backlights are, therefore, required to produce the bright images required in laptop computers.

Despite their proven commercial success, active matrix and STN displays cannot meet the needs of many developing electronic applications. The pager, fax and cellular phone are blending into a single device that allows users to send and receive messages and facsimiles from remote locations. Publishers of magazines, newspapers and books will eventually abandon the printed page for an electronic tablet. These new devices must be very light weight, precluding the heavy batteries and backlights found in today's laptop computers. The displays must operate using ambient light and produce a bright, high-resolution image. Early attempts to commercialize these devices using STN displays, such as the Newton, have had limited success because of the poor optical characteristics of the liquid crystal screen. Only by improving the brightness of the reflected images will the full commercial potential of these new electronic applications

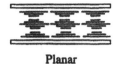

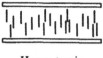

Planar   Focal Conic   Homeotropic

Figure 11.1 Schematic of the cholesteric planar, focal-conic and homeotropic states.

be met. Success will explode the size of the liquid crystal market to well beyond the $20 billion predicted. Clearly, new display technologies must be used in the upcoming portable electronic devices.

11.2 Reflective Cholesteric Displays

For the last 5 years Doane has led a research effort into reflective cholesteric displays (Yang *et al.*, 1991). Unlike earlier research into cholesteric devices, Doane's team has developed cholesteric materials that Bragg-reflect in the visible portion of the spectrum. They have also utilized polymers and surfaces to stabilize the various cholesteric states, producing bistable reflective displays.

Many of the earliest liquid crystal display technologies utilized the variable light scattering properties of cholesteric materials (Adams *et al.*, 1973), such as the cholesteric–nematic phase change mode (Wysocki *et al.*, 1968). Optical metastability and storage effects were also observed in the early 1970s for cholesteric mixtures having either positive or negative dielectric anisotropy and having homogeneous or homeotropic alignment (Gruebel *et al.*, 1973; Heilmeier and Goldmacher, 1968). None of these early cholesteric display devices were commercially developed because of materials limitations, long response times and the commercial interest in twisted-nematic displays developed about the same time.

Yuan and Yang *et al.* have provided excellent reviews of the bistable reflective cholesteric technology in other chapters of this book. Here I will discuss relatively high-polymer-content cholesteric formulations and their use for black on white displays and with plastic substrates. To review briefly, the reflective cholesteric displays developed by Doane utilize three different cholesteric configurations (Figure 11.1). The planar state reflects light over a narrow-wavelength region defined by the pitch length and the average refractive index of the liquid crystal, $\lambda = np$, where λ is the wavelength of maximum reflection, n is the average refractive index of the liquid crystal, and p is the pitch of the cholesteric phase. Reflection in the visible portion of the spectrum, therefore, requires a pitch length between 0.25 and 0.6 micrometres. The focal-conic state breaks the cholesteric helices into small domains. With the small pitch lengths used for visibly reflecting planar states, the focal-conic state is weakly light scattering. Upon application of an electric field sufficient to untwist the cholesteric phase, the liquid crystal takes on homeotropic alignment.

The planar state reflects essentially all light of one circular polarization for cell thicknesses greater than 8 pitch lengths. The opposite circular polarization is transmitted through the display. The maximum reflected intensity therefore approaches 50% for cell thicknesses of 5 μm (Doane *et al.*, 1994). A flat-black backing on the display results in high contrast between the planar and focal-conic states. The viewer observes bright reflected colour from the planar state and the black background from the focal-conic state.

Varying the concentration of the chiral additive in the PSCT formulations varies the pitch length, the wavelength of maximum reflection and therefore the colour of the display. As the wavelength of maximum reflection moves to longer wavelengths the width of the reflection also increases. Red is therefore somewhat difficult to achieve without the addition of dyes (Fritz *et al.*, 1994).

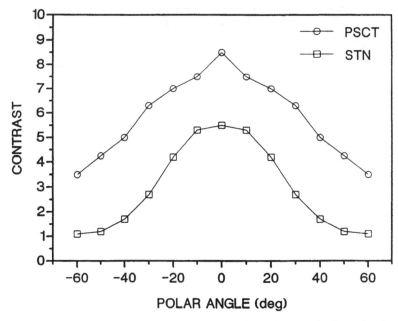

Figure 11.2 Contrast of the PSCT cell and reflective STN cell as a function of polar angle. The incident light is isotropic and unpolarized.

The polymer network of the PSCT materials breaks the material into small domains. The helical axis varies slightly from domain to domain. This variation broadens the width of the reflection peak and reduces the change in wavelength maximum as a function of viewing angle, compared with a well-aligned planar state. The PSCT cells do not use polarizers. The viewing characteristics are, therefore, spherically symmetric (Yang *et al.*, 1994).

Because of the high reflection efficiency of the planar state and because they do not use polarizers, polymer-stabilized cholesteric texture displays are much brighter than TN or STN cells. High brightness of the planar state and the low light scattering of the focal-conic state produce a high-contrast image. The high contrast is maintained even at wide viewing angles. Figure 11.2 compares the contrast of a green-reflecting PSCT cell and an STN cell. The PSCT cell maintains reasonable contrast, even out to 60°.

The bistable switching of the cholesteric materials allows high-resolution displays using simple multiplexing addressing techniques. The images remain until updated by another set of driving pulses. Images have been maintained for over a year with no signs of degradation.

Bistable reflective displays can also be made without the polymer network by using substrates with proper surface treatments (Yang *et al.*, 1994; Wu *et al.*, 1994; Lu *et al.*, 1995). Bare ITO surfaces, or unrubbed polyimide- or PMMA-coated substrates can be used to fabricate bistable, reflective cholesteric displays. Elimination of the polymer network removes the need to photopolymerize during cell fabrication.

The properties of the reflective bistable cholesteric displays meet many of the requirements for the next generation of portable electronic equipment. Consumers may initially accept a display with images of a single colour on a black background (similar to the blue on gold STN screens initially used on portable computers) if no other options are available. Broad acceptance of these new devices requires black-on-white and full-colour displays that are very light weight.

11.3 High-polymer-content Bistable Cholesteric Formulations

High-polymer-content bistable cholesteric formulations have also been reported (West *et al.*, 1993). These formulations combine many of the properties of PDLCs and reflective cholesteric materials. The materials are self-sealing and self-sustaining while maintaining the bistable reflective properties of the lower-polymer-content and neat cholesteric mixtures. Plastic displays are relatively easy to fabricate using the high-polymer-content formulations. Because the optical response of the materials does not require polarized light, birefringent substrates, such as commercially available ITO-coated polyester, can be used. The high-polymer-content materials act as an adhesive for the front and back subtrates, making assembly and thickness control straightforward.

The same techniques used to form polymer-dispersed liquid crystals can be used to form high-polymer-content formulations (West, 1988). Thermoset polymers, such as epoxies and polyuthanes, thermoplastics such as polymethylmethacrylate and polycarbonate, and UV-curable polymers can be mixed with the cholesteric formulations in weight percentages ranging from 10–40%. These formulations maintain the bistable, reflective, electro-optic properties of the cholesteric mixture while introducing the mechanical properties found in a PDLC (West *et al.*, 1994).

When properly cured, the high-polymer-content formulations appear to form a network of small domains. The domain size and structure depends on the type of polymer, the weight percentage of polymer and the polymerization and phase separation conditions. We have found that by utilizing the proper polymer and phase separation conditions, the reflection spectrum of the planar state can be broadened to span the entire visible spectrum, producing a black-on-white display.

We investigated cholesteric formulations dispersed in the UV-cured polymer Masterbond UV 10 (Masterbond Corporation). The cholesteric mixtures consisted of chiral additives, CB15, CE2 and ZLI 4572, dissolved in a nematic host, E48 (EM Chemicals). The chiral additive was then mixed with the nematic host. The concentration of the chiral additive determined the pitch of the resulting mixture. The cholesteric formulation was then dispersed in the Masterbond UV 10 photomonomer. A blank cell is formed using bare ITO-coated glass substrates. The cell is capillary filled with the UV 10/cholesteric dispersion and photocured at a controlled temperature, using an ELC 4000 light curing unit (Electro-Lite Corp.) to expose to UV light for 10 minutes.

The reflectance spectra of the resulting shutters in the planar state were measured using a Perkin Elmer λ 4b spectrophotometer equipped with an integrating sphere. Figure 11.3 shows the reflectance spectra of the completed cells in the planar state as a function of the polymer content.

The bulk chiral nematic has a reflection peak around 518 nm with a peak width at half height of about 65 nm for a 5 μm thick film. The reflection intensity at the peak wavelength for the bulk material approaches the theoretical maximum of 50%. Increasing the weight percentage of UV 10 from 5 to 20 progressively reduces the peak reflection intensity while broadening the reflection spectrum.

Several mechanisms can explain the broadening observed in the reflection spectra of high-polymer-content formulations. The polymer network serves to break the planar state into small domains. Distribution of the helical axis of the domains about the normal broadens the reflection spectrum of the planar state, but only to shorter wavelengths. The longer-wavelength reflections can only be explained by an increase in the refractive index of the liquid crystal mixture or by an increase in the pitch length. A substantial increase in the average refractive index of the liquid crystal is highly unlikely. An increase in the pitch length can be caused by a change in the chiral concentration. This could result from the chiral components dissolving preferentially in the polymer network; however, the breadth of the reflection spectrum could only be explained by having a plurality of domains with different chiral compositions. This

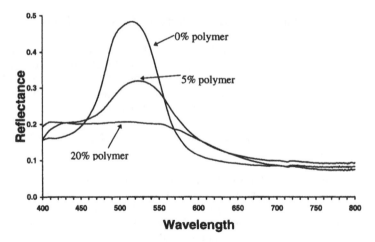

Figure 11.3   Reflectance spectra as a function of weight per cent of Masterbond UV 10.

scenario requires that diffusion be prevented between the domains to maintain different composition. Finally, the polymer domains may distort the pitch of the chiral mixture through surface interactions. We have, therefore, examined the structure of the polymer network using scanning electron microscopy (West *et al.*, 1994).

Samples for scanning electron microscopic examination were prepared using one glass and one sodium chloride substrate. The polymer was irradiated through the glass substrate. The resulting cell was then immersed in water to dissolve the salt plate, exposing the polymer network without mechanically disturbing the structure. Figure 11.4 shows an SEM photograph of the 20% Masterbond UV 10 formulation formed using this technique.

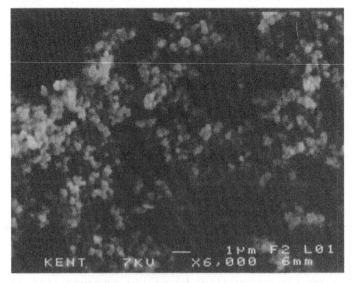

Figure 11.4   Scanning electron microscopic image of the polymer network formed from a 20%, by weight, Masterbond UV10 formulation.

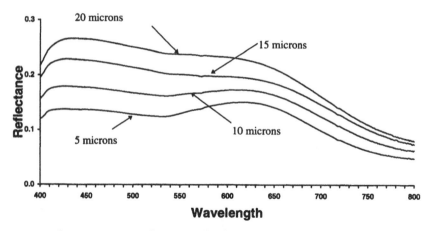

Figure 11.5 Reflection spectras as a function of film thickness for a 20% Masterbond UV 10 formulation.

The 20% Masterbond UV 10 structure shows very small domains of less than a micrometre. The phase separation occurs above the clearing point of the liquid crystal. The alignment of the liquid crystal at the domain boundaries may be determined upon the initial formation of the liquid crystal phase. As the formulation is further cooled, the small domains and surface alignment alter the pitch length of the chiral nematic mixture. This scenario produces different pitch lengths in the various domains, resulting in a broadening of the spectrum.

While the increased polymer content broadens the reflection spectrum it also substantially decreases the reflection intensity. This decrease results from the different pitch lengths in the various domains. Theory predicts that maximum reflection requires at least 8 pitch lengths. For a sample reflecting in the green portion of the visible spectrum, maximum reflection requires a 4 μm thick sample. For a black-on-white display reflecting with blue, green, and red reflecting domains, a thickness of 10-12 μm will be required to produce maximum reflection.

Figure 11.5 shows the reflection spectrum of a 20% Masterbond UV 10 formulation as a function of film thickness. As predicted, the reflection intensity increases with film thickness; however, even for a 20 μm thick cell, the reflection intensity only approaches 25%. This is much greater than the 10–12 μm thickness theory predicted for maximum reflection.

The low reflectivity of the high-polymer-content formulations may result from several factors. Lower reflectivity will result because of the large number of domains. Domains of similar pitch lengths are not in a continuous layer. Some domains may also reflect outside of the visible spectrum and, therefore, do not contribute to the brightness of the display.

Despite the low measured reflectivity of the 20 μm thick cells, they produce high-contrast black-on-white images. Figure 11.6 shows a photograph of a 20 μm thick cell. While producing an acceptable image the increased thickness requires over 100 volts to switch the material into the white reflecting planar state. The increased thickness also increases the amount of scattering from the focal-conic state, turning the black into a dark grey.

An alternate method of producing a black-on-white display utilizes a multiple-layer device. Preliminary research has shown that multiple layers, each with a different chiral composition and pitch length, can be spin coated on glass substrates. A barrier layer is formed between the layers to prevent diffusion of the chiral component between layers. For example, a layer using polymethylmethacrylate as the binder was spin coated on a glass substrate. A water-soluble barrier layer of PVA was spin coated on top of this layer. A second polymethylmethacrylate layer of a different chiral composition was then spin coated on top. The reflection from this cell combined the reflected light from the individual layers, demonstrating the concept.

Figure 11.6   Photograph of a black on white 20 μm thick, 20% by weight Masterbond UV10/cholesteric cell.

High-polymer-content bistable, reflective cholesteric formulation can also be used to form cells using plastic substrates. The resulting polymer network serves to adhere the front and back substrates and maintains uniform cell thickness across the device (West *et al.*, 1995).

Figure 11.7 shows a 320 × 320 pixel high-polymer-content cholesteric display fabricated using polyester substrates. We fabricated all plastic, bistable, reflective cholesteric displays using ITO-coated polyester substrates and a high-polymer-content cholesteric formulation. A 4 inch square (103 cm^2) interdigitated pattern of lines, with a resolution of 80 lines per inch, was etched on pieces of 0.007 in (0.18 mm) thick ITO-coated polyester substrate (Southwall) using standard photolithographic and solvent etching techniques. The substrates were coated with a cholesteric/polymer formulation, E48/CE2/CB15/NOA65, in a weight ratio of 0.49/0.18/0.18/0.15, respectively. The planar state of this formulation reflects light in the green portion of the spectrum. The formulation was heated to 60 °C to assure miscibility of the liquid crystal and monomer. The thickness of the cholesteric film was controlled using 4.5 μm spacers sprayed on the front and back substrates. The cholesteric formulation was poured on one substrate and then the second substrate was pressed over the cholesteric formulation. A roller was used to squeeze the resulting sandwich, assuring uniform thickness across the display. The resulting plastic cell was irradiated with UV light from an Electro-Lite ELC 4000 (25 mW/cm^2 at 365 nm) for 10 minutes resulting in polymerization of the NOA 65 and formation of a uniform network structure throughout the cell. Polymerization serves to adhere the front and back plastic substrates, increasing the durability of the display and maintaining uniform thickness.

Excess cholesteric/polymer formulation was removed from the edges of the cell. The back of the cell was painted black to produce coloured images on a black background. The resulting display was addressed using drive circuitry specifically designed to address the bistable, reflective cholesteric formulations (Yang *et al.*, 1994).

As with the bistable cholesteric formulations formed on glass substrates, the images on the cells formed using plastic substrates are stable with no field applied. Images may be changed by application of an addressing field or by application of pressure. Applied pressure serves to switch the focal-conic to the planar state. The flexibility of the plastic substrates allows pressure to be applied to selective areas of the display. Images can be written on the display using a

Figure 11.7 320 × 320 pixel high-polymer-content cholesteric display fabricated using polyester substrates.

stylus. Written information can be read electronically using different electrical properties of the focal-conic and planar states, providing a method of interfacing with the device.

Bistable, reflective cholesteric formulations produce bright images, viewable in lighting conditions ranging from a dim room to bright sunlight. Plastic substrates greatly reduce the weight of the display (about one-seventh of the weight of comparable glass cells) while increasing their durability. Because the displays do not utilize polarized light, commercially available polyester substrates can be used. The polyester substrates are thermally stable and unaffected by the chemicals and solvents used in the conventional processing of liquid crystal displays. The high polymer content of the cholesteric formulations serves to adhere the front and back substrates, simplifying manufacture and increasing the durability of the cell.

The bistable, reflective cholesteric displays using plastic substrates are particularly well suited for a variety of portable electronic devices, such as electronic books and newspapers, portable communications and personal data assistants. Plastic displays will significantly reduce the weight and increase the durability of the resulting device. The displays are simple to manufacture, using conventional processes and commercially available polyester substrates. Images can be mechanically written on the display, providing a convenient user interface. The plastic displays offer the potential of a low-cost, high-resolution display suitable for the wide variety of portable electronic devices planned for the near future.

11.4 Conclusion

The bistable cholesteric materials serve the unique needs of many of the portable electronic devices under development. The high reflectivity provides bright, high-contrast images under lighting conditions ranging from normal room light to bright sunlight. Eliminating the need for power-hungry backlights greatly reduces the power requirements. Bistable switching allows high-resolution images to be produced using inexpensive multiplex drive schemes. Static images are flicker free, reducing viewer fatigue and eye strain. Driving voltages are only applied when the image is to be changed, further reducing the power requirements of the display.

Conventional plastic substrates such as polyester can be used to fabricate lightweight, flexible and rugged displays.

Bistable, reflective cholesteric materials will not replace active matrix and STN displays in laptop computers in the near future, if ever. Their features, however, offer advantages over these well-established technologies for portable electronic devices.

References

ADAMS, J. E., HAAS, W. E., LEDER, L. B., MECHLOWITZ, B., SAEVA, F. D. and DAILEY, J. L. (1973) *Proc. SID* **14**, 121–126.

AKATSUKA, M., KOH, H., NAAGAWA, Y., MATSUHIRO, K., SOUDA, Y. and SAWADA, K. (1989) *Proc. Japan Display '89*, p. 335.

ALT, P. M. and PLESHKO, P. (1974) Scanning limitations of liquid-crystal displays, *IEEE Trans. Electron. Dev.* **ED-21**, 146.

ASANO, K., ARAI, K. and NISHI, S. (1986) *Proc. Japan Display '86*, p. 392.

DOANE, J. W., ST JOHN, W. D., LU, Z. J. and YANG, D.-K. (1994) *Conference Record of the IDRC, Monterey, CA*, pp. 65–68.

FIRESTER, A. H. (1988) Active matrix liquid crystal display technologies for automotive applications, *SPIE* **958**, 80.

FRITZ, W. J., ST JOHN, W. D., YANG, D.-K. and DOANE, J. W. (1994) *Proc. SID, San Jose, CA*, pp. 841–844.

GRUEBEL, W., WOLFF, U. and KRUBER, H. (1973) Electric field induced texture changes in certain nematical cholesteric liquid crystal mixtures, *Mol. Cryst. Liq. Cryst.* **24**, 103.

HEILMEIR, G. H. and GOLDMACHER, J. E. (1968) A new electric-field-controlled reflective optical storage effect in mixed-liquid crystal systems, *Appl. Phys. Lett.* **13**, 132.

KATOH, K., ENDO, Y., AKZATSUKA, M., OHGAWARA, M. and SAWADA, K. (1987) Application of retardation compensation: A new highly multiplexable black-white liquid crystal display with two supertwisted nematic layers, *Jpn. J. Appl. Phys.* **26**, L1784–L1786.

LEEHOUTS, F. and SCHADT, M. (1986) *Proc. Japan Display '86*, p. 388.

LU, Z.-J., ST. JOHN, W. D., HUANG, X. Y., YANG, D.-K. and DOANE, J. W. (1995) Surface modified reflective cholesteric displays, *SID Dig. Tech. Pap.* **XXVI**, 172–175.

MATSUMOTO, S., HATOH, H., MURAYAMA, A., YAMAMOTO, T., KONDO, S. and KAMAGAMI, S. (1989) A single cell high quality black and white STN liquid crystal display, *IEEE Trans. Electron. Dev.* **36**, 1905.

SCHEFFER, T. J. and NEHRING, J. (1984) A new, highly multiplexable liquid crystal display, *J. Appl. Phys. Lett.* **45**, 1021.

SCHEFER, T. J. and NEHRING, J. (1985) Investigation of the electro-optical propertes of 270 degrees chiral nematic layers in the birefringence mode, *J. Appl. Phys. Lett.* **58**, 3022.

WEST, J. L. (1988) Phase separation of liquid crystals in polymers, *Mol. Cryst. Liq. Cryst.* **157**, 427–441.

WEST, J. L., AKINS, R. B., FRANCL, J. and DOANE, J. W. (1993) Cholesteric/polymer dispersed light shutters, *Appl. Phys. Lett.* **63**, 1471–1473.

WEST, J. L., MAGYAR, G. R. and FRANCL, J. J. (1994) Polymer-stabilized cholesteric texture materials for black-on-white displays, *Proc. SID, San Jose, CA*, pp. 608–610.

WEST, J. L., ROUBEROL, M., FRANCL, J. J., JI, Y., DOANE, J. W. and PFEIFFER, J. (1995) Flexible displays utilizing bistable, reflective cholesteric/polymer dispersions and polyester substrates, *SID 15th Int. Display Research Conf., Asia Display '95* (submitted).

WU, B. G., GAO, J., MA, Y. D., ZHOU, V. and TIMMONS, R. B. (1994) *Conference Record of the IDRC, Monterey, CA*, pp. 476–479.

WYSOCKI, J. J., ADAMS, J. and HAAS, W. (1968) Electric-field-induced phase change in cholesteric liquid crystals, *Phys. Rev. Lett.* **20**, 1024.

YANG, D. K., CHIEN, L.-C. and DOANE, J. W. (1991) Cholesteric liquid crystal/polymer gel dispersion bistable at zero field, *Conference Record of the IDRC SID*, pp. 49.

YANG, D.-K., DOANE, J. W., YANIV, Z. and GLASSER, J. (1994) Cholesteric reflective display: Drive scheme and contrast, *Appl. Phys. Lett.* **64**, 1905–1907.

12

Bistable Reflective Cholesteric Displays

H. YUAN

12.1 Introduction

With the arrival of the information age, electronic displays are playing an ever increasing role as the interface between information processing machines and human beings. Computer monitors, television sets, public information displays are just a few examples. The trend of electronic displays is towards thinner, lighter and more energy-efficient packages owing to the huge demand in portable applications such as notebook and palmtop personal computers, personal digital assistants (PDAs), pagers, portable fax machines, portable information terminals, etc. This is why flat-panel displays, including liquid crystal displays (LCDs), plasma display panels, electroluminescence displays and field emission displays are expanding their applications. In particular, applications of active matrix twisted-nematic LCDs (AM-TN-LCDs or simply AM-LCDs) and super-twisted-nematic LCDs (STN-LCDs) for notebook computers are expanding rapidly. Among the four flat-panel display technologies, LCDs are expected to be the dominating technology for the foreseeable future owing to its maturity and low power consumption. Display industries all over the world have already invested billions of dollars in this technology and its world-wide annual sales are forecasted to triple at least to over $10 billion by the year 2000.

Among LCDs, colour thin-film transistor (TFT) TN-LCDs (or simply colour TFT-LCDs) offer the best contrast and overall image quality. But the cost to produce colour TFT-LCDs is high because of the complicated manufacturing processes to make the huge number of transistors on active display substrates and the colour filters on other substrates. Two polarizers used in these LCDs usually cut out more than 60% of the light in the transmissive (backlit) mode and more than 85% in the reflective mode. Colour filters on the colour substrate and metal lines on the active substrate block even more light, thus giving very low light utilization (or efficiency). Usually the complete panels can only utilize less than 5% of the backlight and are too dim without it. To display the image effectively, both the backlight and driver electronics need to be powered at all times, which consumes a lot of power and requires heavy batteries to operate even for only a few hours.

Passive matrix (PM) LCDs such as STN-LCDs and other LCDs that rely on polarizers to create contrast suffer the same low-light efficiency. Technologies that rely on polarizers also require non-birefringent materials as substrates. That is why most of these displays use glass as a substrate and it is not easy to adopt lightweight and impact-resistant plastic substrates for these LCDs. In fact, the non-birefringent (or at least very low birefringent) plastic substrates

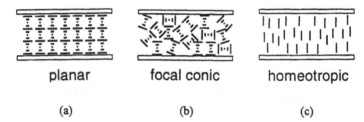

planar   focal conic   homeotropic

(a)   (b)   (c)

Figure 12.1 Illustration of the three textures (or states) of a cholesteric liquid crystal sandwiched between two BRCD substrates.

cost about five times more than glass substrates at present. This has definitely limited the potential for many portable applications where light weight, impact resistance and thin profile are critical.

The other problem that these PM-LCDs face is the cross-talk issue, which limits the maximum number of scanning lines that the display can have without significant degradation in contrast. AM-LCDs (e.g. TFT-LCDs) address this problem with the non-linearity of the active element at each pixel (e.g. the transistors on the active substrate in TFT-LCDs). Unfortunately, this increases the manufacturing cost by one order of magnitude. An alternative is to creat intrinsic display memory so that high-information-content (HIC) images can be displayed without cross-talk. Ferroelectric LCDs (FLCDs) are examples which offer intrinsic display memory and fast switching speed. Many groups and companies have been making efforts to commercialize surface-stabilized FLCDs (SSFLCDs), but they are still not yet available. One of the difficulties in manufacturing these displays is the extremely narrow spacing (1 to 2 μm against 4 to 8 μm for TN-LCDs, STN-LCDs and TFT-LCDs). Another disadvantage with SSFLCDs is the poor impact resistance. A mechanical shock can irreversibly destroy the alignment and make the display disfunctional.

Therefore, a display technology that does not use any polarizers, is reflective, and has intrinsic display memory is very desirable. The bistable, reflective cholesteric display (BRCD) technology invented (Yang *et al.* 1991; Doane *et al.*, 1992; Yang and Doane, 1992) at the Liquid Crystal Institute (LCI) of Kent State University in Kent, Ohio, USA, and further developed by Kent Display Systems (KDS) also in Kent, Ohio (Yuan, 1994; Lu *et al.*, 1995), fulfils all these critical needs. These BRCDs are made in a very similar fashion to other passive LCDs in that two substrates with patterned electrodes are used to sandwich the liquid crystal material. The cholesteric liquid crystals (CLCs) used here are chiral nematics with pitches in the range of around 250 to 500 nm so that the Bragg reflection peaks at a visible wavelength. When these CLCs are sandwiched between the display substrates, there are three possible states (or textures) depending on the surface treatment, bulk treatment and applied field. Figure 12.1 illustrates the three textures: planar, focal conic and homeotropic. Usually, only one texture is fully stable at zero field for a cholesteric liquid crystal. However, the work of Yang *et al.* (1991) and Yuan (1994) has shown that two textures, the planar and focal conic, can be stabilized by either dispersing a low concentration of polymer inside the cholesteric liquid crystal or by specially treating the display substrate surfaces. The former are called polymer-stabilized cholesteric texture LCDs (PSCT-LCDs); the latter are sometimes referred to as polymer-free cholesteric texture LCDs (PFCT-LCDs). In this chapter we will focus on PSCT-LCDs. Related work can be found in the chapters by Yang *et al.*, West and Hikmet in this volume.

The intrinsic display memory of BRCDs makes one-line-at-a-time addressing possible with inexpensive, simple passive substrates (low cost) without contrast degradation and cross-talk for very-high-information-content displays. Owing to the same intrinsic memory effect, the displays need no power to keep the image once it is written. This not only lowers the power consumption but also offers flick-free images. The high reflectivity of these BRCDs eases the lighting (and hence the power consumption) requirement since they are very readable in both normal indoor room lighting and outdoor ambient light. They are also readable in direct sunlight. Their contrast is also very good over a very wide viewing cone. Since these BRCDs do not use polarizers to generate contrast, it is possible to use inexpensive (because birefringence is now allowed in these substrates), lightweight and impact-resistant plastics as substrates. All of these features offer BRCDs great application potentials in areas such as indoor and outdoor signs, pagers, PDAs, electronic books and newspapers (paperless printing).

12.2 Electro-optical Characteristics and Driving Methods

12.2.1 *Cell Preparation*

A PSCT-LCD cell is prepared using two substrates with patterned transparent indium tin oxide (ITO) electrodes. Usually the patterned ITO electrodes are narrow lines with even narrower gaps between them. The ITO lines usually run horizontally (often called rows) on one substrate and vertically (often called columns) on the other. Each overlapping area of a column line and a row line is called a pixel which defines the smallest feature that the display can show. The surfaces of these two substrates are coated with polyimide material, baked and rubbed to generate homogeneous alignment at the liquid crystal–substrate interfaces. A certain percentage of chiral agents are doped to a nematic liquid crystal to get the proper pitch (the pitch P is defined as the length over which the cholesteric liquid crystal molecules twist through $360°$) so that a predetermined colour can be achieved. The spectrum of the Bragg reflection of a cholesteric liquid crystal layer in a planar state with pitch P at normal incident light is peaked at

$$\lambda_{\text{peak}} = \bar{n}P \tag{12.1}$$

where $\bar{n}$ is the average refractive index of the cholesteric liquid crystal. The spectral width of the Bragg reflection at full width at half maximum (FWHM) is

$$\Delta\lambda = P\Delta n \tag{12.2}$$

where Δn is the birefringence of the liquid crystal mixture.

A small percentage of monomer and photoinitiator are then added and mixed uniformly into the cholesteric liquid crystal mixture before it is vacuum filled into an assembled empty cell. The filled cell is then applied with sufficient electrical field to align the liquid crystal molecules into the homeotropic state while ultraviolet (UV) radiation is applied to the cell. Since the monomer is aligned by the homeotropic liquid crystal during the photopolymerization, the polymer network formed after the polymerization is anisotropic and has an aligning effect on the liquid crystal. This aligning effect is a bulk effect and favours homeotropic alignment. The competition between this bulk aligning effect and the surface aligning effect leads to a multidomain structure which stabilizes both the planar and focal-conic textures. The photo-polymerized cell is then coated with an absorbing layer to enhance the contrast between the highly reflecting quasi-planar (multidomains with different helix directions) texture and the weakly scattering (essentially non-reflecting) focal-conic texture.

12.2.2 *Electro-optical Characteristics*

Figure 12.2 shows the typical electro-optical curves of a PSCT-LCD. The curves were measured with a PSCT cell prepared as discussed above. The measurements were carried out with the incident light at an angle of $22.5°$ with respect to the cell normal and detecting angle at the specular angle. The horizontal axis shows the r.m.s. voltage that was applied before the cell was grounded and reflectance was measured. The pulse width in these measurements was fixed

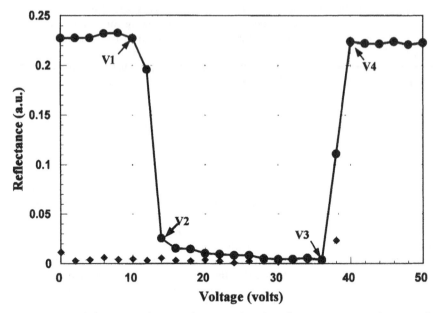

Figure 12.2 Typical electro-optical curves of a PSCT-LCD. The reflectance is measured 800 ms after the corresponding voltage with a 20 ms pulse width is applied and then grounded. The circles represent the reflectance when the initial state was in the quasi-planar state while the diamonds represent the reflectance when the initial state was in the focal-conic state.

at 20 ms. The vertical axis shows the reflectance measured 800 ms after the cell was applied with the corresponding voltage and then grounded to allow the liquid crystal to relax to its stable state. The solid circles represent the measured reflectance when the initial state was in the quasi-planar state while the solid diamonds represent the measured reflectance when the initial state was in the focal-conic state. A refreshing voltage is used before each measurement to ensure that the cell is always in either the reflecting quasi-planar state or in the non-reflecting focal-conic state regardless of the cell's previous state.

From the curve through the solid circles in Figure 12.2 we can see that the initial reflecting state essentially will not change its reflectivity after an r.m.s. voltage equal to or smaller than V_1 at 20 ms pulse width is applied and then grounded. For this particular cell, V_1 is around 10 volts. When the applied voltage is above V_1, the reflectance starts to decrease as the applied voltage increases. When the applied voltage reaches V_2, the decrease in reflectance with the increasing voltage slows down. V_2 is around 14 volts in this case. Observations with an optical microscope indicate that the grey-scale states between V_1 and V_2 are the result of a mixing of reflecting domains with non-reflecting domains since some of the domains are switched from the initially reflecting state to the non-reflecting final state. These domains are micrometre sized and the naked eye sees nice and uniform grey-scale states. When the applied voltage is further increased beyond V_2, the reflectance decreases slightly until another threshold voltage V_3 is reached. Once the applied voltage is increased beyond V_3, the reflectance starts to increase with the increase of the applied voltage because some of the initially reflecting domains are switched into the non-reflecting final state while others are switched into the homeotropic state with the applied voltage and then switched into the reflecting final state after grounding. This continues until the saturation voltage V_4' is reached. Any voltage higher than V_4' essentially turns all the domains into the homeotropic state with the applied pulse and then switches them into the reflecting final state after the grounding and hence switches the display into the reflecting quasi-planar final state with about the same reflectance.

From the curve through the solid diamonds in Figure 12.2 we can see that the non-reflecting focal-conic state is essentially unaltered when the applied voltage is below V_3'. When the voltage is increased beyond V_3', part of the initially non-reflecting domains are switched into the homeotropic state during the application of the 20 ms voltage pulse and then switched to the reflecting final state after grounding at the end of the pulse. This continues until the saturation voltage V_4 is reached. Between V_3' and V_4, the higher the voltage applied, the more domains are switched into the reflecting final state, and hence the higher the grey-scale level of reflectance. Any voltage higher than V_4 essentially turns all the domains into the homeotropic state during application of the 20 ms voltage pulse and then switches them into the reflecting final state after the grounding and hence switches the display into the reflecting quasi-planar final state with about the same reflectance. Depending on the liquid crystal mixture, surface treatment and polymer network, V_3' can be a few tenths of a volt to a few volts higher than V_3, and V_4 can similarly be higher than V_4'.

It is important to note from the two curves that it does not matter what the initial state was: (i) a PSCT cell that has a voltage applied equal to or lower than V_1 with a certain pulse width (20 ms in this example) will not change its initial state; (ii) a voltage higher than V_4 always switches the cell into the reflecting quasi-planar state; and (iii) a voltage between V_2 and V_3 (preferably closer to V_3 for better contrast) always switches the cell into the non-reflecting focal-conic state. All the driving schemes to be discussed in the next section will be based on these electro-optical characteristics.

12.2.3 *Driving Methods*

Based on the unique electro-optical characteristics of BRCDs, a straightforward one-line-at-a-time driving scheme (Yang *et al.*, 1994) can be used to drive the PSCT-LCDs. This simple driving scheme sets only one row (the 'selected' row) at a time to a voltage V_R and all the other rows (the 'non-selected' rows) to ground. At the same time the data signal is applied to all the columns. All the columns that are connected to the pixels on the selected row are set to ground if these pixels are to be displayed as non-reflecting dots. All the other columns that are connected to the pixels (to display reflecting dots) on the selected row are given a voltage V_C that is 180° out of phase with that of the selected row voltage V_R. It is clear from the electro-optical characteristics that in order for this simple driving scheme to work, the characteristic voltages V_1, V_2, V_3 and V_4 as defined above, and the selected driving voltages V_R and V_C, have to satisfy the following relationships:

$$V_2 < V_R < V_3 \tag{12.3}$$

and

$$V_R + V_C > V_4. \tag{12.4}$$

Expression (12.3) is required to ensure that the pixels to be driven to the non-reflecting state will have very low reflectance after being addressed. From the electro-optical curves in Figure 12.2 one can clearly see that a V_R selected closer to V_3 rather than V_2 gives least reflection (and hence highest contrast) and least hysteresis, and in practice V_R is selected a couple of volts lower than V_3 to accommodate manufacturing tolerance. Choosing V_R closer to V_3 also eases the amplitude required for V_C. Expression (12.4) is required to ensure that the pixels to be driven to the reflecting state will have the highest saturated reflection after being addressed. For a non-direct driven (multiplexed) PSCT-LCD, the following condition also has to be satisfied in addition to conditions (12.3) and (12.4):

$$V_C < V_1 \tag{12.5}$$

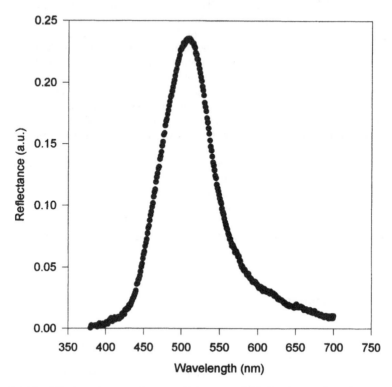

Figure 12.3 The reflection spectrum of a 5 μm thick PSCT cell filled with one of KDS' green production mixtures KDS101. The reflectances were measured at the specular angle with light incident at 22.5°.

This is needed to ensure that the portion of the image that has already been written on a multirow PSCT-LCD will not be altered when a new row is selected and addressed. It is clear from conditions (12.3), (12.4) and (12.5) that for this driving scheme to work, the electro-optical characteristics of a PSCT-LCD must satisfy

$$V_1 > V_4 - V_3 \tag{12.6}$$

In practice, low V_C is very desirable because: (i) inexpensive and readily available CMOS chips can be used for the column drivers; (ii) there are less changes in the reflectivity of the pixels on the columns with the applied voltage V_C; and (iii) there is less unwanted contrast between the reflecting pixels on the columns with the applied voltage V_C and the reflecting pixels on the columns that are grounded.

To improve these conditions, a new driving scheme was proposed, tested and implemented for PSCT-LCD productions by Catchpole *et al.* (to be published). With the new driving scheme, still only one row is selected at a time. But now the voltage applied to the selected row V_R is set at

$$V_3 < V_R < V_4 \tag{12.7}$$

and the amplitude of column voltage V_C is set at

$$V_4 - V_R < V_C < V_1 \tag{12.8}$$

Yellow PSCT Reflective Spectrum

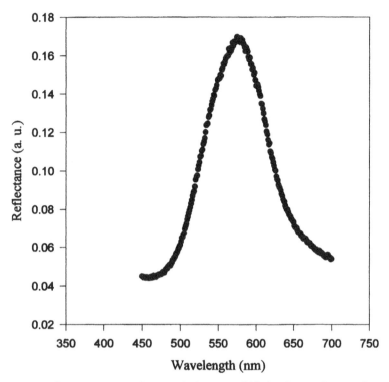

Figure 12.4 The reflection spectrum of a 5 μm thick PSCT cell filled with one of KDS' yellow production mixtures KDS201. The reflectances were measured at the specular angle with light incident at 22.5°.

for all columns, although the phase of the voltage applied to the pixels to be driven into the reflecting state is 180° out of phase with that of the row voltage, while that applied to the pixels to be driven into the non-reflecting state is in phase with that of row voltage. This new driving scheme gives manufacturers of PSCT-LCDs more tolerance and improves the display image quality at the same time. This will become clear with the following example.

Let us, for example, set the selected row voltage at

$$V_R = (V_3 + V_4)/2 \qquad (12.9)$$

Now the amplitude required for the column voltage becomes

$$V_1 > V_C > (V_4 - V_3)/2 \qquad (12.10)$$

instead of $V_1 > V_C > (V_4 - V_3)$ required by the previous driving scheme. This shows that by using the new driving scheme, the requirement for the column voltage can be reduced at least by a factor of two. In other words, the new driving scheme at least doubles the tolerance and makes it possible to drive a bistable reflective cholesteric display with $(V_4 - V_3)/2 < V_1 < (V_4 - V_3)$ when it is impossible to use the previous driving scheme to address the display properly. Using this new driving scheme we have: (i) successfully tested and implemented 5 V CMOS chips for the column drivers to lower the cost of the display electronics;

271

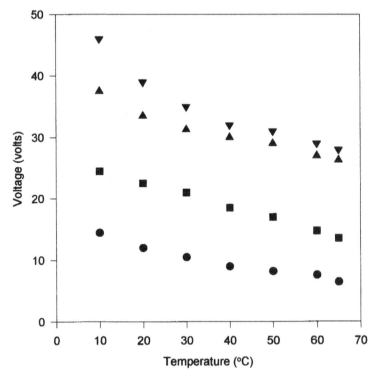

Switching Voltage vs Temperature

Figure 12.5 Measured characteristic voltages V_1, V_2, V_3 and V_4 as a function of cell temperature for a 5 μm thick PSCT cell filled with the KDS001 mixture.

(ii) greatly reduced the change to the reflectivity of the reflecting pixels during and after addressing; and (iii) eliminated any contrast among the reflecting pixels from different columns because all the columns now have the same voltage amplitude.

12.3 Critical Parameters

There are several critical parameters that need to be considered carefully before materials are chosen for bistable reflective cholesteric displays. They are the cholesteric temperature range, the intrinsic pitch P and its variation with temperature dP/dT, birefringence Δn, dielectric anisotropy $\Delta\varepsilon$, elastic constants, especially the twist elastic constant K_{22}, viscosity γ of the cholesteric liquid crystal mixture; the percentage of monomer and photoinitiator added to the mixture; and the cholesteric liquid crystal layer thickness d.

The first step is to choose a nematic liquid crystal with a wide nematic temperature range, low viscosity, large birefringence and dielectric anisotropy, good UV and thermal stability. Once the host nematic liquid crystal is chosen, the next step is to make the cholesteric mixture with a proper pitch because it will determine the reflecting colour of the cholesteric layer according to expressions (12.1) and (12.2). This is achieved by doping a suitable percentage of chiral agents with good UV and thermal stability. If a single chiral agent is used, the pitch is

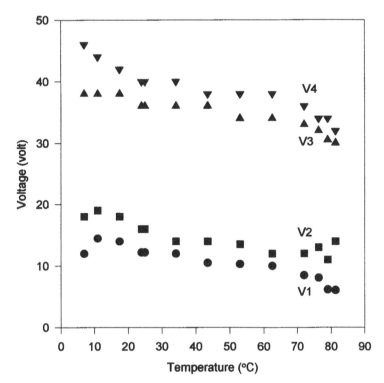

Switch Voltage vs Temperature

Figure 12.6 Measured characteristic voltages V_2, V_2, V_3 and V_4 as a function of cell temperature for a 5 μm thick PSCT cell filled with the KDS101 mixture.

inversely proportional to the product of helical twist power and the concentration of the agent. In practice, more than one chiral agent is used to achieve the desired pitch without reaching the saturation limit of any single agent and to minimize the pitch variation with the temperature.

Figure 12.3 shows, as an example, the reflection spectrum of a 5 μm thick PSCT-LCD cell filled with a green cholesteric liquid crystal mixture KDS101. The reflectances were measured at the specular angle with light incident at 22.5° with respect to cell normal. Using the same components but lowering the concentration of chiral agents we have formulated a yellow cholesteric liquid crystal mixture KDS201. The reflection spectrum of a 5 μm thick PSCT-LCD cell filled with the mixture KDS201 is shown in Figure 12.4. As can be seen from Figures 12.3 and 12.4, when the pitch is elongated by reducing the concentration of the chiral agents, the peak wavelength of the reflection spectrum is shifted from 510 to 575 nm, and the FWHM spectrum width is widened accordingly.

It is important to point out that the pitch is not only a critical parameter for the reflecting colour, but also a critical parameter affecting driving voltage. The threshold field E_{th} (an applied field above E_{th} can turn the cholesteric liquid crystal into the homeotropic state) is related to P by (de Gennes and Prost, 1993)

$$E_{th} = \frac{2\pi^2}{P} \sqrt{\frac{\pi K_{22}}{\Delta \varepsilon}}$$

(12.11)

Driving Voltage vs Temperature

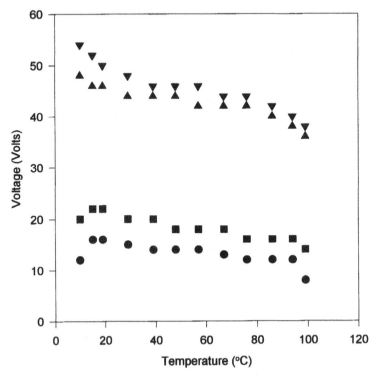

Figure 12.7 Measured characteristic voltages V_1, V_2, V_3 and V_4 as a function of cell temperature for a 5 μm thick PSCT cell filled with the KDS201 mixture.

So, with the same host nematics and cell spacing, the driving voltage is inversely proportional to the pitch. Furthermore, the variation of the pitch with temperature, dP/dT, will not only shift the display colour but also change its characteristic voltages V_1, V_2, V_3 and V_4. Therefore it is important to minimize dP/dT by choosing the appropriate host nematics and the right combination of chiral agents. Figures 12.5, 12.6 and 12.7 show examples of our efforts (Yuan, 1994) to minimize dP/dT and extend the overall operating temperature range. Figure 12.5 shows the characteristic voltages V_1, V_2, V_3 and V_4 as a function of cell temperature for a 5 μm thick green PSCT cell filled with the green cholesteric liquid crystal mixture KDS001. A reasonable operating temperature range is from 15 to 60 °C without a heater. Over this range we measured dP/dT at about 0.5 nm/°C and dV/dT at about 0.2 V/°C.

Figure 12.6 shows the characteristic voltages V_1, V_2, V_3 and V_4 as a function of cell temperature for a 5 μm thick green PSCT cell filled with the green CLC mixture KDS101 with different host nematics and chiral agents from those of the KDS001 mixture. The operating temperature has been extended to $+80$°C, the pitch variation dP/dT has been reduced to about 0.1 nm/°C and the threshold voltage variation dV/dT has been reduced to about 0.05 V/°C. So, mixture KDS101 not only extended the operating temperature range by 20 °C but also reduced the unwanted dP/dT and dV/dT by four times compared with those of KDS001. Figure 12.7 shows an example of further improvements to the operating temperature range and dV/dT. The measurements were taken with a 5 μm thick PSCT cell filled with the KDS R&D mixture

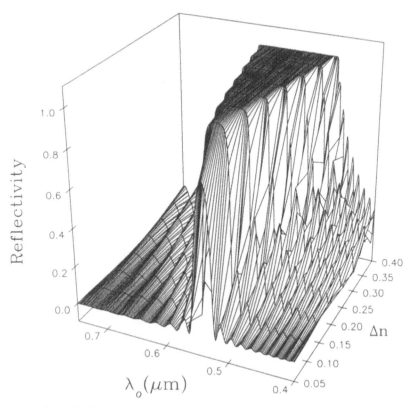

Figure 12.8 Calculated reflection spectra versus the birefringence Δn. $P = 340$ nm and $d = 15P = 5.1$ μm were used in the modelling.

9408. The upper limit of the operating temperature has been extended to over 90 °C with this mixture.

Another two important parameters that affect BRCD performances are the birefringence Δn and cell thickness d. Extensive modelling (St John *et al.*, 1995) has revealed the relationship between the display reflectivity and Δn and d/P. Figure 12.8 shows their modelling results of the reflection spectra against Δn when the pitch $P = 340$ nm and the cell thickness $d = 15P = 5.1$ μm were fixed. It clearly indicates that the FWHM spectral width is linearly proportional to the birefringence Δn while the peak reflectivity increases non-linearly with Δn. So, it is important to choose a mixture with large Δn to make bright BRCDs. Figure 12.9 shows their modelling results of the reflection spectra against the cell thickness d/P when $P = 340$ nm and $\Delta n = 0.228$ were fixed. The results show that the reflectivity increases nonlinearly with increasing d and saturates at about $d = 10P$ with $\Delta n = 0.228$. This is very useful in choosing the right cell thickness. In order to obtain the brightest BRCDs a thicker cell spacing is preferred. But a thicker cell means a higher driving voltage which in turn will increase the cost of the driver electronics and power consumption. So, in practice, the best cell spacing is chosen to be just a few pitch lengths longer than the saturation thickness to ensure the brightest display and enough manufacturing tolerance without too high a driving voltage. The driving voltage can also be reduced by choosing nematics with larger $\Delta \varepsilon$ and smaller K_{22} as indicated by eqn (12.11).

The concentration of the monomer and photoinitiator is also critical because if it is too low some of the domains are too big to be truly bistable and tend to relax with time; if is too high the domain sizes are too small and the large amounts of the defects created by the polymer

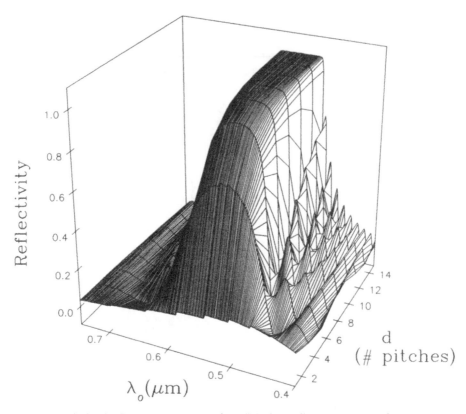

Figure 12.9 Calculated reflection spectra versus the cell thickness d/P. $P = 340$ nm and $\Delta n = 0.228$ were used in the modelling.

network will reduce the reflectivity of the BRCD. Also very high polymer concentration will cause too much scattering in the focal-conic state and hence reduce the contrast of the BRCD. The optimal concentration depends on the surface treatment and the cholesteric mixture used and usually requires many tests of the actual cell and mixture combination.

12.4 Recent Developments and Applications

In addition to their bistability and high reflectivity, these BRCDs also have very high contrast over a very wide and quite symmetrical viewing cone. Figure 12.10 shows a typical isocontrast plot measured at specular angles. These desirable characteristics make the BRCDs suitable for many applications including public information displays. One recent development is the minimization of dP/dT and dV/dT combined with a temperature compensation circuit implemented in the driver electronics (Lu *et al.*, 1995). This extended the operating temperature over $-10\,°C \sim +75\,°C$ without a heater. Heaters have also been applied to PSCT-LCDs to extend the low-end operating temperature limit to below $-35\,°C$ for outdoor applications. These applications include bus signs, wayside signs, and signs used at bus stations, railway stations and airports. Recent colour development work at KDS has expanded the availability

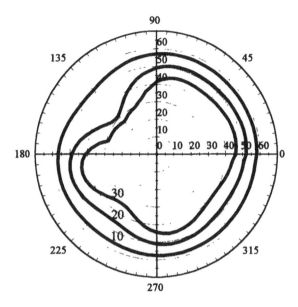

Figure 12.10   A typical isocontrast plot of a PSCT-LCD measured at specular angles.

of colour (Lu *et al.*, 1995). This is achieved by developing different colour mixtures in combination with different colour backgrounds to achieve both the desired active and inactive colours.

The development efforts for document-like displays at KDS in cooperation with the Liquid Crystal Institute and the University of Stuttgart have recently resulted in very high-resolution BRCDs called Viewer (Pfeiffer *et al.*, 1995). Figure 12.11 shows a photograph of the first prototype green/black Viewer at a resolution of over 100 lines per inch (LPI) and an active area of 8.5 × 11 square inches (22 × 28 cm²). Figure 12.12 shows a photograph of a prototype yellow/black Viewer with the same resolution and size.

These high-resolution, high reflective and bistable displays (such as the Viewers) are very attractive for applications such as electronic books, newspapers and magazines to replace printed materials in the future. One of the issues to be resolved before BRCDs can be widely adopted in such applications is the writing speed. Currently, with off-the-shelf chips we have been able to drive BRCDs at around 10 ms per line. For low-to-medium-resolution signs this may not be a problem, but for high-information-content (HIC) displays such as the Viewers it could take more than 10 seconds to write the whole page and this is too slow. Recently the group at the Liquid Crystal Institute has developed a dynamic driving scheme (Huang *et al.*, 1995) which can write to BRCDs at a speed at least one order of magnitude faster. A recent development by Lu and Yuan (to be published) has resulted in a new technique to make BRCDs extremely impact resistant and to offer higher contrast at the same time. Efforts have also been made to produce full-colour BRCDs using colour pixelization through a photo-lithographic process (Chien *et al.*, 1995). The recent efforts by Yuan and Lu (to be published) and coworkers using plastic substrates for BRCDs will certainly move us closer to the realization of ultrathin and very lightweight and high-impact-resistant, high-definition displays.

In conclusion, the BRCD is a relative new technology with many attractive characteristics such as intrinsic display memory, high reflectivity, high contrast and wide viewing angles. With

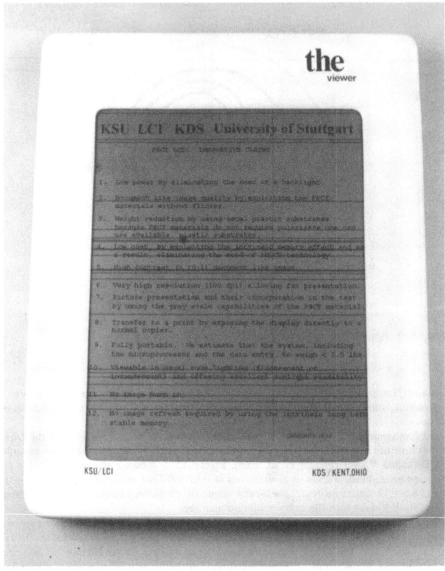

Figure 12.11 A photograph of the first prototype green/black Viewer developed by Kent Display Systems in cooperation with the Liquid Crystal Institute and Stuttgart University. The display has an 8.5 × 11 square inches (22 × 28 cm²) active area and over 100 LPI resolution.

continuous research and development, BRCDs should find ever-increasing applications in the information age.

Acknowledgements

A lot of work presented here was done in collaboration with Drs Minhua Lu, Clive Catchpole and Zvi Yaniv at Kent Display Systems. The author would like to thank Drs Deng-Ke Yang,

Figure 12.12 A photograph of a prototype yellow/black Viewer developed by Kent Display Systems in cooperation with the Liquid Crystal Institute and Stuttgart University. The display has an 8.5 × 11 square inch (22 × 28 cm^2) active area and over 100 LPI resolution.

L.-C. Chien, J. W. Doane, Jack Kelly, Peter Palffy-Muhoray, Phil Bos, John West, Matthias Pfeiffer, Doyle St John and Hefen Lin at the Liquid Crystal Institute for many stimulating discussions. The author would also like to acknowledge fruitful collaborations with Dr E. Luder at Stuttgart University, Dr David Coates at Merck Ltd and Dr Frank Allan at EMI. This work was partially supported by ARPA under Grant N-61331-94-C-0041.

References

CATCHOLE, C., YUAN, H. and LU, M.-H. (to be published).

CHIEN, L. C., MULLER, U., NABOR, M.-F. and DOANE, J. W. (1995) Multi-color reflective cholesteric displays, *SID '95 Dig.*, pp. 169–171.

DE GENNES, P. and PROST, J. (1990) *The Physics of Liquid Crystals*, 2nd edn, Oxford: Clarendon Press, p. 288.

DOANE, J. W., YANG, D.-K. and YANIV, Z. (1992) Front-lit flat panel display from polymer stabilized cholesteric textures, *Japan Display '92*, pp. 73–76.

HUANG, X.-Y., YANG, D.-K., BOS, P. J. and DOANE, J. W. (1995) Dynamic drive for bistable reflective cholesteric displays, *SID '95 Dig.*, pp. 347–350.

LU, M.-H. and YUAN, H. (to be published).

LU, M.-H., YUAN, H., CATCHPOLE, C., WEGER, J. and YANIV, Z. (1995) Study of the temperature dependence of the PSCT display, *SID '95 Application Dig.*, pp. 28–30.

PFEIFFER, M., YANG, D.-K., DOANE, J. W., BUNZ, R., LUDER, E., LU, M.-H., YUAN, H., CATCHPOLE, C. and YANIV, Z. (1995) A high information content reflective cholesteric display, *SID '95 Dig.*, pp. 706–709.

ST JOHN, W. D., FRITZ, W., LU, Z.-J. and YANG, D.-K. (1995) Bragg reflection from cholesteric liquid crystals, *ALCOM Tech. Rep.* **1**, 1–16.

YANG, D.-K. and DOANE, J. W. (1992) Cholesteric liquid crystal/polymer gel dispersion: Reflective display applications, *SID '92 Dig.*, pp. 759–761.

YANG, D.-K., CHIEN, L.-C. and DOANE, J. W. (1991) Cholesteric liquid crystal/polymer gel dispersion bistable at zero field, *Conference Record of Int. Display Research Conf., SID*, pp. 49–52.

YANG, D.-K., DOANE, J. W., YANV, Z. and GLASSER, J. (1994) Cholesteric reflective display: Drive scheme and contrast, *Appl. Phys. Lett.* **64**, 1905–1907.

YUAN, H. (1994) Development of bistable cholesteric reflective displays, Invited presentation at the *ALCOM Symp. on Stabilized and Modified Liquid Crystals, Kent State University, Kent, Ohio, USA, October.*

YUAN, H. and LU, M.-H. (to be published).

13

The Effect of Polymer Networks on Twisted-nematic and Super-twisted-nematic Displays

P. J. BOS, D. FREDLEY, J. LI AND J. RAHMAN

13.1 Introduction

Yang *et al.* (1991) have shown for the case of cholesteric reflective bistable devices that the addition of a low percentage of a polymer additive can stabilize a desired director configuration (for more details regarding bistable cholesteric devices, see the chapters by Yang *et al.* and Yuan in this volume) Here we investigate how this new tool of polymer stabilization (Hikmet, 1991) of desired director configurations is useful in twisted-nematic (TN) and super-twisted-nematic (STN) devices.

A problem to be solved in TN devices is how to reduce the operational voltages. To accomplish this, several approaches have been proposed. One approach is to use high-dielectric-constant materials with low elastic constants. The trade-off in using such materials is the problem of higher ion solubility that results in 'image sticking' in active matrix displays. Another approach is to use a high pretilt alignment (compare curve 1 with curve 2 in Figure 13.1). The higher pretilt angle can be achieved by evaporated alignment, but this process is not commonly consided to be manufacturable.

A problem to be solved in STN devices is also related to the need for a higher pretilt angle than is easily achievable. Super-twisted birefringent effect (SBE) devices (or super-twisted-nematic (STN)) devices have supplanted TN devices in highly multiplexed displays owing to their inherently steeper electro-optical response. In Scheffer's original paper (1984) on SBE devices he demonstrated (for the given conditions) that the steepest curve corresponds to a 270° twist of the director field. However, in highly twisted devices there are competing director configurations that occur in the presence of an electric field applied along the cell normal. As shown in Figure 13.2, in addition to the desired conical deformation, a director configuration can form where the helical axis is along the plane of the cell producing a director configuration that appears under the microscope as stripes. Increasing the pretilt angle of the director at the surface has been shown to be effective in suppressing the stripe deformation (Scheffer and Nehring, 1984; Akatsuka *et al.*, 1987). Pretilts of 5° have been shown to eliminate the stripe deformation in devices having up to about a 240° twist. However, a pretilt high enough to eliminate stripes for a 270° twist are attainable only with obliquely evaporated SiO, which is an elaborate and expensive process, or special polyimides, which often do not give consistent results.

Motivated by the results of Yang *et al.* (1991) the goal of this work is to apply the conncept of polymer stabilization to these problems.

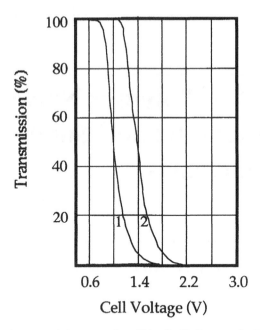

Figure 13.1 Voltage against transmission curves for a TN cell with the same liquid crystal material but with a low pretilt resulting from deposition of PVA (2), and with a high pretilt resulting from deposition of SiO(1). (From BDH.)

13.2 Polymer-stabilized TN Devices

We considered that a polymer network could be useful to lower the operating voltages of a TN device. (Details about polymer network morphologies can be found in the chapters by Hikmet, Yang and coworkers and Crawford and Žumer in this volume.) We found (Bos *et al.*, 1993) that a significant lowering of the operating voltages of a TN device can be achieved by doping a liquid crystal with a photocurable polymer and applying a voltage to the cell during the photocuring process. We describe here the experiments performed and the results obtained.

All the test cells in this study were made with ITO-coated Corning borosilicate glass. The alignment layer was a heat-cured polyimide that provided a pretilt angle in the ~3°–4° range. The cells were assembled as 90° twist cells. The liquid crystal used was ZLI4718 which was doped with a chiral additive to prevent reverse twist disclinations (0.05% ZLI-3786). The photocurable polymer material (DSM-950-044) was added to the liquid crystal in concentrations from 1% to 5% in a semi-dark room with minimum lighting conditions.

The effect of the amount of polymer added to the liquid crystal material on the electro-optical curve of a TN device was investigated. We found that the 1% doped cells showed no reduction in the operating voltages of a TN device independent of the voltage applied during curing. At the other extreme, the 5% doped cells maintained the homeotropic texture at zero volts after curing with the voltage applied. Between these extremes it was found that cells containing 4% polymer remained homeotropic if the curing voltage was above 20 volts, and although a device operated as a TN device and showed some reduction in the operational voltages if the curing voltage was between 5 and 10 volts, the cells had a slight scattering texture. It was only with cells that were doped with approximately 2% of the polymer additive that the desired voltage reduction in the electro-optic response was achieved without any unwanted degradation in the electro-optical properties of the TN device.

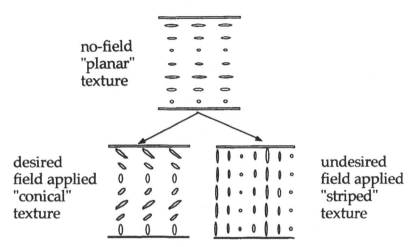

Figure 13.2 The effect of an electric field applied to the 'planar' texture of an SBE device. Shown is an edge view of a liquid crystal cell. The top drawing approximates the director configuration when no field is applied. The bottom drawings approximate the possible director configurations when an electric field is applied along a perpendicular to the cell surfaces (along the vertical direction in the figure). On the left is the desired 'conical' deformation, and on the right is the undesired 'stripe' deformation.

To see the effect of the curing voltage, Figure 13.3 shows the electro-optic curves of a series of 6.0 μm cells that were cured over a month before the measurements were taken. The 0 volt cured cell required higher voltages than the control 5.5 μm undoped cell. The cells cured at 20 volts or higher showed a significant voltage reduction in both the voltage corresponding to 10% transmission (V_{10}) and the voltage corresponding to 90% transmission (V_{90}). However, the cell cured at 50 volts when driven at 0 volts had an effective Δnd that was insufficient to match the first interference minimum condition for a TN cell. To see the effect of thickness on the effect of curing voltage, 7.5 μm cells were made. The results in Figure 13.4 show that in this case the cell cured at 50 volts then appeared to be at the first interference minimum.

The turn-on (t_{on}) and turn-off (t_{off}) times were measured for each of these cells at room temperature by giving a burst of 5 volt r.m.s. pulses to turn on the cell and then driving it to ground. The data, along with those for V_{10} and V_{90}, are shown in Table 13.1. We saw effects similar to that previously observed by Braun *et al.* (1992). The 50 volt cured cell had a slower turn-off response time compared with the control cell; 60 Hz square waves with the r.m.s. values shown were used for curing.

Table 13.1 Summary of electro-optical performance parameters

Cell thickness (μm)	Cure voltage (V)	V_{10} (V)	V_{90} (V)	t_{on} (ms)	t_{off} (ms)
5.5	NA	2.1	3.4	10.0	14.0
6.0	0	2.4	4.0	11.0	11.0
6.0	10	1.9	3.5	9.5	16.0
6.0	50	1.2	2.8	8.6	35.0
7.5	10	2.1	3.7	12.0	17.5
7.5	50	1.0	2.6	8.8	58.0

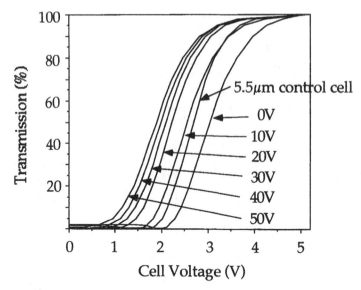

Figure 13.3 The effect of cure voltage. Cells were exposed with UV light to cure the polymer with the indicated voltage applied. All cells were 6 μm thick except the control cell which contained no polymer and was 5.5 μm thick.

To check for the possibility of a short-term memory effect that could result from deformation of the polymer network, two 6.0 μm cells were considered that were cured at 30 volts and 50 volts respectively. Their electro-optic effect was measured for increasing and decreasing applied r.m.s. voltages. The cells were turned on at 10 volts r.m.s. and left on overnight, then turned

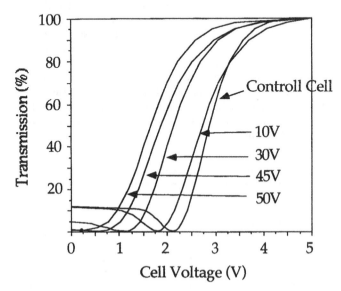

Figure 13.4 The effect of cure voltage. Cells were exposed with UV light to cure the polymer with the indicated voltage applied. All cells were 7.5 μm thick except the control cell which contained no polymer and was 5.5 μm thick.

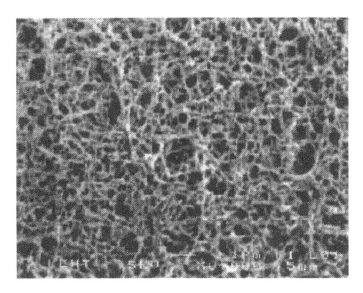

Figure 13.5 SEM of an SBE cell with 1.8% polymer additive. The cell was constructed and one plate was removed to acquire this picture (West and Rouberol, 1994). The actual size of the area photographed is about 12 × 15 μm^2.

off and immediately turned on at 5 volt r.m.s. The electro-optic effect was measured by ramping the r.m.s. voltage to 0 volts and then immediately ramping it back up to 5 volts r.m.s. No difference was observed in the electro-optical curve from those that had been stored with no voltage applied. These results indicate that short-term memory effects are not a problem in these cells.

13.3 Polymer-stabilized STN Devices

We have found (Fredley *et al.*, 1994) that for STN devices the concept of polymer stabilization can be used to control the formation of stripes. We describe here the processes used in building test devices and the results obtained.

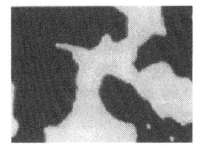 

Figure 13.6 Microscopic pictures of a 270° twist cell with 1.8% polymer additive (left) and a 240° twist cell with no polymer additive (right). Both pictures were taken with a voltage applied to the cell that corresponds to the midpoint of the electro-optical curve. The 240° twist cell shows the stripe texture, while the 270° cell with the polymer additive does not. The distinct black and white regions (rather than a uniform grey) are an indication of a very steep electro-optical curve.

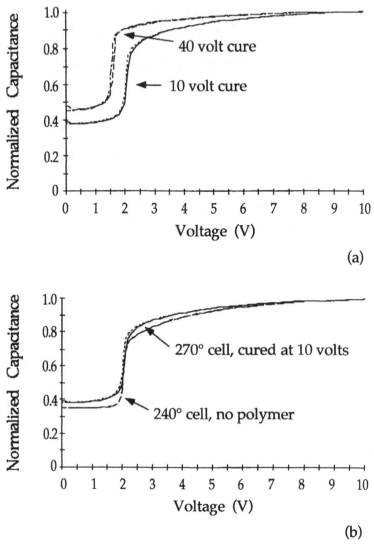

Figure 13.7 The effect of cure voltage in SBE cells. Shown are electro-capacitive curves where the applied voltage was ramped from 0 to 10 volts, and then back down to 0. The top picture shows data acquired for 270° twist cells that contained 1.8% polymer and had been cured while 10 volts and 40 volts was applied. The lower picture shows the cell cured with 10 volts and a 240° twist cell that contained no polymer.

Nissan 610 polyimide was spin coated onto 1.5 inch square (3.8 cm^2) ITO glass substrates and rubbed. Cells were constructed with 5 μm glass spacers obtained from the Merck Co. The polymer was produced from the photocurable diacrylate monomer DSM-950-044 available from the Desotech Co. The liquid crystal was ZLI 1694 and the chiral additive was CB15 from the Merck Co. A cell thickness to an LC material pitch ratio (d/p) of 0.72 was chosen for the 270° cells. The liquid crystal mixture was mixed with the monomer; the cells were then vacuum filled, squeezed and sealed. The resultant cells were then exposed to an ultraviolet light source from O-lite while an electric field was applied.

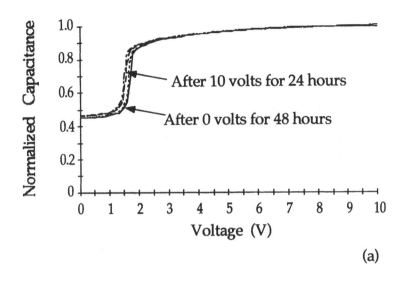

(a)

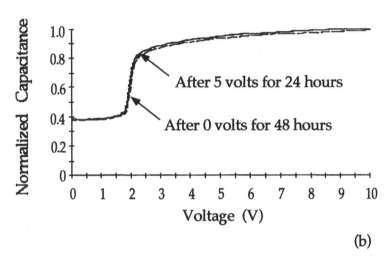

(b)

Figure 13.8 The effect of cure voltage on short-term memory: (a) is for the test cured with 40 volts applied; (b) is for the test cell cured with 10 volts applied.

Cells were analysed with a capacitance bridge from Hewlett-Packard. The capacitance of the cells was measured as a function of voltage by ramping the probe voltage upward and downward at a rate of 3 V/min. The resultant capacitance against voltage scans were normalized at 10 volts to remove the effects of small differences in the thickness of the cells. The reported driving voltage was taken at the midpoint of the electrocapacitance response.

The effect of the amount of polymer was investigated. Cells were constructed with 1, 1.5, 1.75, 1.8, 2 and 3% polymer concentration and cured while 20 volts was applied to the ITO electrodes. It was discovered that cells with polymer percentages below 1.75% demonstrated the formation of the striped texture in a way similar to cells with no polymer. The cell with 3% polymer was homeotropic after curing, and was not operable as an SBE device. At concentrations above 1.8%, small scattering domains could be observed, and at 2% the microscopic texture was somewhat grainy. The best results were obtained within a narrow range of polymer concentration between 1.75 and 1.8%. Figure 13.5 shows an SEM picture taken of the polymer network (West and Rouberol, 1994). Figure 13.6 shows that the cell built with 1.8% polymer does not show the striped director configuration.

Both of the pictures in Figure 13.6 were taken with a voltage applied corresponding to the steepest region of the electrocapacitive curve. This is intended to yield a grey level with about 50% the maximum intensity achieved by optimizing the voltage for maximum transmission. It is noticeable that the picture taken of the 270° twist cell at this voltage actually does not show any grey scale but a distribution of on and off regions. This implies that the electro-optical curve is exceptionally steep. In fact is it likely that the finite slope of the electrocapacitive curves in the transition region is due to slight non-uniformities over the about 1 inch square (6.45 cm^2) active area of the test cells.

The effect of the intensity of the UV light used for curing was investigated. Cells were constructed and exposed at 3 mW/cm^2 and 80 mW/cm^2. Cells exposed at the higher intensity had a granular texture to them, while cells exposed at the low intensity had a smoother texture more characteristic of SBE displays with no polymer dopant. There was no difference in the electro-optical characteristics of the cells built with the two curing intensities.

The curing voltage was also studied. Cells were cured while being held at voltages of 10 and 40 volts by exposing with the UV source at 3 mW/cm^2. As can be seen in Figure 13.7(a) the cell cured with the higher voltage resulted in an electrocapacitive curve that is shifted to lower voltages. Figure 13.7(b) shows that the cell cured with 10 volts had similar operating voltages to an industry standard 240° cell (no polymer added).

The amount of time after curing has an effect on the response curve as well. In the first 3 days after curing the curves increased by about 0.1 volts. The curves were then more stable with a shift of less than 0.1 volts over 1 month.

Short-term memory effects were observed. Figures 13.8(a) and (b) show the electrocapacitive response for cells that had a voltage applied to them for 24 hours just prior to the measurement, and the results of a measurement performed 48 hours later (no voltage applied). It can be seen that in the case of the cell cured at 40 volts and having a 10 volt level applied to it for 24 hours that a shift is noted, but in the cell cured at 10 volts and having 5 volts applied for 24 hours, the shift is smaller than the 0.10 volt resolution of the measurement.

13.4 Conclusions

The polymer-stabilized TN cells are shown to have significantly lowered operational voltages, and polymer stabilization in the STN cells was shown to eliminate the striping texture at high twist states. Reduction of the driving voltage is demonstrated for the case of the STN devices, but in this case we observed a short-term memory effect.

Acknowledgements

The authors wish to acknowledge Mr Doug Bryant for his asistance in cell preparation and Elaine Landry for manuscript editing. This work was supported in part by ARPA under Grant No. MDA972-91-J-1020, and by NSF ALCOM Center under Grant No. DMR 89-20147.

References

AKATSUKA, M., KATOH, K., SAWADA, K. and NAKAYAMA, M. (1987) Electro-optical properties of supertwisted nematic display obtained by rubbing technique, *Proc. SID* **28**, 159–165.
BDH Catalog 1189PP/3.0/0878.

Bos, P., Rahman, J. and Doane, J. W. (1993) A low threshold-voltage polymer network TN device, *SID Dig. Tech. Pap.* **XXIV**, 887–880.

Braun, D., Friek, G., Grell, M., Klemis, M. and Wendorff, J. H. (1992) Liqud crystal/liquid crystalline network composite systems. Structure formation and electro-optic properties, *Liq. Cryst.* **11**, 929–939.

Fredley, D. S., Quinn, B. M. and Bos, P. J. (1994) Polymer stabilized SBE devices, *Conference Record of the 1994 Int. Display Research Conf.*, pp. 480–483.

Hikmet, R. A. M. (1991) Anisotropic gels and plasticized networks formed by liquid cystal molecules, *Liq. Cryst.* **9**, 405–416.

Scheffer, T. J. and Nehring, J. (1984) A new highly multiplexable liquid crystal display, *Appl. Phys. Lett.* **45**, 1021–1023.

West, J. L. and Rouberol, M. (1994) Technique for SEM analysis of PSCT and PDLC polymer matrices (to be published).

Yang, D.-K., Chien, L.-C. and Doane, J. W. (1991) Cholesteric liquid crystal polymer gel dispersion bistable at zero field, *Conference Record of the 1991 Int. Display Research Conf.*, pp. 49–52.

14

Liquid Crystalline Polymer-stabilized Amorphous TN-LCDs

Y. IWAMOTO, Y. IIMURA, S. KOBAYASHI, T. HASHIMOTO, K. KATOH, H. HASEBE and H. TAKATSU

14.1 Introduction

Recent performances of twisted-nematic liquid crystal displays (TN-LCDs) driven by active matrix elements are becoming almost comparable with those of conventional CRT displays. The progress in these LCDs strongly depended on improvements in the device fabrication processes. In general, a rubbing method is used to align liquid crystal molecules in industrial production. However, the rubbing process generates dust and electrostatic charges, which result in polluting the clean environment and damaging the active elements fabricated on a glass substrate. This thus leads to a reduction in the production yield of the active matrix (AM-)LCD panels. Therefore, the development of a 'rubbing-free technology' is a key subject in the mass production of AM-LCDs.

There is also another subject for AM-TN-LCDs to become post-CRT displays: that is, an improvement of the viewing angle characteristics of the LCDs. This is becoming a particularly serious problem in large-size LCD panels.

Amorphous (a-)TN-LCDs (Toko *et al.*, 1993) are one of the display modes and solve the problems mentioned above. This mode is known to use the random alignment of nematic liquid crystal (NLC) molecules, which results from a non-rubbing nature of the display mode. Electro-optical (EO) performance of this mode is almost comparable with that of a conventional TN mode, and the a-TN-LCDs possess the features of wide and uniform viewing angle characteristics and a good grey-scale capability.

However, because of the random orientation of NLC molecules in a-TN-LCDs, generation the pretilt angle is impossible. This leads to the inevitable appearance of reverse tilt disclinations (RTDs) during the grey-scale operation and thus degrades the legibility of the display. This indicates that control of the RTD generation is important to improve the display quality.

In order to control the RTD generation, we tried to fabricate a-TN-LCDs with polymer networks, which were formed by using UV curable liquid crystalline monomers with mesogenic cores. We call this new type of LCD 'liquid crystalline polymer-stabilized amorphous TN-LCD' (LCPS-a-TN-LCD).

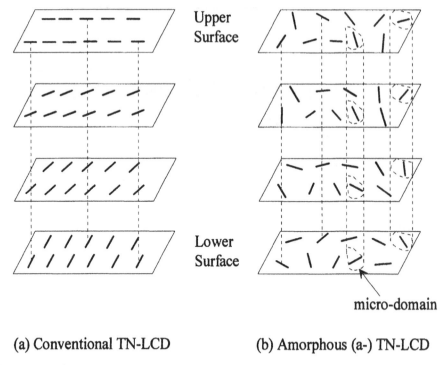

(a) Conventional TN-LCD   **(b) Amorphous (a-) TN-LCD**

Figure 14.1   Schematic structures of (a) a conventional TN- and (b) an amorphous (a-) TN-LCD.

In this chapter, we report on a study of various optical characteristics of the LCPS-a-TN-LCDs compared with those of conventional a-TN-LCDs. From these results, we found that RTDs were frozen and immobilized (stabilized) by forming polymer networks in a-TN-LCDs, and the RTD domain size was controlled by choosing the bias voltage during the UV curing.

14.2 Amorphous TN-LCDs

14.2.1 *Features and an Operating Principle of a-TN-LCDs*

Before discussing LCPS-a-TN-LCDs, we will describe the main features and the operating principle of a-TN-LCDs, which form the basis of the LCPS-a-TN-LCDs.

Schematic structures of conventional TN- and a-TN-LCDs are shown in Figure 14.1. The conventional TN-LCD has the structure where NLC molecules are aligned in one direction on both substrate surfaces and show a 90° twisted configuration. On the other hand, the a-TN-LCD shows that the LC molecular orientations on both substrate surfaces are randomly distributed in such a way that the directors at the substrate surface continuously vary their azimuthal orientation. It is important to note that, as shown in Figure 14.1(b), the a-TN-LCDs can be considered to consist of many, extremely minute TN columns that have a 90° twisted configuration; a multidomain TN structure is formed.

Figure 14.2 shows the luminance against applied voltage curves of a conventional TN- and an a-TN-LCD. As can be seen, the EO characteristic of the a-TN-LCD is almost comparable with that of the conventional one. This good EO performance of the a-TN-LCD can be

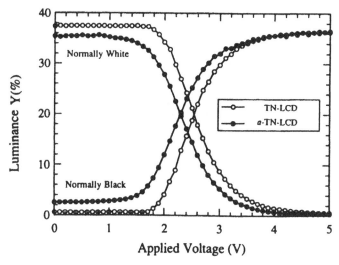

Figure 14.2 Luminance Y(%) against applied voltage characteristics for a conventional TN- and an a-TN-LCD operated in the normally white and black modes.

explained by considering incident light propagation through the TN media. If a linearly polarized monochromatic light whose wavelength satisfies Gooch and Tarry's first minimum condition (Gooch and Tarry, 1975) is incident to the TN cell, the polarization direction of the light passing through the cell is rotated by 90° with respect to the polarization direction of the incident light. This 90° rotation of the transmitted light polarization does not depend on the angle between the incident light polarization and the director at the entrance surface of the cell. Therefore, the luminance at zero applied voltage only degrades about 2% as shown in Figure 14.2. Moreover, it is known that conventional TN-LCDs characterized by a monodomain structure exhibit a strong viewing angle dependence on the contrast ratio. But as shown in Figure 14.3, the viewing angle characteristic of the a-TN-LCD is almost homogeneous owing to its random orientation nature. The a-TN-LCDs are also free from a contrast inversion problem, which is very important in the grey-scale operation.

These features of the EO performance for the a-TN-LCDs were also proved by computer simulations (Sugiyama *et al.*, 1993).

14.2.2 Textures of a-TN Media

As mentioned above, a-TN-LCDs fabricated by a non-rubbing process exhibit excellent viewing angle characteristics. Here we will discuss the textures of the a-TN media.

The textures of the a-TN media are shown in Figure 14.4, which were observed by using an optical microscope under a parallel-Nicols condition. The textures of Figure 14.4(a) were taken at zero applied voltage. It is a schlieren texture and shows $m = \pm\frac{1}{2}$ and $m = \pm 1$ disclinations (de Gennes and Prost, 1993). Information on the molecular orientation of the NLC media can be obtained from the texture. For example, the dark regions indicate that the NLC directors are aligned parallel or perpendicular to the polarizers. Regions that show the same colour are called microdomains, the size of which varies with the cell thickness. The control of the microdomain size of a-TN media is discussed by Hashimoto *et al.*

Figure 14.4(b) shows a texture at an applied voltage (3 V) above the threshold voltage (V_{th}). Since the a-TN-LCDs do not have any pretilt angle, the RTD generation can be seen in the

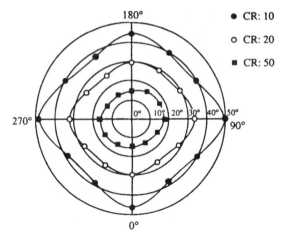

Figure 14.3 Isocontrast ratio contours of an a-TN-LCD operated at 5 V. The contrast is evaluated by the luminance.

figure. The RTD domain size is gradually enlarged increasing applied voltage and then the RTDs disappear when the applied voltage is further increased as shown in Figure 14.4(c). This is because, in the region of low applied voltage, the same molecular tilted direction in the NLC media is obtained in a certain range of the substrate area, which depends on the applied voltage, and when the applied voltage becomes high enough, the director in all the substrate area is oriented almost perpendicular to the substrate surface. Therefore, the a-TN configuration transits from a multidomain to a monodomain structure with increasing applied voltage.

It is important to note that a change of bias voltage induces the transient behaviour of the RTD texture owing to changing the RTD domain size, and this transient behaviour strongly affects the quality of the display.

In conventional TN-LCDs, RTD generation is prevented by selecting an approriate pretilt angle. In a-TN-LCDs, however, the molecular orientation on substrate surfaces is random; so a uniform pretilt angle is impossible to obtain. This indicates the contradiction between the

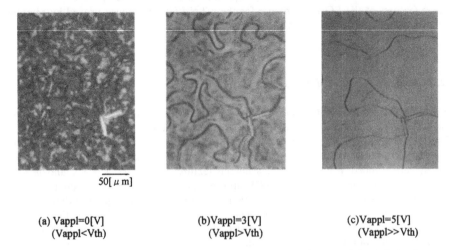

(a) Vappl=0[V]   (b)Vappl=3[V]   (c)Vappl=5[V]
(Vappl<Vth)   (Vappl>Vth)   (Vappl>>Vth)

Figure 14.4 Applied voltage (V_{appl}) dependences of a-TN textures under a parallel-Nicols condition at static state.

wide and homogeneous viewing angle characteristics and the uniform pretilt angle. Therefore, we suppose that it is impossible to eliminate the reverse tilt disclinations perfectly.

14.3 LCPS-a-TN-LCDs

14.3.1 Objective

Reverse tilt disclinations appear at the grey-scale operation of a-TN-LCDs. But the RTDs do not become a serious problem in a static operation mode. In the case of a dynamic operation mode, however, these RTDs show a transient behaviour which strongly influences the quality of the display and degrades the legibility of the a-TN-LCDs.

The following methods were considered to solve the problem:

1 eliminating the RTDs,
2 controlling the RTD domain size, and
3 controlling the generation position of the RTDs.

As discussed in section 14.2.2, it is essentially impossible to eliminate the RTDs in a-TN-LCDs, and thus method 1 does not help to improve the quality of display.

Bos *et al.* (1993) reported the results of TN-LCDs with bulk tilt angle controlled by polymer networks. The LCDs were fabricated by the UV curing of acrylate monomers doped into the NLC with a bias voltage. The LCDs exhibited similar EO characteristics to conventional TN-LCDs with high pretilt angle and were able to reduce the driving voltage significantly. Also, Bos *et al.* reported the results of 270° twisted (STN-)LCDs fabricated using the same method and proved that a stripe domain texture, which was observed in STN-LCDs with low pretilt angle, was eliminated in the LCDs (Fredly *et al.*, 1994; Bos *et al.* this volume). This new method to generate the tilt angle in a bulk region can be utilized to control the RTD generation in a-TN-LCDs.

We used UV curable liquid crystalline diacrylate monomers to form polymer networks in NLC media.

14.3.2 Fabrication of LCPS-a-TN-LCDs

First, we prepared an NLC mixture doped with a chiral material to give $d/p = 0.25$, liquid crystalline diacrylate monomers, a photoinitiator and an inhibitor. The sample used was a sandwich-type cell with non-rubbed polyimide films. The NLC mixture was injected into the empty cell in the isotropic phase, and an amorphous TN configuration was formed by cooling the cell to room temperature. The cell is irradiated with UV light under application of a voltage (V_{cure}) of 130 Hz square wave in the nematic phase. By using this process, we can control the configuration of polymer networks formed in the NLC media.

Table 14.1 shows the material parameters used and experimental conditions for sample preparation. The chemical structure of the liquid crystalline diacrylate monomer and its phase sequence are shown in Figure 14.5. The phase diagram of the NLC mixture is also shown in Figure 14.6.

14.3.3 Observation of the Textures of LCPS-a-TN-LCDs

The texture variation of an LCPS-a-TN-LCD 5 μm thick having the monomer concentration of 0.5 wt.% and $V_{cure} = 2$ V is shown in Figure 14.7 for three different applied voltages. These

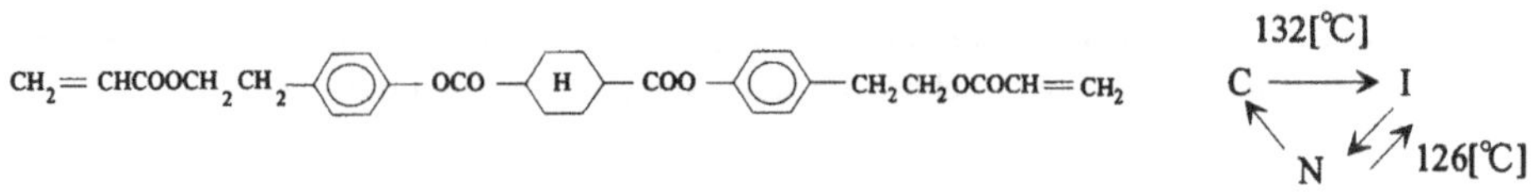

$$CH_2 = CHCOOCH_2 CH_2 - \bigcirc - OCO - \langle H \rangle - COO - \bigcirc - CH_2 CH_2 OCOCH = CH_2 \qquad C \xrightarrow{132[°C]} I$$

Figure 14.5 The chemical structure of a liquid crystalline diacrylate monomer and its phase sequence.

Table 14.1 Materials for the NLC mixture and the sample preparation conditions

Materials:	
NLC	ZLI-4792 (Merck) ($\Delta\varepsilon = 5.2$, $\varepsilon_{\parallel} = 3.1$)
Chiral additive	S-811 (Merck)
Monomer	Liquid crystalline diacrylate monomer (the chemical structure is shown in Figure 14.5)
Others	Photoinitiator and inhibitor
Conditions:	
Alignment firm	Non-rubbed PI (JALS-219) (Japan synthetic rubber)
Cell thickness	5–6 μm
UV lamp	High-pressure Hg lamp, 200
Concentration of photoinitiator	5 wt.% of total monomer concentration
Concentration of doped monomer	0.5–2.0 wt.% (including photoinitiator) in NLC mixture
Curing time	30–60 s
Applied voltage during UC curing (V_{cure})	0–5 V

textures were observed under a cross Nicol condition. In the figure, the long bright lines indicates RTDs, and the short bright ones represent polymer networks formed in the NLC media. It should be noted that, although RTD generation is observed, the transient motion of the RTD domains, which is observed in conventional a-TN-LCDs, cannot be recognized. This indicates that the RTD domain is perfectly frozen and immobilized, and thus the domain size and the generation position of the RTD domain do not change with the applied voltage.

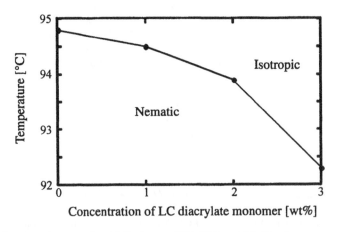

Figure 14.6 The phase diagram of an NLC mixture of ZLI-4792 and LC diacrylate monomer.

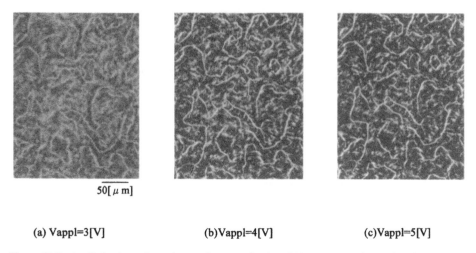

50[μm]

(a) Vappl=3[V]   (b)Vappl=4[V]   (c)Vappl=5[V]

Figure 14.7 Applied voltage dependence of textures for the LCPS-a-TN-LCD observed under a cross Nicol condition. The samples of 5 μm thickness are fabricated at $V_{cure} = 2.0$ V with a monomer concentration of 0.5 wt.%.

This phenomenon may be explained as follows. In a-TN-LCDs, the molecular orientation is only determined by the surface anchoring at the substrate surfaces, and the boundary condition is not controlled by the applied voltage. However, in LCPS-a-TN-LCDs, the application of voltage during the UV curing orients polymer networks along some preferential direction. Therefore, the polymer networks give additional boundary conditions to the bulk LC molecules. This boundary condition may promote the bulk tilt angle and immobilize the reverse tilt disclination.

Textures of the LCPS-a-TN-LCD with a cell thickness of 5 μm fabricated at $V_{cure} = 0$ V and $V_{appl} = 5$ V are shown in Figure 14.8 as a function of the doping concentration of diacrylate

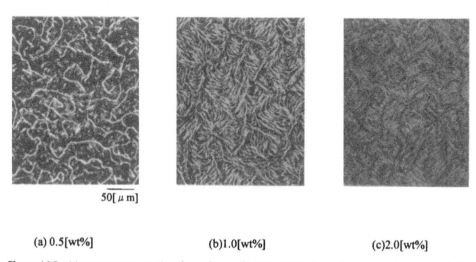

50[μm]

(a) 0.5[wt%]   (b)1.0[wt%]   (c)2.0[wt%]

Figure 14.8 Monomer concentration dependence of LCPS-a-TN textures observed under a cross Nicol condition at $V_{appl} = 5$ V. The samples of 5 μm thickness are fabricated at $V_{cure} = 0$ V and the monomer concentrations of the samples are (a) 0.5, (b) 1.0 and (c) 2.0 wt.%.

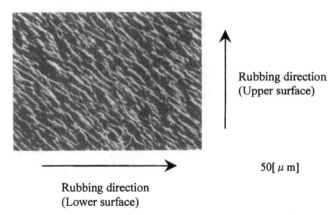

Rubbing direction
(Upper surface)

50[μ m]

Rubbing direction
(Lower surface)

Figure 14.9 Texture of the uniformly aligned TN cell with polymer networks. The observation is done under a crossed Nicols condition.

monomers. At the doping concentration of 0.5 wt.% (Figure 14.8(a)), we can observe well-defined RTDs. With increasing monomer concentration, the disclination arising from the polymer networks becomes closer. In order to investigate the configuration of the polymer networks, we fabricated a well-oriented TN-LCD with polymer networks. The texture obtained is shown in Figure 14.9. In this figure, it can be seen that the orientation of the polymer networks is directed at approximately 45° with respect to the rubbing directions on both substrate surfaces. From this result, we can consider that the orientation of the polymer network is mainly determined by the LC molecular direction in a centre region of the cell.

There is one question about why the polymer network is clearly visualized. In polymer-dispersed liquid crystal displays, the visualization of the polymer matrix is due to the light scattering a refractive-index-mismatched interface between an LC droplet and the polymer matrix. But in our case, the polymer concentration in LC media is so small that the visualization of the polymer network may not be caused by light scattering. In order to clarify the origin of the visualization, we raised the sample temperature above the clearing point, and then observed the change in texture of the cell with decreasing temperature under a polarization

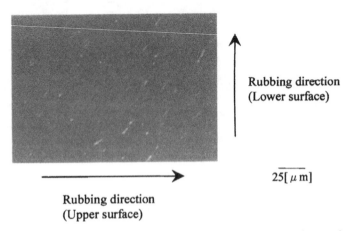

Rubbing direction
(Lower surface)

25[μ m]

Rubbing direction
(Upper surface)

Figure 14.10 Texture of an LCPS-LCD in an isotropic phase. The observation is done under a crossed Nicols condition. Bright lines represent polymer networks covered by LC molecules, and the networks are observed by the birefringence effect.

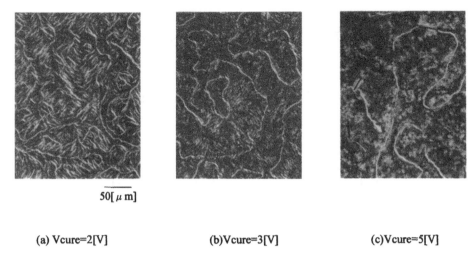

50[μ m]

(a) Vcure=2[V] (b)Vcure=3[V] (c)Vcure=5[V]

Figure 14.11 A curing voltage (V_{cure}) dependence of LCPS-a-TN textures observed under a crossed Nicols condition at $V_{appl} = 5$ V. The samples of 6 μm thickness are fabricated using a monomer concentration of 1.0 wt.%.

microscope with a cross Nicol configuration. In a high-temperature region ($< 95\,°C$), there were no features under the microscope. But as the temperature approached the clearing point, short bright lines representing the polymer networks appeared in the bulk region as shown in Figure 14.10, and they gradually became clear. We also observed that the brightness of the lines was strongly influenced by rotating the sample, and the maximum brightness was obtained when the direction of the line was 45° with respect to the polarizers, at which the maximum effect of the birefringence was expected.

These observations indicate that the visualization of the polymer networks in LC media is caused by the adsorption of LC molecules on the polymer chains and the formation of polymer networks surrounded by the homogeneously aligned LC. This means that we do not observe the polymer network itself, but instead the LC molecular-dressed polymer network.

Figure 14.11 shows the V_{cure} dependence of the texture of an LCPS-a-TN-LCD 6 μm thick, which contained the monomer concentration of 1.0 wt.%. The texture observation was conducted at $V_{appl} = 5$ V. The reverse tilt disclination domains can be clearly seen, and the domain size increases with increasing V_{cure}. This is because, in the LCPS-a-TN cell before the UV curing, the domain size becomes larger with increasing applied voltage V_{cure}, and by the UV curing, the stabilized domain increases in size with increasing V_{cure}. This result shows that the RTD domain size can be controlled by choosing the curing voltage V_{cure}.

14.3.4 EO Characteristics

As shown in Figure 14.8, the texture of the LCPS-a-TN cell depends strongly on the doped monomer concentration in the cell. This indicates that the EO characteristics of the LCPS-a-TN cell also depend on the monomer concentration. Figure 14.12 shows the monomer concentration dependence of the transmittance–voltage ($T-V$) characteristics of the 5 μm thick LCPS-a-TN cell fabricated without the curing voltage, $V_{cure} = 0$ V. All the samples satisfy the first minimum condition at a wavelength of 550 nm. The measurement was done by using a white light source under a normally white mode. In this figure, a 100% level of transmittance corresponded to the zero voltage transmittance of a conventional a-TN-LCD with the same cell parameters.

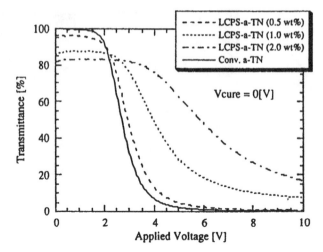

Figure 14.12 Transmittance–voltage (*T–V*) characteristic of LCPS-a-TN-LCDs with different monomer concentrations. The samples of 5 μm thickness are fabricated at V_{cure} = 0 V. For comparison, the *T–V* characteristic of a conventional a-TN-LCD is also shown in the figure.

The transmittance at V_{appl} = 0 V decreases with increasing doped monomer concentration. This may be caused by the formation of a deformed twisted-nematic region near the polymer network as discussed in the last section. It is also observed that the steepness in the *T–V* curve gradually decreases with increasing monomer concentration, which may be due to an increasing bulk anchoring force from the polymer network to the LC molecules. In order to keep the good *T–V* characteristics without losing the stabilization capability of the RTD domain, the optimum monomer concentration should be investigated.

The *T–V* characteristics of the 6 μm thick LCPS-a-TN-LCD cell with a monomer concentration of 1.0 wt.% are shown in Figure 14.13 as a function of V_{cure}. The measurement was done by using an He–Ne laser (632.8 nm) under a normally white mode. A 100% level

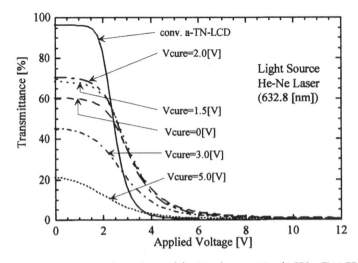

Figure 14.13 Curing voltage (V_{cure}) dependence of the *T–V* characteristic of LCPS-a-TN-LCDs with a cell thickness of 6 μm. The monomer concentration of the cells is 1.0 wt.%.

Table 14.2 *T–V* characteristics of LCPS-a-TN-LCDs with different curing voltages

V_{cure} (V)	V_{90} (V)	V_{10} (V)	V_{10}/V_{90}
0	2.10	5.10	2.43
1.5	2.03	4.83	2.37
2.0	1.97	4.73	2.40
3.0	1.50	4.67	3.11
5.0	0.97	4.23	4.36
Conv. a-TN-LCD	1.90	3.27	1.72

of transmittance is defined as the transmittance of the light through two parallel polarizers. For comparison, the result for a conventional a-TN-LCD is also shown in the figure. The sample does not satisfy the first minimum condition and the sample's condition is situated between the first and the second minimum points. From the figure, the transmittance at $V_{appl} = 0$ V increases with increasing V_{cure} in the range of $V_{cure} < 2$ V, and then decreases at $V_{cure} > 2$ V. This behaviour can be understood by considering that, in the LCPS-a-TN-LCD cured at 2 V, the first minimum condition is satisfied owing to changing the bulk tilt angle and the deviation from the curing voltage of 2 V results in a departure from the first minimum condition. In Table 14.2, we summarize the values of the threshold and saturation voltages and the steepness in the *T–V* curves. The threshold (V_{90}) and the saturation (V_{10}) voltages are defined as voltages where 90% and 10% transmittance levels, respectively, are obtained with respect to the zero voltage transmittance level. The steepness is defined by V_{10}/V_{90}. From the table, it is clear that V_{10} decreases and V_{10}/V_{90} increases with increasing V_{cure}.

The hysteresis in the *T–V* characteristics of the LCPS-a-TN-LCD is also measured, and a typical result is shown in Figure 14.14. The sample used has a cell thickness of 6 μm and was cured at $V_{cure} = 2$ V with a monomer concentration of 1.0 wt.%. In Figure 14.14, it can be seen that a weak hysteresis characteristic appears in the sample. In PDLCDs, it was considered that one of the factors generating the hysteresis in the *T–V* characteristic associates with anchoring properties acting between a liquid crystal and a polymer matrix (Niiyama *et al.*,

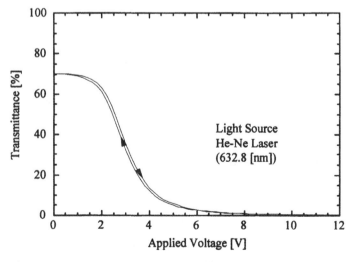

Figure 14.14 A hysteresis characteristic in the *T–V* curve of an LCPS-a-TN-LCD. The cell of 6 μm thickness has a monomer concentration of 1.0 wt.%.

1992, 1993). It was also reported that the hysteresis characteristic can be reduced by matching the material parameters of a liquid crystal (e.g. the dielectric permittivity and the elastic constant) those of a polymer matrix. Therefore, in order to reduce the hysteresis in the LCPS-a-TN-LCD, a proper combination of liquid crystal and polymer materials is important.

14.3.5 Response Time

The results of the optical response time for LCPS-a-TN-LCDs are shown in Table 14.3. τ_{on} and τ_{off}, respectively, represent the rise time and the fall time in the EO response after a pulse voltage is applied. τ_{on} of the LCPS-a-TN cells is almost comparable with that of the conventional a-TN-LCD, but τ_{off} becomes faster than that of the conventional LCD and increases significantly with increasing monomer concentration. Particularly at the monomer concentration of 2.0 wt.%, the fall time is about two times faster than that of the conventional a-TN-LCD. Similar results were also obtained by Hikmet, who used a sample of a homogeneously aligned planar anisotropic gel (Hikmet, 1990; and this volume). This change in the response time may result from the anchoring property of NLC molecules on the polymer network, but the detailed mechanism is not clear at this moment.

Table 14.3 Response times of LCPS-a-TN-LCDs prepared under various monomer concentrations (V_{appl} = 10 V, square wave, 130 Hz)

V_{cure} (V)	Monomer concentration (wt.%)	τ_{on} (ms)	τ_{off} (ms)
0	0.5	9.7	20.2
	1.0	9.2	16.0
	2.0	10.9	12.9
2	0.5	10.3	21.8
	1.0	9.4	17.2
	2.0	10.1	12.2
Conventional a-TN-LCD		11.1	24.3

14.3.6 Viewing Angle Characteristic

Figure 14.15 shows an isocontrast ratio curve for an LCPS-a-TN-LCD operated at V_{appl} = 10 V. This result indicates a wide and uniform viewing angle characteristic, which is almost comparable with those of conventional a-TN-LCDs (Figure 14.3). This means that the fundamental properties of a-TN-LCDs are not disturbed by introducing the polymer networks in the a-TN cell.

14.3.7 Dielectric Permittivity

From the observations of texture, it was clarified that the RTD domain size varies with changing V_{cure}. In order to understand the phenomenon in more detail, we measured the capacitance of LCPS-a-TN cells, from which the average tilt angle of the LC cells can be derived. The measurements were done using a 1 kHz sinusoidal wave with an amplitude of 0.2 V (r.m.s.). Figure 14.16 shows the V_{cure} dependence of the dielectric permittivity for two types of LCPS-a-TN cells with monomer concentrations of 1.0 and 2.0 wt.%. The average tilt angle of the cells gradually increases with increasing V_{cure}. The cell with the higher monomer

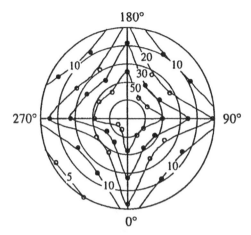

Figure 14.15 Isocontrast ratio contours of an LCPS-a-TN-LCD operated at 10 V. The contrast is evaluated by the transmittance.

concentration (2 wt.%) shows a larger dielectric permittivity than the lower one. These results indicate that an increase in V_{cure} results in the higher average tilt angle and the average tilt angle also becomes large with increasing monomer concentration. Combining these results with the results from the observations of texture, we conclude that the change of the RTD domain size with V_{cure} is caused by the change of the average bulk tilt angle. We also proved that, for a monomer concentration less than 0.5 wt.%, the dielectric permittivity depends weakly on V_{cure}.

In Table 14.4, we summarize the experimental results obtained in this study.

Table 14.4 A summary of the results obtained in this study: (a) the curing voltage V_{cure} dependence; (b) the monomer concentration dependence

(a)	Low	High
Disclination domain size	Small	Large, and then disappears
Average bulk tilt angle	Small	Large

(b)	Low	High
Disclination domain size	Small	Large
Transmittance at zero voltage	High	Low
Average bulk tilt angle	Small	Large

14.4 A Model of Orientation for LCPS-LCDs

In line with the experimental results and the discussions mentioned above, we suggest a model for the configuration of the polymer network in LCPS-a-TN cells. Schematic diagrams of the model are shown in Figure 14.17. (Hasebe *et al.*, 1994). Figure 14.17(a) shows the model for a low-monomer-concentration system, and Figure 14.17(b) for a high-concentration case. In Figure 14.17(a), liquid crystal molecules near the mesogen group of the polymer network are strongly anchored by an interaction with the mesogen group, which is the same effect as the

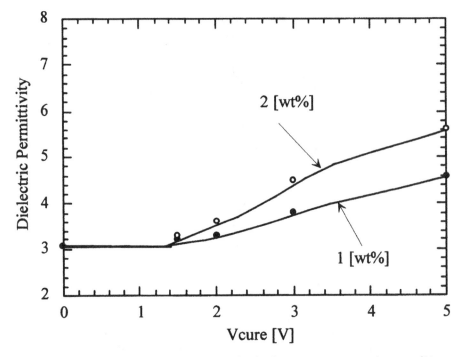

Figure 14.16 Dielectric permittivity variation of two kinds of LCPS-a-TN-LCDs as a function of V_{cure}.

anchoring at the substrate surfaces. The LC molecules near the polymer network are oriented along the direction of the network, and the TN configuration near the network is distorted, which degrades the EO characteristics of the LCPS-a-TN cells. As shown in Figure 14.9, the orientation of the polymer network is mainly determined by the LC molecular orientation in a central region of an LC cell, and the molecular orientation in the central region increases with rising V_{appl}. Therefore, the average bulk tilt angle of the LC molecules increases with increasing V_{cure}. The average bulk tilt angle also depends on the monomer concentration in LCPS-a-TN cells. This is because, if the monomer concentration is increased, the distance between adjacent polymer networks is reduced and the tilt angle of LC molecules between the networks approaches the orientation of the networks. As mentioned above, the orientation of the polymer network determines that of LC molecules near the network, and thus it is considered that the stabilization of the reverse tilt disclination domain in LCPS-a-TN-LCDs is due to the anchoring of LC molecules on the oriented polymer network, which do not change the orientation by the applied voltage.

14.5 Conclusion

In order to improve the degradation in the legibility of a-TN-LCDs arising from the transient motion of the reverse tilt disclination, we tried to control the RTD generation by introducing polymer networks in the a-TN-LCD. It was shown that it is possible to stabilize the reverse tilt disclination and control the disclination domain size. We also showed that the response time became faster when the polymer network was formed in a-TN media. However, the formation of the polymer network in a-TN media degraded the EO performance significantly,

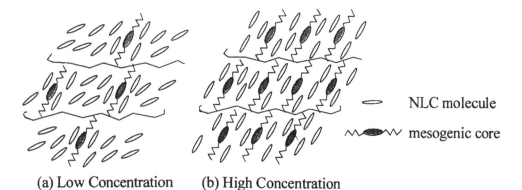

⊖ NLC molecule

ᴡᴧᴧᴧ⬬ᴧᴧᴧ mesogenic core

(a) Low Concentration   (b) High Concentration

Figure 14.17   A model of the orientation of polymer networks in LCPS-LCDs.

and we suggest that these problems will be solved by optimizing the material parameters used. Judging by the experimental results, the optimum monomer concentration in our material system may exist in the range of 0.5–1.0 wt.%.

We applied the LCPS method to improve the legibility of a-TN-LCDs, but the result was not satisfactory enough. Therefore, more studies including ergonomic measurements are needed to characterize LCPS-a-TN-LCDs.

References

BOS, P. J., RAHMAN, J. A. and DOANE, J. W. (1993) A low threshold voltage polymer network TN device, *SID Int. Symp. Dig. Tech. Pap.* **24**, 887–890.

DE GENNES, P. G. and PROST, J. (1993) *Physics of Liquid Crystals*, 2nd edn, Ch. 4, Oxford: Clarendon Press, pp. 163–197.

FREDLY, D. S., QUINN, B. M. and BOS, P. J. (1994) Polymer stabilized SBE device, *Conference Record of the 1994 Int. Display Research Conf.*, pp. 480–483.

GOOCH, C. H. and TARRY, H. A. (1975) The optical properties of twisted nematic liquid crystal structure with twist angle ⩽ 90°, *J. Phys. D: Appl. Phys.* **8**, 1575–1584.

HASEBE, H., TAKATSU, H., IIMURA, Y. and KOBAYASHI, S. (1994) Effect of polymer network mode of liquid crystal diacrylate on characteristics of liquid crystal display device, *Jpn. J. Appl. Phys.* **33**, 6245–6248.

HASHIMOTO, T., TOKO, Y., SUGIYAMA, T., KATOH, K., IIMURA, Y. and KOBAYASHI, S. *Jpn. J. Appl. Phys.* (to be published).

HIKMET, R. A. M. (1990) Electrically induced light scattering from anisotropic gel *J. Appl. Phys.* **68**, 44–4412.

NIIYAMA, S., HIRAI, Y., KUMAI, H., WAKABAYASHI, T. and GUNJIMA, T. (1992) A hysteresis-less LCPC device for a projection display, *SID Int. Symp. Dig. Tech. Pap.* **23**, 575–578.

NIIYAMA, S., HIRAI, Y., OOI, Y., KUNIGITA, M., KUMAI, H., WAKABAYASHI, T., IIDA, S. and GUNJIMA, T. (1993) Hysteresis and dynamic response effects on the image quality in a LCPC projection display, *SID Int. Symp. Dig. Tech. Pap.* **24**, 869–872.

SUGIYAMA, T., TOKO, Y., HASHIMOTO, T., KATOH, K., IIMURA, Y. and KOBAYASHI, S. (1993) Analytical simulation of electrooptical performance of amorphous twisted nematic liquid crystal display, *Jpn. J. Appl. Phys.* **32**, 5621–5632.

TOKO, Y., SUGIYAMA, Y., KATOH, K., IIMURA, Y. and KOBAYASHI, S. (1993) Amorphous twisted nematic-liquid-crystal displays fabricated by nonrubbing showing wide and uniform viewing-angle characteristics accompanying excellent voltage holding ratios, *J. Appl. Phys.* **74**, 2071–2075.

15

Filled Nematics

M. KREUZER and R. EIDENSCHINK

15.1 Introduction

Layers consisting of a uniform nematic phase are the basis of today's liquid crystal display technology. During the last few years stabilized and modified liquid crystals have attracted increasing interest in the field of high-information-content flat-panel displays (West, 1988; Yamaguchi *et al.*, 1993; Fritz *et al.*, 1994). The best-known examples are two-phase systems, like polymer-dispersed liquid crystals (PDLCs) described in the chapter by Crawford and Žumer, which exploit the scattering of light at the interface between nematic droplets and a polymer, and polymer-stabilized cholesteric textures (PSCTs) as pointed out in the chapters by Hikmet, Kitzerow, West and Yuan and in the chapter by Yang *et al.* Recently, a new promising concept has been introduced with filled nematics (Eidenschink and de Jeu, 1991; Kreuzer *et al.*, 1992).

Filled nematics (FNs), which are stable dispersions of inorganic particles formed by the aggregation of spheres with diameters in the range of 10 nanometres, permit switching between a transparent and a scattering state, both of which are stable without an electric field. Taking into account the interactions at the surface between the individual particles and the mesogenic molecules, FNs offer an approach to structure the space of a nematic bulk on a microscopic scale. This makes the addressing by a laser beam especially promising.

15.2 Materials

It has been known for many decades that highly dispersed silica consists of particles of very complex shape. Fumed silica, which is obtained by the hydrolysis of silicon tetrachloride in an oxygen–hydrogen flame, as well as precipitated silica, can form densely packed suspensions in which the silica occupies only 2 to 3% of the volume. The particles of the precipitated silica are characterized by extended caves, whereas fumed silica forms irregularly branched strings (Michael and Ferch, 1993). Figure 15.1 shows the building principle of fumed silica which turned out to be especially useful for FNs. The spherical primary particles of amorphous silicon dioxide obtainable with average diameters between 7 and 16 nm form aggregates via $\equiv$Si—O—Si$\equiv$ moieties. The $\equiv$SiOH groups on the surface of these aggregates can form hydrogen bonds between different particles which lead to the formation of agglomerates. Compared with covalent bonds the energy of hydrogen bonds is relatively low so that they can be unlocked by mechanical interaction and are subsequently available for the formation

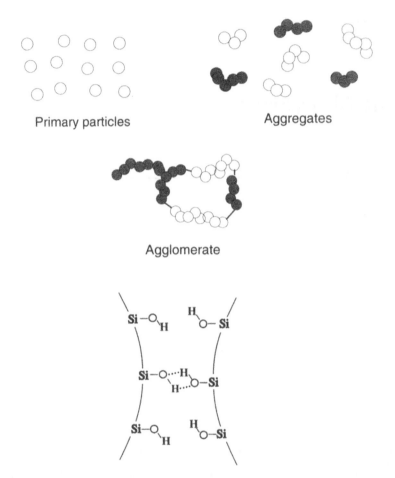

Figure 15.1 Building principle of solid framework of fumed silica depicting the formation of hydrogen bonds between aggregates.

of new bonds thus fixing a new spatial arrangement of the aggregates. This can be seen in analogy to the hydrogen bonds in organic matter. It should be emphasized also that other interactions between the particles can stabilize the solid framework as van der Waals' forces (Gray, 1968) and specific orbital interactions (Michael and Ferch, 1993). The present work has been restricted to highly dispersed silica. Some material parameters such as density and BET surface are similar to the so-called aerogels (see, for example, the chapters by Bellini and Clark and Rappaport *et al.*), which are produced by a sol–gel process followed by supercritical drying of the wet gel. Whereas aerogels have an extended, fixed solid network the particles used in FNs are capable of changing their mutual arrangement. This is an additional possibility of changing physical properties on a microscopic scale.

The high density of silanol groups on the surface of original fumed silica cause their distinct hydrophilic character. By a different silanization reaction hydrophobic materials can be obtained that have only about 10 % of the initial density of silanol groups left on their surfaces. The specific surfaces of commercially available silanized fumed silica which can be detected by the BET method range from 50 to 300 m^2/g with 0.1 to 0.4 silanol groups per square nanometre.

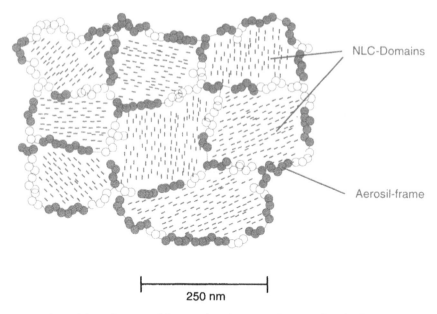

NLC-Domains

Aerosil-frame

250 nm

Figure 15.2 Fictional three-dimensional framework with nematic domains of randomly oriented director.

The silica is dispersed in nematic mixtures by ordinary stirring. After pressing the dispersion between two glass plates carrying ITO electrodes, one can obtain turbid layers of some 10 μm thickness. By applying a voltage – the magnitude of which depends on the kind and the concentration of the silica as well as the value of the (positive) dielectric anisotropy $\Delta\varepsilon$ of the low-molecular-weight nematic phase – at the electrodes the layer becomes transparent as one would expect. The nature of the solid framework does not require any matching of the optical properties of the dispersed particles and the nematic phase.

Surprisingly, the transparent homeotropic state is maintained after the voltage has been switched off despite the fact that the free surface area of the solid particles can exceed the surface area of the electrodes by two orders of magnitude. This feature opens up new fields of application, different from those for monostable systems like PDLC. As will be demonstrated below, a scattering state can be brought back by different methods. Before making hypotheses about this strange behaviour some basic observations must be taken into account.

Obviously the maintenance of the transparent state without an electric field is connected with the ability of the solid frame to rearrange. This can be concluded from the fact that after switching off the voltage a strongly scattering state appears when the particles are impeded by a polymer network that occupies only 2% of the volume (Kreuzer and Tschudi, 1994). The ability to form a stable homeotropic alignment is obviously also lost when the particles are made hydrophobic by introducing particular bulky silyl groups (Yaroshchuk *et al.*, 1994).

The homeotropic state obtained after turning off the electric field can be stabilized by a rearranged solid framework to a different extent. Thus a layer of an FN between two plates, not being treated with a surface agent, containing 2.5 vol.% of the hydrophilic fumed silica Aerosil 200 (Degussa) in the nematic phase E7 (Merck Ltd) upon cooling returns to the transparent state after having been heated a few degrees above its clearing temperature. However, using a hydrophobic silica (e.g. Aerosil R976) the transparent state turns into a scattering one after this procedure. This leaves some doubt about whether the arrangement of the solid framework alone is decisive.

The stabilization of the homeotropic orientation by rebuilding agglomerates lies behind a recent explanation (Kreuzer *et al.*, 1993). Assuming strong anchoring of the nematic director at the surfaces of the particles forming the framework, not only should the surfaces act elastically on the director, but also the director should act on the surfaces and therefore on the particles. Figure 15.2 schematically shows the partition in domains of different director alignments in the FN before the application of an electric field.

The reorientation of the director in an external electric field can then lead to a rearrangement. During this process hydrogen bonds may be broken, and new ones formed with the surface orientation adjusted to the aligned director. Assuming a typical domain size of $l_m \approx 250$ nm the energy density in a Freederiks transition can be estimated to 1×10^{-2}–4×10^{-2} J/m^3. Compared with the stabilization energy density of about 10^{-2} J/m^3, which we know from shear experiments of the thixotropic material (Kreuzer, 1994), we can conclude that the stabilization of the network can be broken during the electric poling process. However, taking into account the irregular form of the aggregates themselves and a strong coupling between the molecules in the nematic phase and the surface, the stabilization of a homeotropic alignment of the good quality found is not fully understandable.

From a kinetic point of view, this passing of the anchoring forces may be favoured by the considerable increase of the bulk viscosity of fluids containing thixotropic additives. However, it has been shown (Kreuzer, 1994) that the homeotropic alignment in displays can be kept for several years. Also the scattering domains generated, for example, by a laser are conserved.

15.3 How to Generate a Scattering State

After the stable transparent state has been created by an electric field it is not immediately obvious how it can be changed. Besides the heating above the clearing a small mechanical shear is also effective. Some preliminary experiments show that the scattering state can be obtained by applying ultrasound *through* the glass plates. This basic method could be especially interesting if the ultrasound could be generated locally.

A purely electric addressing of a display is possible if the nematic phase combines both $\Delta\varepsilon > 0$ and $\Delta\varepsilon < 0$. Using a nematic mixture of $\Delta\varepsilon = 2$ at 400 Hz and $\Delta\varepsilon = -1.9$ at 20 kHz both states could be accomplished by short electrical pulses. The enforcement of a planar molecule alignment by the high-frequency field leads to the formation of randomly oriented domains (Eidenschink and de Jeu, 1991). Because of the special features of the generation of patterns by a laser this topic will be discussed in a separate section.

15.4 Laser-addressed Projection Display

The bistability of FN displays opens up the field of applications where non-local addressing methods like laser addressing can be used to achieve high-resolution and high-information-content displays. As in previously known laser-addressed displays (Dewey, 1984; Eichler *et al.*, 1992; Kakinuma *et al.*, 1992; Kahn, 1973) thermally induced effects seemed to be an effective mechanism to switch back from the transparent homeotropic state to a scattering state. Therefore, appropriate dyes, which convert electromagnetic radiation into heat, were dissolved in the liquid crystal as mentioned above to reduce the laser intensity needed for thermally addressed laser writing.

As we will discuss later in detail, the first experiments have shown that no phase transition process like those in known thermally addressed LCDs is involved in laser-addressed FN displays, but a shock mechanism is responsible for generating randomly oriented domains.

To investigate the laser writing process in detail we used a setup normally used to produce computer-generated holograms (Kreuzer *et al.*, 1992, 1993). With this setup high-information-

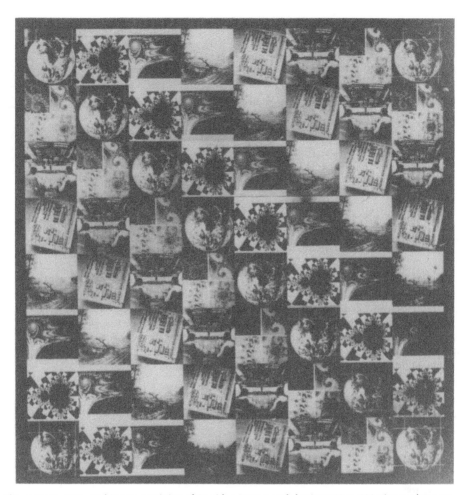

Figure 15.3 Projected image consisting of 64 video images each having 512 × 512 picture elements (pels) with 16 grey levels; 16 million pels with a size of 5 μm are stored on a display area of 4 cm².

content grey-scale images with 4094 × 4096 points were written on an area of 4 cm² (see Figure 15.3). Later on we built a demonstration unit based on a conventional slide projector. A commercial semiconductor laser was used as a reliable and inexpensive laser source delivering about 25 mW ($\lambda = 780$ nm) onto the display. A pair of galvanometer mirrors scans the laser beam, which is focused by a lens via a dichroic mirror over the display while the image is on-line projected onto a screen by a white light source. The galvanometers may be operated in random access mode for vector scanning, or in raster scan mode for grey-scale images. The maximum writing speed was about 1.5 m/s and grey levels could be realized by direct modulation of the injection current (intensity) of the semiconductor laser. Figure 15.4 shows the schematic setup which allows grey-scale images of 1200 × 900 points to be written on a display area of 12 mm × 9 mm.

15.4.1 *Spatial Resolution*

A pixel size of 5 μm, as shown in Figure 15.3, is completely sufficient for high-resolution

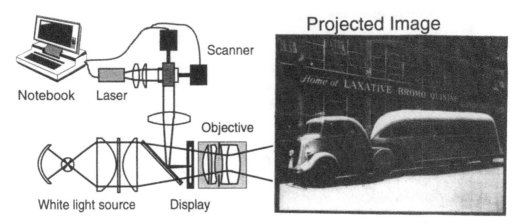

Projected Image

Figure 15.4 Schematic setup of the projection unit.

displays because it matches the resolution of high-quality projection lenses. Nevertheless, applications for optical data storage demand a spatial resolution better than 1 μm. To provide such a resolution in filled nematics, the liquid crystal layers must not be thicker than the Rayleigh length z_0, which is 1 μm for $\lambda = 785$ nm and a focus diameter $\omega_0 = 1$ μm.

With our preparation technique, the minimal cell gap achieved was 7 μm, so that this limit could not be reached, and the minimal pixel size was 2 μm. Obviously the material does offer the potential for optical data storage as well as for information display.

15.4.2 Sensitivity and Grey Levels

For applications in information display, the most important aspects are:

1 Grey-level capacity
2 Sensitivity
3 Writing speed
4 Long-term stability

Figure 15.3 already gives an impression of the grey-level capacity in filled nematics. In fact the measurement depicted in figure 15.5 shows that over a wide range the contrast ratio is a linear function of the absorbed laser energy, until a saturation value is reached for high-energy densities. The contrast is defined here as the ratio of transmitted intensities for the transparent and scattering state. For small energy densities, there is an additional modulation of the contrast caused by the fact that pixel sizes become smaller than the laser focus diameter.

The quantitative analysis of Figure 15.5 shows that there is a threshold value for the writing energy density of $s_t \approx 0.3$ nJ/μm^2. The saturation contrast is typically reached at $s_s \approx 1.4$ J/μm^2. Although these values are largely independent of material, they have to be seen under the following important aspect:

1 The laser writing process is essentially a thermal shock effect. This means that energy has to be delivered within a certain time interval to achieve a permanent effect. A detailed description of the laser writing process, including the possible writing speed, will be given later; it should be mentioned here that laser illumination in the microsecond range for one pixel is easily reached.

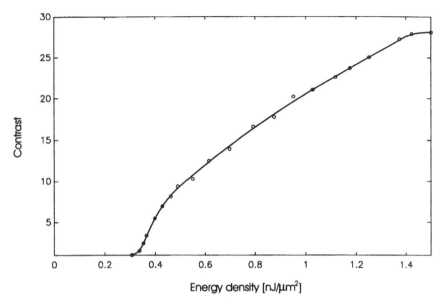

Figure 15.5 Scattering contrast as a function of the absorbed laser energy density. The contrast was measured with a white light source and a projection objective with an F-number of 2.8.

The written information has excellent long-term stability. For example, the photograph of Figure 15.3 was taken 9 months after the image was stored. Experiments conducted over a period of 2 years under normal environmental conditions showed no visible changes in the written images.

15.4.3 Contrast

Information storage in FN is based on the local generation of a scattering state. Therefore the contrast depends on the scattering characteristics of the display and the aperture of the projection system.

In principle, small apertures can improve the contrast, but the divergence of the light source then reduces the brightness. Because the light source is projected into the entry pupil of the objective lens, the light-generating volume should be kept as small as possible to provide high lumen yield even with small apertures. Most projection systems use metal halide lamps to ensure this.

On the other hand, the scattering characteristics are given by the material and display parameters:

1 Refractive index anisotropy Δn
2 Cell thickness
3 Orientation and size of scattering domains

Because of the complexity of the system (multiple scattering, anisotropy, etc.), a quantitative model for light scattering in filled nematics has not yet been found. One may expect an increase in light scattering with increasing domain sizes (Rayleigh scattering), but this quantity is limited by interaction forces of the material that do not allow stable dispersions with less than 1 vol.% of Aerosil. To optimize the material's scattering characteristics, we used liquid crystal mixtures with a high anisotropy of the refractive index ($\Delta n \approx 0.26$).

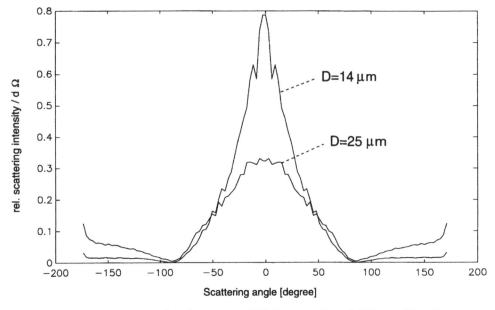

Figure 15.6 Scattering characteristics for two typical FN displays with a cell thickness of $D = 14\ \mu m$ and $D = 25\ \mu m$ ($\lambda = 633$ nm).

The achievable contrast was determined in an automated light scattering measurement site. We measured the angular distribution of scattered intensity for two displays of different cell thickness. Figure 15.6 shows the typical scattering characteristics*. The contrast can be derived by integration over the spatial angle elements. With an aperture of 5.6, one finds contrast ratios of $40:1$ for $D = 14\ \mu m$ and $90:1$ for $D = 25\ \mu m$. These values compare well with the scattering efficiency of PDLC systems (Drzaic and Gonzales, 1992; Whitehead *et al.*, 1993).

Increasing the thickness of the FN layer can yield any desired contrast ratio, but this raises the necessary writing energy as well as the erasing voltage. Therefore cell thicknesses may vary for different applications.

15.4.4 Selective Erasure

The stored information can be erased globally by applying a sufficient voltage to the display. However, optical data storage as well as information display demand a selective erasure process. Figure 15.7 shows that such a process is indeed possible in FN cells. Applying a low bias voltage (which does not change the written information) and simultaneously writing with the laser locally erases the information. The necessary laser energy is comparable with the threshold value for the writing process.

To induce a scattering state in an FN cell with a laser beam, two principal mechanisms have to be initiated:

1   The stabilizing effect of the Aerosil network must be removed.
2   The local orientation of the liquid crystal domains must be disordered.

---

* Both polarization components were determined separately, yet the scattered light is almost fully depolarized at a cell thickness of $D = 14\ \mu m$.

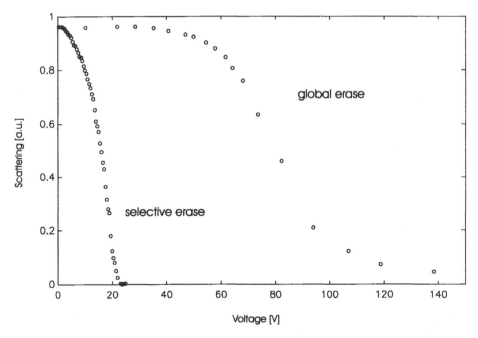

Figure 15.7 Selective erasure applying a small voltage of about 20 V (selective erase), while other information is not influenced (global erase).

The details of laser writing will be discussed in the next section; for selective erasure, only the first aspect has to be explained. Using the model assumptions for the material introduced previously, one may find an explanation for the process without full knowledge of the laser writing mechanism. During selective erasure the low bias voltage cannot induce a global reorientation of the liquid crystal molecules against the molecular interactions, so that the written information remains unchanged. The simultaneous action of the laser *locally* breaks the stabilizing effect of the Aerosil network for a short time interval. During this interval, the electric field can reorientate the molecules until the new transparent state is stabilized by the forming of new hydrogen bonds. Indeed, other experiments have shown that after the hydrogen bonds are broken, it takes about 10 ms until new bonds are formed and the stabilizing effect returns.

We can repeat that the development of FN presents a material for high-resolution optical data storage with grey-scale capacity, excellent contrast and selective erasure. Now we have to take a closer look at the physical aspects of the laser writing process.

15.5 Theory

The mechanism of laser writing to obtain scattering again is evidently thermal in nature. It should be realized that no heating above the nematic–isotropic phase transition is involved or necessary. This distinguishes the display in an important way from older scattering displays that exploit the smectic–nematic phase transition, and thus require thermostatization (Kahn, 1982). Therefore, we assume the following mechanism to be responsible for generating a scattering state by laser writing. The very fast increase in temperature of the illuminated area*

---

* The thermalization time τ^* is much shorter than all other processes to be examined (Batra *et al.*, 1971).

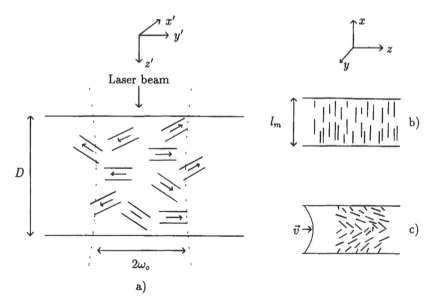

Figure 15.8 (a) Schematic description of the Aerosil network as randomly orientated capillaries. The arrows point in the direction of the induced flow. (b) Orientation in one capillary before the writing process. (c) Hydrodynamic reorientation during the writing process, l_m is a typical capillary radius, ω_o is the laser beam diameter ($1/e^2$ intensity, Gaussian beam TEM$_{00}$) and D is the cell thickness.

leads to hydrodynamic effects due to the change in density. Since the specific volume expansion $\alpha = (1/V)(\partial V/\partial T)_p$ of the Aerosil network is two orders of magnitude smaller than the specific volume expansion of the liquid crystal, a microflow of the liquid crystal in randomly oriented capillaries of the network will be induced (see Figure 15.8). Owing to viscous effects and the strong anchoring of the liquid crystal at the surface of the Aerosil particles this will lead to a typical Poiseuille flow (see Figure 15.8(c)) of the anisotropic fluid resulting in a reorientation of the nematic domains. Finally, the connected shear forces break up the network and after the illumination a new network will be formed which stabilizes the new scattering state. Therefore, hydrodynamically induced reorientation leads to statistically reorientated NLC domains in the illuminated area. The reorientation angle Θ will give a measure for the scattering contrast to be achieved.

Now we have found a plausible description for the coupling between the laser-induced increase in temperature and the reorientation of liquid crystal molecules. Owing to the anisotropy of the liquid crystals and the coupling between temperature, hydrodynamic flow and orientation, the equations governing the system will be extremely complex. For further discussion we will therefore make the following assumptions:*

1 Material constants, which are tensors in general, will be replaced by *effective* scalar quantities. The reaction of reorientation on the flow will be taken into account by introducing effective viscosity coefficients.
2 The Reynolds number of the flow is small, so that convection effects do not contribute. It should be mentioned that owing to the random orientation of capillaries no turbulent flow is necessary to generate a scattering state. Since the relative change in density $|\rho - \rho_0| \ll \rho_0$

---

* Of course, one can use the complex equations for numerical simulations. Yet such calculations have shown that the simplifications are justified.

is very small, the Navier–Stokes equation and the continuity equation can be linearized and combined into one wave equation. The influence of the flow on the heat conduction equation can be neglected.

3   We look at only one capillary shown in Figures 15.8(b) and (c), i.e. a flow with velocity $\mathbf{v} = [0, 0, v_z(x, y)]$.

Now we have three equations. One contains the heat conduction, the second describes the change in density ρ and the third gives the coupling of reorientation angle Θ to the flow via nematodynamics (de Gennes, 1974):

$$\frac{\partial T}{\partial t} = r_t \Delta T + \frac{W}{V \rho_0 C_p} \tag{15.1}$$

$$\frac{\partial^2 \rho}{\partial t^2} = \tilde{c}^2 \Delta \rho + \eta \frac{\partial}{\partial t} \Delta \rho + \rho_0 \tilde{c}^2 \alpha \Delta T \tag{15.2}$$

$$\gamma_1 \frac{\partial \Theta}{\partial t} = K \Delta \Theta - \frac{1}{2} (\gamma_1 - \gamma_2 \cos 2\Theta) \frac{\partial v_z}{\partial x} \tag{15.3}$$

with r_t the heat conductivity coefficient, W the absorbed laser power in volume V, ρ_0 the density, C_p the heat capacity, η the viscosity and the speed of sound $\tilde{c}$. K is the Franck elastic constant in the one-constant approximation and γ_1 and γ_2, which have the dimension of viscosity, are combinations of the Leslie coefficients $\gamma_1 = \alpha_3 - \alpha_2$, $\gamma_2 = \alpha_6 - \alpha_5$ (Blinov, 1983). Typical values are displayed in Table 15.1.

From eqns (15.1) and (15.3) one may extract two time constants. Having a laser beam with radius ω_0, a FN cell of thickness D and a capillary of diameter l_m we get a thermal relaxation time τ_{therm} (heat diffusion) of

$$\tau_{therm} = \frac{1}{2r_t} \left(\frac{1}{\omega_0^2} + \frac{1}{D^2} \right)^{-1} \tag{15.4}$$

and on the other side the director relaxation time τ_{dir}

$$\tau_{dir} = \frac{\gamma_1 l_m^2}{4K\pi^2} \tag{15.5}$$

Equation (15.2) is much more difficult to handle. We cannot deduce a solution or a typical time constant as easily as in the other two equations.

To keep dissipation effects (due to heat diffusion) and saturation effects (due to reorientation) negligible, the laser pulse τ_p should be shorter than the relaxation time τ_{dir} and shorter than τ_{therm} for an effective transformation of energy (laser energy = power × time → to reorientation). Therefore we will restrict ourselves to this parameter range for further calculations. In this case effects of the induced flow after the illumination ($t > \tau_p$) can be neglected. The temperature equation (15.1) is now easily integrated using $W = \sigma I V$:

$$T = \frac{I\sigma}{\rho_0 C_p} t = T_0(x', y')t \quad \text{for } t \leqslant \tau_p \tag{15.6}$$

Table 15.1   Typical data and liquid crystal parameters

$K = 1 \times 10^{-11}$ N	$\gamma_1 = 0.8$ g/cm/s (Poise)
$l_m = 2.5 \times 10^{-5}$ cm	$D = 1.4 \times 10^{-3}$ cm
$r_t = 1.1 \times 10^{-3}$ cm^2/s	$\tilde{c} = 1500$ m/s
$\alpha = 1.2 \times 10^{-3}$ K^{-1}	$\gamma_2 = -0.8255$ g/cm/s (Poise)
$\eta = 1$ cm^2/s	$C_p = 1.8$ J/g/K

where σ (cm^{-1}) is the absorption constant and

$$I(x', y') = I(r) = I_0 \exp(-2r^2/\omega_0^2) \tag{15.7}$$

is assumed for a cylindrically symmetric distribution of intensity $I = I(r)$.

Now we have to find the boundary conditions for ρ in the capillary. One may safely assume that the spatial distribution of velocity in a capillary is given by

$$v_z(x, y) = v_0(t) \sin \frac{\pi}{l_m} x \sin \frac{\pi}{l_m} y \tag{15.8}$$

We have to note that we use local Cartesian coordinates x, y in the capillary, but the calculation of the density distribution has to take place in a global coordinate system (x', y', z') (see Figure 15.8). It seems reasonable to choose the z' direction in the laser beam direction because of the rotational symmetry of the Gaussian beam (see again Figure 15.8). For weak absorption, the z' dependence of the problem may be neglected (then $\partial/\partial z' = 0$; this assumption was already made for eqn (15.6)) and the problem can be reduced to polar coordinates $(x', y', \to \phi, r)$. Since the velocity field $\mathbf{v}$ originates from a change in density, one may reduce eqn (15.2) with the boundary conditions following from (15.8):

$$\ddot{\rho} - \tilde{c}^2 \Delta_T \rho - \tilde{\eta} \frac{\partial}{\partial t} \rho = F \Delta_T T_0 t \tag{15.9}$$

with

$$\Delta_T = \frac{\partial^2}{\partial x'^2} + \frac{\partial^2}{\partial y'^2} = \frac{\partial^2}{\partial r^2} + \frac{1}{r} \frac{\partial}{\partial r} \quad \text{for} \quad \frac{\partial}{\partial \phi} = 0 = \frac{\partial}{\partial z'} \tag{15.10}$$

and the substitutions

$$F = \rho_0 \tilde{c}^2 \alpha \quad \text{and} \quad \tilde{\eta} = \frac{2\eta \pi^2}{l_m^2} \tag{15.11}$$

To find a solution, this partial differential equation is Fourier transformed in its spatial coordinates and we get an ordinary differential equation

$$\ddot{\rho} + \tilde{c}^2 k^2 \rho(k) + \tilde{\eta} \dot{\rho} = -k^2 F T_0(k) t = P(k) t \tag{15.12}$$

$$k^2 = k_r^2 = k_{x'}^2 + k_{y'}^2 \quad (k \text{ are the spatial frequencies}) \tag{15.13}$$

In the present case the wave equation is overdamped, so that for the solution of eqn (15.12) we get

$$\rho(k) = -\frac{F T_0(k)}{\tilde{c}^2} \left\{ \frac{\tilde{\eta}}{\tilde{c}^2 k^2} \left[\exp\left(-\frac{\tilde{c}^2 k^2 t}{\tilde{\eta}}\right) - 1 \right] + t \right\} \tag{15.14}$$

We obtain the solution for $\rho(r)$ by retransforming to spatial coordinates. For the density distribution $\rho(r)$ this means that

$$\rho(r) = 2\pi \int_0^\infty k\rho(k) J_0(2\pi rk) \, dk \tag{15.15}$$

with $J_0(x)$ the Bessel function of first kind and zero order. For a Gaussian beam eqns (15.6) and (15.7) we conclude that

$$\rho(r, t) = \rho_0 - t \frac{FI\sigma}{\rho_0 C_p \tilde{c}^2} \left[\exp\left(-\frac{2r^2}{\omega_0^2}\right) + S(r, t) \right] \quad \text{for } t \leqslant \tau_p \tag{15.16}$$

with

$$S(r, t) = \frac{\pi^2 \omega_0^2 \tilde{\eta}}{2\tilde{c}^2 t} \left[\text{Ei}\left(-\frac{\pi^2 r^2}{s_2^2}\right) - \text{Ei}\left(-\frac{\pi^2 r^2}{s_1^2}\right) \right] \tag{15.17}$$

and the substitutions and the exponential integral

$$s_1^2 = \frac{\omega_0^2 \pi^2}{2} \qquad s_2^2 = \frac{\omega_0^2 \pi^2}{2} + \frac{\tilde{c}^2 t}{\tilde{\eta}} \qquad -\text{Ei}(-z) = \int_z^\infty u^{-1} \exp(-u) du$$

Looking at the generalized solution for $\rho(r)$, one can see that the last term $S(r, t)$ (eqn (15.17)) in eqn (15.16) describes a time-dependent 'perturbation' of the equilibrium density distribution $\lim_{t \to \infty} \rho(r, T)$. To obtain the time constant for the decay of this perturbation, we look again at the solution $\rho(k)$ in the Fourier space (eqn (15.12)). For each spatial frequency the time constant $\tau_\rho(k)$ is

$$\tau_\rho(k) = \frac{\tilde{\eta}}{\tilde{c}^2 k^2} \tag{15.18}$$

To find $\tau_\rho(r)$, the average over all values of k is taken regarding a 'driving force'*

$$\tau_\rho = 2 \frac{\int_0^\infty \tau(k)P(k)\,dk}{\int_0^\infty P(k)\,dk} = \frac{4\pi^2 \eta}{l_m^2 \tilde{c}^2} \frac{\int_0^\infty \exp(-k^2\pi^2\omega_0^2/2)\,dk}{\int_0^\infty k^2 \exp(-k^2\pi^2\omega_0^2/2)\,dk} = \frac{4\pi^4 \eta \omega_0^2}{l_m^2 \tilde{c}^2} \tag{15.19}$$

From these calculations we may draw one first important conclusion: as mentioned before, the laser illumination time τ_p should be shorter than the relaxation time constants of the director (τ_{dir}) and the temperature distribution (τ_{therm}). With τ_p we find the first complete condition

$$\tau_\rho < \tau_p < \min(\tau_{dir}, \tau_{therm}) \tag{15.20}$$

which means that there is a limited parameter range for optimal writing conditions (see Figure 15.9).

With the continuity equation

$$\frac{\partial \rho}{\partial t} + \text{div}\,\rho \mathbf{v} = 0 \tag{15.21}$$

we can now determine the velocity v_0 of eqn (15.8)

$$v_0(r) = \frac{\pi^2}{4r\rho_0} \frac{\partial}{\partial t} \int_0^r \rho(r')r'\,dr' \tag{15.22}$$

and calculate the dynamical behaviour of the reorientation angle with eqn (15.3). In general, one will obtain rather complex expressions. But if the condition (15.20) is fulfilled, we can estimate the time development of the density distribution with the equilibrium distribution

$$\rho(r,t) \approx -\frac{FI\sigma}{\rho_0 C_p \tilde{c}^2} t \exp\left(-\frac{2r^2}{\omega_0^2}\right) + \rho_0 \quad \text{for } t \leqslant \tau_p \tag{15.23}$$

(the change in density follows the change in temperature adiabatically) and we obtain

$$v_0(r) = \frac{FI\sigma\pi^2\omega_0^2}{16\rho_0^2 C_p \tilde{c}^2} \left\{\frac{1}{r}\left[1 - \exp\left(-\frac{2r^2}{\omega_0^2}\right)\right]\right\} \tag{15.24}$$

and $v_0 = 0$ for $t > \tau_p$. So in this case the effects of flow and reorientation after laser illumination are negligible. For the maximum velocity we find that with $F = \rho_0 \tilde{c}^2 \alpha$ and the laser power $P = I\pi\omega_0^2$

$$v_{0,\text{max}} \approx \frac{0.9\pi^2 \sigma \alpha}{16\rho_0 C_p \omega_0} P \tag{15.25}$$

---

*The right-hand side of eqn (15.12) with the Fourier transform $T(k)$ of eqn (15.6).

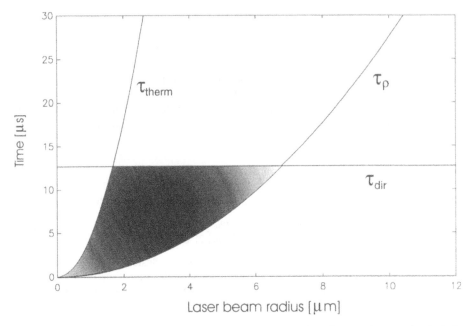

Figure 15.9 A plot of the three constants τ_p, τ_{dir} und τ_{therm} against laser beam radius ω_0. The shaded area shows the optimal range for the illumination time after (15.20).

during laser illumination. To find an estimate for the dynamical reorientation $\Theta(t)$ one can use $v_{0,max}$ together with eqn (15.8) for the integration of the reorientation equation (15.3). Since $\gamma_1 \approx -\gamma_2$ (see for example de Jeu, 1980), eqn (15.3) is linearized as follows:

$$\gamma_1 \frac{\partial}{\partial t} \Theta - K\Delta\Theta + \gamma_1 \frac{\partial v}{\partial x} = 0. \qquad (15.26)$$

Using this, we obtain an approximation for the reorientation angle Θ_0

$$\Theta_0 \approx \Theta_m \left[1 - \exp\left(-\frac{4K\pi^2}{l_m^2 \gamma_1} t \right) \right] \quad \text{for } \tau_p \gg \tau_\rho \qquad (15.27)$$

with

$$\Theta_m = \frac{\gamma_1 l_m v_{0,max}}{4K\pi} = \frac{0.9}{64} \frac{\gamma_1 l_m \pi \sigma \alpha}{K\rho_0 C_p \omega_0} P \qquad (15.28)$$

or, if $\tau_p < \tau_{dir}$,

$$\Theta_0 = \frac{\pi v_{0,max}}{l_m} t = \frac{0.9\pi\alpha\sigma P}{16\rho_0 C_p \omega_0 l_m} t \quad \text{for } 0 \leqslant t \leqslant \tau_p \qquad (15.29)$$

The first thing of interest in the solution of the system of differential equations (15.1)–(15.3) for the present conditions is the estimation of writing sensitivity s ($=$ necessary writing energy per area A). For such an estimation we assume that for good contrast a reorientation angle of $\Theta_0 \approx 60°$ has to be achieved, and using the definition of sensitivity we find with eqn (5.29) that

$$s = \frac{P\tau_p}{A} = s(60°) = \frac{60}{180}\pi \frac{16\rho_0 C_p l_m}{0.9\alpha\sigma\omega_0 \pi^2} \qquad (15.30)$$

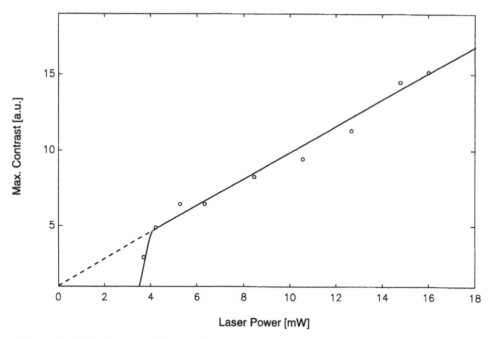

Figure 15.10 Maximum possible (saturation) contrast as a function of laser power P. The dashed part of the line marks the range of purely transient orientation. Experimental parameters: beam diameter $d = 2\omega_o \approx 50~\mu m$, absorption $\approx 65\%$.

relative to the energy delivered onto the display. For a beam radius of $\omega_0 = 10~\mu m$ and the values of Table 15.1 the sensitivity is

$$s \approx 70~\text{mJ/cm}^2 = 0.7~\text{nJ}/\mu m^2 \tag{15.31}$$

relative to the absorbed energy ($\sigma = 500~\text{cm}^{-1} \hat{=} 50\%$ absorption). This agrees very well with the experimental data $s_{\text{exp}} = 0.5\text{–}1.5~\text{nJ}/\mu m^2$.

Looking more closely at eqn (15.30), one can see that the sensitivity seems to improve with increasing beam diameter ω_0. This is a consequence of the fact that an increasing beam diameter raises the maximum flow velocity, which is directly connected to the reorientation angle. But, as shown in Figure 15.9, the optimal writing conditions cannot be fulfilled for large beam diameters. To determine the optimal writing parameters, eqns (15.22) and (15.3) have to be solved numerically.

One of the most important consequences of the theoretical model is the relation between the maximum reorientation angle Θ_m and laser power P in eqn (15.28)

$$\Theta_m \propto P \tag{15.32}$$

This proves that the *maximum possible* contrast is not proportional to the absorbed laser energy but to the laser power.

To investigate this theoretical statement we measured the maximum contrast depending on the laser power, i.e. for each value of the laser power P the illumination time was increased until we achieved a saturation contrast. The results shown in Figure 15.10 confirm the theoretical predictions. In addition, the good linearity justifies the assumption that the contrast is a linear function of reorientation angle Θ.

Now that there is such a good agreement of theoretical predictions and experimental results, we could carry out optimization calculations for the parameters ω_0, P and τ_p. The results are

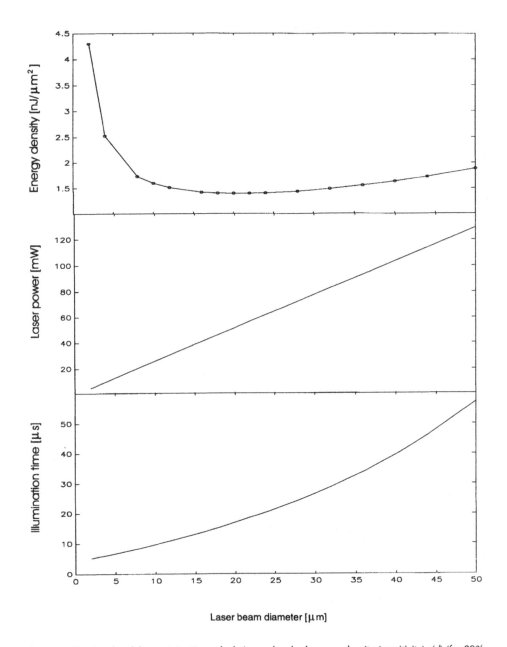

Figure 15.11 Results of the optimization calculations: absorbed energy density (sensitivity) $s(d)$ (for 80% of maximum contrast) against laser beam diameter d. The necessary laser power $P(d)$ and the illumination time $\tau_p(d)$ are depicted below. For the calculations we assumed a typical absorption of 50%. Varying the absorption only changes the necessary laser power (in a first-order approximation).

shown in Figure 15.11 and confirm the experimental value of the energy density, 1.4 nJ/μm^2, for maximum contrast under optimum writing conditions.

A couple of important conclusions can be drawn from the calculations:

1   For a certain range 10 μm $< d <$ 35 μm of laser beam diameter d the sensitivity has a nearly constant good value.

2   To achieve minimum writing time one will try to use small laser beam diameters. Since the written information has to be projected by an optical system and the f-number should be chosen larger than 4 for sufficient contrast, usually 75 lines/mm can be displayed well. This corresponds to a beam diameter of $d \approx 13$ μm, an illumination time of $\tau_p \approx 12$ μs and to a laser power of $P \approx 40$ mW. For an image with 512 $\times$ 512 pixels this means a writing time of about 3 seconds.

Now that hydrodynamic flow has been identified as causing the reorientation process, it has to be ensured that the accompanying forces are large enough to break the stabilization by hydrogen bonds. Without a doubt, the largest forces appearing here are the shear forces connected with the flow. From the quantities calculated earlier, we can estimate the elastic stress τ on the Aerosil network (the energy density) as

$$\tau = \frac{\eta \rho v_{0, \max}}{l_m} \approx 10^3 \text{ J/m}^3 \qquad (15.33)$$

With the results mentioned above we can see that this is much larger than the energy density needed to break hydrogen bonds. This means that only a small part of the absorbed laser energy is responsible for the destabilization of the Aerosil network (break up of hydrogen bonds) while the major part is needed to induce a reorientation of the liquid crystal.

To exploit this fact in recent experiments we have added a small amount of a chiral dopant to the nematic liquid crystal. The *stored* pitch energy in a stable homeotropic state (in contrast to the bistable cholesteric systems described in the chapters by Yang *et al.* and by Yuan, where the homeotropic state is not stable without an applied electric field) should be set free to help create a scattering state during the laser writing while reducing the needed laser energy.

Using 3.5% of the chiral dopant CB15 (Merck) in the liquid crystal E48 (BDH) we have found an increase in sensitivity by a factor of about three. Also a remarkable increase in the maximum achievable contrast was found to the same order, which is not fully understood up to now. Further investigations of the influence of chiral dopants on the behaviour of FN displays are under way.

15.6 Conclusion

During the last few years stabilized and modified liquid crystals have attracted increasing interest in the field of high-information-content flat-panel displays. In this area a new promising concept has been introduced with filled nematics.

Based on a rebuildable framework of highly dispersed silica in a nematic phase a bistable scattering type of display has been developed. The bistability allows laser addressing with high resolution, good contrast and brightness, continuous grey scale and good sensitivity. Selective erasure can be reached by repeating the writing process in combination with a moderate electric field.

A theoretical modelling of the laser writing process has been presented showing that an optically induced hydrodynamic flow in randomly oriented microcapillaries is responsible for creating scattering states. First experiments have shown that the addition of chiral dopants can enhance the effectiveness of this laser writing process resulting in higher sensitivity and better contrast ratio. Therefore, we may conclude that filled nematics have considerable potential for applications in projection displays, optical data storage and photolithography.

Acknowledgements

We would like to thank W. H. de Jeu (FOM Amsterdam), T. Tschudi, N. V. Tabiryan and M. Bittner (TH Darmstadt) and the participants of the DALI Project (Institut für Mikrotechnik Mainz) for scientific advice and fruitful discussions.

References

BATRA, I. P., ENNS, P. H. and POHL, D. (1971) Stimulated thermal scattering of light, *Phys. Status Solidi (b)* **48**, 11–63.

BLINOV, L. M. (1983) *Electro-optical and Magneto-optical Properties of Liquid Crystals*, New York: John Wiley.

DE GENNES, P. G. (1974) *The Physics of Liquid Crystals*, Oxford: Clarendon Press.

DE JEU, W. H. (1980) *Physical Properties of Liquid Crystalline Materials*, New York: Gordon and Breach.

DEWEY, A. G. (1984) Laser addressed liquid crystal displays, *Opt. Eng.* **23**, 230–240.

DRZAIC, P. S. and GONZALES, A. M. (1992) Light scattering mechanisms in polymer-dispersed liquid crystal films, *Proc. 12th Int. Display Research Conf., SID Japan Display '92*, pp. 687–690.

EICHLER, H. J., HEPPKE, G., MACDONALD, R. and SCHMID, H. (1992) Storage of erasable laser induced holographic gratings in low molar mass cholesteric liquid crystals, *Mol. Cryst. Liq. Cryst.* **223**, 159–168.

EIDENSCHINK, R. and DE JEU, W. H. (1991) Static scattering in filled nematic: New liquid crystal display technique, *Electron. Lett.* **27**, 1195.

FRITZ, W. J., ST JOHN, W. D., YANG, D.-K. and DOANE, J. W. (1994) *SID 94 Dig.*, pp. 841–844.

GRAY, W. A. (1968) *The packing of solid particles*, London: Chapman and Hall.

KAHN, F. J. (1973) IR-laser-addressed thermo optic smectic liquid crystal storage displays, *Appl. Phys. Lett.* **22**, 111.

KAHN, F. J., TAYLOR, G. N. and SCHONHORN, H. (1973) *Proc. IEEE* **61**, 823.

KAKINUMA, T., KOSHIMIZU, M. and TESHIGAWARA, T. (1992) Laser-addressed liquid crystal-light valve employing polymer dispersed smectic liquid crystals, *Proc. 12th Int. Display Research Conf., SID Japan Display '92*, pp. 851–854.

KREUZER, M. (1994) Grundlagen und Anwendungen von Flüssigkristallen in der optischen Informations- und Kommunikationstechnologie, PhD Thesis, Technische Hochschule Darmstadt.

KREUZER, M. and TSCHUDI, T. (1994) New materials for optical data storage and high information content displays, *Lasers in Engineering*, ed. W. Waidelich, Berlin: Springer-Verlag.

KREUZER, M., TSCHUDI, T. and EIDENSCHINK, R. (1992) Erasable optical storage in bistable liquid crystal cells, *Mol. Cryst. Liq. Cryst.* **223**, 219–227.

KREUZER, M., TSCHUDI, T., DE JEU, W. H. and EIDENSCHINK, R. (1993) A new bistable liquid crystal display with bistability and selective erasure using scattering in filled nematics, *Appl. Phys. Lett.* **62**, 1712–1714.

MICHAEL, G. and FERCH, H. (1993) *Grundlagen und Anwendungen von AEROSIL*, Schriftenreihe Pigmente, Heft 11, 5th edn, Frankfurt: Degussa AG.

WEST, J. L. (1988) Phase separation of liquid crystals in polymers, *Mol. Cryst. Liq. Cryst.* **157**, 427–441.

WHITEHEAD, JR, J. B., ŽUMER, S. and DOANE, J. W. (1993) Light scattering from a dispersion of aligned nematic droplets, *J. Appl. Phys.* **73**, 1057–1065.

YAMAGUCHI, R., OOKAWARA, H., ISHIGAME, M. and SATO, S. (1993) Thermally addressed and erasable displays by polymer-dispersed nematic liquid crystals with memory properties, *J. SID* **1**, 347–352.

YAROSHCHUK, O. V., REZNIKOV, Yu.A., HLUSHCHENKO, A. and LOPUKHOVICH, Yu.N. (1994) The storage mode control in filled nematics, Presentation at the *15th Int. Liquid Crystal. Conf.*, Budapest, July.

16

Phase Transitions in Restricted Geometries

D. FINOTELLO, G. S. IANNACCHIONE and S. QIAN

16.1 Introduction

Studies of the influence of confinement and random substrate disorder on physical properties are topics of current interest which are under close scrutiny (Fisher et al., 1989; Krauth et al., 1991). Such studies are particularly important for liquid crystal systems since their weak orientational and translational order is considerably influenced by the presence of surfaces. In addition, because of the existence of long-range correlations near a phase transition and the presence of different order phase transitions, liquid crystals are a unique and rich system to test in restricted geometries. With the recent availability of new porous substrates, the competition between surface ordering and disordering effects on properties like the director configuration and configurational transitions within cavities, as well as thermodynamic responses like the specific heat, can now be addressed. The behaviour of the orientational order and changes in director configurations that include wetting transitions in submicrometre-size cavities are best probed via ^{2}H-NMR (deuterium nuclear magnetic resonance) (Allender et al., 1991; Crawford et al., 1991a,b, 1993; Žumer and Crawford, this volume). The translational order can be addressed via X-ray measurements (Clark et al., 1993; Iannacchione et al., 1994), while phase transtions are best studied through measurements of the specific heat (Bellini et al., 1992; Iannacchione and Finotello, 1992, 1994; Iannacchione et al., 1992b, 1993a,b, 1994; Wu et al., 1995).

In this chapter, we describe specific heat measurements on the homologous alkyl-cyanobiphenyl (nCB) series ($5 \leqslant n \leqslant 10$) confined to the untreated and treated cylindrical pores of Anopore membranes. This work unveiled effects at the weakly first-order nematic to isotropic (NI), the continuous smectic A to nematic (AN), and the first-order smectic A to isotropic (AI) transitions. Results for nCB liquid crystals confined to the extremely small and interconnected pores of Vycor glass are also presented. New results for nCB confined to the randomly connected pores, comparable in size to Anopore, of fibrous Millipore filter paper are also presented. A comparison with similar specific heat studies in silica aerogel (Bellinic and Clark, this volume; Wu et al., 1995), is also included.

To summarize the major results, in Anopore, the specific heat is strongly dependent on the liquid crystal configuration within the pore; specific heat peaks at depressed transition temperatures are rounded, broadened and suppressed. The pore surface plays the dual role of inducing orientational order near the pore wall while disordering effects dominate through the core of the pore. Smectic ordering is inhibited for liquid crystals with a nematic phase; in contrast, the formation of smectic layers is not strongly affected if the nematic phase is absent in the parent bulk liquid crystal. For a particular surface treatment yielding a stronger surface–liquid crystal interaction, a quantized growth of the smectic phase is induced in the isotropic phase. Despite the randomness of the Millipore confinement, the AN transition remains quite prominent, while the specific heat peak at the NI transition is round and broad. In Vycor, no evidence of any phase transition is found; the orientational order continuously grows with decreasing temperature. A simple, phenomenological Landau–de Gennes type of model, applied to a set of independent pores, quantitatively explains the experimental results.

16.2 Experimental Technique

16.2.1 AC Calorimetry Technique

The specific heat measurements were performed with an AC calorimetry technique (Sullivan and Seidel, 1969). A heating voltage is applied sinusoidally to a sample; the amplitude of the resulting temperature oscillations is inversely proportional to the sample's heat capacity. When the oscillating voltage is applied to a resistive wire heater, Joule's heat is generated at a rate $Q = Q_0 \cos^2(\omega_v t) = \frac{1}{2}Q_0[1 + \cos(2\omega_v t)]$, where $\omega_v = 2\pi f_v$ is the angular voltage frequency and Q_0 is its heating amplitude. The induced temperature oscillations are at twice the voltage frequency, $\omega = 2\omega_v$. For our experimental setup the heat capacity C is given by (Steele *et al.*, 1993a)

$$C = \frac{V_{pp}^2 T^2 I R_{th}}{32\sqrt{2f_v} R_h} [A_1 + 2A_2 \log(R_{th}) + 3A_3 \log(R_{th})^2 + ...]^{-1} \tag{16.1}$$

where V_{pp} is the applied peak-to-peak voltage, T is the average sample temperature, R_h is the heater resistance, R_{th} is the thermometer resistance, and I is the DC current biasing the resistive thermometer. The coefficients A_i arise from the temperature calibration (polynomial fit) of the thermometer.

The experimental cell consists of a 10 kΩ carbon flake thermobead and a 50 Ω Evanohm wire (28 Ω/cm, 0.038 mm cross-section) heater attached to the same side of a sapphire disk of 10 mm diameter and 0.1 mm thick, chosen for its rigidity, flatness, high thermal conductivity, and low heat capacity. The thermobead (a glass-encapsulated carbon flake) has the advantage of extremely small size, less than or equal to 0.05 mm on the side, with a short internal time constant.

16.2.2 Substrates Properties

Anopore membranes (Anopore from Whatman Laboratories, Clifton, NJ) are made from an inorganic aluminium oxide matrix using an electrochemical anodizing process (Furneaux *et al.*, 1989). The anodizing voltage controls the pore size and density, yielding a highly reproducible structure. The cylindrical pores so formed extend parallel to one another through the 60 μm thickness. Anopore has a sharply centred pore size distribution with over 90% of the pores at the rated 0.2 μm diameter. From nitrogen adsorption isotherms and scanning electron microscopy (SEM) (Crawford *et al.*, 1992a), their porosity was estimated at 40%.

Vycor thirsty glass (Levitz *et al.*, 1991) (Corning Vycor brand 7930) is prepared from a melt of 75% SiO_2, 20% B_2O_3 and 5% Na_2O, which is quenched below but near its consolute temperature, inside the miscibility gap. During heat treatment it separates into an SiO_2-rich phase and a B_2O_3 alkali-oxide-rich phase. The boron-rich phase is removed by leach leaving an almost pure SiO_2 skeleton network of 3D randomly connected pore segments with relative uniform diameter $d \sim 70$ Å and average length $l \sim 300$ Å.

Millipore fibrous filter paper (Millipore Corporation, Bedford, MA) was chosen for several reasons: (i) studies can take place as a function of pore size (and thus porosity), (ii) the pore size is comparable with that in the cylindrical Anopore pores, yet there is random interconnection like in Vycor, and (iii) 2D helium films exhibit an essentially identical behaviour in Anopore and Millipore (Steele *et al.*, 1993b; Yeager *et al.*, 1994). Millipore MF-type membranes are thin (125 μm mean thickness) polymeric structures with a high porosity between 50 and 75% composed of pure, biologically inert mixtures of cellulose acetate and cellulose nitrate, and are compatible with dilute acids and bases, aliphatic and aromatic hydrocarbons. They are available with pore sizes ranging from 0.025 to 8 μm. However, SEM studies on several membranes indicated that the typical pore size is a factor of three–six times larger than the nominal size. A broad pore size distribution exists in Millipore.

16.2.3 Confined Sample Preparation

Anopore membranes, used with and without pore treatment with surfactants, were cut into disks less than or equal to 10 mm in diameter. The small disks were cleaned in an acetone or ethyl alcohol ultrasonic bath and dried in an oven. For lecithin-treated Anopore, the disks were immersed in a 2% solution of egg-yolk lecithin and hexane or chloroform. After the treatment, the disks were placed in a vacuum oven at 70 °C for total solvent removal. Anopore was also treated using a similar procedure with a variable length aliphatic acid chain C_mH_{2m+1}—COOH to change the surface anchoring energy. Treated or untreated, all disks were immersed in an isotropic bath of the desired nCB liquid crystal. To avoid unwanted signals, the excess material on the outer surfaces of the disks were removed by squeezing them between Whatman filtration paper. Millipore filters were cut to similar-size disks without cleaning and were not treated with any surfactant.

A chip of Vycor glass approximately $3 \times 7 \times 0.3$ mm^3 was made. After thorough cleaning in an ultrasonic 30% hydrogen peroxide bath and pumping while heating, the sample was left overnight in isotropic 5CB to ensure complete filling of the pores. The 10.6 mg Vycor piece contained approximately 2.5 mg of 5CB. Similar Vycor glass sizes and liquid crystal amounts were used with 7CB and 8CB.

For the calorimetry measurements, a *single* filled Anopore (Millipore) disk, containing 2 to 2.5 mg of liquid crystal, or the Vycor piece, was rested on the experimental cell and cycled through the bulk isotropic transition several times. This annealing process is performed to ensure that the director field had reached a final stable configuration.

16.3 Results and Discussion

Since three different porous substrates, two pore surface treatments and three different order phase transitions were probed, the results are discussed separately and according to substrate.

16.3.1 Anopore Membranes

For untreated Anopore, the director is aligned parallel to the pore axis; after lecithin or aliphatic acid treatment (longer chains), the alignment is homeotropic or radial (Crawford *et al.*, 1991b, 1992b, 1993). For a cylindrical geometry there exist several possible radial configurations; in Anopore, the nematic director configuration is most likely an escaped radial with point defects; the director starts being aligned perpendicular at the pore wall, eventually (and gradually) bending parallel to the pore axis near its centre. For the radial configuration, both the order parameter and the director field have a spatial dependence, while for the axial configuration, the order parameter is spatially dependent with a uniform director field parallel to the pore axis.

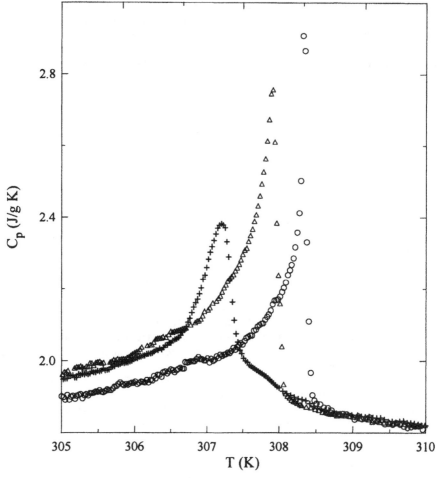

Figure 16.1 Specific heat as a function of temperature near the NI transition for bulk (○), axial (△) and radial (+) 5CB configurations.

16.3.1.1 *Nematic to isotropic (NI) transition*

Specific heat results near the NI transition for bulk, axial and radial confined samples of 5CB, 7CB and 8CB are presented in Figures 16.1–16.3. The magnitude of the bulk specific heat in the isotropic phase and the transition temperatures agree with known results. For all liquid crystals and configurations, a suppressed, rounded and downward shift in the temperature specific heat peak is found. Comparing the confined alignments, transition shifts, peak height suppressions and broadening are more noticeable for the radial case. Compared with the bulk, the axial T_{NI} is shifted down by 0.44, 0.87 and 0.75 K for 5CB, 7CB and 8CB respectively. In the radial case, larger shifts are measured: 1.15, 2.36 and 1.41 K for 5CB, 7CB and 8CB respectively, which are comparable with those in other confined systems (Kuzma and Labes, 1983; Wu *et al.*, 1995). The transition width, the full width at half maximum (FWHM) of the specific heat peaks, broadens from the axial to radial alignment. The axial specific heat peak retains the sharp, bulk-like decrease on the isotropic side while the radial peak is quite symmetric about the transition temperature T_C ($T_C = T_{NI}$ for the bulk) and is somewhat reminiscent of a second-order phase transition.

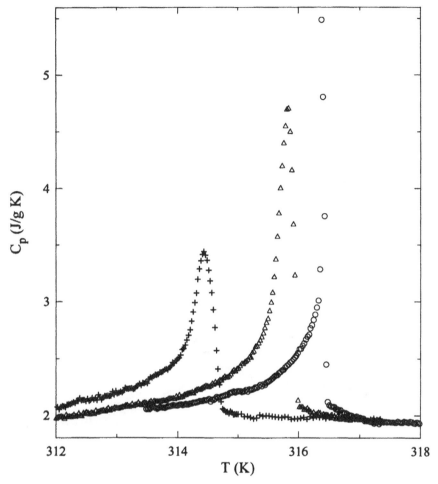

Figure 16.2 Specific heat as a function near the NI transition for bulk ($\bigcirc$), axial ($\triangle$) and radial ($+$) 7CB configurations.

The suppression of the specific heat peak maximum at T_C can be attributed to the effect of surface anchoring. As T increases above T_C, only the liquid crystal near the pore centre disorders, the rest remaining ordered (pinned) by the surface, deep in the isotropic phase. There is no clear boundary between these two regions and the order evolves gradually across the pore diameter. Surface pinning prevents the liquid crystal from undergoing the transition at T_C (surface immobilization) and higher temperatures are needed to free the molecules; the transition is consequently rounded and broad. The surface coupling constant, which is a measure of the surface anchoring strength, is an order of magnitude larger for the lecithin-treated than for the untreated Anopore (Crawford *et al.*, 1991b). Stronger pinning exists in the radial case so less material undergoes the transition at T_C. The specific heat calculated from the measured heat capacity divided by the total mass is accordingly smaller.

The effect of pinning can be estimated from the specific heat peak for axial 5CB which is suppressed by 12%. Assuming that 12% of the molecules are pinned, it would be equivalent to a surface-induced nematic cylindrical shell about 62 Å thick. Such nematic pinning lengths are not unreasonable considering the results of dielectric spectroscopy and diffusion studies

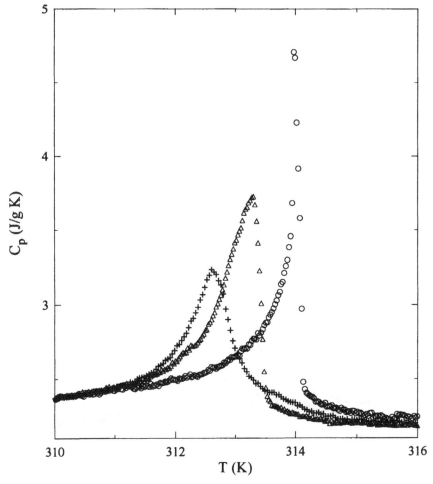

Figure 16.3 Specific heat as a function near the NI transition for bulk (O), axial (△) and radial (+) 8CB configurations.

that have been interpreted in terms of similarly thick surface-induced nematic (polar ordered) shells (Aliev and Breganov, 1989). Additional evidence for a pinning model can be extracted from the transition enthalpy ΔH, obtained from the area under the specific heat curve using a similar temperature range enclosing T_C. For homeotropic 5CB where the peak is a factor of two broader than the bulk and clearly rounded, the enthalpy is smaller than the bulk by about 17%, corresponding to a surface-induced nematic pinning length of about 78 Å. A related discussion regarding the influence of confinement on dynamic and interfacial properties can be found here in the chapter by Aliev.

When dealing with confined systems on these length scales, finite size effects must be taken into account. However, the correlation length at $T_{NI}(\sim 170$ Å) is much shorter than the pore 1000 Å radius, and transition shifts cannot be due to finite size effects. Alternatively, owing to the rigid pore walls, another cause for transition shifts are surface tension effects introduced via the density difference between the nematic and the isotropic phase. At constant volume, an interface within the pore must form to accommodate the density change at T_C. The effect on T_C may be calculated from (Kuzma and Labes, 1983)

$$\Delta T = \frac{2\Delta\sigma T_{NI}}{R\Delta H v} \tag{16.2}$$

$\Delta\sigma$ is the difference in surface tension between the nematic and isotropic phases, ΔH is the NI transition enthalpy, v is the number density of the liquid crystal and R is the confining size. Using typical values for bulk 5CB ($\Delta\sigma \approx 0.26 \times 10^{-3}$ N/m, $\Delta H \approx 2.085 \times 10^{-21}$ N m/M, $v \approx 0.5 \times 10^{28}$ M/m^3 and $T_{NI} \approx 308$ K, where M is the number of molecules) and $R = 1000$ Å, a downward shift of 0.16 K would be introduced. Such a shift is still an order of magnitude smaller than the typical T_C depression.

Downward shifts in these systems most likely arise from elastic deformations of the nematic director, which are particularly severe for the radial configuration. An estimate can be obtained from

$$\Delta T = \frac{K_0}{2a_0} \left(\frac{2\pi}{R}\right)^2 \tag{16.3}$$

Using 5CB values for the Frank elastic constant $K_0 \approx 5.5 \times 10^{-11}$ N, and the Landau–de Gennes free-energy expansion constant $a_0 = 2a/3 = 9.0 \times 10^4$ J/m^3 K, the downward shift is $\Delta T \cong 1.2$ K, in good agreement with the measurements. Thus, elastic constraints play a dominant role in these systems.

Finally, for all samples at the same reduced temperature t in the isotropic phase, $t = |T/T_C - 1|$, the magnitude of the specific heat is higher for the radial case than the axial or bulk case, reflecting the residual surface-induced nematic order near the pore wall (Crawford *et al.*, 1991a,b). As determined by NMR, surface-induced order in the first layers, extending as far as 20 K deep in the isotropic phase, is fairly typical in these porous substrates. The specific heat for the axial case is slightly above the bulk, a reflection that the thermal fluctuations for this system are similar to those of the bulk. The nematic order at the untreated pore wall is small and temperature independent while for the lecithin-treated pores it is large but decreases with increasing temperature.

In contrast, for all samples at the same t in the nematic phase, the magnitude of the bulk specific heat is intermediate between the (lower) radial and the (higher) axial. This is understood by considering the spatial dependence of both the order parameter S and the director n. For the radial alignment, the thermal fluctuations of S may be suppressed by the additional spatial dependence of n, yielding the lowest nematic specific heat magnitude. For the axial configuration, the director is uniform across the pore, yet the degree of order near the pore wall is less than that near the pore centre. This increases fluctuations between nematic ordered domains near the core centre and those near the wall, resulting in a higher than bulk specific heat.

16.3.1.2 Smectic A to nematic (AN) transition

The specific heat at the AN transition, more strongly affected by confinement than the NI transition, is shown in Figures 16.4 and 16.5 for 8CB and 9CB. The specific heat peak is greatly suppressed, nearly disappearing for the radial case. For 8CB, the maximum for the axial case is 16% that of the bulk while for the radial alignment it is only 7.5%; these are 3.5 and five times larger than the corresponding suppressions at T_{NI}. The enthalpy ΔH decreases by 25% for the axial and by 67% for the radial case (over the same narrow temperature range) to be compared with bulk $\Delta H_{AN} = 0.48$ J/g and $l_{AN} = 0.0014$ J/g (Thoen *et al.*, 1982; Thoen, 1992). Identical effects are found for 9CB. Thus, given the earlier discussion regarding the NI transition, it would appear that an even smaller fraction of the material undergoes the AN than the NI transition, an indication of the pore surface preference for a nematic. In the axial case, although having a larger specific heat peak, the transition temperature is lower than in the radial alignment. The axial nematic range ($T_{NI} - T_{AN}$) is slightly wider than the bulk; the radial nematic range is 1 K narrower.

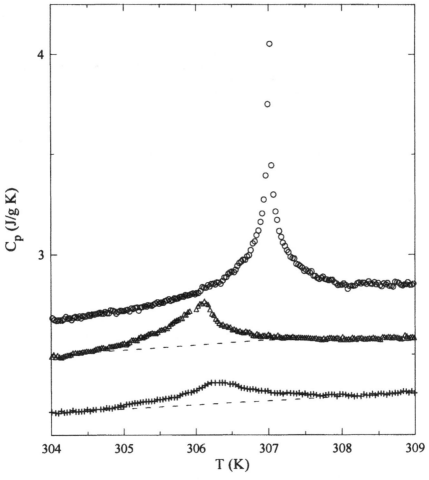

Figure 16.4 Specific heat as a function near the AN transition of 8CB for bulk (○), axial (△) and radial (+) configurations. Dashed lines are guides to the eye. Axial and bulk data have been shifted for clarity.

Because of pore wall heterogeneities, the axial case is influenced by weak surface anchoring at diametrically opposed surfaces of the cylindrical pore. Smectic layers that uniformly span the entire diameter of the pore are difficult to form. A shell of nematic material is left near the pore wall and does not participate in the AN transition, leading to a smaller peak maximum than the bulk. This shell of nematic material, which completely surrounds the smectic core, acts as a disordering surface for the enclosed smectic phase and lowers T_C.

Radially confined 8CB to the larger pores of lecithin-treated glass capillary arrays (25 μm in diameter) was studied by X-rays which showed that distorted smectic layers of variable thickness are formed only near the pore wall (Mang *et al.*, 1992). These effects should be present and be more pronounced for the lecithin-treated Anopore since its pore size is 100 times smaller. Consequently, a thin cylindrical shell of liquid crystal material comprising only a few smectic layers undergoes the AN transition, leaving a nematic core in a nearly escaped-radial configuration. The smectic region is bordered on one side by a strong orientationally ordering interface (the pore surface); the transition would not be shifted as much as for the axial case. Fewer layers are formed in the radial case, and thus the specific heat peak maximum would

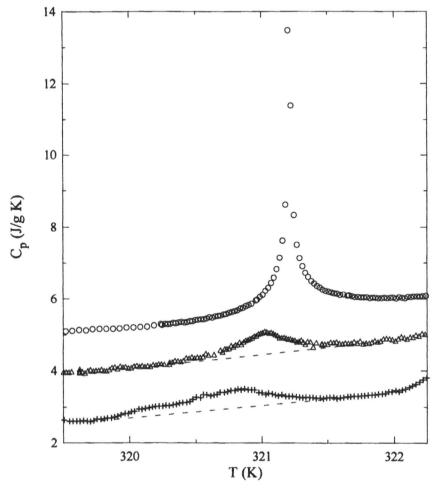

Figure 16.5 Specific heat as a function near the AN transition of 9CB for bulk (○), axial (△) and radial (+) configurations. Dashed lines are guides to the eye. Axial and bulk data have been shifted for clarity.

be even smaller. The smectic correlation length may not grow beyond a small fraction of the pore size (Bellini *et al.*, 1992; Clark *et al.*, 1993); a translationally disordering nematic is the preferred liquid crystal phase in these pores.

Because of diverging correlation lengths, finite size effects could play more of a role at the AN transition. For smectics there is a distinct difference between the correlation length parallel ($\parallel$) and perpendicular ($\perp$) to the layers. For 8CB (Litster *et al.*, 1979) $\xi_\perp \approx |t|^{-n(\perp)}$ and $\xi_\parallel \approx |t|^{-n(\parallel)}$ where $n_\perp = 0.51$ and $n_\parallel = 0.67$; the parallel component dominates with decreasing system size. For instance, the perpendicular correlation length is comparable with the pore size (1000 Å) only within 0.05 K of the transition, while the parallel correlation length is comparable with the pore diameter at twice this ΔT, or 100 mK. For bulk 8CB at 0.11 K from $T_c(t = 3.5 \times 10^{-4})$, $\xi_\parallel \approx 900$ Å and $\xi_\perp \approx 126$ Å. For the axial confinement, $\xi_\parallel$ must be compared with the pore radius, while the radial correlation length $\xi_\parallel$ must be compared with the 60 μm pore length. Thus, finite size effects may play a role in the axial case yet only very close to the transition. The AN specific heat suppression is too large to be completely attributed to finite size effects.

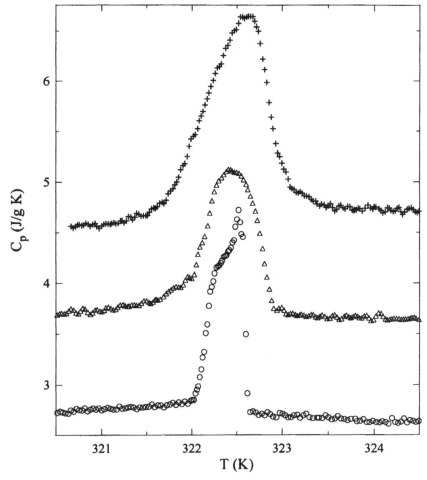

Figure 16.6 Specific heat against temperature near the AI transition of bulk ($\bigcirc$), axial ($\triangle$) and radial (+) 10CB. Confined data were shifted for clarity. Note the insignificant temperature shift as compared with those at the NI or AN transition.

16.3.1.3 Smectic A to isotropic (AI) transition

Measurements of the specific heat for bulk, axial and radial 10CB are shown in Figure 16.6. The confined AI transition is not shifted to lower temperatures and, surprisingly, the specific heat peak is not suppressed. The transition region is quite broad, becoming more rounded for the confined cases. In the radial case, the transition is broader and has a small upward temperature shift. This is similar to observations at the AN transition in the sense that T_C is higher for the radial than the axial alignment, with both T_C below the bulk transition. Under confinement, step-like features apparent in the bulk specific heat (also seen for 12CB (Iannacchione *et al.*, 1993b) disappear. Because of their disappearance and considering the experimental setup consisting of a sapphire with a few milligrams of liquid crystal material residing on top, these features have been attributed to interfacial effects and a temperature-dependent surface anchoring energy; no such behaviour was observed in identical measurements with nCB liquid crystals with $n < 10$. Alternatively (see below), such phenomena may be a consequence of the underlying sapphire inducing smectic layering transitions.

The broad transition region contains large and highly energetic fluctuations. This is evident from the calculated enthalpy, $\Delta H \approx 1.2$ J/g, which is of the same order of magnitude as the enthalpy at the NI (1.2 to 2 J/g) or at the AN (0.5 to 0.8 J/g) transitions, but almost completely due to the two-phase coexistence. As a comparison, the latent heat at the 10CB AI transition is $l_{AI} = 8.86$ J/K (Thoen, 1992). ΔH increased upon confinement: 1.4 J/g for the axial and 2 J/g for the radial case. This increase in ΔH may result from enhanced fluctuations between regions within the pore having a different degree of order, a consequence of how the latent heat is released in the confined system, or a combination of both. If surface anchoring energies are to play a significant role, as they do for the NI and AN transitions, then they must be comparable in magnitude with the energy of direct smectic fluctuations in an isotropic phase. Integrating these results with those at the AN transition, if an intermediate nematic phase is absent from the bulk liquid crystal, a smectic phase is formed within the pores.

An interesting scenario develops when the Anopore pores are treated with an aliphatic acid, C_mH_{2m+1}—COOH, whose length can be varied; in this way, the surface–liquid crystal interaction is changed. When 12CB is confined to these Anopore-treated pores, and for long aliphatic acid lengths, it was shown through X-ray and calorimetry measurements (Iannacchione *et al.*, 1994) that the translational order exhibits a stepwise growth at temperatures well into the isotropic phase. In addition to a considerable broadening and rounding of the AI transition, there are smaller yet unmistakably clear specific heat peaks which arise from this surface-induced smectic order. The transition enthalpy of the small peaks is consistent with the growth of a single bilayer. These results are shown in Figure 16.7 for $m = 12$ aliphatic acid treatment of the pores where as many as nine steps in the X-ray intensity and three distinct specific heat peaks are seen. The peaks' location can be identified with the position where the X-ray results show discrete increases. As the AI transition is approached, layering transitions occur near T_{AI}; since they occur over a wide temperature range, they cannot be distinguished from and thus contribute to the width of the AI transition. No discrete increases in the translational order are seen for $m < 10$ or for the untreated or lecithin-treated pores; the strength and extent of the surface interactions, chemical in nature, changes with the length of the aliphatic acid and it is optimum for $m = 12$. A bilayer-by-bilayer growth, observed for the first time in a cylindrical geometry, has been previously observed in planar systems (Ocko, 1990).

16.3.2 *Vycor Glass*

The most extensively employed confining substrate for a variety of physical systems is porous Vycor glass. In contrast to the well-defined Anopore cylindrical geometry, Vycor offers random and severe confinement as the typical pore size, 70 Å diameter, is a factor of two smaller than the NI correlation length. NMR and calorimetry results for 5CB (and also 8CB) in Vycor (Iannacchione *et al.*, 1993a) shown in Figure 16.8 reveal that severe constraints introduced by the random porous network yield an inhomogeneous and depressed nematic orientational order which is intermediate between the nematic and isotropic phases. The NI transition is substituted by the gradual increase of local orientational order with decreasing temperature.

The Vycor results are well understood in terms of a simplified model of independent subsystems, liquid crystals in pore segments, where the orientational order is described by an effective scalar order parameter Q from a free-energy expansion

$$f = f_0 + \tfrac{1}{2}a(T - T^*)Q^2 - \tfrac{1}{3}bQ^3 + \tfrac{1}{4}cQ^4 + hQ^2 - gQ \tag{16.4}$$

where f_0, a, T^*, b and c are known parameters in a Landau expansion (de Gennes, 1974).

The hQ^2 term in the Landau expansion describes the disordering effects of surface-induced deformations which strongly depend on pore shape and the roughness of its surface; h, whose effect is that of renormalizing T^*, is expected to vary from pore to pore. The $-gQ$ term, whose effect is similar to that of an external magnetic field on a spin system (Bellini *et al.*, 1992;

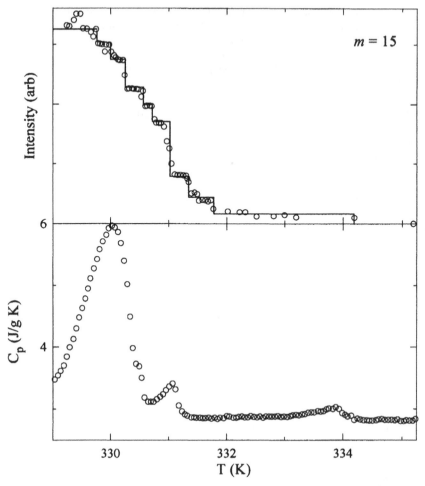

Figure 16.7 X-ray intensity and specific heat as a function of temperature for 12CB confined to $m = 15$ aliphatic-acid-treated Anopore to show the discrete nature of surface-induced bilayer growth in the isotropic phase. Full lines are guides to the eye.

Iannacchione *et al.*, 1993a; Kralj *et al.*, 1991; Poniewerski and Sluckin, 1986; Sheng, 1976, 1982), accounts for the ordering effect of the surface interactions; g should be shape independent and so is taken as constant throughout the system. The model is expected to be valid because of the Vycor pore's relatively elongated shape; the order is mostly determined by local or intraporous constraints. Collective or interporous effects are expected to be weak enough that the interaction between subsystems is negligible (Iannacchione *et al.*, 1993a). Effects due to the interconnections become noticeable in porous glasses similar to Vycor but with a slightly larger pore size (Tripathi *et al.*, 1994). A complete discussion of the influence of the interconnections can be found in the chapter by Tripathi and Rosenblatt.

Considering the above approximations, the Vycor system is represented by a Gaussian distribution characterized by an average disorder parameter and a certain width; by averaging over this, response functions including the NMR spectral pattern and the orientational order contributrion to the heat capacity can be calculated; a comparison of the model prediction with the experimental results is presented in the inset to Figure 16.8. Evidently, the inclusion

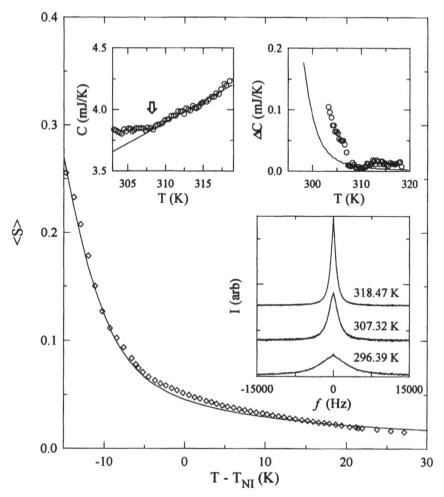

Figure 16.8 Average order parameter as a function of temperature for 5CB in Vycor. Also shown are NMR lineshapes at several temperatures bracketing the bulk T^{NI}. Left inset: heat capacity results as a function of temperature; the arrow indicates the bulk transition temperature. Right inset: comparison of the model-predicted orientational contribution to the heat capacity (full line) with the experimental results.

of ordering and disordering terms in a Landau free energy yields a satisfactory and quantitative description of the data.

The Vycor results should be contrasted with those in silica aerogel (Bellini *et al.*, 1992; Clark *et al.*, 1993; Kralj *et al.*, 1993; Wu *et al.*, 1995) that are also described here in the chapters by Bellini and Clark. For 8CB in the 175 Å aerogel pores, there is a pronounced and only marginally shifted NI transition peak; the AN transition is considerably suppressed. By comparing both random structures, particularly the pore length to diameter aspect ratio, which is four in Vycor while it is nearly unity in aerogel, in the latter the interporous interaction should be stronger. Thus, to understand the aerogel results, a model where subsystems are affected by both interporous and intraporous interactions must be developed. Interestingly, certain features of the aerogel results are reminiscent of those for homeotropically aligned liquid crystals in Anopore (Wu *et al.*, 1995).

In summary, the study of phase transitions in porous glasses is a fascinating research area. The experimental results have stimulated theoretical efforts (Maritan *et al.*, 1994) which are fully described in this book in the chapter by Maritan *et al.*, and in that by Cleaver *et al.*

16.3.3 *Millipore Filters*

The last substrate that we wish to consider is Millipore filter paper. Here, we discuss specific heat and NMR measurements for *n*CB liquid crystals in the 0.05 μm pore size Millipore. Owing to the large void size and isotropic nature of the filters, no surface-induced preferred orientation should be expected. However, in the NMR studies also described and because of the larger than nominal pore sizes becoming on the order of the magnetic coherence length for the 4.7 T field used, a magnetic-field-induced partial alignment may take place.

16.3.3.1 *Nematic to isotropic transition*

Results for specific heat against temperature for 5CB, 8CB and 9CB are shown in Figure 16.9. Comparing with the bulk, the transition is shifted to lower temperatures by approximately 0.5 K for 5CB and 0.8 K for 8CB. Possibly due to its narrow nematic range and smectic influence, T_{NI} is only shifted by 0.23 K for 9CB. The NI specific heat peak is slightly broad and round; like in Anopore, the divergent nature of the transition disappears under confinement. The broadening and rounding may be attributed to the existing distribution of void sizes; however, since even the smallest nominal size of 0.05 μm is at least three times larger than the NI correlation length, these effects are more likely a reflection of the underlying random disordering nature of the substrate.

For all liquid crystals, the FWHM is about 0.15 K, which is a factor of two or three narrower than in Anopore, and it is not too different from the bulk. This is partly a consequence of the fact that the magnitude of the confined specific heat is larger than that of the bulk, a result of how the latent heat is released in randomly confining systems. A larger than bulk maximum was also measured in aerogel (Wu *et al.*, 1995) but not in the well-defined Anopore geometry. Note that the AC calorimetry technique used here and in the aerogel work provides a measure of C_P and not the enthalpy; thus a quantitative determination of the latent heat is not available. The delta function that corresponds to the bulk latent heat (not detectable with our technique) is not included in the above comparisons.

According to NMR measurements in progress (Qian *et al.*, 1995), the broadening of the NI transition may also be understood in terms of coexistence of two paranematic domains with isotropic liquid. Two distinct NMR quadrupolar splittings, which are seen in the nematic phase, merge into a single broad absorption peak in the isotropic phase. The broad isotropic absorption peak which is attributed to a distribution of weakly ordered nematic areas has an FWHM of about 300 Hz near T_{NI}, decreasing to less than 100 Hz 16 K above T_{NI}, again a reflection of the existence of residual surface-induced nematic order. The larger quadrupolar splitting is approximately a factor of two smaller than the bulk and arises from nematic material in the voids. A 1/2 bulk splitting is typical of a radial configuration, i.e. a configuration where LC molecules are aligned perpendicular to the external magnetic field (4.7 T); here, it most likely reflects the competition between the surface anchoring that prevents a complete alignment by the field. The smaller quadrupolar splitting, an order of magnitude less than that of magnetic-field-aligned bulk material, is due to liquid crystal molecules in much closer proximity to the substrate fibres; the liquid crystal molecules are strongly anchored and thus only weakly affected by the NMR field. Both quadrupolar splittings are weakly dependent on the angle between the field and the NMR probe, which can be understood in terms of a small anisotropy of the filters.

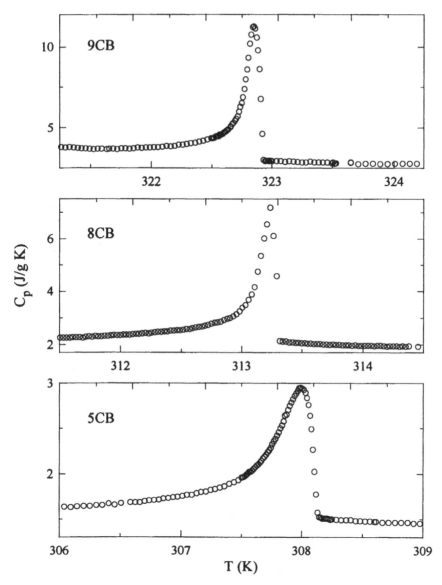

Figure 16.9 Specific heat as a function of temperature at the NI transition for 5CB, 8CB and 9CB in the 0.05 μm Millipore pores.

16.3.3.2 *Smectic A to nematic transition*

The AN transition in porous materials is substantially modified. Studies of the 0.2 μm diameter pores of Anopore membranes found a strongly suppressed AN transition, dependent on the director orientation in the pores; the specific heat peak was at most 15% that of the bulk. In silica aerogel (Bellini *et al.*, 1992; Wu *et al.*, 1995), the AN transition is broad and suppressed, these effects becoming stronger with increasing aerogel density (decreasing porosity). In Millipore, as evident from the specific heat plots for 8CB and 9CB shown in Figure 16.10, the AN transition is manifest via a prominent and symmetric specific heat peak.

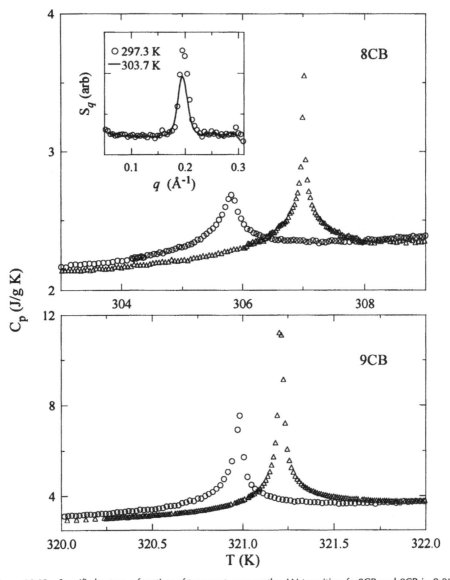

Figure 16.10 Specific heat as a function of temperature near the AN transition for 8CB and 9CB in 0.05 μm Millipore. Bulk results (○) are also shown. Inset: small-angle neutron scattering results for 8CB in Millipore.

Despite the interconnection of voids in Millipore, smectic layers are easily formed; the smectic spacing, determined from small-angle neutron scattering (SANS), is estimated to be 31.4 Å, suggestive of the presence of a bilayered structure (Qian *et al.*, 1994). The SANS results, plotted in the inset to Figure 16.10, clearly show an intensity peak that grows by 30% with decreasing temperature. The existence of such a peak to temperatures as low as 297 K indicates considerable supercooling, as also found in silica aerogel (Bellini *et al.*, 1992; Clark *et al.*, 1993). The melting/freezing LC transition in porous materials is an interesting topic that requires further investigation.

16.4 Summary

The investigation of phase transitions in restricted geometries is in general a rich and exciting area. This is particularly true for liquid crystals owing to their strong interactions with surfaces. In this chapter, we described specific heat studies at liquid crystal (nCB) phase transitions confined to the 0.2 μm diameter cylindrical pores of Anopore membranes, with and without pore surface treatment, to the severely restrictive and random connected network of pores of Vycor glass, and to the 0.05 μm voids of Millipore fibrous filter paper. Through complementary NMR studies that were briefly presented, we discussed effects on the orientational order. The translational order existent in these systems was studied via X-ray and small-angle neutron scattering. This work explored confining and surface effects at the NI, AN and AI transitions. Results were contrasted to the bulk and to other confining substrates, in particular silica aerogel.

At the weakly first-order NI transition, the confined specific heat peak for both untreated and lecithin-treated Anopore membranes is suppressed, rounded, broadened and shifted to lower temperature. These effects are severe for a homeotropic alignment; although weaker for the axial configuration they are none the less surprising. The axial specific heat peak retains a bulk-like sharp decrease into the isotropic phase. The transition broadening is attributed to a two-phase coexistence. The radial NI transition is shifted to even lower temperatures with a greater suppression and broadening of the specific heat peak. The divergent nature of the bulk is lost and the peak shows a symmetric appearance reminiscent of a continuous transition. The radial configuration is strongly influenced by elastic distortions and the transition temperature shift is well explained by a simple elastic model. A large two-phase coexistence region is also present for this configuration. It is, however, very possible that under certain conditions the order of the phase transition is changed.

The continuous 8CB AN phase transition in Anopore is significantly more affected than the NI transaction. The specific heat peak at T_{AN} is extremely broad and greatly suppressed. This is accompanied by a decrease in the transition enthalpy a direct indication of the difficulty of forming smectic layers within these pores whenever a (confinement-)preferred orientationally ordered nematic phase exists in the bulk liquid crystal. A nematic phase provides a mechanism for lowering the free energy of the confined liquid crystal in the presence of elastic distortions and surface disordering. The smectic correlation length is probably much shorter than the pore size which is too large to understand these phenomena in terms of finite size effect theories.

The AI transition is very broad; most of the effects observed at the NI and AN transitions are absent. The specific heat peak position does not shift significantly in temperature. The transition is broader for the axial case and even broader for the radial alignment, resulting in an increase in the transition enthalpy from bulk to axial to radial. The absence of a nematic phase in the bulk permits smectic layers to form within the pores, possibly with the exception of a nematic-like shell at the pore wall which is not allowed to grow. A discrete bilayer-by-bilayer smectic growth in the isotropic phase is induced by the Anopore pore surface when it is coated with an aliphatic acid chain.

In Vycor, all phase transitions are completely suppressed. Owing to severe constraints on the liquid crystal imposed by the random porous network, the NI transition is replaced by a continuous evolution of local orientational order. A simple model of independent pores, where the orientational order is described by an effective scalar order parameter and the inclusion of ordering and disordering terms in the Landau free energy, leads to a quantitative understanding of the experimental results. For aerogel, the interconnection must be incorporated.

Results in the 0.05 μm interconnected pores of fibrous Millipore filter paper find a slightly rounded NI transition which suggests the coexistence of a mixture of paranematic domains. In contrast to the similar-size and well-defined Anopore geometry, the AN transition in Millipore is quite prominent; the specific heat peak, although somewhat suppressed, retains the sharpness and symmetric features of a second-order phase transition. The translational order is present down to temperatures below the bulk freezing.

Acknowledgements

We have benefited from multiple discussions with D. Allender, G. Crawford, D. Johnson, S. Kumar, M. Lee, P. Palffy-Muhoray, P. Sokol, S. Žumer, and especially J. W. Doane. This work was supported by NSF through the ALCOM-STC Grant DMR 89-20147.

References

ALIEV, F. M. and BREGANOV, M. N. (1989) Electric polarization and dynamics of molecular motion of polar liquid crystals in micropores and macropores, *Zh. Eksp. Teor. Fiz.* **95**, 122 [*Sov. Phys. JETP* **68**, 70 (1989)].

ALLENDER, D. W., CRAWFORD, G. P. and DOANE, J. W. (1991) Determination of the liquid crystal surface elastic constant K_{24}, *Phys. Rev. Lett.* **567**, 1442.

BELLINI, T., CLARK, N. A., MUZNY, C. D., WU, L. GARLAND, C. W., SCHAEFER, D. W. and OLIVER, B. J. (1992) Phase behavior of the liquid crystal 8CB in a silica aerogel, *Phys. Rev. Lett.* **69**, 788.

CLARK, N. A., BELLINI, T., MALZBENDER, R. M., THOMAS, B. N., RAPPAPORT, A. G., MUZNY, C. D., SCHAEFER, D. W. and HRUBESH, L. (1993) X-ray scattering study of smectic ordering in a silica aerogel, *Phys. Rev. Lett.* **71**, 3505.

CRAWFORD, G. P., YANG, D.-K., ŽUMER, S., FINOTELLO, D. and DOANE, J. W. (1991a) Ordering and self-diffusion in the first molecular layer at a liquid crystal-polymer interface, *Phys. Rev. Lett.* **66**, 723.

CRAWFORD, G. P., STANNARIUS, R. and DOANE, J. W. (1991b) Surface-induced orientational order in the isotropic phase of a liquid crystal material, *Phys. Rev. A* **44**, 2558.

CRAWFORD, G. P., STEELE, L. M., IANNACCHIONE, G. S., ONDRIS-CRAWFORD, R., YEAGER, C. J., DOANE, J. W. and FINOTELLO, D. (1992a) Characterization of the cylindrical cavities of Anopore and Nuclepore membranes, *J. Chem. Phys.* **96**, 7788.

CRAWFORD, G. P., ALLENDER, D. W. and DOANE, J. W. (1992b) Surface elastic and molecular anchoring properties of nematic liquid crystals confined to cylindrical cavities, *Phys. Rev. A* **45**, 8693.

CRAWFORD, G. P., ONDRIS-CRAWFORD, R. ŽUMER, S. and DOANE, J. W. (1993) Anchoring and orientational wetting transitions of confined liquid crystals, *Phys. Rev. Lett.* **70**, 1838.

FISHER, M. P. A., WEICHMAN, P. B., GRINSTEIN, G. and FISHER, D. S. (1989) Boson localization and the superfluid–insulator transition, *Phys. Rev. B* **40**, 546; and references therein.

FURNEAUX, R. C., RIGBY, W. R. and DAVIDSON, A. P. (1989) The formation of controlled-porosity membranes from anodically oxidized aluminum, *Nature* **337**, 147.

DE GENNES, P. G. (1974) *The Physics of Liquid Crystals*, Oxford: Clarendon Press.

IANNACCHIONE, G. S. and FINOTELLO, D. (1992) Calorimetric study of phase transitions in confined liquid crystals, *Phys. Rev. Lett.* **69**, 2094.

IANNACCHIONE, G. S. and FINOTELLO, D. (1994) Specific heat dependence on orientational order at cylindrically confined liquid crystal phase transitions, *Phys. Rev. E* **50**, 4780.

IANNACCHIONE, G. S., CRAWFORD, G. P., DOANE, J. W. and FINOTELLO, D. (1992) Orientational effects on confined 5CB, *Mol. Cryst. Liq. Cryst.* **222**, 205.

IANNACCHIONE, G. S., CRAWFORD, G. P., ŽUMER, S., DOANE, J. W. and FINOTELLO, D. (1993a) Randomly constrained orientational order in porous glass, *Phys. Rev. Lett.* **71**, 2595.

IANNACCHIONE, G. S., STRIGAZZI, A. and FINOTELLO, D. (1993b) Interface influenced phase transitions of alkylcyanobiphenyl liquid crystals. A calorimetric study, *Liq. Cryst.* **14**, 1153.

IANNACCHIONE, G. S., MANG, J. T., KUMAR, S. and FINOTELLO, D. (1994) Surface-induced discrete smectic order in the isotropic phase of 12CB in cylindrical pores, *Phys. Rev. Lett* **73**, 2708.

KRALJ, S., ŽUMER, S. and ALLENDER, D. W. (1991) Nematic-isotropic phase transition in a liquid crystal droplet, *Phys. Rev. A* **43**, 2943.

KRALJ, S., LAHAJNAR, G., ZIDANŠEK, A., VRBANČIČ-KOPAČ, N., VILFAN, M., BLINC, R. and KOSEC, M. (1993) Deuterium NMR of a pentylcyanobiphenyl liquid crystal confined in a silica aerogel matrix, *Phys. Rev. E* **48**, 340.

KRAUTH, W., TRIVEDI, N. and CEPERLEY, D. (1991) Superfluid–insulator transition in disordered boson systems, *Phys. Rev. Lett.* **67**, 2307.

KUZMA, M. and LABES, M. M. (1983) Liquid crystals in cylindrical pores: Effects on transition temperatures and singularities, *Mol. Cryst. Liq. Cryst.* **100**, 103; and references therein.

LEVITZ, P., EHRET, G., SINHA, S. K. and DRAKE, J. M. (1991) Porous Vycor glass: The microstructure as probed by electron microscopy, direct energy transfer, small-angle scattering, and molecular adsorption, *J. Chem. Phys.* **95**, 8.

LISTER, J. D., ALS-NELSON, J., BIRGENAU, R. J., DANA, S. S., DAVIDOV, D., GARCIA-GOLDING, F., KAPLAN, M., SAFINYA, C. R. and SCHAETZING, R. (1979) High resolution x-ray and light scattering studies of bilayer smectic compounds, *J. Phys. C: Solid State Phys.* **3**, 339.

MANG, J. T., SAKAMOTO, K. and KUMAR, S. (1992) Smectic layer orientation in confined geometries, *Mol. Cryst. Liq. Cryst.* **223**, 133.

MARITAN, A., CIEPLAK, M., BELLINI, T. and BANAVAR, J. R. (1994) Nematic–isotropic transition in porous media, *Phys. Rev. Lett.* **72**, 4113.

OCKO, B. M. (1990) Smectic layer growth at solid interfaces, *Phys. Rev. Lett.* **64**, 2160; and references therein.

PONIEWERSKI, A. and SLUCKIN, T. J. (1986) *Fluids Interfacial Phenomena*, ed. C. Croxton, New York: John Wiley.

QIAN, S., IANNACCHIONE, G. S., FINOTELLO, D., STEELE, L. M. and SOKOL, P. E. (1994) Smectic order in a porous interconnected substrate, *Proc. 15th Int. Liquid Crystal Conf.*

QIAN, S., IANNACCHIONE, G. S. and FINOTELLO, D. (1995) Orientational order and the nematic to isotropic phase transition in random media, unpublished.

SHENG, P. (1976) Phase transition in surface aligned nematic films, *Phys. Rev. Lett.* **37**, 1059.

SHENG, P. (1982) Boundary layer phase transition in nematic liquid crystals, *Phys. Rev. A* **26**, 1610.

STEELE, L. M., IANNACCHIONE, G. S. and FINOTELLO, D. (1993a) AC calorimetry technique. Applications to liquid helium films and liquid crystals, *Rev. Mex. Fís.* **39**, 588.

STEELE, L. M., YEAGER, C. J. and FINOTELLO, D. (1993b) Precision specific heat studies of thin superfluid films, *Phys. Rev. Lett.* **69**, 3673.

SULLIVAN, P. and SEIDEL, G. (1969) Steady state, ac-temperature calorimetry, *Phys. Rev.* **173**, 679.

THOEN, J. (1992) Calorimetric studies of liquid crystal phase transitions: Steady state adiabatic techniques, *Phase Transitions in Liquid Crystals*, ed. S. Martellucci and A. N. Chester ed., Ch. 10, New York: Plenum Press.

THOEN, J., MARYNISSEN, H. and VAN DAEL, W. (1982) Temperature dependence of the enthalpy and the heat capacity of the liquid crystal octylcyanobiphenyl 8CB, *Phys. Rev. A* **26**, 2886.

TRIPATHI, S., ROSENBLATT, C. and ALIEV, F. M. (1994) Orientational susceptibility in porous glass near a bulk nematic–isotropic phase transition, *Phys. Rev. Lett.* **72**, 3674.

WU, L., ZHOU, B., GARLAND, C. W., BELLINI, T. and SCHAEFER, D. W. (1995) Heat capacity study of nematic–isotropic and nematic–smectic-A transitions for octylcyanobiphenyl in silica aerogel, *Phys. Rev. E,* **51**, 2157.

YEAGER, C. J., STEELE, L. M. and FINOTELLO, D. (1994) Calorimetric study of localized helium films in disordering substrates, *Phys. Rev. B* **49**, 9782.

Liquid Crystals and Polymers in Pores

The influence of confinement on dynamic and interfacial properties

F. M. ALIEV

17.1 Introduction

Investigations of condensed matter in porous matrices have revealed various new properties and effects not observed in the same substances when they are in the bulk. The non-triviality of physical phenomena that occur in pores is manifested, for example, when the influence of confinement on phase transition and phase separation in liquid binary mixtures was studied (Aliev *et al.*, 1993; Awschalom and Warnock, 1987; Dierker and Wiltzius, 1991; Drake *et al.*, 1993; Frisken and Cannell, 1992; Goldburg *et al.*, 1995; Maher *et al.*, 1984). The confinement of fluids can lead to such prominent changes in their properties that even in the case of one-component isotropic fluids the physical picture of these changes is far from well understood (Klafter and Drake, 1989; Korb *et al.*, 1993; Pissis *et al.*, 1994; Schuller *et al.*, 1994; Warnock *et al.*, 1986). The difference between the surface and bulk properties (Jerome, 1991; Zhuang *et al.*, 1994), as well as finite size effects, are manifested most strikingly in the case of liquid crystals (LCs) (Aliev and Breganov 1988, 1989; Aliev and Pojivilko, 1988; Aliev, 1992, 1994; Armitage and Price, 1976, 1978a,b; Bellini *et al.*, 1992, 1995; Clark *et al.*, 1993; Crawford *et al.*, 1991, 1992, 1993; Dadmun and Muthukumar, 1993; Goldburg *et al.*, 1995; Iannachione *et al.*, 1993; Iannachcione and Finotello, 1992, 1994; Kralj *et al.*, 1993; Kralj and Žumer, 1995; Kuzma and Labes, 1983; Martian *et al.*, 1994; Sheng *et al.*, 1992; Tripathi *et al.*, 1994; Vrbančič *et al.*, 1993; Wu *et al.*, 1992). This is due to the fact that LCs are 'soft' systems because the energy responsible for long-range orientational order is fairly small. The substrates between which a layer of LC is confined can exert their influence on the LC up to distances L which may reach several thousand angstrom (Jerome, 1991). Investigations of LCs in pores of sizes less than L should in principle, provide information on the interfacial properties of these LCs. From the pioneering investigations of Armitage and Price (1976, 1978a,b) it is well known that restricted geometries have a significant effect on the ordering, structure and phase transition of nematic liquid crystals (Aliev and Breganov, 1988, 1989; Aliev and Pojivilko, 1988; Aliev, 1992, 1994; Armitage and Price, 1976, 1978a,b; Bellini *et al.*, 1992; Clark *et al.*, 1993; Crawford *et al.*, 1991, 1992, 1993; Dadmun and Muthukumar, 1993; Iannacchione *et al.*, 1993; Iannacchione and Finotello, 1992, 1994; Kralj *et al.*, 1993; Kralj and Žumer, 1995; Kuzma and Labes, 1983; Maritan *et al.*, 1994; Tripathi *et al.*, 1994; Vrbančič *et al.*, 1993; Yokoyama, 1988; Žumer and Kralj, 1992) or on the isotropic phase of LCs (Aliev *et al.*, 1991; Schwalb *et al.*, 1994; Schwalb and Deeg, 1995). There has been little work performed to characterize the influence of the confinement on the dynamics of LCs (Aliev and Breganov, 1988, 1989; Aliev and Kelly, 1994; Bellini *et al.*, 1995; Goldburg *et al.*, 1995; Schwalb *et al.*, 1994; Schwalb and Deeg, 1995; Wu *et al.*, 1992) or the second-order phase transition (Bellini *et al.*, 1992; Clark *et al.*, 1993), namely the smectic C–smectic A transition. The dynamic aspects of the influence of the confinement on the behaviour of LCs can be studied by using dielectric spectroscopy and photon correlation spectroscopy (also called dynamic or quasi-elastic light scattering). The second method makes it possible to study modes which are not active dielectrically.

Ferroelectric liquid crystals (FLCs) are good candidates to study the influence of the confinement on the dynamics of collective modes and on the second-order phase transition. In the smectic A phase the director, describing the time-averaged orientation of the long molecular axes, is oriented parallel to the smectic layer normals. Within the layers there is no positional ordering of the molecules. The spontaneous polarization in the SmA phase is zero. In the smectic C* (SmC*) phase the director tilts with respect to the layer normals and precesses from one layer to another so that the helicoidal structure is formed with a period of about 10^3 layer thicknesses. If chiral molecules have a permanent dipole moment transverse to their long molecular axes the in-plane component of the spontaneous polarization becomes non-zero.

The dynamics of different excitations in bulk FLCs has been studied by dielectric spectroscopy (Filipič *et al.*, 1988; Levstik *et al.*, 1987, 1990, 1991; Gouda *et al.*, 1991). In the SmC* phase there are two dielectrically active modes: the Goldstone mode, which corresponds to the collective director reorientation around the smectic cone, and the soft mode connected with tilt fluctuations. In the SmA* phase there is one collective process – the soft mode. Along with investigations of the bulk samples FLCs were studied in the free-standing film geometry (Young *et al.*, 1978; Rosenblatt *et al.*, 1979) and at a solid–LC interface (Patel *et al.*, 1991; Brandow *et al.*, 1993). Measurement of polarization, elastic constants and viscosities as functions of film thickness as well as studies of the FLC–solid interface have revealed many interesting features of FLCs. For FLCs confined in porous media several important questions arise immediately. Does smectic ordering exist in restricted geometries? What is the influence of confinement on the SmC*–SmA* phase transition? What are the dynamics of the Goldstone and the soft modes in pores if they exist in confined geometry? And what is the temperature dependence of the rotational viscosities associated with the Goldstone and the soft modes if they are found in the pores? We try to answer these questions in this chapter.

We performed dielectric, DSC and X-ray scattering measurements to study the influence of the cofinement on the dynamics, phase transitions and structure of an FLC infused into silica porous glasses with thoroughly interconnected and randomly oriented pores with average pore sizes of 1000 Å (volume fraction of pores 40%) and 100 Å (27%) respectively. Below, for simplicity, we will treat the porous matrices with the 1000 Å pore size as macroporous glasses (macropores), and the matrices with 100 Å pores as microporous glasses (micropores). We found that the smectic C and smectic A phases are formed in macropores and the SmC–SmA phase transition temperature is reduced in the pores by about 15°C compared with the bulk value. The Goldstone and the soft modes are found in macropores and the rotational viscosity associated with the soft mode is about 10 times higher than in the bulk. In order to describe the experimental temperature dependencies of the dielectric permittivities and relaxation times of the soft mode in pores by equations based on a Landau-type theory, we suggest that modulated structures whose characteristic length is about 85 Å consisting of periodically bent msmectic layers with conserved local symmetry could be formed in macropores.

In micropores X-ray measurements show the existence of a frozen layered structure but no temperature dependence of the layer spacing and no phase transitions are observed. The Goldstone and the soft modes are not detected in micropores although low-frequency relaxation is present. The last is probably associated with surface polarization effects typical of two-component heterogeneous media.

In the case of polar molecules, the substrate may induce a polar order and give rise to polarization effects, which can also be due to a gradient of the order parameter and inhomogeneity of orientation. Clearly, the bent nature of the pore may induce these effects in porous matrices. The influence of substrate (pore wall) and confinement on the orientational mobility of nematic LC molecules is another topic under consideration in this chapter. We investigate this topic by dielectric spectroscopy. We show that the difference between the dynamics of orientational motion of the polar molecules of NLCs in confined geometries and in the bulk is qualitatively determined by the total energy F_s of the interaction between molecules and the surface of the pore wall, which is found to be $F_s \approx 10^2$ erg/cm^2.

Another insufficiently investigated area is polymers dispersed in pores. The synthesis and investigation of polymers in porous matrices with a developed pore surface are of considerable interest because the difference between the surface and bulk properties of polymers may be great in these systems (Aliev *et al.*, 1990; Aliev and Zgonnik, 1991; Li *et al.*, 1994). In microheterogeneous polymer–porous glass systems, the impregnating polymer may be in the colloidal degree of dispersity, and the role of the interface becomes very great. Therefore, it is possible to vary the characteristics and structure of the material over a wide range. From a practical point of view the investigations of polymers in pores are encouraged by the possibility of using the results for the solution of such important problems as the surface modification, the preparation of new composites and microencapsulation. The synthesis of the polymer–porous glass systems is promising for the preparation of new microcomposites for optical applications (Aliev and Zgonnik, 1991; Li *et al.*, 1994). The difference between the surface and bulk poperties of the substance may be of fundamental character and may be revealed in the appearance near the surface of new phases and transitions between them not found in the volume. In order to establish the relationship between the length of the side group in the homologous series of poly(alkyl methacrylates) and their physical properties in pores, poly(methyl methacrylate) (PMMA), poly(ethyl methacrylate) (PEMA), poly(butyl metha-crylate) (PBMA), poly(octyl methacrylate) (POMA), poly(decyl methacrylate) (PDMA) and poly(2-ethyl hexyl acrylate) (PEHA) were synthesized in porous silicate glass matrices. The macromolecules of the last four polymers exhibit comb-like structure. Polystyrene (PS) was synthesized in the porous glasses as a typical glassy polymer with an aromatic side group for comparson with comb-like polymers. We performed optical, differential scanning calorimetry (DSC) and dielectric measurements to study the influence of the confinement on the physical properties of synthesized polymers in porous silica glasses with thoroughly interconnected and randomly oriented pores.

In macroporous matrices which contain poly(alkyl methacrylate) (PAM) with long side groups, an anomalous change in light scattering, which switches the composite from an opaque state to transparent, was observed in a narrow temperature range. Taking into account the results of thermal and dielectric measurements, this thermally controlled light scattering is explained by the assumption of the orientational ordering of the side groups induced by the inner pore surface at low temperatures. The transition into a state with disordered arrangement of these groups is a first-order phase transition which is absent in the bulk PAM under investigation.

17.2 Samples

Porous matrices of different types have been effectively used to investigate the finite size effects in condensed matter. Porous silicate glasses are particularly promising for use as porous matrices (Aliev *et al.*, 1993; Aliev, 1994; Tripathi *et al.*, 1994; Tripathi and Rosenblatt, this volume), since they are distinguished by their uniformity of chemical composition, purity and mechanical strength. Porous silicate glass can be used as an ideal matrix to study the influence of temperature on the surface effects that occur at the interface between the glass and some other material. Since the structural characteristics of these matrices are nearly independent of the temperature, all observable effects when the temperature is changed can be attributed to the change in the physical properties of the second component (LC). Moreover, the dielectric permittivity of the silica porous glass matrix is independent of the temperature and frequency for a wide range of frequencies. This fact and the practically negligible electrical conductivity of the matrix greatly simplify the interpretation of the results of the dielectric measurements and make it possible to avoid a number of difficulties that were encountered earlier in studies of dielectric properties of heterogeneous systems (Van Beek, 1967; Hikmet and Zwerver, 1991; Rout and Jain, 1992).

Macroporous and microporous matrices with thoroughly interconnected and randomly oriented pores with average pore sizes of 1000 Å (volume fraction of pores 40%) and 100 Å (27%), respectively, were prepared from the original sodium borosilicate glasses. The sodium borate phase was removed by leaching, and the matrix framework consisted of SiO_2. The characteristics of the matrices were determined by small-angle X-ray scattering (Aliev and Pojivilko, 1988). These porous matrices are similar to porous Vycor glass, except that the pore size of macroporous glasses is significantly greater. Like Vycor glass (Drake and Klafter, 1990; Levitz *et al.*, 1991) the porous glasses used in our experiments are characterized by very narrow distributions of pore sizes (for details of the quantitative characterization of the macroporous and microporous matrices as well as the image of the structure of microporous glass obtained by electron microscopy see Aliev and Pojivilko (1988). The FLC we used was SCE 12 synthesized by BDH Ltd through EMI. This FLC has been investigated thoroughly in the free state (see for example Li *et al.*, 1990, 1991). The phase transition temperatures of SCE 12 in the bulk are: SmC*66°C, SmA*81°C, N 121°C I. Nematic liquid crystals were 5CB and cyanophenyl ester of heptilbenzoic acid, i.e. cyanophenyl heptylbenzoate (CPHB). The dipole moment μ of the 5CB molecule is $5D$, and it is parallel to the longitudinal axis of the molecule ($\beta = 0$, where β is the angle between the dipole and the long axis of a molecule); and for CPHB $\mu = 6.1D$ and $\beta = 16°$. The temperatures of the phase transitions of 5CB and CPHB in the free state are $T_{CN} = 295$ K and $T_{NI} = 308.18$ K (5CB), and $T_{CN} = 316.5$ K and $T_{NI} = 328.1$ K (CPHB). In dielectric measurements the samples were porous glass plates, of dimension $2 \times 2 \times 0.1$ cm^3. They were heated to 450°C and pumped out; this was followed by impregnation with the LCs from an isotropic melt. In dynamic light scattering experiments we used microporous glass plates of dimensions $1 \times 1 \times 0.2$ cm^3, which were optically transparent. In the case of macroporous matrices, which are opaque, in order to reduce the contribution from multiple scattering, the thickness of the samples was 0.2 mm. All surfaces of the matrices were optically polished.

The polymers were synthesized directly in the pores at the Institute of Macromolecular Compounds, Academy of Sciences of Russia, St Petersburg, and kindly provided by L. V. Zamoiskaya and V. N. Zgonnik. All polymers were synthesized in evacuated matrices by free-radical polymerization of the monomers at 30°C for 4 days with AIBN as the initiator at a concentration of 0.1 wt.%. The molecular weights of external PMMA, PEMA, PBMA and PS were 4.9×10^6, 5×10^6, 6.9×10^6 and 6×10^5, respectively. Other polymers were partially cross-linked and did not dissolve in the free state, so it was impossible to determine their molecular weights.

17.3 Phase Transitions and Structure of FLC in Pores

In investigating phase transitions in finite systems, fundamental difficulties arise in that the transition region occupies a wide temperature interval and it is unclear, in general, what should be regarded as the phase transition temperature. The set of methods of X-ray scattering, DSC and dielectric spectroscopy we used to study FLCs in pores permits adequate judgements regarding the phase transition temperature. The main purpose of this chapter is to study dielectric relaxation in confined FLCs, and we use X-ray and DSC as additional methods to establish the existence of layered structures (X-ray scattering) and the phase transition (X-ray and DSC). Using X-ray scattering it was found that at temperatures below 52°C, the distance d_p between the layer spacing of FLC in pores depends on temperature and increases from

$d_p \approx 28.3$ Å at 25 °C up to $d_p \approx 29.7$ Å at ≈ 52 °C. In the temperature range 52 °C $\leqslant T \leqslant 80$ °C, $d_p \approx 29.7$ Å and is constant. For the bulk FLCs in the temperature range 25 °C, d_b varies within the limits 28.1 Å $\leqslant d_b \leqslant 31.2$ Å and is characteristic of the SmC* phase. In the temperature range 67 °C $\leqslant T \leqslant 80$ °C, $d_b = 31.2$ Å and corresponds to the SmA* phase. For FLCs confined in pores there is not enough information to make a conclusion about the types of smectic phases: are they SmC* and SmA* or SmC and SmA? Therefore, below we will call smectic phases in pores simply SmC and SmA.

DSC measurements show a bump covering a broad temperature region 42 °C $\leqslant T \leqslant 52$ °C in the plot of heat flow against temperature. This bump is connected with a smeared phase transition and a temperature of about 50 °C could be attributed to the phase transition from SmC to SmA. This transition temperature shows a significant shift ($\Delta T \approx 15$ °C from the bulk transition temperature). Since the investigated FLC is a mixture (Li *et al.*, 1990, 1991) there are two possible explanations for this shift: first, preferential adsorption of one component of the mixture to the matrix which can result in a change in the various concentrations and hence alter the transition temperature; and second, the finite size effects due to the confining geometry tend to lower the transition temperature. These mechanisms for shifting the transition temperature hold for all of the transitions: smectic A–nematic, and nematic–isotropic. One can estimate (Iannacchione *et al.*, 1993; Iannacchione and Finotello, 1994; Finotello *et al.*, this volume) the expected shift in the nematic–isotropic transition due to finite size effects. We find this expected shift to be approximately 0.2 °C and consistent with the shift we measure, assuming that changes in the concentration play a smaller role. Because of the nature of the SmC–SmA transition, one expects the confining geometry (its curvature and so forth) to have an even greater effect than it does on the N–I transition.

From the X-ray scattering measurements we find a layered structure with a distance between layer spacing of $d_p = 30.5$ Å is formed in micropores. This structure is thermally stable and does not change even at temperatures corresponding to the bulk isotropic phase. DSC measurements show that there is no evidence of the existence of phase transitions for FLCs in micropores.

17.4 Dielectric Properties of FLC in Pores

Having identified the existence of layered structures of SmC and SmA types and a very broad phase transition between these phases in pores we turn our attention to study the dynamics of the Goldstone and the soft modes by dielectric spectroscopy. Measurements of the real ε' and imaginary ε'' parts of the complex dielectric permittivity in a frequency range of 1 Hz to 5 MHz at different temperatures were carried out using a computer-controlled Schlumberger Technologies 1260 impedance/gain-phase analyser. The quantities measured directly were the permittivities ε'_{syst} and the dielectric loss factors ε''_{syst} of the two-phase heterogeneous systems comprising of a matrix and a liquid crystal. The determination of the permittivity of the disperse phase (liquid crystal) calls for a theory allowing for anisotropy and local inhomogeneity of liquid crystal dispersed in a matrix and valid at high concentrations of liquid crystal. Such a theory is not known to us, so we had to calculate the permittivity of the second phase (liquid crystal) using Bottcher's theory (Van Beek, 1967):

$$\varepsilon_{syst} = \varepsilon_m + 3\omega\varepsilon_{syst} \frac{\varepsilon - \varepsilon_m}{2\varepsilon_{syst} + \varepsilon} \tag{17.1}$$

where ε_m and ε are the dielectric constants of the matrix material and the liquid crystal; ω is the volume fraction of pores. A satisfactory theory may first of all alter the absolute values of dielecric permittivity of the liquid crystal but it should have less effect on the nature of the temperature and frequency dependencies which are discussed below. In accordance with our expectations for FLCs confined in macropores we did find two regions of dielectric dispersion.

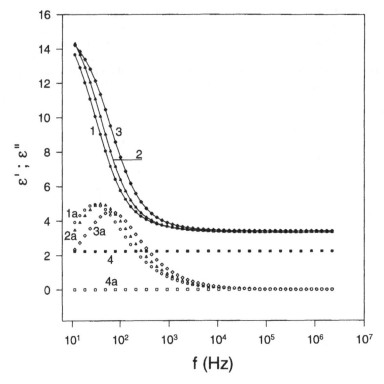

Figure 17.1 Frequency dependencies of ε' (1, 2, 3) and ε'' (1a, 2a, 3a) of FLC in macropores due to the Goldstone mode at different temperatures: (1, 1a), 21 °C; (2, 2a), 25 °C; (3, 3a), 35 °C; 4 and 4a, ε' and ε'' of empty macroporous glass.

The first (low-frequency) region we attribute to the Goldstone mode and the second to a soft mode. Plotting ε'' as a function of ε' (the Cole–Cole diagrams) enables us to separate the dispersion regions and determine the values of the permittivities which in the second dispersion region were static and at the same time corresponded to the 'high-frequency' plateau for the first dispersion region. The Cole–Cole diagrams indicate that for LCs in pores the parameter α, which represents empirically the spectrum of the relaxation times, varies from 0.2 to 0.35, and hence there is a spectrum of relaxation times. It should be noted that in dielectric investigations of bulk FLC at temperatures far from the SmC*–SmA* phase transition temperature the relaxation process follows Debye-type behaviour ($\alpha = 0$), but close to the phase transition temperature α slightly deviates from zero. It was suggested (Gouda *et al.*, 1991) that this increasing of α may be connected with the divergence and slowing down of the fluctuations typical of second-order phase transition point. The Cole–Cole plot was performed for all the samples and permits the use of the Debye equation for complex permittivity ε^*, modified by Cole and Cole (see for example Scaife, 1989). According to Cole and Cole the frequency dependence of the complex dielectric permittivity of the system which has more than one relaxational process is described by the equation

$$\varepsilon^* = \varepsilon_\infty + \sum_{j=1} (\varepsilon_{js} - \varepsilon_\infty)/(1 + i2\pi f \tau_j)^{1-\alpha_j} - i\sigma/2\pi\varepsilon_0 f^n \tag{17.2}$$

where ε_∞ is the high-frequency limit of the permittivity, ε_{js} the low-frequency limit, τ_j the mean relaxation time, and j the number of the relaxational process. The term $i\sigma/2\pi\varepsilon_0 f^n$ takes into account the contribution of the conductivity σ and n is a fitting parameter ($n \simeq 1$). In our

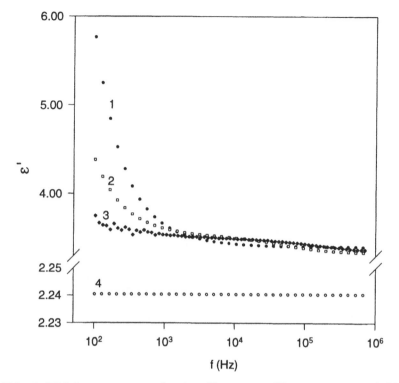

Figure 17.2 ε' of FLCs in macropores as a function of frequency at different temperatures: 1, 21 °C; 2, 50 °C; 3, 70 °C; 4, empty macroporous glass.

experiments we obtain (for frequencies $f < 1$ kHz) $0.9 \leqslant n \leqslant 0.96$ and 8×10^{-10} $(\Omega \text{ m})^{-1} \leqslant \sigma \leqslant 2 \times 10^{-9}$ $(\Omega \text{ m})^{-1}$ for all temperature ranges and for both macropores and micropores. The conductivities of the matrix material (σ_m) and empty matrix are negligibly small compared with the obtained conductivity for the matrix–liquid crystal system, and both ε'_m and ε''_m do not depend on frequency.

A graphical fitting enables us to determine ε_{js} and τ_j. Thus, using this fitting procedure and eqn (17.2) we have determined quasi-static ε_{As} and high-frequency ε_∞ values for the soft mode, and the temperature dependence of the soft mode dielectric strength $\Delta\varepsilon_A = \varepsilon_{As} - \varepsilon_\infty$.

As an example of low-frequency relaxation, Figure 17.1 shows the real and imaginary parts (the contribution of conductivity is taken into account) of permittivity against frequency in the SmC phase. The positions of the low-frequency peaks of $\varepsilon''(f)$ as well as the positions of the inflection points of the $\varepsilon'(f)$ curves increase in frequency as temperature increases. The contributon to the dielectric constant from the soft mode is negligible in the SmC phase, and thus this relaxation can be attributed to the Goldstone mode (we do not consider here the polarization modes which are beyond our experimental range). At first sight, one might be tempted to attribute low-frequency dispersion to the Maxwell–Wagner mechanism, which arises in heterogeneous materials: for a mixture of two or more components the accumulation of charges at the interfaces between phases gives rise to polarization which contributes to relaxation if at least one component has non-zero electric conductivity. This phenomenon is known as the Maxwell–Wagner (M–W) effect (Van Beek, 1967) (the M–W equation for the relaxation time of the interfacial polarization for spherical particles is given below). These two possibilities (the Goldstone mode or the Maxwell–Wagner mechanism) can be distinguished

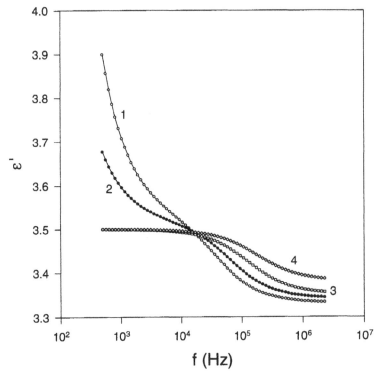

Figure 17.3 Dispersion curves ε' due to the soft mode as a function of frequency at different temperatures: 1, 45 °C; 2, 55 °C; 3, 70 °C; 4, 80 °C.

by studyng the temperature dependence of the dielectric permittivity. The low-frequency contribution to the dielectric permittivity decreases with increasing temperature and at a temperature of about 50 °C is much smaller than in the temperature region below about 40 °C (see Figure 17.2). This behaviour is consistent with the interpretation of the feature as the Goldstone mode (which vanishes at the critical temperature) and would be inconsistent with its interpretation as resulting from the M–W mechanism. The contribution from the M–W mechanism should exist at all temperatures. It is characterized by a single relaxation time, and the corresponding relaxation time should not strongly depend on the temperature. Possibly at the low frequencies we observe the overlapping of the Goldstone mode and the relaxation of interfacial polarization, not of M–W origin but rather due to the formation of the surface layer on the pore wall with the relaxation mechanism which does not exist in the bulk liquid crystal.

The overlapping of the Goldstone mode and another low-frequency process does not make possible the quantitative analysis of the Goldstone mode, and the data in Figures 17.1 and 17.2 are qualitative indications of the existence of the Goldstone mode in macropores connected with tilted layered structure identified from X-ray measurements. But it should be noted that the characteristic frequencies determining the relaxation times of the slow process of FLC in pores are in a frequency range much lower than is predicted for the Goldstone mode by the theory of the bulk FLC (Blinc and Zĕkš, 1978; Martinot-Lagard and Durand, 1981).

At the same time, when temperature approaches about 50 °C the contribution from the second relaxation process, corresponding to the soft mode, increases, and at the temperature above the phase transition temperature determined from X-ray scattering and DSC measurements this mechanism dominates. Again, like in X-ray scattering and DSC measurements, dielectric measurements do not show the existence of a sharp phase transition between the

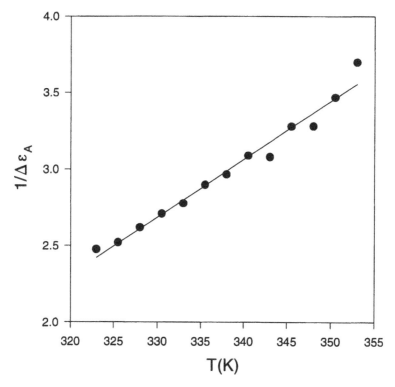

Figure 17.4 Temperature dependence of the reciprocal of the soft mode dielectric strength.

SmC and SmA phases and the Goldstone mode contribution almost completely vanishes (Figure 17.3) at high enough temperatures. The expressions determining the temperature dependencies of the soft mode dielectric strength $\Delta\varepsilon_A$ and the relaxation time τ_A of the bulk SmA* phase are well known (Blinc and Žekš, 1978; Martinot-Lagard and Durand, 1981)

$$(\Delta\varepsilon_A)^{-1} = (a/\varepsilon^2 C^2)(T - T_c) + Kq^2/\varepsilon^2 C^2 \tag{17.3}$$

$$\tau_A^{-1} = a(T - T_C)/\eta_A + Kq^2/\eta_A \tag{17.4}$$

In these equations a is the coefficient contained in the temperature-dependent term of the Landau expansion of the free-energy density, C is the linear coupling between polarization and tilt angle, ε is the dielectric constant of the system in the high-frequency limit, $K = K_3 - \varepsilon\mu^2$, where K_3 is the bend elastic constant, and μ is the flexoelectric coupling constant, For the bulk FLC, with a helix pitch p, q is the wavevector of the pitch: $q = 2\pi/p$. For the liquid crystal confined in pores, distortions of FLC can exist, and it is natural to assume that in this case q corresponds to the wavevector of the distortions, characterized by linear size $\approx$ pore size. The results show that the temperature dependencies of $1/\Delta\varepsilon_A$ (Figure 17.4) and $1/\tau_A$ (Figure 17.5) are linear in accordance with eqns (17.3) and (17.4). The dielectric strength and the relaxation time of the soft mode are related (through eqns (17.3) and (17.4)) to each other by (Gouda *et al.*, 1991; Levstik *et al.*, 1991)

$$\Delta\varepsilon_A \tau_A^{-1} = \varepsilon^2 C^2/\eta_A \tag{17.5}$$

and this relation makes it possible to determine the rotational soft mode viscosity. The slope of the experimental $\Delta\varepsilon_A^{-1}(T)$ dependence, $d(1/\Delta\varepsilon_A)/dT = a/\varepsilon^2 C^2$, is equal to about 0.04 K^{-1}

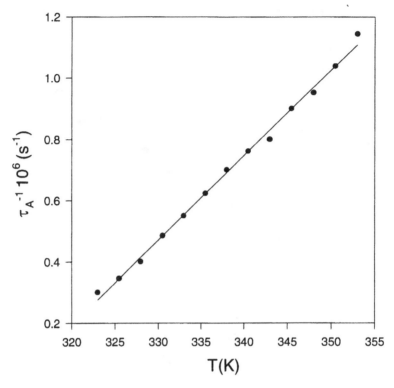

Figure 17.5 Temperature dependence of the reciprocal of the soft mode relaxation time.

(Figure 17.4) and assigning a typical value of $a = 3.5 \times 10^4$ N/m^2 K we can estimate the rotational soft mode viscosity. The temperature dependence of η_A, calculated using eqn (17.5) and experimental data on the temperature dependence of $(\Delta\varepsilon_A)^{-1}$ (Figure 17.4) and τ_A^{-1} (Figure 17.5), is presented in Figure 17.6 together with the temperature dependence of $\tau_A/\Delta\varepsilon_A$ which is proportional to η_A. This viscosity varies within the limits 3 N s/m$^2 \leqslant \eta_A \leqslant 8$ N s/m^2 in the temperature range $50\,°C \leqslant T \leqslant 80\,°C$, and its temperature dependence obeys the Arrhenius equation

$$\tau = \tau_0 \exp(U/RT) \tag{17.6}$$

with activation energy $U_A = 0.31$ eV. Note that estimated values of viscosity are about 8–10 times higher than the typical corresponding viscosities of the bulk FLC.

It is possible to estimate q from the formulas (17.3) and (17.4) at $T = T_c$ using the measured values of $\Delta\varepsilon_A$, τ_A and assigning a typical value of $K = 3.4 \times 10^{-12}$ N. It is surprising that both formulae give approximately the same value of $q \approx 7.4 \times 10^{10}$ m^{-1}. This value corresponds to the linear size $l \approx 85$ Å and is close to the thickness of three molecular layers. This result is at first sight surprising as one expects the natural length scale to be of the pore size. One possible explanation for this result could be the existence of a modulated structure which consists of periodically bent smectic layers with conserved local symmetry. In our system this structure may arise from the surface-induced deformations. The characteristic wavelength of this structure (Clark and Meyer, 1973; Ribotta *et al.*, 1973) can be estimated roughly as

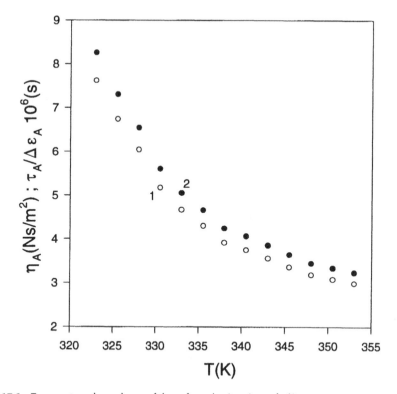

Figure 17.6 Temperature dependence of the soft mode viscosity and $\tau/\Delta\varepsilon_A$.

$\lambda \sim (\pi d\xi)^{1/2}$, where d is the pore size and $\xi \approx 5$–10 Å is the penetration length. The result is $\lambda \approx 10^2$ Å. It should be noted the calculation of this smaller length scale ($l \approx 85$ Å) arises from a theory that does not contain any contributions from the surface.

In micropores the low-frequency dispersion of the dielectric permittivity is observed at all temperatures under investigation (Figure 17.7) and does not vanish even at temperatures corresponding to the bulk isotropic phase. The relaxation process is not described by single relaxation time. The characteristic relaxation times are temperature dependent and this dependence obeys the Arrhenius relation with activation energy $U = 0.58$ eV. The soft mode was not detected in dielectric measurements. These facts suggest that the observed low-frequency relaxation is not connected with the Goldstone mode.

On the other hand it is very difficult to attribute the observed low-frequency relaxation process to the M–W relaxation. In fact, since $\sigma_m \ll \sigma$, the relaxation time due to the M–W effect is given by (Van Beek, 1967)

$$\tau_{MW} = 2\pi\varepsilon_0 \frac{2\varepsilon_m + \varepsilon + \omega(\varepsilon_m - \varepsilon)}{\sigma(1 - \omega)}$$

and the estimation for τ_{MW} is $\tau_{MW} \geq 0.1$ s which is slower than the measured τ. This is also correct for the behaviour of FLCs in macropores which was discussed above. Possibly at low frequencies we observe the relaxation of interfacial polarization not of M–W origin but rather due to the formation of a surface layer on the pore wall.

355

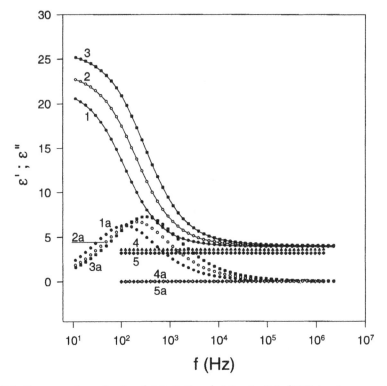

Figure 17.7 Frequency dependencies of ε' (1, 2, 3) and ε'' (1a, 2a, 3a) of FLCs in micropores at different temperatures: (1, 1a), 78 °C; (2, 2a), 87 °C; (3, 3a), 96 °C; 4 and 4a, ε' and ε'' of empty microporous glass; 5 and 5a, ε' and ε'' for CCl$_4$ in micropores at 20 °C.

17.5 Dynamic Properties of Nematic Liquid Crystals in Pores

17.5.1 *Dielectric Spectroscopy*

Dielectric properties of NLCs were studied in the frequency range 100 Hz to 30 MHz. The dependencies of ε on the frequency of the external electric field are shown in Figure 17.8 for 5CB in micropores (curve 1) and in macropores (curve 2) and for CPHB in micropores (curve 3). It is evident that all the curves of $\varepsilon(f)$ are relaxational in character and exhibit two regions of the permittivity dispersion. The low-frequency (first) region of the dispersion is not present in the bulk NLCs.

Again, as in the investigations of FLCs, using the Cole–Cole diagrams and fitting procedure we were able to separate the dispersion regions and determine the values of the permittivities which in the second dispersion region were 'static' and at the same time corresponded to the 'high-frequency' plateau for the first dispersion region. These values are identified by the dotted lines in Figure 17.8. The Cole–Cole diagrams indicate that for LCs in pores the parameter α, which represents empirically the spectrum of the relaxation times, varies from 0.2 to 0.35. For the bulk nematic phase, the Cole–Cole parameter is equal to zero. For LCs inside the pores, α is not equal to zero, and there is then a spectrum of relaxation times. This is due to the fact that the properties of the surface layers begin to vary at distances of the order of molecular dimensions, and depending on the distance from the surface of a pore, the various layers of LC can have different relaxation times.

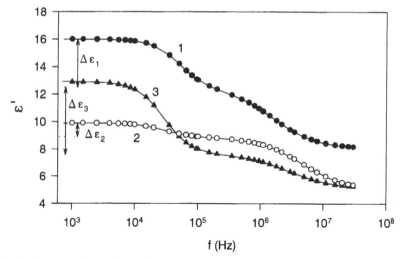

Figure 17.8 Frequency dependence of ε' at $T = 290$ K: (1) 5CB in micropores; (2) 5CB in macropores; (3) CPHB in micropores. The dotted horizontal lines are the values of ε_s for the second dispersion region.

The main difference between the behaviour of the permittivity of LC in pores and their behaviour in the bulk is the existence of a low-frequency dispersion region, absent in the free state and characterized by the relaxation time $\tau \sim 10^{-6}$ s, which could not be attributed to the orientational motion of the molecules, but represents some collective process. The relaxation time τ_2 for the second region is close to τ of the bulk LC, but the temperature dependence of τ_2 is weaker in both micropores and macropores. The values of τ_2 and their temperature dependencies in macropores are close to the bulk behaviour than that observed in micropoes, in agreement with expectations.

All these properties and the temperature hysteresis of ε' measured at $f = 1$ kHz (Aliev and Breganov, 1989; Aliev, 1994) can be explained using the assumption that the wall induces in an NLC polar order of smectic type. In fact, it is known that the $\varepsilon'(T)$ dependence for bulk LCs in the smectic A and smectic C phases exhibits a hysteresis, and the temperature dependence of the relaxation times governing the rotational mobility of molecules around the short axis becomes weaker than in the nematic phase (Druon and Wacrenier, 1982).

We found that the dependence of $\ln\tau_2$ on T^{-1} is linear and is described by the relation (17.6). The activation energy is equal, in barrier theories, to the difference in potential energy of the stable orientations $\theta = 0$ or π and the highest potential energy of intermediate orientation ($\theta = \pi/2$) (where θ is the angle between the director and the dipole). There is no polar ordering in the bulk; therefore, the orientations $\theta = 0$ and $\theta = \pi$ are equally probable. The values of the activation energies of the investigated NLCs in the bulk were found to be $U_{1f} = 8.8 \times 10^{-13}$ erg for 5CB and $U_{2f} = 14 \times 10^{-13}$ erg for CPHB.

The interaction of NLC molecules with the surface (characterized by the interaction energy U_0) is equivalent to the interaction with an external field producing a polar order close to the surface, stabilizing one orientation ($\theta = 0$) and correspondingly changing U_f to $U_f + U_0$. For $\theta = \pi$ the potential becomes $U = U_f - U_0$ and corresponds to the activation energy (Druon and Wacrenier, 1982).

The activation energies U_i, corresponding to the dependences $\tau_2(T)$ in micropores, are $U_1 = 1.7 \times 10^{-13}$ erg (5CB) and $U_2 = 8.5 \times 10^{-13}$ erg (CPEH). A comparison of U_f and U_{0i} gave the energy of the interaction of molecules with the surfaces of the pores, $U_{0i} = U_f - U_i$, and these values were $U_{01} = 7.1 \times 10^{-13}$ erg (5CB) and $U_{02} = 5.5 \times 10^{-13}$ erg (CPHB). These estimates are only qualitative; it would be reasonable to assume that $U_0 \approx 5 \times 10^{-13}$ erg.

Taking into account that the number of molecules per square centimetre is $n_s \approx (2–3) \times 10^{14}$, we found that the surface energy of the nematic liquid crystal $F_s = U_0 n_s$ should be $F_s \sim 10^2$ erg/cm^2.

Another argument in support of the existence of polar ordering at the pore wall is the low-frequency dispersion. It is natural to assume that the ratio of the low-frequency increments $\Delta\varepsilon_2'$ and $\Delta\varepsilon_1'$ corresponding to curves 2 and 1 in Figure 17.8 is proportional to the volume fraction g of the polar ordered surface layer (thickness l) in the macropores, if all the molecules in the micropores belong to this layer. Then, if the macropore is modelled by a cylinder of radius R, $\Delta\varepsilon_2/\Delta\varepsilon_1 \approx 1 - (R - l)^2/R^2$. Using the experimental values $\Delta\varepsilon_1 = 3.5$ and $\Delta\varepsilon_2 = 1$ determined at the same temperature and $R \approx 500$ Å, we find that $l \approx 75$ Å. The ratio of $\Delta\varepsilon_3$ to $\Delta\varepsilon_1$ is equal to 1.65. On the other hand $\Delta\varepsilon_3/\Delta\varepsilon_1 \simeq \mu_3^2/\mu_1^2 = 1.65$. Thus the differene between the dynamics of dielectrically active modes corresponding to the orientational motion of the polar molecules of NLCs in confined geometries and in the bulk can be qualitatively explained by assuming the formation of a surface layer with polar ordering on the solid–LC interface (the thickness of this layer is about 10^2 Å).

17.5.2 *Dynamic Light Scattering*

In the above consideration of the dynamic properties of liquid crystals confined in porous glasses with randomly oriented pores, studied by dielectric spectroscopy, we did not use the concept of the random field Ising model (Brochard and de Gennes, 1983; de Gennes, 1984; for details see also the chapters by Maritan *et al.* and Cleaver *et al.* in this book). These results are described by taking into account interfacial phenomena arising at the pore wall–liquid crystal interface and ordering effects of surface interactions. At the moment of completion of this chapter only two original papers (Wu *et al.*, 1992; Bellini *et al.*, 1995) on the investigations of confined liquid crystals by dynamic light scattering had been published. In studies of the nematic ordering of the liquid crystal 8CB in sintered porous silica (Wu *et al.*, 1992) it was found that the liquid crystal shows orientational glass-like dynamics near the nematic–isotropic phase transition. It was shown (Goldburg *et al.*, 1995; Wu *et al.*, 1992) that although a proper theory that can explain the dynamics of liquid crystals in porous media (dynamic light scattering) is still lacking, some features of dynamic behaviour of these systems can be explained on the basis of the model in which the porous medium imposes a random uniaxial field on the liquid crystal. In the most recent investigation (Bellini *et al.*, 1995) of the dynamic properties of 8CB in an aerogel host by dynamic light scattering, the observed dynamic behaviour was different from this in sintered porous silica (Wu *et al.*, 1992). Nevertheless according to Bellini *et al.* (1995) the spin-glass interpretation given in Bellini *et al.* (1995) and the random field interpretation given by Wu *et al.* (1992) are consistent if the geometrical differences between two matrices are taken into account. The matrices used in both the Wu *et al.* and Bellini *et al.* experiments had a mean pore size of 200 Å and wide pore size distribution.

Our 5CB–microporous glass samples are optically transparent, so that there is no multiple scattering for the samples with thickness 5 mm, and they are very convenient for optic investigations. We performed dynamic light scattering measurements using a 6328 Å He–Ne laser and the ALV-5000/Fast Digital Multiple Tau Correlator operating over delay times from 12.5 ns up to 10^4 s with the Thorn EMI 9130/100B03 photomultiplier and the ALV preamplifier. The depolarized component of scattered light at a scattering angle of $\Theta = 30°$ was investigated. Observation of the depolarized component of the scattered light makes it possible in the isotropic phase of bulk liquid crystal to detect the contribution with order parameter fluctuations only, and for liquid crystals in pores blocks out the scattering from the fixed matrix structure. The difference in the behaviour of isotropic bulk 5CB, bulk nematic multidomain 5CB and 5CB in macropores as well as in micropores can be seen by comparing Figure 17.9 and 17.10. In the dynamic light scattering experiment, one measures the intensity–intensity autocorrelation function $g_2(t) = \langle I(t)I(0)\rangle/\langle I(0)\rangle^2$. The plots in Figures 17.9 and 17.10 represent the autocorrelation function $G_2(t) = g_2(t) - 1$. In an isotropic phase of

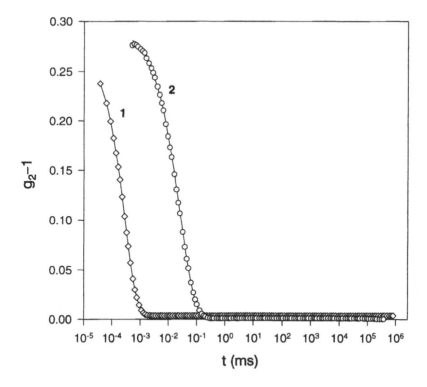

Figure 17.9 Intensity–intensity autocorrelation functions for the bulk 5CB: 1, isotropic phase, 308.33 K; 2, nematic phase, 295.82 K (full lines – fitting).

the bulk liquid crystal the intensity–intensity autocorrelation function of the depolarized component of scattered light is determined by order parameter fluctuations and the corresponding decay is single exponential. The experimental data for the isotropic phase of 5CB ($T = 308.33\ ^\circ$C) are perfectly fitted (Figure 17.9, curve 1) by a single-exponential decay function ($G_2(t) = g_1^2(t)$, $g_1(t) = a\ \exp(-t/\tau)$) with the relaxation time $\tau = 5.59 \times 10^{-7}$ s. In the nematic phase (Figure 17.9, curve 2) the relaxation process due to the director fluctuations is described by the macroscopic equations of nematodynamics (Groupe Orsay, 1969), and the relaxation time in the dynamic light scattering experiment is determined by the viscoelastic properties of nematic liquid crystals, the geometry of an experiment and light polarization. Since the investigation of the dynamics of the bulk liquid crystal is not our purpose and corresponding data are presented only in order to stress the difference between dynamics in pores and in the bulk, we do not need a rigorous consideration of the dynamic light scattering in the nematic phase, which is given by Groupe Orsay (1969). If, according to Groupe Orsay, we assume for simplicity that six Leslie coefficients have the same order of magnitude and are approximately η (η is an average viscosity), and three elastic constants are equal (K), then the relaxation time $\tau = \eta/Kq^2$ ($q = 4\pi n\ \sin(\Theta/2)/\lambda$, where n is the refractive index). The relaxation time $\tau = 5.9 \times 10^{-5}$ s which corresponds to curve 2 (Figure 17.9) is in agreement with the theory (Groupe Orsay, 1969).

The relaxation processes of 5CB in both microporous and macroporous matrices (Figure 17.10) are highly non-exponential and the correlation functions of confined 5CB are in contrast to G_2 of the bulk 5CB (the data for 5CB in macropores are given only for illustration since there was a contribution from multiple scattering in this sample). We are not able to find the correlation function (or superposition of correlation functions) known from previous publica-

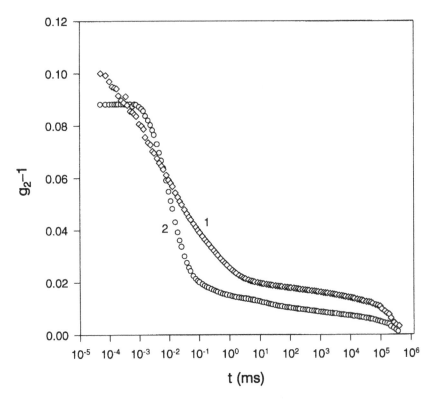

Figure 17.10 Intensity–intensity autocorrelation functions for 5CB in pores: 1, 5CB in 100 Å pores, 295.84 K; 2, 5CB in 1000 Å pores, 294.77 K.

tions which would satisfactorily describe the whole experimental data from $t = 10^{-4}$ ms up to $t = 10^5$ ms. However, we found that in the time interval 10^{-3}–10^3 ms (six decades on the time scale) and the temperature range 290–303 K the autocorrelation function

$$g_1(t) = a \exp(-x^z) \tag{17.7}$$

where $x = \ln(t/\tau_0)/\ln(\tau/\tau_0)$, and in our case $\tau_0 = 10^{-5}$ ms provides the best fitting for 5CB in micropores (Figure 17.11) compared with other conventional decay functions. In the expression (17.7) $z = 3$ corresponds to the activated scaling theory for random field systems with conserved order parameter (Huse, 1987) (which is not our case since liquid crystals are systems with no conserved order parameter). In the theory of Randieria *et al.* (1985) for d-dimensional short-range Ising spin glasses z is determined by the dimensionality of the system, $z = d/(d - 1)$, and correlation function (17.7) describes the slow relaxation of large isolated clusters above the spin-glass transition. Randieria *et al.* suggested that this is the signature for an intermediate Griffiths phase between the spin-glass and the paramagnetic phases.

Examples of fitting the data for 5CB in micropores by function (17.7) are presented in Figures 17.11 and 17.12. The parameter z was about 2, i.e. it was 2.3 for low temperatures and 1.7 for temperatures close to 303 K. The correlation functions for 5CB in micropores corresponding to different temperatures are presented in Figure 17.12. The temperature dependence of the relaxation times obtained using correlation function (17.7) obeys the Vogel–Fulcher expression (see Williams, 1991)

$$\tau = \tau_0 \exp[B/(T - T_0)] \tag{17.8}$$

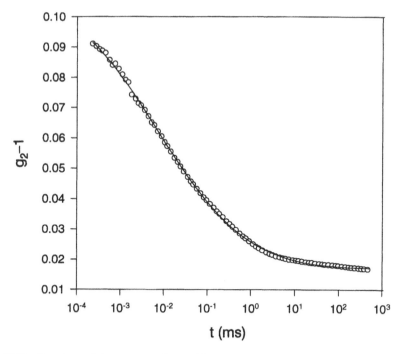

Figure 17.11 Experimental data (open circles) and fitting (full line) according to eqn (17.7) for 5CB in 100 Å pores, 295.84 K.

which is the characteristic of glass-like behaviour. The parameters for 5CB in micropores are $\tau_0 = 3 \times 10^{-9}$ s, $B = 560$ K and $T_0 = 253$ K. We found that the autocorrelation functions describing our data, as in the experiments by Wu *et al.* (1992) (see also Aliev *et al.*, 1993; Goldburg *et al.*, 1995) obey activated dynamical scaling with the scaling variable $x = \ln t/\ln \tau$, but experimental correlation functions are not described by the scaling function in the form $g_1(t) = 1/(1 + x^n)$ with $n = 2$ for liquid crystals in pores (Wu *et al.*, 1992) or $n = 3$ for binary liquid mixtures in pores (Aliev *et al.*, 1993; Goldburg *et al.*, 1995). Our data cannot be more or less satisfactorily described using a standard form of dynamical scaling variable (t/τ) and stretched exponential form of the correlation function as was done for the slow nematic relaxation observed in Bellini *et al.* (1995). Since the microporous glass used in our experiments has a narrow pore size distribution the observed wide spectrum of relaxation times cannot be attributed to the wide pore size distribution, which was characteristic of porous matrices in Bellini *et al.* (1995). It should be mentioned that in the investigations of the orientational dynamics of 5CB confined in nanometre-length-scale porous silica glass by the time-resolved transient grating optical Kerr effect (Schwalb *et al.*, 1994; Schwalb and Deeg, 1995) in the temperature range corresponding to the bulk deep isotropic phase, non-exponential relaxation and the distribution of relaxation times were observed. The pore size dependence of relaxation times was explained from the point of view of the Landau model based on independent pore segments. However, this model and the distribution of the pore sizes do not describe the non-exponentiality of the decay. In the opinion of Schwalb and Deeg (1995) this suggests that it is necessary to include consideration of the interporous interaction in order to describe the dynamics satisfactorily. The results of dynamic light scattering (Bellini *et al.*, 1995; Goldburg *et al.*, 1995; Wu *et al.*, 1992), dielectric (Aliev and Breganov, 1989; Aliev and Kelly, 1994) and NMR (Crawford *et al.*, 1993; Kralj *et al.*, 1993; Vrbančič *et al.*, 1993) investigations show that the dynamics of liquid crystals in confined geometries is an extremely fertile field to study.

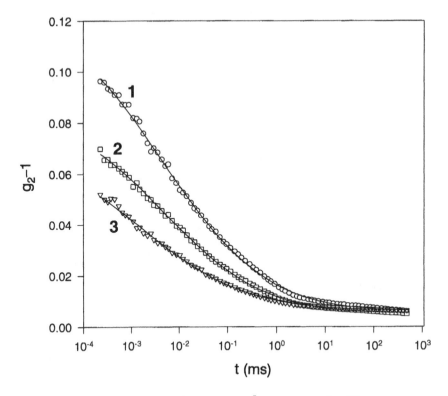

Figure 17.12 Autocorrelation functions for 5CB in 100 Å pores measured at different temperatures; 1, $T = 296.08$ K; 2, $T = 297.17$ K; 3, $T = 299.506$ K. Full lines show fitting using the correlation function according to eqn (17.8).

17.6 Physical Properties of Polymers in Pores

In the investigations of light transmission through porous glass plates, the pores of which contain poly(alkyl methacrylates) with a long side group (PEHA, POMA, PDMA), anomalous changes in the light transmission in a relatively narrow temperature (T) range were observed. Figure 17.13 shows the temperature dependencies of the transmission coefficients (A) measured at the wavelength $\lambda = 550$ nm for POMA in macroporous glass.

It is clear that the light transmission for POMA increases by two orders of magnitude when the temperature increases from 65 to 80 °C and the transmission decreases to the initial value with decreasing T. Since the light absorption by the matrix and the polymers is absent at $\lambda = 550$ nm, the observed dependence $A(T)$ is caused by the change in intensity of the scattered light with T. This behaviour of A is characteristic of polymers with sufficiently long side groups (PEHA, POMA, PDMA) and cannot be explained by differences in the temperature dependencies of the refractive indices of a silicate matrix ($\mathrm{d}n_1/\mathrm{d}T \approx 10^{-6}\ °\mathrm{C}^{-1}$) and polymers filling the pores ($\mathrm{d}n_2/\mathrm{d}T \approx 10^{-4}\ °\mathrm{C}^{-1}$) as in two-component systems consisting of isotropic substances. This point can be seen by comparing the shapes of curves 1 and 2 in Figure 17.13 with the expression for the transmission coefficient, given by a formula valid for the temperature dependence of the transmission coefficient caused by refractive index mismatch:

$$A = \exp(-\kappa l) = \exp\{-\mathrm{constant}[n_1 - n_{02}(1 - \alpha_2 T)]^2 l\}$$

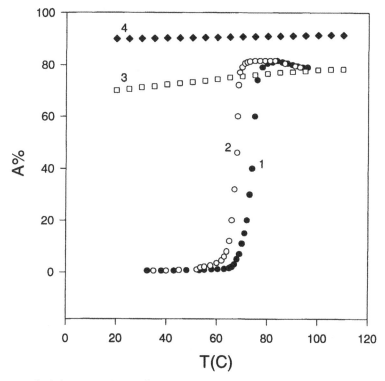

Figure 17.13 The light transmission coefficient against T: POMA in macropores, (1) heating, (2) cooling; (3) PEMA in macropores; (4) PEMA in micropores.

where n_1 is the refractive index of the matrix framework, n_{02} is the polymer refractive index at $T = 0\,°C$ and α_2 is the temperature coefficient of the polymer refractive index. The shapes of curves 1 and 2 are completely different from the behaviour due to refractive index mismatch. In contrast to PEHA, POMA and PDMA in systems containing PMMA, PEMA and PS, the change in transmission with varying temperature is determined by the temperature dependence of the refractive index of the polymer. The typical dependence of $A(T)$ for these systems is shown in Figure 17.13 (curve 3) for PEMA.

In microporous matrices the above effects are not observed. The typical temperature dependencies of the transmission coefficients for polymers in micropores are shown in Figure 17.13 for PEMA (curve 4). The spectral dependencies of the extinction coefficients for these samples obey the relationship $\sigma \sim \lambda^{-4}$ in accordance with the Rayleigh theory of light scattering. This result is natural because the characteristic size of the optical inhomogeneity in microporous matrices (equal to pore size) is much less than the wavelength of the light.

One possible explanation for the observed anomalous light scattering in the macropores with polymers having sufficiently long side groups may be the formation of the orientational ordering of these groups induced by an inner pore surface at low T. The assumption of the formation of orientationally ordered regions in PEHA, POMA and PDMA near the inner pore surface is in agreement with the molecular structure of these polymers and with that of the surface layer of silica. The surface of the pore wall contains a large number of SiOH groups. The molecules of poly(alkyl methacrylates) contain ester groups including the C=O bond. The oxygen atom of the carbonyl group of the macromolecule forms a hydrogen bond with the hydrogen atom of the silanol group, and the polymer chain should be arranged on the silica surface. Hence, relatively long side groups on the pore surface should be mutually oriented.

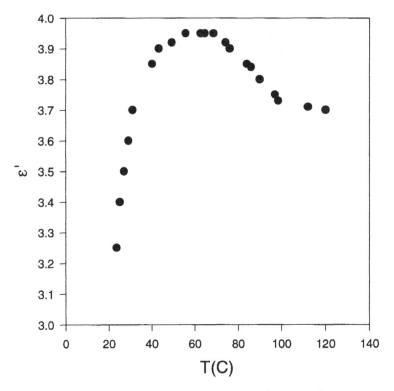

Figure 17.14 Temperature dependence of dielectric permittivity of POMA in the macropores ($f = 1$ kHz).

The orientation of the side groups of poly(octadecyl methacrylate) normal to the silica suface has been detected (Mumby *et al.*, 1985) by infrared spectroscopy in a monolayer of this polymer.

The character of the temperature dependence of the light transmission coefficient, and the temperature hysteresis, suggest that in the temperature range in which the transmission coefficient changes abruptly, a smeared-out first-order phase transition is observed. This suggestion is confirmed by DSC measurements.

Calorimetric measurements for POMA in the macropores show the endothermic and exothermic peaks upon heating and cooling. The latent heat corresponding to these peaks for POMA in micropores is approximately 3.5 J/g. Similar calorimetric behaviour is observed for PEHA and PDMA. The temperatures at which the endothermic and exothermic peaks appear coincide with those at which the samples change their transparency. The values of the latent heat obtained for our samples are typical for the weak first-order phase transition from the liquid crystalline to isotropic phase. The transparency of the system changes abruptly at this transition. The observed behaviour for comb-like polymers synthesized in macropores is similar to that of nematic liquid crystals at the nematic–isotropic phase transition.

The synthesized microcomposites with short side groups (PMMA, PEMA and PBMA) are transparent in the wavelength range 450–1000 nm, so that for sample thickness of 1 mm the light transmission coefficient is greater than 85 %. It is important to determine the temperatures T_d at which the thermal degradation starts because this value determines the maximum temperature at which transparency is retained. The temperature T_d governs the possibility of practial applications of the transparent composites. We found from thermogravimetric measurements that the values of T_d for PMMA, PEMA and PBMA in micropores are 292 and 272 °C respectively. For bulk polymers the values of T_d are 217 °C (PMMA), 217 °C (PEMA)

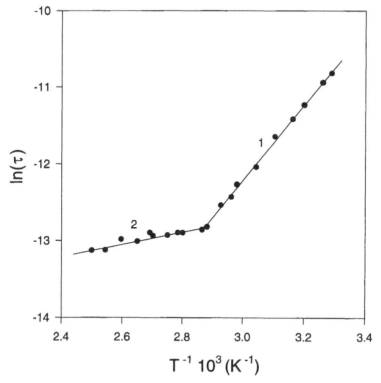

Figure 17.15 Temperature dependence of the dielectric relaxation times of POMA in the macropores; full lines 1, 2 correspond to the Arrhenius plot.

and 212 °C (PBMA). Thus the temperatures T_d are higher for the polymers in pores than for these polymers in the bulk by 75 °C. The factors that can increase thermal stability of the polymers in micropores are: the interaction between the macromolecules and the surface; the possibility of grafting molecular fragments onto the surface; the cage effect, which leads to considerable recombination of primary radicals formed in degradation; and the hindrance of the diffusion. These factors prevent chain free-radical degradation processes and increase thermal stability (Aliev and Zgonnik, 1991).

We performed dielectric measurements in addition to optical and calorimetric measurements in order to obtain information on the influence of confinement on the dynamics of POMA in the macropores.

The real part of the measured dielectric permittivity of the POMA–macroporous glass system has a frequency dependence in the wide frequency range $f = 500$ Hz to 2 MHz. The frequency dependence of the complex dielectric permittivity for POMA in the macropores was satisfactorily described by eqn (17.2) with the Cole–Cole parameter α varying from 0.4 to 0.6, which is characteristic of a wide spectrum of relaxation times. The temperature dependence of the volume-averaged real part of the permittivity of POMA in the macropores is presented in Figure 17.14. We found that there is a pronounced change in the temperature dependencies of ε and the relaxation times (Figure 17.15) in the temperature interval, which was interpreted as the phase transition temperature range from optical and DSC measurements. The plot in Figure 17.15 shows that the dependence of ln τ on $1/T$ falls reasonably well on two straight lines of different slopes, corresponding to different values of the activation energies in the Arrhenius equation (17.6). In the low-temperature region $T \leqslant 80$ °C, corresponding to the opaque state of POMA in the macropores, the activation energy is $U_1 = 40.5$ kJ/mol (line 1), and at $T \geq 80\,°C$ (transparent state) the activation energy is $U_2 = 7$ kJ/mol (line 2). The temperature dependence of the relaxation times presented in Figure 17.15 is similar to that observed previously (Wang *et al.*, 1983) in the investigations of reorientational motion of α-phenyl *o*-cresol in the normal liquid and supercooled liquid state, but the expression (17.8) does not fit the data as well as the Arrhenius expression (17.6) with two activation energies.

17.7 Conclusion

Investigations of FLC confined in pores show that the smectic C and smectic A phases are formed in 1000 Å pores, and the smectic C–smectic A phase transition temperature is reduced in macropores by about 15 °C. The Goldstone and the soft modes are found in macropores, and the rotational viscosity associated with the soft mode is about 10 times higher in pores than in the bulk. The layered structure is also formed in micropores. This structure is thermally stable and does not change even at temperatures corresponding to the bulk isotropic phase. No phase transition or critical phenomena, neither the soft nor the Goldstone modes, are observed for FLCs in micropores.

The difference between the dynamics of orientational motion of the polar molecules of NLCs in confined geometries and in the bulk can be qualitatively explained by assuming the formation of a surface layer with polar ordering on the solid–LC interface. The thickness of this layer is about 10^2 Å and is quantitatively determined by the total energy F_s of the interaction between molecules and the surface of the pore wall, which is found to be $F_s \approx 10^2$ erg/cm^2.

The combination of optical, thermophysical and dielectric data, and the available information on the molecular structure of the bulk polymer and the pore surface of silica porous matrices, make it possible to explain the properties of comb-like polymers in the macropores, by using an assumption about the surface-induced orientational order in relatively long, linear, aliphatic side groups. The transition into the state with a disordered arrangement of these groups is a first-order phase transition from the phase with orientational order to the isotropic phase. Poly(alkylmethacrylates) in microporous glasses form the microcompositions transparent in the optical wavelength range. The thermal stability of the polymers in the micropores is higher than that of polymers in the free state.

Although it is possible to explain (at least qualitatively) the physical properties of confined liquid crystals (and other complex fluids) using the concepts of the surface-induced ordering effects and the disordering effects due to surface-induced deformations, there are nevertheless many unsolved problems and further systematic investigations are needed. The slow dynamics detected by dynamic light scattering and the extremely wide spectrum of relaxation times still remain unexplained.

Acknowledgements

I have benefited from the discussions with M. Anisimov, L. Blinov, N. Clark, B. Doane, G. Durand, D. Finotello, W. I. Goldburg, E. Kats, J. Kelly, P. Palffy-Muhoray, C. Rosenblatt,, X.-L. Wu and V. Zgonnik. The investigations of FLCs were performed in collaboration with J. Kelly and L. Chen. I wish to thank V. Nadtotchi for assistance with the dynamic light scattering measurements.

This work was supported by the NSF Center for Advanced Liquid Crystalline Optical Materials grant DMR89-20147, EPSCoR-NSF grant EHR-9108775 and DOE-EPSCoR grant DE-FG02-94ER75764.

References

ALIEV, F. M. (1992) Optical properties of liquid crystals in porous glasses, *Mol. Cryst. Liq. Cryst.* **222**, 147–163.

ALIEV, F. M. (1994) Liquid crystals – porous glasses heterogenous systems as materials for investigation of interfacial properties and finite-size effects, *Mol. Cryst. Liq. Cryst.* **243**, 91–105.

ALIEV, F. M. and BREGANOV, M. N. (1988) Temperature hysteresis and dispersion of the dielectric constant of a nematic liquid crystal in micropores, *Sov. Phys. JETP Lett.* **47**, 117–120.

ALIEV, F. M. and BREGANOV, M. N. (1989) Dielectric polarization and dynamics of molecular motion of polar liquid crystals in micropores and macropores, *Sov. Phys. JETP* **68**, 70–79.

ALIEV, F. M. and KELLY, J. (1994) Dynamics, structure, and phase transitions of ferroelectric liquid crystal confined in a porous matrix, *Ferroelectrics* **151**, 263–268.

ALIEV, F. M. and POJIVILKO, K. S. (1988) Determination of the structural properties of porous glasses by small-angle X-ray diffractometry and electron microscopy, *Sov. Phys. Solid State* **30**, 1351–1354.

ALIEV, F. M. and POJIVILKO, K. S. (1989) Critical behavior of an interphase layer and the surface properties of a liquid crystal in micropores, *Sov. Phys. JETP Lett.* **49**, 308–312.

ALIEV, F. M. and ZGONNIK, V. N. (1991) Thermooptics and thermal stability of poly(alkyl methacrylates) in porous matrices, *Eur. Polym. J.* **37**, 969–973.

ALIEV, F. M., POJIVILKO, K. S. and ZGONNIK, V. N. (1990) SAXS and DSC studies of surface and size effects for poly(vinyl stearate), *Eur. Polym. J.* **26**, 101–104.

ALIEV, F. M., VERSHOVSKAYA, G. Yu. and ZUBKOV, L. A. (1991) Optical properties of the isotropic phase of a liquid crystal in pores, *Sov. Phys. JETP* **72**, 846–948.

ALIEV, F. M., GOLDBURG, W. I. and WU, X.-I. (1993) Concentration fluctuations of a binary liquid mixture in a macroporous glass, *Phys. Rev. E* **47**, R3834–R3837.

ARMITAGE, D. and PRICE, F. P. (1976) Size and surface effects on phase transitions, *Chem. Phys. Lett.* **44**, 305–308.

ARMITAGE, D. and PRICE, F. P. (1978a) Supercooling and nucleation in liquid crystals, *Mol. Cryst. Liq. Cryst.* **44**, 33–44.

ARMITAGE, D. and PRICE, F. P. (1978b) The differential scanning dilatometer and the phase transitions in liquid crystals, *J. Polym. Sci.: Polym. Symp.* **63**, 95–107.

AWSCHALOM, D. D. and WARNOCK, J. (1987) Supercooled liquids and solids in porous glass, *Phys. Rev. B* **35**, 6779–6785.

BELLINI, T., CLARK, N. A., MUZNY, C. D., WU, L., GARLAND, C. W., SCHAEFER, D. W. and OLIVER, B. J. (1992) Phase behavior of the liquid crystal 8CB in a silica Aerogel, *Phys. Rev. Lett.* **69**, 788–791.

BELLINI, T., CLARK, N. A. and SCHAEFER, D. W. (1995) Dynamic light scattering study of nematic and smectic A liquid crystal ordering in silica aerogel, *Phys. Rev. Lett.* **74**, 2740–2743.

BLINC, R. and ZĔKŠ, B. (1978) Dynamics of helicoidal ferroelectric smectic-C liquid crystal, *Phys. Rev. A* **18**, 740–745.

BRANDOW, S. L., HARRISON, J. A., DiLELLA, D. P., COLTON, R. J., PHEIFFER, S. and SHASHIDHAR, R. (1993) Scanning tunnelling microscopic study of the interfacial order in a ferroelectric liquid crystal, *Liq. Cryst.* **13**, 163–170.

BROCHARD, F. and DE GENNES, P. G.1 (1983) Phase transitions of binary mixtures in random media, *J. Phys. Lett.* **44**, L785–L791.

CLARK, N. A. and MEYER, R. B. (1973) Strain-induced instability of monodomain smectic A and cholesteric liquid crystals, *Appl. Phys. Lett.* **22**, 493–494.

CLARK, N. A., BELLINI, T., MALZBENDER, R. M., THOMAS, B. N., RAPPAPORT, A. G., MUZNY, C. D., SCHAEFER, D. W. and HRUBESH, L. (1993) X-ray scattering study of smectic ordering in a silica aerogel, *Phys. Rev. Lett.* **71**, 3505–3508.

CRAWFORD, G. P., YANG, D.-K., ŽUMER, S., FINOTELLO, D. and DOANE, J. W. (1991) Ordering and self-diffusion in the first molecular layer at a liquid-crystal–polymer interface, *Phys. Rev. A* **43**, 2943–2952.

CRAWFORD, G. P., ALLENDER, D. W. and DOANE, J. W. (1992) Surface elastic and molecular-anchoring properties of nematic liquid crystals confined to cylindrical cavities, *Phys. Rev. A* **45**, 8693–8708.

CRAWFORD, G. P., ONDRIS-CRAWFORD, R., ŽUMER, S. and DOANE, J. W. (1993) Anchoring and orientational wetting transitions of confined liquid crystals, *Phys. Rev. Lett.* **70**, 1838–1841.

DADMUN, M. and MUTHUKUMAR, M. (1993) The nematic to isotropic transition of a liquid crystal in porous media, *J. Chem. Phys.* **98**, 4850–4851.

DE GENNES, P. G. (1984) Liquid-liquid demixing inside a rigid network. Qualitative features, *J. Phys. Chem.* **88**, 6469–6472.

DIERKER, S. B. and WILTZIUS, P. (1991) Statics and dynamics of a critical binary fluid in a porous medium, *Phys. Rev. Lett.* **66**, 1185–1188.

DRAKE, J. M., KLAFTER, J., KOPELMAN, R. and AWSCHALOM, D. D. (eds) (1993) *Dynamics in Small Confining Systems*, Materials Research Society Symposium Proceedings, Vol. 290, Pittsburgh: Materials Research Society.

DRUON, C. and WACRENIER, J. M. (1982) A study of 4 nonaoate, 4'cyanobiphenyl using dielectric relaxaton method, *Mol. Cryst. Liq. Cryst.* **88**, 99–108.

FILIPIČ, C., CARLSSON, T., LEVSTIK, A., LEVSTIK, I., ŽĒKŠ, B., BLINC, R., GOUDA, F., LAGERWALL, S. T. and SKARP, K. (1988) Dielectric properties near the smectic-C*– smectic-A phase transition of some ferroelectric liquid-crystalline systems with a very large spontaneous polarization, *Phys. Rev. A* **38**, 5833–5839.

FRISKEN, B. J. and CANNELL, D. S. (1992) Critical dynamics in the presence of a silica gel, *Phys. Rev. Let.* **69**, 632–635.

GOLDBURG, W. I., ALIEV, F. and WU, X.-L. (1995) Behavior of liquid crystals and fluids in porous meda, *Physica A* **213**, 61–70.

GOUDA, F., SKARP, K. and LAGERWALL, S. T. (1991) Dielectric studies of the soft mode and Goldstone mode in ferroelectric liquid crystals, *Ferroelectrics* **113**, 165.

GROUPE, ORSAY (1969) Dynamics of fluctuations in nematic liquid crystals, *J. Chem. Phys.* **51**, 816–822.

HIKMET, R. A. M. and ZWERVER, B. H. (1991) Dielectric relaxation of liquid crystal molecules in anisotropic confinements, *Liq. Cryst.* **10**, 835–847.

HUSE, D. A. (1987) Critical dynamics of random-field Ising systems with conserved order parameter, *Phys. Rev. B* **36**, 5383–5397.

IANNACCHIONE, G. S. and FINOTELLO, D. (1992) Calorimetric study of phase transitions in confined liquid crystals, *Phys. Rev. Lett.* **69**, 2094–2097.

IANNACCHIONE, G. S. and FINOTELLO, D. (1994) Specific heat dependence on orientational order at cylindrically confined liquid crystal phase transitions, *Phys. Rev. E* **50**, 4780–4795.

IANNACCHIONE, G. S., CRAWFORD, G. P., ŽUMER, S., DOANE, J. W. and FINOTELLO, D. (1993) Randomly constrained order in porous glass, *Phys. Rev. Lett.* **71**, 2595–2598.

JEROME, B. (1991) Surface effects and anchoring in liquid crystals, *Rep. Prog. Phys.* **54**, 391–451.

KLAFTER, J. and DRAKE, J. M. (eds) (1989) *Molecular Dynamics in Restricted Geometries*, New York: John Wiley.

KORB, J.-P., XU, S. and JONAS, J. (1993) Confinement effects on dipolar relaxation by translational dynamics of liquids in porous silica glasses, *J. Chem. Phys.* **98**, 2411–2422.

KRALJ, S. and ŽUMER, S. (1995) Saddle-splay elasticity of nematic structures confined to a cylindrical capillary, *Phys. Rev. E* **51**, 366–379.

KRALJ, S., LAHAJNAR, G., ZIDANŠEK, A., VRBANČIČ-KOPAČ, N., VILFAN, M., BLINC, R. and KOSEC, M. (1993) Deuterium NMR of a pentylcyanobiphenyl liquid crystal confined in a silica aerogel matrix, *Phys. Rev. E* **48**, 340–349.

KUZMA, M. and LABES, M. M. (1993) Liquid crystals in cylindrical pores: Effect on transition temperatures and singularities, *Mol. Cryst. Liq. Cryst.* **100**, 103–110.

LEVITZ, P., EHRET, G., SINHA, S. K. and DRAKE, J. M. (1991) Porous vycor glass: The microstructure as probed by electron microscopy, direct energy transfer, small-angle scattering, and molecular adsorbtion. *J. Chem. Phys.* **95**, 6151–6161.

LEVSTIK, A., CARLSSON, T., FILIPIČ, C., LEVSTIK, I. and ŽEKŠ, B. (1987) Goldstone mode and soft mode at the smectic-A–smectic-C* phase transition studied by dielectric relaxation, *Phys. Rev. A* **35**, 3527–3534.

LEVSTIK, A., KUTNJAK, Z. A., FILIPIČ, C., LEVSTIK, I., BREGAR, Z., ŽEKS, B. and CARLSSON, T. (1990) Dielectric method for determining the rotational viscosity in thick samples of ferroelectric chiral smectic-C* liquid crystals, *Phys. Rev. A* **42**, 2204–2210.

LEVSTIK, A., KUTNJAK, Z. A., FILIPIČ, C., LEVSTIK, I., ŽEKŠ, B. and CARLSSON, T. (1991) A dielectric method for determination of the rotational viscosity in ferroelectric liquid crystals, *Ferroelectrics* **113**, 207–217.

LI, X., KING, T. A. and PALLIKARI-VIRAS, F. (1994) Characteristics of composites based on PMMA modified gel silica glasses, *J. Non-Cryst. Solids* **170**, 243–249.

LI, Z., DILISI, G. A., PETSCHEK, R. G. and ROSENBLATT, C. (1990) Nematic electroclinic effect, *Phys. Rev. A* **43**, 1997–2004.

LI, Z., AKINS, R. B., DILISI, G. A., ROSENBLATT, C. and PETSCHEK, R. G.(1991) Anomaly in the dynamic behavior of the electroclinic effect below the nematic–smectic-A phase transition, *Phys. Rev. A* **43**, 852–857.

MAHER, J. V., GOLDBURG, W. I., POHL, D. W. and LANZ, M. (1984) Critical behavior in gels saturated with binary liquid mixtures, *Phys. Rev. Lett.* **53**, 60–63.

MARITAN, A., CIEPLAK, M., BELLINI, T. and BANAVAR, J. R. (1994) Nematic–isotropic transition in porous media, *Phys. Rev. Lett.* **72**, 4113–4116.

MARTINOT-LAGARD, Ph. and DURAND, G. (1981) Dielectric relaxation in a ferroelectric liquid crystal, *J. Phys.* **42**, 269–275.

MUMBY, S. J., RABOLT, Y. F. and SWALEN, Y. D. (1985) Structural characterization of a polymer monolayer on a solid surface, *Thin Solid Films*, **133**, 161.

PATEL, J. S., LEE, S.-D. and GOODBY, J. W. (1991) Nature of smectic ordering at a solid–liquid crystal interface and its influence on layer growth, *Phys. Rev. Lett.* **66**, 1890–1893.

PISSIS, P., DAOUKAKI-DIAMANTI, D., APEKIS, L. and CHRISTODOULIDES, C. (1994) The glass transition in confined liquids, *J. Phys.: Condens. Matter* **6**, L325–L328.

RANDIERIA, M., SETHNA, J. and PALMER, R. G. (1985) Low-frequency relaxation in Ising spin-glasses, *Phys. Rev. Lett.* **54**, 1321–1324.

RIBOTTA, R., DURAND, G. and LITSTER, D. (1973) Rayleigh scattering induced by static bends of layers in smectic A liquid crystal, *Solid State Commun.* **12**, 27–29.

ROSENBLATT, C., PINDAK, R., CLARK, N. A. and MEYER, R. B. (1979) Freely suspended ferroelectric liquid-crystal films: Absolute measurements of polarization, elastic constants and viscosities, *Phys. Rev. Lett.* **42**, 1220–1223.

ROUT, D. K. and JAIN, S. C. (1992) Dielectric properties of a polymer dispersed liquid crystal film, *Mol. Cryst. Liq. Cryst.* **210**, 75–81.

SCAIFE, B. K. P. (1989) *Principles of Dielectrics*, Oxford: Clarendon Press.

SCHULLER, J., MEL'NICHENKO, Yu. B., RICHERT, R. and FISCHER, E. W. (1994) Dielectric studies of the glass transition in porous media, *Phys. Rev. Lett.* **73**, 2224–2227.

SCHWALB, G. and DEEG, F. W. (1995) Pore-size-dependent orientational dynamics of a liquid crystal confined in a porous glass, *Phys. Rev. Lett.* **74**, 1383–1386.

SCHWALB, G., DEEG, F. W. and BRAUCHLE, C. (1994) Influence of surface interaction on the reorientational dynamics of pentylcyanobiphenyl confined in a porous glass, *J. Non.-Cryst. Solids* **172–174**, 438–352.

SHENG, P., LI, B.-Z., SHOU, M., MOSES, T. and SHEN, Y. R. (1992) Disordered-surface-layer transition in nematic liquid crystals, *Phys. Rev. A* **46**, 946–950.

TRIPATHI, S., ROSENBLATT, C. and ALIEV, F. M. (1994) Orientational susceptibility in porous glass near a bulk nematic–isotropic phase transition, *Phys. Rev. Lett.* **72**, 2725–2728.

VAN BEEK, L. K. H. (1967) Dielectric behaviour of heterogenous systems, *Progress in Dielectrics*, Vol. 7, ed. J. B. Birks, London: Heywood, pp. 69–114.

VRBANČIČ, N., VILFAN, M., BLINC, R., DOLINŠEK, J., CRAWFORD, G. P. and DOANE, J. W. (1993) Deuteron spin relaxation and molecular dynamics of a nematic liquid crystal (5CB) in cylindrical microcavities, *J. Chem. Phys.* **98**, 3540–3547.

WANG, C. H., MA, R. J., FYTAS, G. and DORFMULLER, Th. (1983) Laser light scattering studies of the dynamics of molecular reorientation of a viscoelastic liquid: α-phenyl *o*-cresol, *J. Chem. Phys.* **78**, 5863–5873.

WARNOCK, J., AWSCHALOM, D. D. and SHAFER, M. W. (1986) Orientational behavior of molecular liquids in restricted geometries, *Phys. Rev. B* **34**, 475–478.

WILLIAMS, G. (1991) Molecular motion in glass-forming systems, *J. Non-Cryst. Solids* **131–133**, 1–12.

WU, X.-L., GOLDBURG, W. I., LIU, M. X. and XUE, J. Z. (1992) Slow dynamics of isotropic–nematic phase transition in silica gels, *Phys. Rev. Lett.* **69**, 470–473.

YOKOYAMA, H. (1988) Nematic–isotropic transition in bounded thin films, *J. Chem. Soc. Faraday Trans.* **84**, 1023–1040.

YOUNG, C. Y., PINDAK, R., CLARK, N. A. and MEYER, R. B. (1978) Light-scattering study of two-dimensional molecular orientation fluctuations in a freely suspended ferroelectric liquid-crystal film, *Phys. Rev. Lett.* **40**, 773–776.

ZHUANG, X., MARRUCCI, L. and SHEN, Y. R. (1994) Surface-monolayer-induced bulk alignment of liquid crystals, *Phys. Rev. Lett.* **73**, 1513–1516.

ŽUMER, S. and KRALJ, S. (1992) Influence of K_{24} on the structure of nematic liquid crystal droplets, *Liq. Cryst.* **12**, 613–624.

18

Magnetic Field Effects on Liquid Crystalline Order in Porous Media

S. TRIPATHI and C. ROSENBLATT

Condensed matter in confined geometries is a field which has taken many directions in the past decade, and will undoubtedly continue to evolve in the future. In some cases the host matrix is a porous glass or polymer which is subsequently filled with a fluid. In other cases one relies on phase separation of liquids or liquid crystals in a polymer matrix, giving rise to individual, disconnected droplets of the confined fluid. A detailed history of confined liquid crystals can be found in the chapter by Crawford and Žumer. Professor J. William Doane, who has contributed so much to the field of liquid crystals, has made numerous seminal contributions to the physics of confined geometries and, in fact, is largely responsible for the discovery and development of polymer-dispersed liquid crystals; this field is described in more detail in the chapter by Kitzerow. We therefore view this opportunity as both a privilege and a pleasure to contribute to this book which honours one of the giants of liquid crystal research.

The physics of fluids in confined geometries is an extraordinarily rich topic. For example, structure and ordering, phase transitions, order parameter fluctuations, dynamics of molecular motion, and dynamics of collective modes have received considerable attention during the past decade. Details may be found in the chapters by Finotello et al. and Aliev. All of these subjects have bulk analogues in which one may observe, for example, the divergence of the correlation length near a second-order phase transition. What happens, however, when the walls of the container are so closely spaced that the bulk correlation length would ordinarily extend beyond these walls? In very large systems one may recall the squeezing walls of the trash compactor in the 1977 science fiction film *Star Wars*. In that case the heroes escaped before the container had shrunk to a size comparable with the relevant length in the problem – approximately 0.5 m. In the physics of condensed matter, of course, the relevant length scale is considerably smaller, generally of the order of tens to hundreds of ångströms. Very specialized containers are needed and, indeed, are available, facilitating a variety of experiments. These studies usually involve the impregnation of porous glasses and aerogels with substances such as liquid helium, binary liquid mixtures (BLMs), and liquid crystals, generally having a second-order or weakly first-order phase transition. A few years ago it seemed that the fundamental physical problems of liquid crystals and BLMs in pores might be solved in the very near future, and that the most important aspects of their physical behaviour would be understood. These hopes were connected with the fact that experimental results (Goh et al., 1987; Dierker and Wiltzius 1987; Wiltzius et al., 1989; Dierker et al., 1990; Dierker and Wiltzius, 1991) apparently lend support to the random field approach (Brochard and de Gennes, 1983; Fisher, 1986; de Gennes, 1984), which is based on the argument that the randomness of the pore structure gives rise to random field Ising-like behaviour near the phase transition temperature (see the chapter by Aliev). Recently an alternative theoretical approach was introduced (Liu *et al.*, 1990; Liu and Grest, 1991; Monette *et al.*, 1992), whereby confinement in small pores slows down domain growth in certain regions of the wetting phase diagram. In this 'single pore model' it was suggested that the random field picture is unlikely to apply to porous media with randomly distributed pores. Moreover, many features observed in experiments on BLMs in porous media which had been interpreted in terms of a random field model were instead shown (Monette *et al.*, 1992) to be consistent with wetting in a confined geometry, and with no randomness. In more recent experiments using silica gels (Frisken and Cannell, 1992) it was shown that the dynamics is consistent with the random field model, whereas a spontaneous ordering just above the critical temperature is not accounted for by the random field model. Finally, for BLMs in pores with pore size greater than the correlation length (Aliev *et al.*, 1993), a limit in which random field theory is inapplicable, a logarithmic decay was still observed and the slow dynamics remains unexplained.

In the case of a nematic liquid crystal the situation is even more complicated. In these investigations the emphasis is no longer on finite size and surface effects which are always present in porous media. Rather, attention has been focused on the effects of disorder on phase transitions and dynamics of fluctuations which are due to geometrical and chemical inhomogeneities. Elastic and quasi-elastic light scattering results (Wu *et al.*, 1992) obtained at the nematic–isotropic phase transition in silica gel were explained on the basis of the gel imposing a random uniaxial field on the liquid crystal. The equilibrium phase transition is smeared out by the randomness, and dynamically the system exhibits (Wu *et al.*, 1992) the kind of self-similarity that is associated with the conventional random field behaviour (Ogielski and Huse, 1986). Previously (Bellini *et al.*, 1992) some features of nematic ordering in the aerogel porous matrix were qualitatively explained in terms of the random field Ising model with an asymmetric distribution of random fields (Maritan *et al.*, 1991). Later it was mentioned (Kralj *et al.*, 1993) that the same random field argument (Bellini *et al.*, 1992; Maritan *et al.*, 1991) is qualitatively consistent with results of NMR investigations of liquid crystals in aerogels. The absence of a nematic–isotropic phase transition and a gradual increase of the local orientational order was observed (Iannacchione *et al.*, 1993) in NMR and calorimetric studies on pentylcyanobiphenyl (5CB) in Vycor porous glass with an average pore size about 75 Å. These results were satisfactorily described by taking into account the ordering effects of surface interactions and the disordering effects due to surface-induced deformations. A Landau theory was used which included an effective scalar order parameter, and a random field theory Hamiltonian was not needed. Many of these issues are discussed in more detail in the chapters by Cleaver *et al.* and Maritan *et al.*

It has unfortunately become apparent that even more questions seem to arise with each new attempt to clarify the physics of BLMs and liquid crystals in random porous media. Thus, before a definitive explanation of the nature of the physical behaviour of liquid crystals in pores can be achieved, still further systematic investigations are needed; many of these are described in detail in chapters by Finotello *et al.* and Clark. In this chapter we discuss magnetically induced birefringence measurements which were carried out in collaboration with Professor Fouad M. Aliev of the University of Puerto Rico (Tripathi *et al.*, 1994). The experiments were performed in the nematic liquid crystal 5CB in porous glass around the *bulk* nematic–isotropic phase transition temperature T_{NI}. (In porous glass there is no well-defined phase transition (Iannacchione *et al.*, 1993), and orientational order exists even above the bulk T_{NI}.) We find that the Cotton–Mouton coefficient $C(\equiv d\Delta n/dH^2)$ exhibits S-shaped behaviour with temperature in the vicinity of T_{NI}. This is in contrast to the algebraic behaviour observed in the bulk isotropic phase above T_{NI} (Stinson *et al.*, 1972), wherein $C \propto (T - T_{NI}^*)^{-\gamma}$, where T_{NI}^* is the supercooling limit of the isotropic phase and γ is the susceptibility critical exponent – generally $\gamma = 1$ for mean-field-like behaviour. Additionally, the S-shaped behaviour in porous glass is different from the non-analytic behaviour which is found in the bulk nematic phase, wherein $C \propto |H|$ (Malraison *et al.*, 1980). In that case the director fluctuations are quenched for all wavevectors up to some q_{max} comparable with an inverse molecular length. We discuss our results in terms of nematic order induced by the walls of the randomly oriented pores, such that the liquid crystal director is subsequently reoriented by the magnetic field. Additionally, our results indicate that in addition to field-induced changes of the director orientation, the magnetic field may significantly affect the magnitude of the nematic order parameter as well.

A porous matrix with thoroughly interconnected and randomly oriented pores was prepared from the original sodium borosilicate glass Na 7/23 consisting of 7% Na_2O, 23% B_2O_3 and 70% SiO_2. The sodium borate phase was removed by acid leaching, similar to the process used to create porous Vycor glass (Levitz *et al.*, 1991; Levitz and Tchoubar, 1992). The matrix framework consisting of SiO_2 is almost chemically pure and distinguished by its uniformity of chemical composition and mechanical strength. The characteristics of the porous matrix were determined by small-angle X-ray scattering (Aliev and Pozhivilko, 1988) using a Porod analysis. The average chord length of a pore – a chord is a linear path which correlates two distinct points on the interface – was found to be 100 Å, the volume fraction of pores was 0.27, and the specific area of the pores' surface was 108 m^2/cm^3. Since the structural characteristics of this matrix are independent of the temperature, at least in the temperature region used in our experiment, all observable temperature-dependent effects can be attributed to the change in the physical properties of the liquid crystal which impregnates the pores. In this experiment we used a porous glass plate of dimensions 1 cm × 1 cm × 0.21 cm, all surfaces of which were optically polished. The polishing procedure was performed in two stages: the glass plate was first polished before leaching, and then it was polished additionally after the porous structure was formed. Since the linear size of optical inhomogeneities as determined by the pore size is much smaller than the wavelength of visible light, the matrix was optically transparent. The porous matrix was then heated to 400°C and placed in a vacuum so as to remove any additional impurities; this was followed by impregnation with 5CB from the isotropic melt at 40 C over a period of 72 hours.

A problem endemic to most liquids, including liquid crystals, in porous glass involves temperature-driven density changes, resulting in leakage of the liquid crystal from the pores and associated optical complications. For example, when heated above the bulk nematic–isotropic phase transition temperature T_{NI}, liquid crystal tends to leak from the matrix, forming small droplets on the surface of the matrix. If a solid glass plate is used to cover the matrix, the liquid crystal will leak from the matrix and uniformly fill the narrow gap – typically a few micrometres – between the glass cover and the porous matrix. Above T_{NI} this layer of liquid crystal does not contribute substantially to the measured Cotton–Mouton coefficient. When cooled below T_{NI}, however, some of the liquid crystal will not be reabsorbed by the matrix and will remain at the surface, contributing a large, bulk-like (and unwanted) signal. These difficulties were obviated by applying a thin layer of optically transparent epoxy to the two large faces of the porous glass sample. Microscope cover slides were then placed over the epoxy, resulting in a pair of optically smooth and transparent surfaces; the epoxy effectively sealed the liquid crystal inside the porous glass. Observations with a polarizing microscope below the bulk T_{NI} revealed neither a surface layer of nematic nor any birefringence associated with the curing of the epoxy.

The porous glass was placed in a glass tube for mechanical support, and inserted into an oven temperature controlled to better than 10 mK. The oven was in turn placed into the bore of an 82 kG superconducting magnet. Light from an He–Ne laser was incident onto a polarizer, the sample, a Pockels cell modulated at frequency v, an analyser, and finally a photodiode detector. The output from the detector was input to a lock-in amplifier, which was referenced to the Pockels cell modulation frequency. The output from the lock-in amplifier was proportional to the total phase retardation (sample + Pockels cell) and was used in a feedback circuit to adjust automatically the total retardation to zero. Thus, a measurement of the voltage applied to the Pockels cell yielded the sample retardation and thus the birefringence. A schematic representation of the apparatus is shown in Figure 18.1, and more details can be found elsewhere

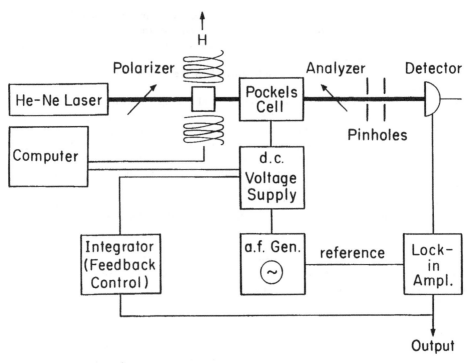

Figure 18.1 Experimental arrangement.

(Rosenblatt, 1984). The sample was oriented such that the light passed through the short (0.21 cm) axis of the porous glass. The sample was first equilibrated at a temperature approximately 2.5°C above the bulk T_{NI}, where T_{NI} was found to be 33.91°C. The magnetic field was ramped up at a rate of 75 G/s to a maximum of 55 kG, and the birefringence Δn was computer recorded. A sample trace is shown in Figure 18.2. The birefringence was also recorded on the downward sweep of the field. Measurements were made in this manner on decreasing the temperature to approximately 29°C, although at lower temperatures the improved signal-to-noise ratio facilitated the use of smaller maximum fields, as low as 35 kG. The birefringence Δn was found to be linear in H^2, and the Cotton–Mouton coefficient C was determined from a two parameter linear least-squares fit of Δn against H^2. C is plotted as a function of temperature in Figure 18.3 for both increasing and decreasing magnetic field sweeps. A small systematic difference between the two sets of data is apparent, although we are unable to offer an explanation for this behaviour. Nevertheless, the differences are small and do not affect our interpretation.

In order to understand this result we first examine the magneto-optic response of *bulk* 5CB above the nematic–isotropic phase transition. Muta *et al.* (1979) obtained a Cotton–Mouton coefficient $C \approx 1 \times 10^{-14}$ G^{-2} just above T_{NI} for bulk 5CB. In a field of 50 kG this would correspond to an induced birefringence of $\Delta n = CH^2 \approx 2.5 \times 10^{-5}$. Since the saturated birefringence Δn_0 of 5CB is approximately 0.19 (Karat and Madhusudana, 1976), the magnetically induced scalar nematic order parameter $S(=\Delta n/\Delta n_0)$ at $H = 50$ kG would be approximately 1.3×10^{-4} in the bulk. This is more than two orders of magnitude smaller than the nematic order induced by the pores in porous glass in the absence of an external field, as obtained by NMR measurements (Kralj *et al.*, 1993; Iannacchione *et al.*, 1993). Given the substantial order in a random porous medium, one might expect that the principal effect of a

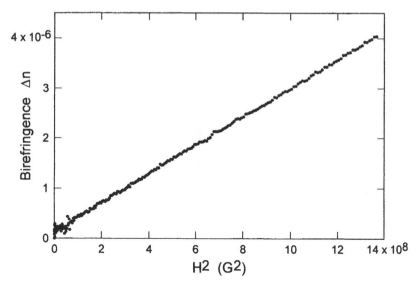

Figure 18.2 Typical computer-recorded trace of birefringence against H^2. Data were taken at temperature $T = 30.54\,°C$ as the field was increasing.

magnetic field is to reorient the director, such that the director orientation is biased along the field direction. This would clearly result in a non-zero spatially averaged birefringence. We shall first adopt this picture and examine its consequences; field-induced changes in the *magnitude* of the order parameter will be discussed later.

One difficulty with implementing this model, of course, is that to date there is no firm evidence for the equilibrium director profile within the very small pore. Kralj *et al.* (1993) have pointed out that the profile depends on both the surface interactions and the bulk elasticity. In small cavities where the average deformation for a radial director profile is large, a planar director orientation is favoured. (See the chapters by Bellini and Clark and Finotello *et al.*) Experimental confirmation of this behaviour was alluded to by Kralj *et al.* (1993) and references therein. In smaller cavities, however, additional factors become important as well. For example, on entropy grounds one might expect a planar director in a cavity which possesses an eccentric shape. The problem of completely filling space can also be achieved if gauche bonds were introduced in the alkyl tails (Gruen, 1985). Nevertheless, the general evidence seems to point towards a planar alignment of the liquid crystalline molecules.

By means of deuterium NMR Iannacchione *et al.* (1993) measured the scalar nematic order parameter $\langle S \rangle$, spatially averaged over a pore, of 5CB in Vycor glass. They found $\langle S \rangle \sim 0.025$ in the vicinity of the bulk nematic–isotropic phase transition temperature T_{NI}, and increasing smoothly as the temperature is lowered. Owing to the pores' elongated shape, similar to that of our experiment, the orientation of the director $\hat{n}$ was taken to be uniform. Additionally, because of the small pore diameter, the *magnitude* of the nematic order parameter was also treated as being spatially uniform. Using these ideas, we now consider a cylindrical correlated pore region. We assume that the axis which characterizes the pore's orientation is initially parallel to the $\hat{z}$ axis in the laboratory frame. Moreover, we assume that a field $\mathbf{H}$ is applied along the $\hat{z}$ axis. If the liquid crystal director $\hat{n}$ makes a polar angle θ with respect to the pore axis – for the moment this is the $\hat{z}$ axis – and an azimuthal angle ϕ with respect to the xz plane, the components of the director are given by

$$n_x = \sin \theta \cos \phi$$

$$n_y = \sin \theta \sin \phi$$

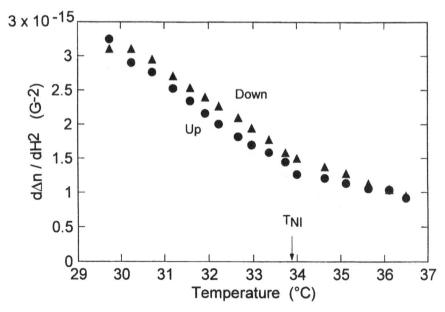

Figure 18.3   Cotton–Mouton coefficient against temperature. T_{NI} corresponds to the bulk nematic–isotropic phase transition temperature. Circles (●) correspond to data taken as the field was swept upward, and triangles (▲) to data as the field was swept back down. Results account for 0.27 filling factor.

$$n_z = \cos \theta$$

If the pore and liquid crystal therein are now rotated by an angle θ_0 in the xz plane, the component of the director along the magnetic field ($\hat{z}$) axis becomes

$$n_z = -\sin \theta_0 \sin \theta \cos \phi + \cos \theta_0 \cos \theta \equiv \cos \alpha$$

where α is the angle between the local director and the magnetic field. The field exerts a torque on the liquid crystal, subject to both thermal fluctuations and a restoring force determined by an effective anchoring strength coefficient W. The magnetic energy $U_M = -\frac{1}{2}\Delta\chi H^2 4 \cos^2 \alpha$ is therefore given by $U_M = -\frac{1}{2}\Delta\chi H^2 4(-\sin \theta_0 \sin \theta \cos \phi + \cos \theta_0 \cos \theta)^2$. Here V is the volume of a correlated pore region and $\Delta\chi$ is the effective magnetic susceptibility anisotropy, where $\Delta\chi = \langle S \rangle \Delta\chi_0$ and $\Delta\chi_0$ is the susceptibility anisotropy for saturated ($\langle S \rangle = 1$) order. The restoring energy due to surface anchoring is $U_R = \frac{1}{2}WA \sin^2\theta$, where A is an appropriate lateral area of the pore segment. Note that W includes the usual anchoring strength (appropriately averaged around the cylinder), as well as an entropy part which comes about from space-filling considerations ($\hat{n}$ prefers to be parallel to the pore's walls). (For more details on anchoring and surfaces, see the chapter by Barbero and Durand.) Additionally, the parallel orientation tends to increase the translational component of entropy (Onsager, 1949). By thermally averaging over the director orientation within the pore (θ and ϕ) and the orientation of the pore with respect to the magnetic field (θ_0), we obtain the average square component of the director along the $\hat{z}$ axis in the laboratory frame, i.e. $\langle \cos^2\alpha \rangle = \langle (-\sin \theta_0 \sin \theta \cos \phi + \cos \theta_0 \cos \theta)^2 \rangle$:

$$\langle \cos^2\alpha \rangle = \frac{1}{Z} \iiint \exp\left(-\frac{U_M + U_R}{k_B T} \right)$$

$$\times (-\sin \theta_0 \sin \theta \cos \phi + \cos \theta_0 \cos \theta)^2 \sin \theta_0 \, d\theta_0 \sin \theta \, d\theta \, d\phi \qquad (18.1)$$

where k_B is Boltzmann's constant and T is temperature. The partition function Z is given by

$$Z = \iiint \exp\left(-\frac{U_{\mathrm{M}} + U_{\mathrm{R}}}{k_{\mathrm{B}} T}\right) \sin \theta_0 \, d\theta_0 \sin \theta \, d\theta \, d\phi$$

Thus, we can extract $\langle Q \rangle = \langle \frac{3}{2} \cos^2\alpha - \frac{1}{2} \rangle$, an appropriately averaged quantity for the director orientation with respect to the $\hat{z}$ axis. Note that $\langle Q \rangle$ is *not* the nematic order parameter, but is rather a quantity proportional to the magnetically induced birefringence.

In order to evaluate $\langle Q \rangle$ we treat the pores to a first approximation as worm-like cylinders. We use a figure for a suitable surface area of a correlated region characteristic of pores in Iannacchione *et al.* (1993) and, we believe, of our sample as well, i.e. $A = 1.3 \times 10^{-11}$ cm^2. Moreover, V can then be obtained from the specific area of our material (108 m^2/cm^3) giving $V = 1.2 \times 10^{-17}$ cm^3, and $\Delta\chi_0 = 1.2 \times 10^{-7}$ (Malraison *et al.*, 1980). Additionally, we note that the Cotton–Mouton coefficient $d\Delta n/dH^2$ is given by

$$\frac{d\langle Q \rangle}{dH^2} \langle S \rangle \Delta n_0$$

where $\langle S \rangle \Delta n_0$ is the maximum achievable field-oriented birefringence at a given temperature, i.e. if the directors of all correlated volumes are oriented along the $\hat{z}$ axis. This indicates that $d\Delta n/dH^2$ should vary as $\langle S \rangle^2$, with one factor of $\langle S \rangle$ appearing explicitly and the other appearing implicitly as part of $d\langle Q \rangle/dH^2$ through the susceptibility anisotropy. This behaviour is in qualitative agreement with Figure 18.3 using the data of Iannacchione *et al.* (1993). Taking $\langle S \rangle = 0.03$, a value found by Iannacchione *et al.* (1993) a few degrees below the bulk transition temperature, we find after numerically evaluating eqn (18.1) that our model predicts $d\Delta n/dH^2 \approx 4 \times 10^{-16}$ G^{-2} at the same temperature. Although not unreasonable, this figure is nevertheless too small to account completely for the experimental results of Figure 18.3, and we therefore need to look for a complementary mechanism behind the magneto-optic response. (We note in passing that the results of the model are virtually independent of W for W in the range 10^{-5}–10^1 erg/cm^2 (Blinov *et al.*, 1989). The reason for this behaviour is the very small magnetically induced deviation of the director from its initial alignment in the pore, resulting in only a tiny restoring torque. For this range of W, $U_{\mathrm{R}} \ll k_{\mathrm{B}} T$. For larger deviations, of course, U_{R} can become comparable with thermal energies. Thus, the surface interactions – both isotropic and anisotropic – are responsible for inducing an equilibrium alignment of the liquid crystal even above the bulk T_{NI} and suppressing the phase transition.)

The model presented above is quite simple, and neglects an important physical feature of the pores, i.e. the interconnectedness of the pore structure. This interconnectedness gives rise to 'end' effects, inasmuch as an 'end' can be ascribed to a pore. Since the 'ends' are washed out in space, and the appropriate volume is probably larger than the volume V used in the analysis, neglect of the 'ends' is likely to result in an underestimate of the magnitude of the magnetic susceptibility. Nevertheless, it is not clear that 'end' effects can completely account for the deficiency in our calculated value of $d\Delta n/dH^2$.

It is apparent from the model calculations that we need to look for an additional contribution to account for our experimental results. Although we considered only reorientation of an already extant non-zero tensor order parameter, the magnetic field may also induce changes in the *magnitude* of the order parameter. We note in Figure 18.3 that $d\Delta n/dH^2$ above the bulk nematic–isotropic transition temperature is comparable with, or even smaller than, field-induced values in the bulk (Muta *et al.*, 1979). As noted earlier, $C \approx 1 \times 10^{-14}$ G^{-2} in the bulk just *above* T_{NI} (Muta *et al.*, 1979); this is already larger than our experimental result. Moreover, since the pore size is greater than or equal to the nematic correlation length, especially well above T_{NI}, the uniform order parameter approximation used in our model may not be completely appropriate. Thus, despite the significant degree of orientational order already present in the pores, even well above the bulk T_{NI}, it is not unreasonable that field-induced changes in the magnitude of the nematic order parameter may measurably contribute to our observed signal. The temperature and order parameter dependencies of this contribution, however, are not as transparent as that of our model calculation, for which $d\Delta n/dH^2 \propto \langle S \rangle$. Moreover, since **H** and the local director are not necessarily parallel, a change in magnitude would also involve a concomitant change in direction, with the additional possibility of some degree of magnetically induced biaxiality.

The combination of magnetically induced changes in both the magnitude and orientation of the order parameter, specifically in the presence of surface-related effects due to the large interfacial area, is quite complex. Issues of this sort are dealt with in the chapter by Barbero and Durand. An appropriate model would require a full Landau calculation of the tensor order parameter to fourth order, coupled to a surface ordering term, a disordering term and the magnetic field. Additionally, gradients in the tensor order parameter would also be needed, especially in regions where the nematic correlation length becomes comparable with or smaller than the pore diameter. The subject of liquid crystals in confined geometries is an extremely fertile field for study, and will clearly require an extensive battery of tools to achieve a more comprehensive understanding.

Acknowledgements

The authors are indebted to Dr Fouad Aliev for providing samples and support throughout this project. Additionally, the authors thank D. Finotello, G. P. Crawford and R. G. Petschek for useful conversations. This work was supported by the National Science Foundation under grant DMR-9020751 and by the NSF's Advanced Liquid Crystalline Optical Materials Science and Technology Center (ALCOM) under grant DMR-8920147.

References

ALIEV, F. M. and POZHIVILKO, K. S. (1988) Critical behavior of an interphase layer and the surface properties of a liquid crystal in micropores, *Sov. Phys. Solid State* **30**, 1351–1312.

ALIEV, F. M., GOLDBURG, W. I. and WU, X.-L. (1993) Concentration fluctuations of a binary liquid mixture in a macroporous glass, *Phys. Rev. E* **47**, R3834–R3837.

BELLINI, T., CLARK, N. A., WU, L., GARLAND, C. W., SCHAEFER, D. and OLIVIER, B. (1992) Phase behavior of the liquid crystal 8CB in a silica aerogel, *Phys. Rev. Lett.* **69**, 788–791.

BLINOV, L. M., KABAYENKOV, A. Yu. and SONIN, A. A. (1989) Experimental studies of the anchoring energy of nematic liquid crystals, *Liq. Cryst.* **5**, 645–661.

BROCHARD, F. and DE GENNES, P. G. (1983) Phase transitions of binary liquids in random media, *J. Phys. Lett.* **44**, 785–791.

DE GENNES, P. G. (1984) Liquid–liquid demixing inside a rigid network: Qualitative features, *J. Phys. Chem.* **88**, 6469–6472.

DIERKER, S. B. and WILTZIUS, P. (1987) Random field transition of a binary liquid in a porous medium, *Phys. Rev. Lett.* **58**, 1865–1868.

DIERKER, S. B. and WILTZIUS, P. (1991) Statics and dynamics of critical binary fluid in a porous medium, *Phys. Rev. Lett.* **66**, 1185–1188.

DIERKER, S. B., DENNIS, B. S. and WILTZIUS, P. (1990) The use of Raman scattering to study wetting and fluid flow of binary liquids in porous media, *J. Chem. Phys.* **92**, 1320–1328.

FISHER, D. S. (1986) Scaling and critical slowing down in random field Isling systems, *Phys. Rev. Lett.* **56**, 416–419.

FRISKEN, B. J. and CANNELL, D. S. (1992) Critical dynamics in the presence of a silica gel, *Phys. Rev. Lett.* **69**, 632–635.

GOH, M. C., GOLDBURG, W. I. and KNOBLER, C. M. (1987) Phase separation of a binary liquid mixture in a porous medium, *Phys. Rev. Lett.* **58**, 1008–1011.

GRUEN, D. W. R. (1985) A model for the chains of amphiphilic aggregates: 1. Comparison with a molecular dynamic simulation of a bilayer, *J. Phys. Chem.* **89**, 1456–153; (1985) A model for the chains of amphiphilic aggregates: 2. Thermodynamic and experimental comparisons for aggregates of different shapes and sizes, *J. Phys. Chem.* **89**, 153–163.

IANNACCHIONE, G. S., CRAWFORD, G. P., ŽUMER, S., DOANE, J. W. and FINOTELLO, D. (1993) Randomly constrained orientational order in porous glass, *Phys. Rev. Lett.* **71**, 2595–2598.

KARAT, P. P. and MADHUSUDANAS, N. V. (1976) Elastic and optical properties of some 4'-n-alkyl-4-cyanobiphenyls, *Mol. Cryst. Liq. Cryst.* **36**, 51–64.

KRALJ, S., LAHAJNAR, G., ZIDANŠEK, Z., VRBANČIČ-KOPAČ, N., VILFAN, M., BLINC, R. and KOSEC, M. (1993) Deuterium NMR of a pentylcyanobiphenyl liquid crystal confined in a silica aerogel matrix, *Phys. Rev. E* **48**, 340–349.

LEVITZ, P. and TCHOUBAR, D. (1992) Disordered porous solids – From chord distributions to small angle scattering, *J. Phys. I* **2**, 771–790.

LEVITZ, P., EHRET, G., SINHA, S. K. and FRAKE, J. M. (1991) Porous Vycor glass: The microstructure as probed by electron microscopy, direct energy transfer, small angle scattering, and molecular adsorption, *J. Chem. Phys.* **95**, 6151–6161.

LIU, A. J. and GREST, G. S. (1991) Wetting in a confined geometry – A Monte-Carlo study, *Phys. Rev. A* **44**, R7894–R7897.

LIU, A. J., DURIAN, J., HERBOLZHEIMER, E. and SAFRAN, S. A. (1990) Wetting transitions in a cylindrical pore, *Phys. Rev. Lett.* **65**, 1897–1900.

MALRAISON, B., POGGI, Y. and GUYON, E. (1980) Nematic liquid crystals in high magnetic field: Quenching of the transverse fluctuations, *Phys. Rev. A* **21**, 1012–1024.

MARITAN, A., SWIFT, M. R., CIEPLAK, M., CHAN, M. H. W., COLE, M. and BANAVAR, A. (1991) *Phys. Rev. Lett.* **67**, 1821.

MONETTE, L., LIU, A. J. and GREST, G. S. (1992) Wetting and domain growth kinetics in confined geometries, *Phys. Rev. A* **46**, 7664–7679.

MUTA, K., TAKEZOE, H., FUKUDA, A. and KUZE, E. (1979) Cotton–Mouton effect of alkyl- and alkoxy-cyanobiphenyls in isotropic phase, *Jpn. J. Appl. Phys.* **18**, 2073–2080.

OGIELSKI, A. T. and HUSE, A. (1986) Critical behavior of the three dimensional dilute Ising antiferromagnetic in a field, *Phys. Rev. Lett.* **56**, 1298–1301.

ONSAGER, L. (1949) The effects of shape on the interactions of colloidal particles, *Ann. NY Acad. Sci.* **51**, 627–659.

ROSENBLATT, C. (1984) Temperature dependence of the anchoring strength coefficient at a nematic liquid crystal-wall interface, *J. Phys.* **45**, 1087–1091.

STINSON, T. W., LISTER, J. D. and CLARK, N. A. (1972) Static and dynamic behavior near the order disorder transition of nematic liquid crystals, *J. Phys. Coll.* **33**, C1-69–C1-75.

TRIPATHI, S., ROENBLATT, C. and ALIEV, F. M. (1994) Orientational susceptibility in porous glass near a bulk nematic-isotropic phase transition, *Phys. Rev. Lett.* **72**, 2725–2728.

WILTZIUS, P., DIERKER, S. B. and DENNIS, B. S. (1989) Wetting and random field transition of binary liquids in a porous medium, *Phys. Rev. Lett.* **62**, 804–807.

WU, X.-L., GOLDBURG, W. I., LIU, M.-X. and XUE, J.-Z. (1992) Slow dynamics of isotropic-nematic phase transition in silica gels, *Phys. Rev. Lett.* **69**, 470–473.

19

Light Scattering as a Probe of Liquid Crystal Ordering in Silica Aerogels

T. BELLINI and N. A. CLARK

19.1 Introduction

The study of the influence of externally imposed disorder on the structure and dynamical properties of condensed phases has been a rich area of statistical physics. The introduction of randomness via externally applied fields, site dilution and intraparticle interactions has been found to alter critical behaviour, suppress long-range order, and lead to non-ergodicity and glass-like dynamics (Fisher *et al.*, 1988; Binder and Young, 1986). Because of the detailed understanding of phase transitions and critical behaviour developed for fluid media over the past few decades, fluids incorporated into disordered porous solid media are interesting systems for the quantitative study of the effects of imposed disorder (Brochard and de Gennes, 1983; Dierker and Wiltzius, 1987; Wong and Chan, 1990; Kim *et al.*, 1993). Liquid crystals exhibit a rich collection of transitions, some extensively studied in the bulk, between phases having orientational and/or translational order in one, two, and three dimensions (de Gennes and Prost, 1993; Pershan, 1988), and are susceptible to disordering by incorporation into disordered hosts (Chow and Martire, 1969; Armitage and Price, 1976; Aliev *et al.*, 1984; Aliev, 1994; Crawford and Žumer, this volume). The experimental accessibility of liquid crystal structure and dynamics, coupled with the readiness with which quenched disorder can be introduced via disordered surfaces, suggest that liquid crystals will be an important testing ground for our understanding of disorder.

In this chapter we discuss the analysis of light scattering by liquid crystal orientational fluctuations when a liquid crystal in the isotropic, nematic and smectic A phases is confined to silica aerogel. This work extends recent efforts aimed at quantitative characterization of the thermodynamics, structure and dynamics of liquid crystal transitions using aerogels as disordering porous hosts. The elimination of long-range order and bulk-like phase behaviour has been observed for liquid crystal–aerogel systems by calorimetry (Bellini *et al.*, 1992; Wu *et al.*, 1995), X-ray scattering (Clark *et al.*, 1993; Rappaport, 1995; Rappaport *et al.*, 1995), and static light scattering (Bellini *et al.*, 1992). Dynamic light scattering (Wu *et al.*, 1992; Bellini *et al.*, 1995) shows that in the aerogel, orientational confinement appears gradually with decreasing temperature via a glass-like orientational freeze-out to non-ergodicity, in contrast to bulk liquid crystal systems where confinement appears abruptly at the isotropic–nematic/smectic phase transitions. The optical properties of liquid crystals in silica aerogel (LCSA) systems are thus dramatically different from the bulk properties. In particular:

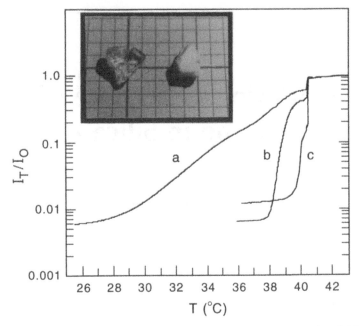

Figure 19.1 Ratio of transmitted to incident intensity against *T* for 8CB–LCSA samples of different aerogel density: 0.69 g/cm³ (curve a), 0.36 g/cm³ (curve b), 0.08 g/cm³ (curve c). Inset shows that the 0.36 g/cm³ aerogel is glass-like when filled with isotropic phase (left), and chalk-like when filled with nematic phase (right). The small squares are 0.1 × 0.1 inches (2.5 × 2.5 mm).

1   LCSA samples have the appearance of glass when the liquid crystal is isotropic, but the appearance of chalk when the liquid crystal is nematic (Figure 19.1).

2   In contrast to what is found in the bulk, where light scattered by the nematic is entirely dynamic, for nematic confined in aerogel, the scattered light is almost entirely static.

3   Both in the nematic and in the smectic phase, the residual fluctuating contribution to the scattered intensity yields a correlation function showing distinct relaxation processes, including very slow glass-like non-exponential dynamics.

4   Smectic ordering produces a drastic slowing down of orientational relaxation, in contrast to the bulk where the suppression of bend and twist by the layering accelerates orientational relaxation.

Thus the relatively few experiments to date suggest that the study of liquid crystals in porous media will offer new insights into cooperativity in the presence of disorder. While several of these novel LCSA characteristics can be understood in terms of the properties of the bulk liquid crystal (e.g. the optical turbidity of the LCSA samples and their fast relaxation component are closely connected, respectively, to the bulk birefringence and to the bulk pretransitional dynamics of paranematic clusters in the isotropic phase), the orientational freeze-out and glassy behaviour require a new understanding of the role of collective dynamics and defect motion in condensed phases. Additionally, interesting new phase diagrams as a function of increasing induced disorder have indeed been predicted (Maritan *et al.*, 1994 and this volume), but not yet observed. In this chapter we briefly review the relevant characteristics of the aerogel hosts, discuss the analysis of the static light scattering (turbidity) data, and present a general discussion and results of dynamic light scattering from LCSA systems.

In the experiments presented here we study the phase behaviour of the single component liquid crystal octyloxycyanobiphenyl (8CB), incorporated in the silica aerogels. The compound has been used as purchased from BDH. The bulk phase diagram is as follows (Karat and Madshusudana, 1976; Kasting *et al.*, 1980):

I $(T_{\text{NI}} = 40.5°C)$ N $(T_{\text{NA}} = 33.7°C)$ A $(T_{\text{AX}} = 21.5°C)$ X

19.2 Structure and Optical Properties of Silica Aerogel

19.2.1 *Description of Silica Aerogel*

The aerogels used in the experiments discussed here are described in more detail in the following chapter. The growth procedure, a polymerization under basic conditions, makes them 'colloidal' aerogels; that is, aerogels where the network-like solid phase is a structure of multiply connected solid silica strands having a typical thickness $\langle s \rangle$, in our case around 50 Å, as opposed to other silica aerogel structures having large silica agglomerates connected by 'bridges' (Schaefer, 1994; Schaefer *et al.*, 1994). Another consequence of this polymerization process is that while the resulting aerogel is quite uniformly dense, over length scales ranging from the particle size (~ 50 Å) to the fractal correlation length (colloidal aggregate size of a few hundred angstroms) the structure is fractal (Emmerling and Fricke, 1992), with a fractal dimension d_f relating mass to length in the range $d_f \approx 2$ (Schaefer and Keefer, 1986). This is in agreement with the experimental finding that the pore size distribution is not simply a power law (Ferri *et al.*, 1991; Sorenson and Lu, 1993).

The basic characteristic properties of the silica aerogels we have used, such as density, mean pore chord $\langle p \rangle$ and mean solid chord $\langle s \rangle$, are reported in Table 19.1. As can be seen, the 'empty' space within these gels, the pore volume (volume fraction ϕ) which will be filled by liquid crystal, ranges from $\phi = 0.70$ to 0.95 of their total volume, while the mean pore size ranges from 100 to 700 Å.

Table 19.1 Parameters relevant to the aerogels and the 8CB aerogel system. Aerogel data (mass density, ρ; pore volume fraction, ϕ; solid mass density, ρ_s; average solid chord, $\langle s \rangle$; average pore chord, $\langle p \rangle$) are from Wu *et al.* (1995). V/S is the inverse surface-to-volume ratio of the pore region of the aerogel. $\xi_{\text{dyn, sat}}$ is the mean pore size extracted from the fast intrapore relaxation observed in the DLS correlation functions. The low-temperature correlation length for smectic ordering, ξ_s, is from Rappaport *et al.* (1995) and Rappaport (1995)

ρ (g/cm^3)	ϕ	ρ_s (g/cm^3)	$\langle s \rangle$ (Å)	$\langle p \rangle$ (Å)	V/S (Å)	$\xi_{\text{dyn, sat}}$ (Å)	ξ_s (Å)
0.08	0.95	1.48	42	700 ± 100	159	410	1400
0.17	0.90	1.59	50	430 ± 65	137	250	520
0.36	0.79	1.73	47	180 ± 45	51	210	180
0.60	0.73	2.2	45	120 ± 25	30	130	95

These aerogels are nearly invisible. When empty they are as transparent as ordinary glass and they have very low surface reflectivity for visible light. This is because all the inhomogeneities within the silica structure are on a length scale sufficiently short relative to optical wavelengths that the optical mean free path is many centimetres. The silica strands themselves have mean diameters of about 50 Å, which are quite small indeed, and their collective structure is random for sizes larger than the fractal correlation length. Specifically, if we model the aerogel as a 10% by volume collection of randomly placed 50 Å diameter silica spheres, the optical mean free path we calculate is $\ell \approx 25$ cm. If the pores are filled with an isotropic liquid, $n_{\text{liq}} - n_{\text{sil}}$, the refractive index difference between pores and solid decreases, and the mean free path,

$\ell \sim (n_{\mathrm{liq}} - n_{\mathrm{sil}})^{-2}$, becomes even longer. *The observed light scattering properties of the LCSA samples are therefore due only to the refractive index fluctuations associated with disordering and defects of the liquid crystal induced by the aerogel.* This is confirmed by the fact that the LCSA samples in the isotropic phase are also perfectly transparent (Figure 19.1).

The 8CB has been introduced in the empty cavities of the silica aerogels by capillary action. To ensure that no air bubbles remain within the aerogels, the filling has been performed under vacuum, and pressure subsequently slowly increased. The liquid crystal has finite surface tension which may alter the aerogel structure during filling. To assess this we have compared the X-ray intensity $I_{\mathrm{ag}}(q)$ scattered by the aerogel before and after filling (see the next chapter). Filling lowers the X-ray index mismatch between solid and pore, thereby reducing $I_{\mathrm{ag}}(q)$, but otherwise, for $q > 0.006$ Å^{-1}, tends only to lower the aggregate fractal dimension by 5 to 10%. It appears, therefore, that the silica structure has not been modified on length scales $l \lesssim 1000$ Å by the insertion of the liquid crystal. Filling does produce cracks in the lowest-density aerogel used here (Table 19.1).

19.2.2 Silica Aerogel Modelled as a Porous Material

The mean size $\langle p \rangle$ of the empty spaces between silica strands can be determined by small-angle X-ray diffraction on the empty aerogels. The principles through which this extraction is performed are detailed by Schaefer (1984) and Schaefer *et al.* (1994) which specifically deal with the geometrical characterization of one of the same aerogels used in these light and X-ray scattering experiments, namely the one having a density of 0.36 g/cm^3. The basic idea is to evaluate S, the total silica surface per unit volume, from the Porod invariant, which in turn is calculated from absolute measurements of the scattered X-ray intensity. S is then simply connected to the length $d = \langle p \rangle + \langle s \rangle$ through the relationship $d = 4/S$, provided that the surface is locally randomly oriented. This is the way these quantities reported in Table 19.1 have been obtained. An independent estimate of $\langle s \rangle$ can be obtained directly from the position of the elbow of the structure factor of the empty silica aerogel. As can be seen from Figure 19.2 of Schaefer *et al.* (1994), which shows the scattered intensity from a sample of silica aerogel, $\langle s \rangle \approx 50$ Å, in agreement with $\langle s \rangle$ as calculated from the Porod invariant.

An alternative measure of $\langle p \rangle$ is the mean size of smectic domains ξ_{S}, extracted from X-ray scattering at temperatures well below the nematic–smectic A transition (see the next chapter). Table 19.1 shows that the values of ξ_{S} obtained for the different aerogels correspond fairly well to $\langle p \rangle$ obtained as described above; the identification of $\langle p \rangle$ with ξ_{S} seems to be appropriate. The X-ray scattering analysis also provides information on the pore eccentricity and shows that the pores are roughly spherical in shape: $R < 2$, where $R = \xi_{\mathrm{S}\parallel}/\xi_{\mathrm{S}\perp}$, the ratio of the values of ξ_{S} parallel and perpendicular to the local mean long molecular axis given by the unit vector director $\mathbf{n(r)}$ (Rappaport, 1995; Rappaport *et al.*, 1995).

As will become clear in the analysis, if we want to calculate experimentally accessible quantities, we need to know not only $\langle p \rangle$ and $\langle s \rangle$ but the pore size distribution $P_p(L)$ as well. The condition that the pores fill the empty space within the aerogel leads to the following normalization requirement:

$$\int P_p(L) L^3 \, \mathrm{d}L = \phi$$

To analyse $P_p(L)$ we imagine a function $f(r)$ which is constant inside each pore and, upon moving from pore to pore, assumes random values between 0 and -1. The autocorrelation of such a function is simply determined by $P_p(L)$ because it contains no cross-correlations between different pores, but 'self' or 'single-pore' correlations only. As discussed in the next chapter, because of the lack of phase coherence among the different smectic domains, the same can be said of the smectic correlation function $g_{\mathrm{S}}(r)$ measured by X-ray scattering: it contains smectic domain self-correlations only. The X-ray scattering shows unambiguously that $g_{\mathrm{S}}(r)$ is simple exponential, and assuming the pores to be shapes with small eccentricity as noted above, we have $g_s(r) \sim \exp(-r/\xi_s)$ (Clark *et al.*, 1993; Rappaport, 1995; Rappaport, *et al.*, 1995). Since from Table 19.1 $\langle p \rangle \sim \xi_s$ we will use $g_s(L)$ to extract $P_p(L)$. The next step is then to determine which pore size distribution $P(L)$ yields a simple exponential for the single-pore correlation. The autocorrelation function of $f(r)$, $g_f(R) = \langle f(r)f(r+R)\rangle$ is given by

$$g_f(R) = \langle f^2 \rangle \int_R^\infty P_p(L)L^2(L-R)\,dL + \langle f \rangle^2 \int_R^\infty P_p(L)L^2 R\,dL + \langle f \rangle^2 \int_0^\infty P_p(L)L^3\,dL \quad (19.1)$$

and contains three terms:

1   The 'self' term, present when the pores are larger than R, and corresponding to the fact that a pore shifted by a distance R partially overlaps with its original volume in the unshifted position.
2   An incoherent contribution from pores larger than R, corresponding to the non-overlapping volume of shifted and unshifted pores.
3   The incoherent term from the pores smaller than R.

By assuming the size distribution to be

$$P_p(L) = (\langle p \rangle L)^{-2} \exp(-L/\langle p \rangle) \quad (19.2)$$

and inserting it in eqn (19.1) we obtain the desired exponential correlation function

$$g_f(R) = (\langle f^2 \rangle - \langle f \rangle^2) \exp(-R/\langle p \rangle) + \langle f \rangle^2 \quad (19.3)$$

We will adopt hereafter the functional form of eqn (19.2) for $P_p(L)$ when a distribution is required to model the void structure within the silica aerogel. Equation (19.1) is exact when the pores are cubes. We have checked that the precise shape of the pores (spheres, cubes, low-eccentricity spheroids) does not change appreciably the result of eqn (19.3). The volume-averaged mean pore size of this distribution is

$$\langle L \rangle = \int P_p(L)L^4\,dL \Big/ \int P_p(L)L^3\,dL^3\,dL = 2\langle p \rangle$$

The expression 'mean pore size' can be applied to many different averaging procedures, depending on the context. We will use it to indicate $\langle p \rangle$.

The lineshape of the powder-averaged Bragg scattering from layer ordering within the pores of the aerogel is Lorentzian (Clark *et al.*, 1993; Rappaport, 1995; Rappaport *et al.*, 1995) This indicates that before powder averaging the mean lineshape of domains about the layering wavevector q_0 is (Lorentzian)^2 (Clark *et al.*, 1993). The X-ray lineshape for the above $P_p(L)$, since the domains are uncorrelated, can be obtained by summing the scattering of single-sized domains, properly weighted by $P_p(L)$. If the domains are spherical, their form factor $F(qL) = 4\pi q^{-3}[qL\cos(ql) - \sin(qL)]$, so we can easily evaluate the scattering intensity

$$I(q) \sim \int_0^\infty F^2(qL)P(L)\,dL \sim [1 + (aq\langle p \rangle)^2]^{-2} = [1 + (0.96q\langle p \rangle)^2]^{-2}$$

This $I(q)$ is essentially perfectly fit by the (Lorentzian)^2 form given with $a = 0.96$. Thus, as expected, $a \cong 1$.

19.3 Static Light Scattering

In a previous paper (Bellini *et al.*, 1992), we have reported the strong temperature dependence of the optical transparency of an 8CB–LCSA sample. In this section we describe in more detail the results and how they have been obtained, and introduce two different ways to model the origin of the huge observed turbidity. Although the models are simple, we believe they grasp the real physical origin of the phenomenon. They enable a successful comparison of the measurements with known static optical properties of the bulk nematic and smectic phases.

19.3.1 Turbidity Experiment and Results

We show in Figure 19.1 the ratio of transmitted to incident intensity for a collimated laser beam, I_T/I_0, as a function of temperature for three 8CB–LCSA 1 mm thick samples having different aerogel density. In Figure 19.1 the bulk T_{NI} is marked by a discontinuous jump in transmittance due to the transformation into nematic of a small amount of remnant bulk 8CB on the sample external surfaces. This step serves as a reference for the bulk transition temperature of the 8CB in the aerogel which occurs at slightly different temperatures in different aerogels because of differing impurity concentrations. Thus, as in the rest of this chapter, we have made a small shift in the temperature scales for the different aerogels to make the steps coincide with the bulk $T_{NI} = 40.5\,°C$.

Inspection of Figure 19.1 reveals that by increasing the aerogel density we increasingly depress the nematic ordering at a given temperature. On the other hand Figure 19.1 also shows the intrinsic difficulty in using the transmitted intensity as a tool to study the optical transparency of turbid samples. Because of the diffuse multiple scattering into the forward direction, which grows upon lowering the temperature, the apparent I_T saturates at about 1% of I_0 when the temperature is reduced. The diffuse background eventually becomes larger than the transmitted intensity so that simple measurement of I_T cannot be used to obtain turbidity. Multiply scattered light can be observed by monitoring the intensity exiting the LCSA sample as a function of the distance from the beam propagation axis at fixed temperatures. The results are shown in Figure 19.2 for a 0.5 mm thick slab of the 0.36 g/cm³ aerogel, at two different temperatures: when $T = 36\,°C$, the transmitted laser beam profile is clearly distinguishable from the diffuse background, while $T = 34\,°C$ only the smooth position dependence of the multiply scattered light exiting the sample can be detected. For this sample we can measure the turbidity directly via I_T for $T > 35.5\,°C$.

For lower temperature, we have extracted the turbidity by comparing the measured diffusive profile of a 1 mm thick sample with that calculated by assuming photon diffusion behaviour for the light propagation through the aerogel. Such an assumption is acceptable when the mean number of scattering events per photon path through the sample is larger than 10 (Bellini *et al.*, 1991), and requires solving the equation

$$\frac{\partial U(r,\ t)}{\partial t} = \frac{c\ell^*}{3}\ \nabla^2 U(r,\ t) \tag{19.4}$$

where $U(r,\ t)$ is the photon density and ℓ^* is the transport mean free path (Pine *et al.*, 1990). Because of the small size of the scattering objects with respect to the optical wavelength, we have assumed ℓ^* to be equal to the photon scattering mean free path ℓ. Perfect-sink boundary conditions are also needed to ensure that the photons exiting the sample do not re-enter it (Pine *et al.*, 1990). The solution of eqn (19.4) in an infinitely wide scattering slab, with U having a Gaussian entering profile and exponential extinction, has been obtained as in Bellini *et al.* (1991). The desired diffuse intensity against position away from the beam axis is obtained by integrating over time the solution for U in the selected position. The unknown absolute amplitude of the incident U has been set by fitting the profiles, also enabling the measurement of the photon mean free path via the transmitted intensity, as for $T = 36\,°C$ in Figure 19.2. At lower temperatures, we have determined the turbidity $\tau = 1/\ell$ deducing ℓ from the best fit of the diffuse background. The good quality of the fits is shown in Figure 19.2 for $T = 34\,°C$. The overall result is shown in Figure 19.3. The measured τ is very large compared with that of bulk nematic 8CB for which $\tau \approx 2$ mm⁻¹. For example, at $T = 7\,°C$ for 8CB in the

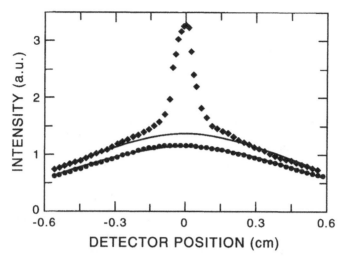

Figure 19.2 Intensity exiting the 0.36 g/cm³ sample as a function of the distance in the sample plane from the incident beam axis. $T = 36\,°C$ (diamonds) and $T = 34\,°C$ (circles). The lines represent the fit to the data by the model of photon diffusion through the aerogel.

0.36 g/cm³ aerogel, $\tau \approx 40\,000$ mm⁻¹, corresponding to a transmittance through a 1 mm thick sample of about $I_T/I_0 \approx 10^{-18}$, making the transmitted beam negligible in comparison with the diffuse scattering.

Figure 19.3 also shows the measured specific heat C_p in the 8CB 0.36 g/cm³ sample (Bellini *et al.*, 1992; Wu *et al.*, 1995). The main peak in C_p is due to the isotropic to nematic transition of liquid crystal within the aerogel pores. The temperature of the peak is depressed by about 1.3 °C with respect to the bulk peak position, T_{NI}, located by the small sharp feature on the right side of the figure, a result of remnant bulk 8CB on the aerogel external surfaces. The T scales are shifted to overlap this feature with the turbidity step due to bulk 8CB, as discussed above.

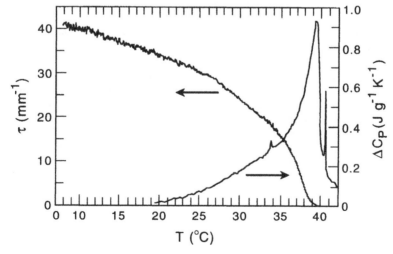

Figure 19.3 Turbidity τ (left axis) and specific heat ΔC_p (right axis) against T for the 8CB–LCSA sample having aerogel density of 0.36 g/cm³.

19.3.2 *Model 1: Single Uniaxial Domains*

The purpose of this and the next section is to discuss the origin of the large measured τ, making use of the optical properties of the bulk nematic and smectic. We will present two different models which yield explicit analytical expressions for τ. In both cases we model the liquid crystal as a multidomain nematic structure, with the gradual increase in τ a consequence of a gradual increase in mean domain size. At low T the nematic domain size distribution saturates to be the pore size distribution, $P_p(L)$, given in eqn (19.2), and the mean pore size saturates at $\langle p \rangle$. At this point we have no information on the nature of the domain size distribution when the mean size is smaller than $\langle p \rangle$. In the case of smectic ordering, the ubiquitous Lorentzian lineshape of the powder-averaged Bragg scattering indicates that for the appearance of smectic order, the size distribution is the same as that of eqn (19.2), but with a T-dependent mean size. We assume here that a similar condition applies to the nematic ordering, namely that, in general, the nematic domain size distribution, $P_\zeta(L)$, is of the form of eqn (19.2), but with a mean size $\zeta < \langle p \rangle$: $P_\zeta(L) = \exp(-L/\zeta)/(\zeta L)^2$. Furthermore, we assume that $\mathbf{n}(\mathbf{r})$ has a different but random orientation in each domain and that the domain boundaries are sharp. Since in the nematic and smectic phases the domains are optically uniaxial and have large intrinsic birefringence $\Delta n = n = n_\parallel - n_\perp$, the local optical properties change abruptly in going from one domain to another. Both models show that the *optical discontinuities at the domain boundaries are sufficient to explain the observed large optical turbidity*. Since the maximum domain size encountered in these experiments is 1400 Å (in the lowest-density aerogel), we can safely assume we are in the Debye–Rayleigh–Gans regime. In fact, since $\Delta n < 0.25$, the system satisfies the condition $<p>\Delta n/\lambda \ll 1$, where $\langle p \rangle$ is the domain size. We will make use of this approximation in many steps of the following discussion.

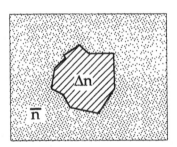

In model 1 we suppose every uniaxial domain is immersed in a medium that is optically isotropic having an effective refractive index $\bar{n} = (n_\parallel + 2n_\perp)/3$. Model 1 is a 'mean field' approach, in the sense that it considers every single domain by itself, the scattering being determined by its optical mismatch with respect to an average of all the other domains. For the equivalent case of uniaxial colloids index matched with the solvent, it has been rigorously proved that the amplitude of both the polarized and depolarized field scattered from a single domain are functions of the director field $\mathbf{n}(\mathbf{r})$ and that they can assume both positive and negative values (Berne and Pecora, 1990; Degiorgio *et al.*, 1994). Coherent summations of light scattered from different domains are therefore, on the average, cancelled by the random director orientation, and hence the overall scattering is an incoherent sum of single-domain contributions. If the incident intensity is linearly polarized along the z axis, the amplitudes of polarized (E_{VV}) and depolarized (E_{VH}) field scattered by a single domain in the Debye–Rayleigh–Gans regime are (Berne and Pecora, 1990)

$$E_{VV} = \frac{E_0 k^2}{4\pi\varepsilon D} (\cos^2\delta - \tfrac{1}{3})\beta \sin \gamma_1 \tag{19.5}$$

$$E_{VH} = i \frac{E_0 k^2}{4\pi\varepsilon D} (\cos \delta \sin \delta)\beta \sin \gamma_2 \tag{19.6}$$

where $k = 2\pi\varepsilon^{1/2}\lambda$, is the wavelength, ε is the average dielectric constant at optical frequency, δ is the angle between the incident optical polarization and $\mathbf{n(r)}$, γ_1 the angle between the scattered wavevector and the component of the induced dipole normal to $\mathbf{z}$ and D the distance at which the scattered field is detected. $\beta = \alpha_\| - \alpha_\perp \sim V (n_\|^2 - n_\perp^2) = L^3(n_\|^2 - n_\perp^2)$ is the anisotropy of the domain polarizability. This total scattered intensity from one domain is

$$I_{tot} = \int 2\pi D^2 \sin\gamma_1 \, d\gamma_1 \langle E_{VV}^2 \rangle + \int 2\pi D^2 \sin\gamma_2 \, d\gamma_2 \langle E_{VH}^2 \rangle \tag{19.7}$$

where the average has to be taken over all the possible director orientations. Note that the integral and the average are independent operations so that, with $V = L^3$,

$$I_{tot} = \frac{I_0 k^4}{\varepsilon^2} \frac{1}{27\pi} 4\bar{n}^2(n_\|^2 - n_\perp^2)L^6 \tag{19.8}$$

We have then to sum up all the contributions coming from the domains within a unit of volume:

$$\tau = \int \frac{I_{tot}}{I_0} P(L) \, dL = \phi \frac{32}{9\pi} \frac{\bar{n}^2(n_\|^2 - n_\perp^2)}{\varepsilon^2} k^4 \zeta^3 = \phi \frac{8}{9\pi} \frac{\Delta\varepsilon^3}{\varepsilon^2} k^4 \zeta^3 \tag{19.9}$$

where we have made use of $\varepsilon = n^2$. In this equation we have not included any form factor, the domains being small with respect to the optical wavelength.

19.3.3 Model 2: Random Scalar Refractive Index

An alternative way to approach the problem is one already proposed (Bellini *et al.*, 1992). It entails reducing the uniaxial nature of the optical properties of each domain to an effective scalar refractive index. This is done by assuming that the effective refractive index in every domain equals the intersection of the index ellipsoid along a direction fixed in the laboratory frame. This approximation has the advantage of yielding the scattering properties not only of one domain, but of the whole multidomain structure at once. Within this second model, the correlation function $g_\varepsilon(R)$ for the dielectric constant fluctuations is obtained by replacing $f(r)$ by $\varepsilon(r)$ in eqns (19.1) and (19.3). The amplitude $\langle\varepsilon^2\rangle - \langle\varepsilon\rangle^2$ of the optical dielectric constant fluctuation has to be evaluated by randomly cutting the index spheroid. We obtain

$$\langle\varepsilon^2\rangle - \langle\varepsilon\rangle^2 \sim \frac{1}{6.5}(\varepsilon_\| - \varepsilon_\perp)^2 \tag{19.10}$$

The scattered intensity $I(q)$ from the whole multidomain structure can be calculated from $g_\varepsilon(R)$ as

$$I(q) = \frac{\zeta^3 k^4 I_0 V}{13\pi D^2(1 + q^2\zeta^2)^2} \frac{(\Delta\varepsilon)^2}{\bar{\varepsilon}^2} \tag{19.11}$$

To calculate the turbidity we integrate over all scattered light directions. In order to do this, we neglect the $q^2\zeta^2$ term in the denominator of eqn (19.11):

$$\tau = \phi \int_0^{2\pi} d\varphi \int_0^\pi \sin^3\gamma \, d\gamma D^2 \frac{1}{I_0 V} \frac{\zeta^3 k^4 I_0 V}{13\pi D^2} \frac{(\Delta\varepsilon)^2}{\bar{\varepsilon}^2} = \phi \frac{8}{36} \zeta^3 k^4 \frac{(\Delta\varepsilon)^2}{\bar{\varepsilon}^2} \tag{19.12}$$

where the additional $\sin^2\gamma$ factor takes into account the dipole radiation intensity dependence of the polarized scattered light. Note that this Model 2 result is quite similar but not identical to that obtained previously in equation (1) of Bellini *et al.* (1991). In particular the dependence of τ on the bulk Δn and on the mean domain size ζ is the same. The main difference comes from neglecting the form factor q^{-4} cut-off in eqn (19.11) and from a different approximation in passing from eqn (19.11) to (19.12). When applied to an LCSA sample having $\xi \approx 200$ Å, eqn (19.12) gives a result which is 1.4 times smaller than equation (1) of Bellini *et al.* (1992). Note that the results of the Models 1 and 2 are very similar. The only difference is in the numerical constants: in eqn (19.9) we have $8/9\pi = 0.283$ while in eqn (19.12) we have obtained $8/36 = 0.222$.

If we want to use these results to discuss the turbidity of 8CB 0.08 g/cm^3 aerogel samples, we would have to improve the modelling by including the q dependence of the single-domain form factor in eqns (19.5) and (19.6) or eqn (19.12). The effect would be to reduce the contributions to τ from the large domains. Both models neglect any correlation between $\mathbf{n}(\mathbf{r})$ in the different domains, and therefore neglect any coherent summation of the scattered field from different domains. This is similar to the assumption made in the analysis of the *x*-ray diffraction by smectic layering in the aerogel which probes single-domain properties only, the coherent contribution being cancelled by the random phase of the smectic order parameter in adjacent domains.

19.3.4 *LCSA Turbidity from Bulk Optical Properties*

In this section we relate the measured aerogel turbidity and bulk properties via eqns (19.9) and (19.12). The temperature-dependent terms in the equations are ξ and Δn. In the bulk, because of the first-order nature of the nematic to isotropic phase transition, Δn changes abruptly when $T = T_{NI}$. In the aerogel upon lowering T, the measured $\tau(T)$ instead grows continuously from zero, indicating that $\langle p \rangle$ grows continuously from zero. Calorimetry (Wu *et al.*, 1995) shows that indeed in the isotropic phase the nematic ordering process is identical to the bulk pretransitional growth of paranematic fluctuations, provided that a temperature shift δT is introduced. The nematic correlation length then grows in the LCSA sample as it does in the bulk isotropic phase until it saturates because of the silica aerogel disordering effect. In the present discussion we use ζ to indicate both the correlation length $\zeta < \langle p \rangle$ controlled by thermally driven fluctuations which, at low T, becomes $\zeta_l \approx \langle p \rangle$, and the nematic mean domain size determined by the aerogel-induced quenched disorder.

In comparing eqns (19.9) and (19.12) with $\tau(T)$ we assume $\Delta n(T) = n_B(T)$, the birefringence measured in the bulk. According to the calorimetry data, we have also introduced a temperature shift δT in $\Delta n_B(T)$ matching the observed depression of the C_p peak. For the 0.36 g/cm^3 aerogel $\delta T \approx 1.3°$C (Bellini *et al.*, 1992; Wu *et al.*, 1995). A problem in using $\Delta n(T)$ as measured in the bulk (Janossy and Bata, 1978) is that results are available in a temperature range of only about 15°C below T_{NI}, i.e. in the whole nematic phase and in part of the smectic phase. Since the smectic to crystal transition is lowered to about 0 °C in the aerogel (Clark *et al.*, 1993), our aerogel data extend in the nematic and smectic A phases for more than 35°C below T_{NI}. By inserting $\Delta n_B(T - \delta T)$ in eqns (19.9) and (19.12) we adjust the value of ζ to fit the experimental data at the lowest temperatures, as shown in Figure 19.4 (open circles). We obtain for $\zeta_\tau \approx \langle p \rangle \approx 209$ Å using eqn (19.9) and $\zeta_\tau \approx 227$ Å using eqn (19.12). This is in good agreement with both the mean smectic domain size as measured by *x*-ray scattering in the same aerogel ($\xi_S \approx 215$ Å (Clark *et al.*, 1993)) and with the mean pore size calculated using the small-angle X-ray diffraction from the empty sample, $\langle p \rangle \approx 180$ Å (see section 19.2.2).

To check better whether the $\Delta \varepsilon_B^2$ temperature dependence is actually sufficient to explain the observed behaviour, we need to extend the range of Δn_B. We have performed this extrapolation by exploiting the fact that with good approximation $\Delta \varepsilon_B(T)$ is proportional to the nematic order parameter $S(T)$ (de Jeu, 1978). We have then proceeded in two different ways. First we have compared the results with S^2 as calculated from computer simulations of the Lebwohl–Lasher model (Fabbri and Zannoni, 1986). The comparison is performed by stretching the T axis to make T_{NI} from the simulation coincide with experimental $T_{NI} - \delta T$, and by scaling the S^2 amplitude to overlap with Δn_B. The result is shown in Figure 19.4. (open

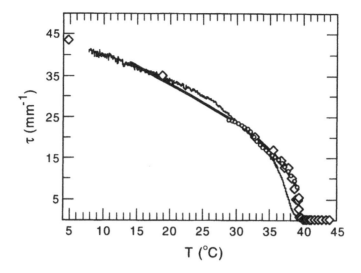

Figure 19.4 Comparison between the *T* dependence of the measured turbidity (thin line) and bulk Δn^2 shifted in temperature so as to overlap the bulk T_{NI} with the LCSA specific heat peak position. Δn^2 obtained in different ways: measured bulk 8CB birefringence from Janossy (circles); mean field theoretical calculation of the nematic order parameter for Humphries (thick line); nematic order parameter from Lebwohl–Lasher computer simulation from Zannoni (diamonds).

diamonds). Second we have similarly compared τ with the S^2 theoretically calculated from an extension of the classical Maier–Saupe theory (Humphries *et al.*, 1972); this is also shown in Figure 19.4 (thick line). We conclude that:

1 The $\tau \sim \Delta\varepsilon_B^2$ scaling is clearly confirmed.
2 In analogy to the calorimetry results, the nice agreement of the temperature dependence of $\tau(T)$ and $\Delta\varepsilon_B{}^2(T)$, together with the numerical agreement on the value of ζ, indicate that, except for a temperature interval close to the I–N transition, *the liquid crystal in the silica aerogel locally behaves as the bulk, having the same growth of the nematic order parameter and therefore the same birefringence.*
3 No sign of smectic order is observed. In part, this is certainly due to the small effect that the N–A transition has on the bulk birefringence itself. It is possible that the small bump in the experimental results between 20 °C and 30 °C is connected to the growth of smectic order within the pores.
4 By reversing the argument we can extract from the turbidity data an extension to the measured $\Delta\varepsilon_B$, as shown in Figure 19.5.
5 Finally, this aerogel experiment enables a (successful) comparison of $\Delta n(T)$ with that of the Lebwohl–Lasher computer simulation extending over a temperature range much wider than normally accessible by study of the bulk.

Figure 19.5 also shows the fit of $\Delta n(T)$ with the phenomenological equation already used to fit successfully the birefringence of the bulk cyanobiphenyl 5CB as well as of other nematogens (Wu and Cox, 1988):

$$\Delta n(T) = \Delta n_0 \left(1 - \frac{T}{T_{NI}}\right)^{\beta} \tag{19.13}$$

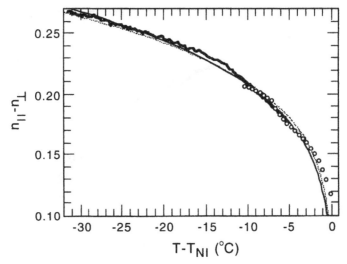

Figure 19.5 Bulk 8CB birefringence against T: measured in bulk 8CB (circles); extracted from LCSA turbidity (thick line); fit with an empirical function (Wu and Cox, 1988) of the LCSA data only (thin line, $\beta = 0.23$) and of the LCSA together with the bulk data (dotted line, $\beta = 0.21$).

where again we have used the LCSA T_{NI} as defined by the calorimetry peak position. Our best fit yields $\Delta n_0 \approx 0.42$ and $\beta \approx 0.21$, quite close to what is found in bulk materials and by fitting simulation results (Berardi *et al.*, 1993).

Both model 1 and model 2 are clearly more appropriate to discuss the low-temperature ($T < 36\,°C$) behaviour where the domain size distribution is controlled by the aerogel structure. At higher temperatures the size of the nematic clusters depends on a mix of bulk pretransitional properties and aerogel-induced disorder and may have a size distribution different from $P_{\langle p \rangle}(L)$. Nevertheless, we assume that eqns (19.9) and (19.12) hold in the whole temperature range, and use them to deduce the temperature dependence of in the proximity of T_{NI}. In order to do this, we have set again $\Delta n(T) = \Delta n_B(T)$ and calculated ζ through eqn (19.9). The result is shown in Figure 19.6 (diamonds). An improvement of model 1, making it more suitable to describe the high-temperature behaviour, is possible by assuming that in each pore of size L there is a nematic domain of smaller size $L' = aL$, where $a \le 1$ is the same for all pores. The total intensity scattered by each domain is then given by eqn (19.8) provided that V is substituted with $V' = a^3 V$. The domain size distribution in eqn (19.9) is instead still the same, being determined by the host pore and not by the actual domain size. The turbidity which results is

$$\tau = \int \frac{I_{tot}}{I_0} \frac{V'^2}{V^2} P(L)\, dL = a^6 \phi\, \frac{8}{9\pi} \frac{\Delta \varepsilon^2}{\varepsilon^2}\, k^4 \zeta^3 \tag{19.14}$$

In Figure 19.6 we show $\zeta' = a\zeta_s$ extracted from the turbidity data by means of eqn (19.14), where ζ_s is the saturated value of ζ. As is clearly visible, at the temperature corresponding to the C_p peak (Bellini *et al.*, 1992) (marked in Figure 19.6 by an arrow), the nematic correlation length is still much shorter than the saturated value. This is analogous to what is found for the N–A transition by comparing X-ray scattering and calorimetry and is a clear proof of the disordering effect of the silica aerogel. As the temperature is lowered, the increased elastic constants, together with the reduced thermal energy, push the domain boundaries closer to the silica strands.

Finally we can calculate the turbidity under the assumption that the aerogel surfaces are disordering, resulting in an isotropic layer of thickness λ on the pore surface. The effective

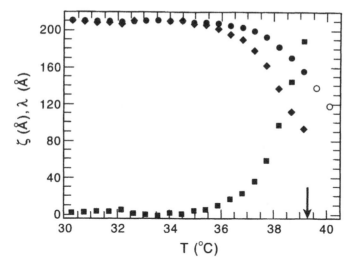

Figure 19.6 Mean pore size ζ extracted from the LCSA turbidity measurement using eqn (19.9) (diamonds) and using eqn (19.14) (circles). The open circles have been obtained by linearly extrapolating the bulk 8CB Δn into the isotropic phase. The arrow indicates the specific heat peak position. The squares give $\lambda(T)$, the thickness of the isotropic layer at the pore surfaces, under the assumption that the pore surfaces are disordering. Pores of dimension $L < 2\lambda$ are isotropic.

ordered volume within a pore of size L then has dimension $L - 2\lambda$ and the turbidity can be obtained as

$$\tau = \int_{2\lambda}^{\infty} I_{\text{tot}}[(L - 2\lambda)^3]/I_0 P(L) \, \mathrm{d}L$$

so that $\lambda(T)$ can be extracted from the measured τ. The resulting $\lambda(T)$ is shown as the squares in Figure 19.6. As T increases order disappears first in the smallest pores, i.e. those for which $L < 2\lambda$.

19.4 Dynamic Light Scattering – General Features

So far two different studies of the phase behaviour of LCSA by dynamic light scattering have appeared (Wu *et al.*, 1992; Bellini *et al.*, 1995). They both find strongly non-exponential decay of the orientational correlation functions, having a relaxation extending over many decades in time. The slowing down, observed upon lowering the temperature in the nematic phase, has been discussed in both cases and different interpretations offered. It is difficult to compare the two results because, although they have been obtained making use of aerogels having nominally similar pore size, the aerogels have quite different structure. Here we extend our earlier results (Bellini *et al.*, 1995) and compare them, where interesting, with the data of Wu *et al.* (1992). Both experiments show that the dynamics of the LCSA exhibits novel phenomenology: the slowing which appears in the bulk abruptly at the ordering transitions is spread out in the aerogel over a wide range of T. In addition study of the dynamics makes possible the observation of phase behaviour, such as glass transitions, not detectable otherwise.

19.4.1 DLS from LCSA Samples: Contrast and Fluctuations

We have performed measurements of dynamic light scattering (DLS) on 8CB – aerogel samples of varying aerogel density for 10 °C $< T <$ 42 °C, i.e. in the nematic and smectic phases, the freezing transition within the aerogel being depressed to about $T = 0$ °C (Clark *et al.*, 1993). The measurements have been performed by using an ordinary DLS setup having a 25 mW He–Ne laser as a light source. The collection optics was a standard two pin-hole scheme enabling selection of the detected intensity well within one coherence area. Time correlation functions were obtained using the multiple sample timebase Brookhaven 9000 digital correlator, enabling measurement over a wide time range, from a fraction of a microsecond to a few seconds. We have measured the intensity correlation function $G_2(\tau) = \langle I(t) I(t + \tau) \rangle$ of light exiting the aerogel slabs as a function of scattering angle for both polarized ($G_2^{VV}(\tau)$) and depolarized ($G_2^{VH}(\tau)$) components.

Because of the high turbidity which develops rapidly as T is lowered below T_{NI}, for nearly the entire T range studied, the DLS probes light which has undergone many scattering events. This multiple scattering potentially increases the number of phase-changing fluctuations per photon path, so we expected to find in these experiments a significant speeding up of the scattering intensity fluctuations with increasing turbidity, as is observed in diffusing wave spectroscopy (Pine *et al.*, 1990). To our great surprise, no sign of increasingly more rapid phase fluctuations was found upon decreasing temperature. On the contrary a progressive slowing down was always observed. Another very notable feature of the measured $G_2(\tau)$ is the low value of its contrast $C = [G_2(0) - G_2(\infty)]/G_2(\infty)]$, corresponding to an almost static speckle pattern which we can directly observe after the first pin-hole. In Figure 19.7 we show the average number $\langle n \rangle$ of scattering events per photon path reaching the detector as obtained by the approximation $\langle n \rangle \approx 1 + d^2/\ell^2$, where $d = 1$ mm is the LCSA sample thickness, and $\ell = \tau^{-1}$ is the photon scattering mean free path. In Figure 19.7(a) we also show the temperature dependence of the mean value of C obtained by averaging C over different positions in the speckle pattern.

The low value of C has two important meanings: first it indicates that the scattering events are principally due to static fluctuations of the refractive index, such as frozen domains or groups of domains strongly correlated in orientation; second it indicates that our dynamical measurements have been performed in the heterodyne regime, the fluctuating intensity being mixed with a static background acting as local oscillator. When $\langle n \rangle$ is large, i.e. when the probability of having photon paths built up by scattering events which are all perfectly elastic is small, the low value of C clearly indicates that upon lowering the temperature, 'dynamical' scattering events become less and less probable. Knowing C and $\langle n \rangle$ enables us to give an estimate of the fraction of the scattering events due to fluctuating sites. This can be done as follows.

The total scattered intensity I is a sum of a static part I_S, emerging from photon paths in which all the scattering events are elastic, and a fluctuating part I_D, due to those photon paths in which at least one scattering event is quasi-elastic. For heterodyne DLS the measured intensity correlation function is

$$G_2(\tau) = \langle I(t)I(t + \tau) \rangle \sim I_S^2 + 2I_S I_D G_1(\tau) \tag{19.15}$$

where $G_1(\tau)$ is the field correlation function. Therefore, for heterodyne scattering, $C \sim 2I_D/(I_S + 2I_D)$. We suppose that p is the probability that a single scattering event is quasi-elastic. The probability of having a whole photon path of n scattering events made of only elastic events is $P = I_S/(I_S + I_D) = (1 - p)^n$. Therefore

$$(1 - p)^n = \frac{2(1 - C)}{2 - C} \tag{19.16}$$

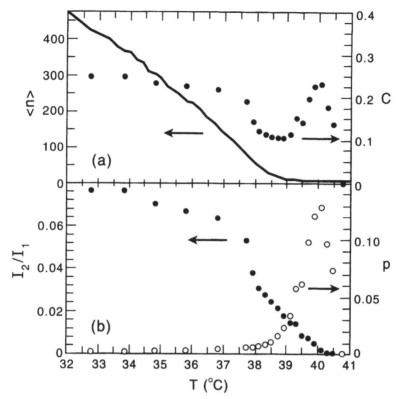

Figure 19.7 (a) Average number of scattering events per photon path ⟨n⟩ and contrast C of the intensity correlation function. (b) Ratio between the intensities carried by photon paths having one only non-elastic scattering event I_1 and two non-scattering events I_2, and probability p of non-elastic scattering events.

Using eqn (19.6) and the data in Figure 19.7(a) we can extract p agaisnt T. We show the result in Figure 19.7(b). Since we are mainly interested in understanding the absence of the typical multiple scattering fast dynamics, it is worth evaluating the ratio I_2/I, where I_1 and I_2 are the intensities of light carried by all the photon paths having, respectively, one and two inelastic scattering events among the total n scattering events per path. By assuming, as in diffusing wave spectroscopy, that the intensity is proportional to the number of photon paths, we can estimate

$$\frac{I_2}{I_1} = \frac{(n-1)p}{2(1-p)} \tag{19.17}$$

When $p \ll 1$ and $n \gg 1$, the typical experimental situation, eqns (19.6) and (19.7) give

$$\frac{I_2}{I_1} \approx \frac{1}{2} \ln\left(\frac{2-C}{2(1-C)}\right) \tag{19.18}$$

That is, when $n \gg 1$, the ratio between the numbers of scattering paths having two dynamic events and the number of scattering paths having a single dynamic event is independent of n. When $C \ll 1$, $I_2/I_1 \approx C/4$. We show in Figure 19.7(b) the ratio I_2/I_1 as determined by eqns (19.16) and (19.17): I_2/I_1 is always less than 0.1 and therefore in our experiment the residual dynamics is dominated by paths having only one non-elastic scattering event. We therefore

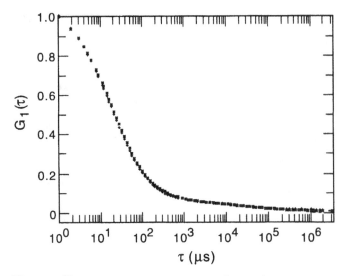

Figure 19.8 $G_1^{VV}(\tau)$ and $G_1^{VH}(\tau)$ measured in the 8CB 0.36 g/cm³ aerogel at $T = 39.6\,°C$.

conclude that, *despite the large average number of scattering events per photon path, the dynamics of the intensity fluctuations is basically the same we would have had in a standard single scattering experiment in the heterodyne regime.* As expected, the dependence on scattering angle of the scattered intensity dynamics, already weak at the highest temperatures, vanishes in the multiple scattering regime.

19.4.2 Modelling the Dynamics of a Multidomain Nematic Structure

We show in Figure 19.8 the normalized $G_2^{VV}(\tau)$ and $G_2^{VH}(\tau)$ measured at $T = 39.6\,°C$, in the 8CB 0.36 g/cm³ sample, a temperature where the single scattering contribution is dominant. As is clear from the figure, $G_1^{VV}(\tau)$ and $G_1^{VH}(\tau)$ are identical. Both relaxations show a faster decay mechanism, completed in about 100 μs, and a slower decay, widely spread over time, completed at this temperature in about 100 ms. The dynamics of the LCSA system is quite likely due to local reorientations of the director or to movements of defects in the nematic orientation pattern. It is useful to extend the multidomain nematic model to include the reorientational dynamics of the domains, since it provides a simple context in which to discuss the results, and because it, in fact, explains some of the experimental observations.

We suppose that the microscopic structure of the LCSA is of nematic-like domains having fixed positions, the open cavities of the silica aerogel, but with a certain freedom to reorient. Let us also adopt the same point of view as model 1 of section 19.3.2: the scattering properties of the domains within the LCSA sample are those of independent anisotropic domains, index matched with the solvent. The consequences are the following:

1. Since the intensity scattered by the different domains adds incoherently, the overall correlation function is just the summation of the correlation functions of the intensity scattered by the single domains. Therefore, assuming that each domain contributes to the overall relaxation with a single-exponential decay having correlation time $\tau(L)$, we obtain

$$G_{1,S}(\tau) = A_S \int_0^\infty P(L) \exp[-t/\tau(L)]L^6 \, dL \qquad (19.19)$$

2 $G_1^{VV}(\tau) \propto G_1^{VH}(\tau)$, as experimentally observed at all temperatures. This result can be obtained for the particular case of free rotators by explicitly performing the correlations of the scattered field in eqns (19.5) and (19.6), or, more generally, by noting that E_{VV} and E_{VH} can be transformed into each other by proper rotations of the reference axis (Berne and Pecora, 1990). In the case of free identical rotators having fixed position $G_1^{VV}(\tau) \propto G_1^{VH}(\tau) \propto \exp(-6D_R\tau)$, where D_R is the rotational diffusion coefficient.

3 If the single domains are free to rotate in the entire 4π solid angle, their constrained position does not give rise to any static background of the scattered light, and the intensity fluctuations should eventually become uncorrelated. This is again a consequence of eqns (19.5) and (19.6), where $\langle E_{VV} \rangle = \langle E_{VH} \rangle = 0$.

4 If, on the other hand, the accessible director orientations within a domain are restricted within a finite solid angle, the correlation function of the intensity scattered by that domain does not decay to the uncorrelated value $\langle I \rangle^2$, but rather saturates to a higher value. Although this also leads to a low contrast C_1 in $G_2(\tau)$ for a single scattering event, the observed low value of C for multiply scattering still cannot be explained by simply assuming a multidomain structure where all domains fluctuate but have restricted rotational freedom. In such a situation, the many scattering processes per photon path augment both the net contrast and rate of the intensity fluctuations produced by each single scattering event, as can be shown as follows. We assume that each scattering event induces a time-dependent phase shift $\delta(t)$ in the scattered field, which is confined within a limited angle $\pm \delta_\infty$, so that, when $t \to \infty$, $\langle \delta(t)^2 \rangle \to \delta_\infty^2$. This is what happens, for example, when scattering from a Brownian oscillator. Hence, after one scattering event, $C_1 = [1 - \exp(\delta_\infty^2)]/[1 + \exp(\delta_\infty^2)]$, while after n scattering events $C = [1 - \exp(n\delta_\infty^2)]/[1 + \exp(\delta_\infty^2)]$. Thus, upon lowering the temperature into the multiple scattering regime, we expect C to be governed by $(1 - C)/(1 + C) \propto [(1 - C_1)/(1 + C_1)]^{\langle n \rangle}$ and the first cumulant Γ of $G_2(\tau)$ to increase as $\Gamma \propto \langle n \rangle$, i.e. make the decay more rapid. This is clearly in disagreement with both the low C of Figure 19.7 and the observed weak temperature dependence of the correlation times. This affirms the conclusion of section 19.4.1 that a large fraction of the domains are actually frozen.

19.4.3 Non-ergodic Behaviour

In section 19.5 we will develop the simplest way to extract predictions from the independent nematic rotator model introduced above, i.e. we will suppose that the domains can be split in two different ensembles: frozen domains, accounting for the static background, and freely rotating domains, producing the observed dynamics. Of course all the intermediate situations have to be present and it is quite likely that the observed dynamics is also due to partially frozen domains.

The dynamic light scattering from non-ergodic media, and in particular the problem of how to extract meaningful dynamical information from scatterers having restricted configurational freedom, has been already widely discussed (Pusey and van Megen, 1989; Pusey, 1994). The discussion usually deals with positional non-ergodicity, but it can easily be extended to the present situation. The way to extract the correct rotational diffusion coefficient D_R of uniaxial domains index-matched with the solvent and having restricted rotational dynamics would be to measure the homodyne $G_2(\tau)$ in different positions in the speckle pattern, and to average them. Such an 'ensemble-averaged' homodyne $G_2(\tau)_E$ has a short-time behaviour which is usually the free diffusional dynamics of the system, and saturates at longer time to a plateau

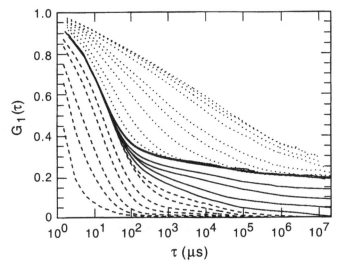

Figure 19.9 Evolution of $G_1(t)$ for an 8CB 0.36 g/cm³ aerogel sample against temperature. The three different regimes discussed in the text are identified by different kinds of lines: dashed, intrapore 'fast nematic' relaxation regime ($T = 40.32, 40.21, 40.11, 39.89, 39.71, 39.48, 39.11$ °C); full, interpore 'nematic glass' transition regime ($T = 38.8, 38.69, 38.31, 37.58, 35.42, 34.45, 32.96$ °C); dotted, intrapore 'smectic slowing down' regime ($T = 30.96, 29.01, 27.01, 25.04, 23.55, 22.09, 21.11, 20.13, 18.19$ °C).

higher than the square of the average scattered intensity. The first cumulant Γ of $G_2(\tau)_E$ yields the 'true' dynamics of the system in the same way it would for scattering from totally free rotators (Berne and Pecora, 1990): $\Gamma = 6D_R$. This method has usually been discussed for homodyne measurements, where the observed static speckle pattern is a consequence of the restricted dynamics only and not of the presence of other static scatterers. It is probably extendable to heterodyne scattering experiments where the local oscillator intensity is known. On the contrary, because of the intrinsic static background present in all our measurements, we cannot evaluate a meaningful ensemble-averaged $G_2(\tau)$. Therefore, *in our experimental situation it is impossible to discern the apparent rotational dynamics of partially frozen domains from the true dynamics of freely rotating domains.*

In this introductory discussion to the DLS results we have indicated that the complexity of quasi-elastic multiple scattering from possibly non-ergodic scatterers makes it difficult to define an unambiguous interpretative scheme, so that the interpretation will necessarily be model dependent. On the other hand, within the context of the model to be described in the next two sections, the results are internally self-consistent and show simple and smooth trends as the temperature is changed.

19.5 DLS Results – Nematic Phase

In Figure 19.9 we show an overview of the temperature dependence of the correlation functions for the 8CB 0.36 g/cm³ LCSA sample. They form a monotonic sequence, characterized by a progressive slowing down as the temperature is lowered. The full line correlation functions have been recorded in the temperature range of the bulk nematic phase ($T > 33$ °C) while the dotted ones represent the growth of the smectic phase.

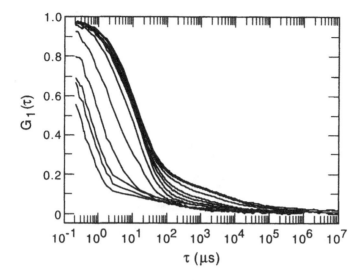

Figure 19.10 Evolution of $G_1(t)$ for the 8CB 0.36 g/cm³ aerogel sample of Figure 19.9 at $T = 40.36$, 40.21, 40.15, 40.05, 39.90, 39.80, 39.69, 30.40, 39.320, 38.19, 35.70, 33.71 °C (from the lowest curve up), but at a different sample position. Here non-ergodic behaviour, evidenced in Figure 19.9 by the growing baseline plateau, is masked by slow drift of the background speckle pattern.

The data sets for the other aerogel densities are similar, each exhibiting several well-defined features which are immediately recognizable:

1 As the temperature drops below the bulk T_{NI} (about 40.5°C), the static scattered field grows and, on the top of it, the first signs of dynamics appear.
2 The correlations measured at the highest temperatures ($T > 40$°C) have an evident bimodal shape, as can be seen in Figure 19.10, where the delay time interval has been reduced down to 0.2 μs.
3 Upon lowering the temperature, the fast relaxation slows down considerably (from 0.2 μs to 20 μs for the 0.36 g/cm³ aerogel), until it saturates at about 39.5°C to a temperature-invariant shape.
4 Upon lowering the temperature into the nematic phase, the slow relaxation mode grows smoothly in amplitude and mean relaxation time.
5 Between 33.5 and 36.5°C the relaxation curves do not depend on the temperature.
6 Upon further temperature reduction a drastic slowdown over many decades is observed. This aspect will be discussed in section 19.6.

A clue crucially motivating our interpretation of the observed fast relaxation dynamics is that the relaxation times are different in the different aerogels. Figure 19.11 shows four relaxation curves recorded at $T \approx 36$°C, from 8CB LCSA samples prepared with the aerogels of Table 19.1, having four different densities and therefore different mean pore sizes. *The larger the cavities within the aerogel, the slower the dynamics of the scattered intensity. Hence, the slowing down observed while entering the nematic phase is due to the continuous growth of the nematic domain size within the aerogel.*

The fast relaxation is non-exponential, indicating a process with a distribution of relaxation times, in agreement with eqn (19.19). The simplest way to extract a semi-quantitative comparison with the experiment from eqn (19.19) is to assume that the orientational dynamics of a given domain within the aerogel is the same as that of the bulk pretransitional paranematic

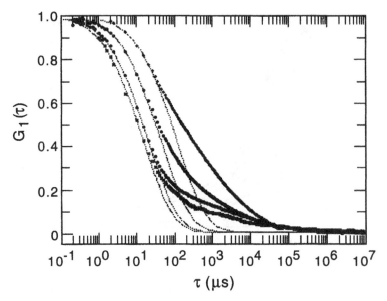

Figure 19.11 $G_1(\tau)$ measured at $T \approx 36\,°C$ for 8CB in four different density aerogels and fitted with eqn (19.20): 0.08 g/cm³ (plus), 0.17 g/cm³ (solid dots), 0.36 g/cm³ (open dots), 0.69 g/cm³ (crosses).

fluctuations in the isotropic phase. This analogy is suggested by the disordering effect of the silica aerogel, which can be viewed as creating an isotropic-like environment around each nematic domain. The orientational dynamics of pretransitional nematic clusters have been measured using the optical Kerr effect (Pouligny *et al.*, 1983). The relaxation of the orientational correlation function is exponential, $\exp(-6D_R t)$, and the rotational diffusion coefficient D_R is proportional to the inverse square of the pretransitional nematic correlation length ξ_N: $D_R = C/\xi_N^2$, where C is a constant. At the transition temperature, $\xi_N \approx 85$ Å in 8CB (Chen *et al.*, 1989) and the orientational correlation time $\tau_N \approx 0.4\ \mu s$ (Pouligny *et al.*, 1983) so that $C = \xi_N^2/6\tau_N$. We assume that in the aerogel D_R depends on pore dimension L in the same way that it depends on ξ_N, even when the paranematic fluctuations turn into nematic domains, having mean size ζ larger than the maximum bulk ξ_N. Under this assumption eqn (19.19) transforms into

$$G_{1,S}(t') = A_S' \int \exp(-y)\, \exp(-t'/y^2)\, y^4\, \mathrm{d}y \tag{19.20}$$

where

$$t' = \frac{\tau}{\zeta^2} \frac{\xi_N^2}{\tau_N} \quad \text{and} \quad A_S' = A_S \zeta^3$$

It is therefore easy to calculate a fit to the faster decay component of the measured $G_1(\tau)$ where the only free parameters are A_S' and ζ_{dyn}, the baseline being determined by the level at large delay times. Examples of fitting curves are given in Figure 19.11. For $T < 40\,°C$ the fits are quite good over delay times extending to about 50 μs. For $T < 40\,°C$, it is difficult to establish the real quality of the fit. By forcing it anyway, we have extracted from the temperature evolution of $G_1(\tau)$ the temperature dependence of ζ_{dyn}, which is shown in Figure 19.12 for the 0.36 g/cm³ aerogel, together with ζ_t, as obtained by the turbidity analysis (section 19.3.4). Differences in the behaviour are evident, the main one being that the temperature interval of nematic growth before saturation at low T is reached is about 2 °C for ζ_{dyn}, and about 3 °C

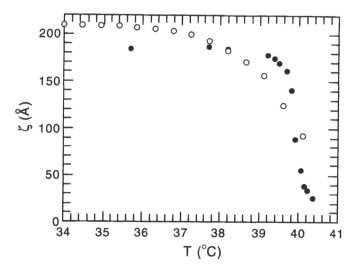

Figure 19.12 Mean nematic domain size ζ as extracted from dynamic scattering data (solid dots) and from turbidity data (open dots).

for ζ_τ. The saturation value of ζ_{dyn} at low T has been extracted for the different density aerogels and the results are: $\zeta_{dyn,\,sat} = 130$ Å for the 0.60 g/cm³ aerogel, 250 Å for 0.17 g/cm³ aerogel and 410 Å for the 0.08 g/cm³ aerogel, as shown in Table 19.1.

Because of the L dependence of the intensity scattered from each domain which varies as L^6 and the dependence of the correlation time which varies as L^2, the dynamics is much more sensitive to the large domains than to the small ones. This can be quantified by calculating the average relaxation time $\langle \tau \rangle$:

$$\langle \tau \rangle \equiv \frac{\int G_{1,s}(t)\,dt}{G_{1,s}(0)} = \frac{\int dt \int P(L)L^6 \exp(-6G_R t)\,dL}{\int P(L)L^6\,dL} = 30\,\frac{\zeta^2}{6A} \qquad (19.21)$$

Thus $\langle \tau \rangle$ is 30 times larger than the relaxation time of a single domain having $L = \zeta$. To compare the size sensitivity of $\langle \tau \rangle$ with that of the turbidity, we should compare the integrands eqns (19.21) and (19.9): their maxima are at $L = 6\zeta$ and $L = 4\zeta$ respectively, and when $L = 6\zeta$ the turbidity has already reached 72% of its saturated value, while $\langle \tau \rangle$ is at about 39% of its saturated value. Since the turbidity is more sensitive to small pores than the average relaxation time, if the pores order via faster filling of the larger pores, ζ_{dyn} should indeed grow faster than ζ_τ extracted from static scattering.

Figures 19.9 and 19.10 show two sets of DLS data on nominally equivalent 8CB 0.36 g/cm³ LCSA samples. One sees that, upon lowering the temperature, the long-time tail in the nematic phase grows in amplitude and slows down. Although this trend has been clearly observed in all the measurement sets on all of the 8CB–LCSA samples, details of the evolution can be dependent on the position of the incident beam on the sample. For example, in the measurement run of Figure 19.9 a significant long time relaxation develops and leads to clear non-ergodic behaviour, whereas in Figure 19.10 the long-time relaxation is significantly smaller. Collection of such data requires long runs to obtain adequate statistics at long delay times (>1 s), and observation of low-contrast fluctuations on top of a high-contrast static speckle pattern. At some places in the sample the distribution of liquid crystal in the bulk cracks or at its air interface in pores near the surface changes slowly even under long-term temperature control,

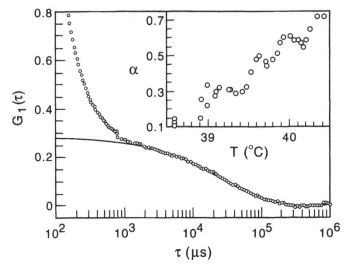

Figure 19.13 Stretched-exponential fit (line) to the long-time component of the correlation function $G_1(\tau)$ measured at $T = 39.93\,°C$ (dots) in the data set of Figure 19.9. Inset: stretching exponent α against T.

producing small transformations of the photon path ensemble and a slow evolution of the speckle pattern that generates decorrelation, masking the long-time non-ergodic behaviour (Frisken and Cannell, 1992).

There are, however, general and interesting features and trends in the long-time tail. First of all, they can all be well represented by stretched exponentials:

$$G_{1,L}(\tau) = A_L \exp[-(t/\tau_L)^\alpha] \tag{19.22}$$

The quality of a typical $G_1(\tau)$ fit for the data set of Figure 19.9 is shown in Figure 19.13 for $T = 39.93\,°C$. The vertical scale has been greatly enlarged to show the tail, whose amplitude is about 5% of the total drop of the intensity correlation function, and about 0.5% of the baseline. All three parameters in eqn (19.22) depend on the temperature. There is an intrinsic connection between the amplitude A_L of $G_{1,L}(\tau)$ and the rotational dynamics of the domains: the very presence of a long-time tail indicates that the fast process of intrapore orientational diffusion does not completely decorrelate the director orientation. The fact that there is little change in the fast relaxation dynamics over a large temperature range in which the long-time tail grows indicates that the appearance of the long-time behaviour does not affect the intrapore dynamics. Rather, long-time behaviour is more likely to be connected with interpore director coupling. The physical picture we offer is therefore as follows: upon lowering T the coupling between the directors of the different domains, or of the director to the silica walls, increases both because of the enlarging domain contact through the openings in the aerogel structure, and because of the increased elastic constants. The rotational diffusion is therefore hindered by an increasingly hilly landscape, which does not modify the short-time free dynamics, but has the effect of saturating the faster relaxation at progressively earlier stages. This is an orientational analogue to the translational dynamics of a Brownian oscillator, freely diffusing on short times and giving rise to a correlation function of the scattered field $G_{1,BO}(\tau)$ which relaxes to a non-zero value: $G_{1,BO}(\tau) = G_{1,FBP}(\delta[1 - \exp(-\tau/\delta)])$, where $G_{1,FBP}(\tau)$ is the correlation function of light scattered from the corresponding free Brownian particles (Pusey, 1994). In order to perform a simultaneous fit of $G_{1,S}(\tau)$ and $G_{1,L}(\tau)$ to the experimental $G_1(\tau)$ we have borrowed the same saturation mechanism:

$$G_1(\tau) = G_{1,S}(\delta[1 - \exp(-\tau/\delta)]) + G_{1,L}(\tau) \tag{19.23}$$

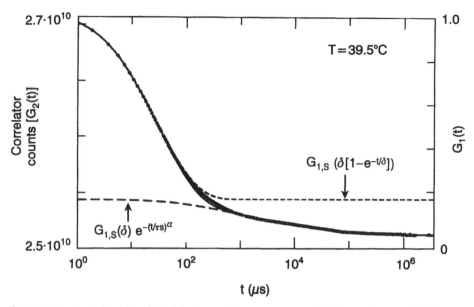

Figure 19.14 Typical combined fit of the fast and slow component of $G_1(\tau)$ by using eqn (19.23).

where the amplitudes are connected through the relation $A_L = G_{1,S}(\delta)$. An example of the composite fit is given in Figure 19.14.

Particularly interesting is the behaviour of the stretching exponent α, shown in the inset of Figure 19.13. The stretched-exponential relaxations together with the decrease of α observed upon lowering the temperature are typical of a system approaching a glass transition temperature (Rammal, 1987), the non-exponential decay being a superposition of hopping modes between configurations equivalent in energy (Binder and Young, 1986). In the LCSA system, the basis for a glass-like behaviour may be the disorder in the interpore interactions, the nematic directors of the different domains being coupled via various channel structures to neighbouring pores. Our system would then resemble a system of spins (the domain directors, placed on the disordered pore location sites) having disordered interaction and approaching a spin glass transition. In such a 'nematic glass' the complete loss of orientational correlation is achieved by a sequence of configurational hops, involving coupled rotations of many domains.

There is an alternative possibility, suggested by the DLS study of 8CB in a sintered porous silica of Wu *et al.* (1992), to explain the observed behaviour: the growth of the nematic domain size enhances the coupling with the surrounding silica branches, so that the long-time dynamics is activated. Stretched-exponential relaxation and divergent decay time would then be a consequence. Wu *et al.* (1992) find that the correlation curves obtained in the nematic phase can be overlapped by stretching the log τ axis, which suggests activated dynamics. It is interesting to note that a stretched exponential having exponent α, when scaled by stretching the log τ axis by a multiplicative constant $\beta > 1$, transforms into a new stretched exponential having a lower exponent $\gamma = \alpha/\beta$. Therefore, although the functional form of the decays are different, it is indeed possible that the long-time relaxation we observe and the whole correlation curves reported by Wu *et al.* arise from the same phenomenon. A comparison between the correlation functions reported by Wu *et al.* for 8CB in sintered porous glass and those in Figure 19.9 for the 8CB 0.36 g/cm³ LCSA system (which has nominally comparable pore size) is made in Figure 19.15. The short-time relaxation which we observe is absent in the sintered silica data, and its presence in the aerogel data makes it difficult to compare the scaling properties of the

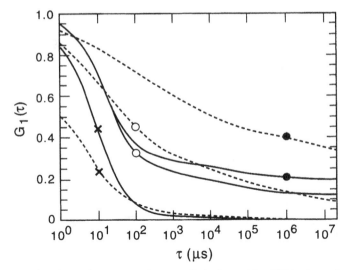

Figure 19.15 Overplot of $G_1(t)$ from Wu *et al.* (1992) (dashed curve) for 8CB in sintered porous glass and from our aerogel data (full curve) at $T = 40\,°C$ (×), 38 °C (○), and 36 °C (●). The 8CB– sintered porous glass system does not show the distinct, fast intrapore decay of the 8CB aerogel.

long-time correlations. The absence of the short-time relaxation may be a consequence of the different structure of the sintered porous silica used by Wu *et al.*, which is only 30% volume fraction pore, and likely to have a smoother pore surface. The larger portion of silica surrounding each pore in this sintered structure may give rise to an effective stronger local field, reducing the angular range of free director reorientations and therefore suppressing the short-time behaviour which we see in the aerogels. In our system, where the pore interconnectivity is very large, it is likely that both pore–surface and pore–pore interactions are operative.

A third way to interpret the data would be to compare them with the dynamics expected in a random field model. Simulated or calculated dynamics of a Lebwohl–Lasher model with added random fields are not available yet, but simulation of a three-state Potts model with added random field yields spin correlation functions having a decay time which increases with decreasing temperature and eventually saturates to a non-zero plateau (see the chapter by Maritan *et al.*).

An important experimental observation substantiating our interpretation is the fact that for a large temperature interval in the nematic phase, between 32 and 36°C, the measured correlation functions are temperature independent. First, this means that the substantial increase in the number of scattering events per photon path in this T interval (see Figure 19.7(a)) does not affect the intensity fluctuation dynamics, in agreement with our discussion of section 19.4.1. Second, it shows that the temperature dependence of the viscosity, which we have neglected in extracting ζ from the decay time, does not play much of a role in slowing down the intrapore dynamics.

19.6 DLS Results – Smectic Slowing Down

As visible in Figures 19.9, 19.16 and 19.17, beginning for $T \approx 32\,°C$, near the bulk T_{NA}, the measured correlation functions exhibit a drastic slowing down of the intrapore relaxation process with decreasing T. Such a 'smectic slowing down' changes the shape of the correlation

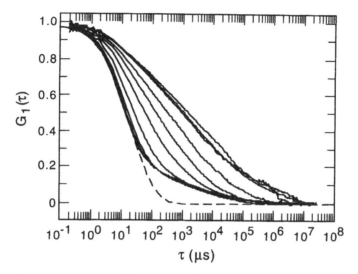

Figure 19.16 $G_1(\tau)$ measured on the LSCA sample having 0.36 g/cm³ density aerogel at $T = 33.41$, 32.22, 30.3, 28.75, 26.77, 24.78, 22.82, 21.83, 20.85, 18.87 °C, from the lowest curve up. Dashed line: fit to short-time decay by eqn (19.20).

functions: they cannot be fitted at the lower temperatures with the same $G_{1,s}(\tau)$ used for $T > 33$ °C. Figure 19.18 shows, however, that they can be made to overlap very well by stretching the log τ axis, and Figure 19.19 gives a measure of the slowing down by showing $\tau_{SA}(T)$, the time needed for $G_1((\tau)$ to reach 0.5, for the four different aerogel densities. $\tau_{SA}(T)$ has a sigmoidal dependence on decreasing T, tending to increase less rapidly at lower T.

This 'smectic slowing down' is an evolution of the feature of $G_1(\tau)$ which we have identified as arising from *intrapore* nematic dynamics, i.e. it is not from the *intrapore* slow relaxation.

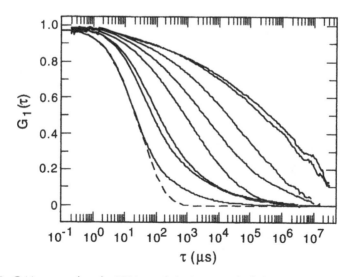

Figure 19.17 $G_1(\tau)$ measured on the LSCA sample having 0.17 g/cm³ density aerogel at $T = 32.17$, 31.3, 30.64, 29.31, 28.40, 26.46, 24.48, 22.51 °C, from the lowest curve up. Dashed line: fit to short-time decay by eqn (19.20).

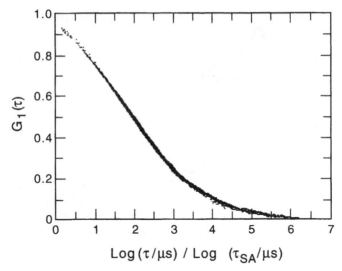

Figure 19.18 Superposition of $G_1(t)$ for the 0.36 g/cm³ aerogel at different temperatures in the 'smectic slowing down' regime, obtained by appropriate scaling of the various log τ axes. The data are for $T = 18.19, 20.13, 21.11, 22.09, 23.55, 25.04, 26.01, 27.01$ °C.

This indicates that smectic slowing down is essentially *intrapore*, a change of local director fluctuation dynamics within the solid-angle excursion allowed by the intrapore coupling. The log τ scaling which we find generally appears for activated kinetics with barriers over a broad range in energy, such as in the random field (Binder and Young, 1986) or vortex glass systems (Fisher *et al.*, 1991), the slowing down being a result of a general increase in the mean barrier height. The only available pore-level fluctuation mechanism of this type in the present case appears to be the thermal motion of defects, e.g. disclination lines, which encounter increasing energy barriers as the smectic order develops. Comparison with X-ray scattering results (see

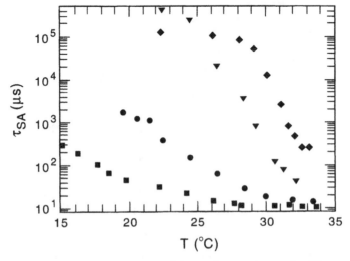

Figure 19.19 Temperature dependence of the half-decay relaxation time τ_{SA} for different aerogel densities: 0.08 g/cm³ (diamonds), 0.17 g/cm³ (triangles), 0.36 g/cm³ (circles), 0.69 g/cm³ (squares).

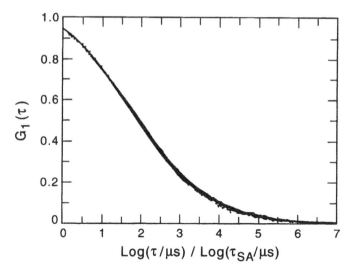

Figure 19.20 Scaled intrapore $G_1(\tau)$ relaxations at several temperatures in the smectic slowing down regime for the 8CB 0.36 g/cm³ system, along with a fit to the form
$$G_1(\tau) \sim \exp\{-[\ln(\tau/\tau_0)]^2/A\} \sim (\tau/\tau_0)^{-[\ln(\tau/\tau_0)/A]}.$$

the next chapter) shows that the temperature dependence of the smectic layering correlation lengths is qualitatively similar to that of $\ln(\tau_{SA}(T))$. This is suggestive of the role of growing smectic order in the 'smectic slowing down', but at this time there is no predictive theory of this effect.

The shape of the 'smectic slowing down' relaxation function is of interest in addition to its time scaling properties. Figure 19.20 shows the scaled relaxation curves for the 8CB ·0.36 g/cm³ sample and a fit to a relaxation Gaussian in ln τ, $G_1(\tau) \sim \exp\{-[\ln(\tau/\tau_0)]^2/A\} \sim (\tau/\tau_0)^{-[\ln(\tau/\tau_0)/A]}$, which is excellent. These correlation functions bear a strong resemblance to those obtained for the staggered magnetization correlation function in a diluted antiferromagnet (Ogielski and Huse, 1986), which also exhibit data collapse with ln scaling. The ln τ scaling has its origin in the exponential dependence of the relaxation time on correlation length for fluctuations in disordered random field systems (Fisher, 1986), $\tau \sim \exp(C/\xi^\theta)$, as ξ becomes large, a consequence of the anomolous growth of free energy in a correlation volume (determined by the exponent θ). Recently Zhang and Chakrabarti (1995), via computer simulation of orientation fluctuations, have found exactly this functional dependence and ln τ scaling for the orientational correlation function in a Lebwohl–Lasher system with a strong external random field.

Acknowledgements

The authors acknowledge many useful conversations with C. W. Garland, L. Wu, B. Zhou, D. W. Schaefer, V. Degiorgio, X.-L. Wu and C. D. Muzny. This work was supported by NSF Solid State Chemistry Grant Grant DMR 92-23729 and a NATO Collaborative Research Grant.

References

ALIEV, F. M. (1994) Liquid crystals – porous glasses heterogeneous systems as materials for investigation of interfacial properties and finite size effects, *Mol. Cryst. Liq. Cryst.* **243**, 91–105.

ALIEV, F. M., MESHKOVSKII, I. K. and KUZNETSOV, V. I. (1984) Change in the phase transition temperatures of liquid crystalline substances in pore of different sizes, *Sov. Phys. Dokl.* **29**, 1009.

ARMITAGE, D. and PRICE, F. P. (1979) Size and surface effects on phase transitions, *Chem. Phys. Lett.* **44**, 305–308.

BELLINI, T., GLASER, M. A., CLARK, N. A. and DEGIORGIO, V. (1991) Effects of finite laser coherence in quasielastic multiple scattering. *Phys. Rev. A* **44**, 5125.

BELLINI, T., CLARK, N. A., MUZNY, C. D., WU, L., GARLAND, C. W., SCHAEFER, D. W. and OLIVIER, B. J. (1992) Phase behaviour of the liquid crystal 8CB in a silica aerogel, *Phys. Rev. Lett.* **69**, 788–791.

BELLINI, T., CLARK, N. A. and SCHAEFER, D. W. (1995) Dynamic light scattering study of nematic and smectic A liquid crystal ordering in silica aerogel, *Phys. Rev. Lett.* **74**, 2740–2743.

BERARDI, R., EMERSON, A. P. J. and ZANNONI, C. (1993) Monte Carlo investigations of a Gay–Berne liquid crystal, *J. Chem. Soc. Faraday Trans.* **89**, 4069.

BERNE, B. J. and PECORA, R. (1990) *Dynamic Light Scattering*, Malabar: Krieger.

BINDER, K. and YOUNG, A. P. (1986) Spin glasses: Experimental facts, theoretical concepts, and open questions, *Rev. Mod. Phys.* **58**, 801.

BROCHARD, F. and de GENNES, P. G. (1983) Phase transitions of binary mixtures in random media, *J. Phys.* **44**, L785.

CHEN, W., MARTINEZ-MIRANDA, L. J., HSIUNG, H. and SHEN, Y. R. (1989) Orientational wetting behavior of a liquid-crystal homologous series, *Phys. Rev. Lett.* **62**, 1860.

CHOW, L. C. and MARTIRE, D. E. (1969) Thermodynamics of solutions with liquid crystal solvents II. Surface effects with nematogenic compounds, *J. Phys. Chem.* **73**, 1127.

CLARK, N. A., BELLINI, T., MALZBENDER, R. M., THOMAS, B. N., RAPPAPORT, A. G., MUZNY, C. D., SCHAEFER, D. W. and HRUBESH, L. (1993) X-ray scattering study of smectic ordering in a silica aerogel, *Phys. Rev. Lett.* **71**, 3505–3508.

DEGIORGIO, V., PIAZZA, R., BELLINI, T. and VISCA, M. (1994) Static and dynamic light scattering of fluorinated polymer colloids with a crystalline internal structure, *Adv. Colloid Interface Sci.* **48**, 61.

DIERKER, S. and WILTZIUS, P. (1987) Random field transition of a binary liquid in a porous medium, *Phys. Rev. Lett.* **58**, 1865.

EMMERLING, A. and FRICKE, J. (1992) Small angle scattering and the structure of aerogels, *J. Non-Cryst. Solids* **145**, 113.

FABBRI, U. and ZANNONI, C. (1986) A Monte Carlo investigation of the Lebwhol–Lasher lattice model in the vicinity of its orientational phase transition, *Mol. Phys.* **58**, 763.

FERRI, F., FRISKEN, B. J. and CANNELL, D. S. (1991) Structure of silica gels, *Phys. Rev. Lett.* **67**, 3626.

FISHER, D. S. (1986) Scaling and critical slowing down in random field Ising systems, *Phys. Rev. Lett.* **56**, 416.

FISHER, D. S., GRINSTEIN, G. M. and KHURANA, A. (1988) Theory of random magnets, *Phys. Today* **41**, 56.

FISHER, D. S., FISHER, M. P. A. and HUSE, D. A. (1991) Thermal fluctuations, quenched disorder, phase transitions, and transport in type-II superconductors, *Phys. Rev. B* **43**, 130.

FRISKEN, B. J. and CANNELL, D. S. (1992) Critical dynamics in the presence of a silica gel, *Phys. Rev. Lett.* **69**, 632.

DE GENNES, P. G. and PROST, J. (1993) *The Physics of Liquid Crystals*, Oxford: Clarendon Press.

HUMPHRIES, R. L., JAMES, P. G. and LUCKHURST, G. R. (1972) Molecular field treatment of nematic liquid crystals, *J. Chem. Soc., Faraday Trans. 2* **68**, 1031.

JANOSY, J. and BATA, L. (1978) Study of elastic properties near a nematic–smectic-A transition, *Acta Phys Pol.* **A54**, 643.

DE JEU, W. H. (1978) The dielectric permittivity of liquid crystals, *Liquid Crystals*, Suppl. 14 of *Physics. Advances in Research and Applications*, ed. L. Lievert, New York: Academic Press.

KARAT, P. P. and MADSHUSUDANA, N. V. (1976) Elastic and optical properties of some 4′-n-alkyl-4-cyanobiphenyls, *Mol. Cryst. Liq. Cryst.* **36**, 51.

KASTING, G. B., GARLAND, C. W. and LUSHINGTON, K. J. (1980) Critical heat capacity of octylcyanobiphenyl (8CB) near the nematic–smectic A transition, *J. Phys.* **41**, 879.

KIM, S. B., MA, J. and CHAN, M. H. W. (1993) Phase diagram of $^3He-^4He$ mixture in aerogel, *Phys. Rev. Lett.* **71**, 2268.

MARITAN, A., CIEPLAK, M., BELLINI, T. and BANAVAR, J. R. (1994) Nematic–isotropic transition in porous media, *Phys. Rev. Lett.* **72**, 4113.

OGIELSKI, A. T. and HUSE, D. A. (1986) Critical behavior of the three dimensional dilute Ising antiferromagnet in a field, *Phys. Rev. Lett.* **56**, 1298.

PERSHAN, P. S. (1988) *Structure of Liquid Crystals*, Singapore: World Scientific.

PINE, D. J., WEITZ, D. A., MARET, G., WOLF, P. E., HERBOLZHEIMER, E. and CHAIKIN, P. M. (1990) Dynamical correlations of multiply scattered light, *Scattering and Localization of Classical Waves in Random Media*, ed. P. Shen, Singapore: World Scientific.

POULIGNY, B., MARCEROU, J. B., LALANNE, J. R. and COLES, H. J. (1988) Pretransitional effects in the isotropic phase of some alkyl cyanobiphenyls. A study by optical Kerr effect and flexoelectricity, *Mol. Phys.* **49**, 583.

PUSEY, P. N. (1994) Dynamic light scattering by non-ergodic media, *Macromol. Symp.* **79**, 17.

PUSEY, P. N. and VAN MEGEN, W. (1989) Dynamic light scattering by non-ergodic media, *Physica A* **157**, 705.

RAMMAL, R. (1987) Spin dynamics in dilute system, *Time Dependent Effects in Disordered Materials*, ed. R. Pynn and T. Risk, New York: Plenum Press.

RAPPAPORT, A. G. (1995) PhD Thesis, University of Colorado.

RAPPAPORT, A. G., CLARK, N. A., THOMAS, B. N. and BELLINI, T. (1995) Pore size dependence of smectic ordering in silica aerogels, to be published.

SCHAEFER, D. W. (1994) Structure of mesoporous aerogels, *MRS Bull.* **19-4**, 49–53.

SCHAEFER, D. W. and KEEFER, K. D. (1980) Structure of random porous materials: Silica aerogel, *Phys. Rev. Lett.* **56**, 2199.

SCHAEFER, D. W., BROW, R. K., OLIVER, B. J., RIEKER, T., BEAUCAGE, G., HRUBESH, L. and LIN, J. S. (1994) Characterization of porosity in ceramic materials by small-angle scattering: Vycor glass and silica aerogel, *Modern Aspects of Small-Angle Scattering*, ed. H. Brumberger, Dordrecht: Kluwer.

SORENSEN, C. M., CAI, J. and LU, N. (1993) Comment on 'Structure of silica gels', *Phys. Rev. Lett.* **71**, 1474.

WONG, A. P. Y. and CHAN, M. H. W. (1990) Liquid-vapor critical point of ^{4}He in aerogel, *Phys. Rev. Lett.* **65**, 2567.

WU, L., ZHOU, B., GARLAND, C. W., BELLINI, T. and SCHAEFER, D. W. (1995) Heat capacity study of nematic–isotropic and nematic–smectic-A transitions for octylcyanobipyl in silica aerogels, *Phy. Rev. E* **51**, 2157–2164.

WU, S. T. and COX, R. J. (1988) Optical and electro-optical properties of cyantolanes and cyanostibenes: Potential infrared liquid crystals, *J. Appl. Phys.* **64**, 821.

WU, X.-I., GOLDBURG, W. I., LIU, M. X. and XUE, J. Z. (1992) Slow dynamics of isotropic–nematic phase transition in silica gel, *Phys. Rev. Lett.* **69**, 470.

ZHANG, Z. and CHAKRABARTI, A. (1995) Dynamics of nematic ordering in porous media, Preprint.

X-ray Scattering as a Probe of Smectic A Liquid Crystal Ordering in Silica Aerogels

A. G. RAPPAPORT, N. A. CLARK, B. N. THOMAS AND T. BELLINI

20.1 Introduction

Phase behaviour in the presence of externally induced disorder is an area of current research in statistical physics and liquid crystals (LCs) are of particular interest in this connection. Because they possess rather weak orientational and partial translational ordering, which is readily affected by surface interactions, the phase behaviour of LCs can be strongly influenced by incorporation into porous solid media of high surface-to-volume ratio (Chow and Martire, 1969; Armitage and Price, 1976; Aliev et al., 1984; Aliev, 1994; Crawford and Žumer, this volume). Liquid crystals can possess partial (one- and two-dimensional) translational order which appears via phase transitions which are continuous, or nearly so, and which can be effectively studied using X-ray diffraction (Pershan, 1988). Liquid crystals thus offer uniquely direct opportunities to study the effect of externally induced disorder on continuous translational freezing processes. The most basic LC translational ordering is the nematic–smectic A (NA) transition, wherein a one-dimensional layering of molecules into a stack of 2D liquid-like layers appears. In this chapter we present a quantitative study of translational ordering in LCs in the presence of induced disorder, probing via X-ray diffraction the smectic A layer structure of the LC 8CB incorporated into the pores of silica aerogel. This is part of an effort to characterize LC phase behaviour in the presence of disorder by combining results of equilibrium (X-ray diffraction and calorimetry) and dynamic (light scattering) probes (Bellini et al., 1992, 1995; Clark et al., 1993; Wu et al., 1995). The static and dynamic light scattering results are detailed in the previous chapter in this volume.

Silica aerogels, highly porous solids formed by annealing-free dehydration of suspensions of aggregates of nanometre dimension silica particles, are attractive as hosts. The aerogel pores form a continuously connected volume which can occupy a volume fraction ϕ approaching unity ($\phi \leqslant 0.95$ in the experiments reported here), of high surface-to-volume ratio. The aerogel internal solid surface is disordered and locally fixes LC molecular orientation and position, directly influencing its orientational and translational order parameters. Various disordering effects of rough surfaces on LC structure are known (Yokoyama et al., 1987), so that the aerogel incorporation can be expected to alter smectic ordering significantly. The aerogels of the kind used here represent the limit of maximum internal surface roughness.

It should be emphasized at the outset that the effect of the host medium on LC phase behaviour will depend strongly on not only pore dimension, but also other characteristics of the host. Pore connectivity, structure dependence on length scale, degree of surface smoothness, surface-to-volume ratio, chemical properties, etc., all play a role in determining the LC structure and vary radically among possible host media. These differences need to be accounted for in any comparison of LC behaviour found in structurally distinct media.

20.2 The Nematic–Smectic A Phase Transition

There are now extensive experimental studies of the bulk nematic–smectic A transition and a well-developed theoretical description which appears to capture its essential thermodynamic and structural features and thus should be a reasonable starting point for understanding the phase behaviour in the aerogels. Here we summarize the aspects of the bulk data and of this description relevant to the phase behaviour in the aerogels, and discuss the specific character-istics of 8CB.

20.2.1 *Phenomenological Description and Critical Behaviour*

At the NA transition the orientationally ordered nematic (mean molecular long axis given by the director, $\mathbf{n}$) becomes modulated by a one-dimensional (1D) mass or electron density wave $n(\mathbf{r}) = \Psi(\mathbf{r}) \exp(i\mathbf{Q}_0 \cdot \mathbf{r})$, of period $\langle d \rangle = 2\pi/Q_0$, describing the 2D fluid layer structure of the smectic A. The complex amplitude $\Psi(\mathbf{r})$ is the order parameter for the transition to the smectic phase. This 1D layering is a phenomenon found in fluid smectic liquid crystals and other systems which have equivalent order parameters, including superfluids and superconductors, in which Ψ represents the wave function of a coherent quantum-mechanical ground state, as well as dynamical instabilities which are spatially periodic. The relationship between smectic A ordering and superconductivity was developed by de Gennes (1972) and is the basis for the phenomenological description of the NA transition (de Gennes and Prost, 1993), employing the two-component order parameter $\Psi(\mathbf{r}) = \psi(\mathbf{r}) \exp(i\phi(\mathbf{r}))$ to characterize the layer ordering. For the smectic A $\Psi(\mathbf{r})$ gives the local magnitude of the density modulation associated with the layering and the phase $\phi(\mathbf{r}) = 2\pi u(\mathbf{r})/d$ is determined by $u(\mathbf{r})$, the displacement of the layers along the layer normal from their positions in a perfectly ordered domain having the equilibrium layer spacing, $\langle d \rangle$. If the layers are disordered then $\nabla\phi = \hat{\mathbf{l}}(\mathbf{r})/l(\mathbf{r})$, where $\hat{\mathbf{l}}$ is the local layer normal and $l(\mathbf{r})$ the local layer thickness. For continuous deformation of the layers, the net number of layers crossed for integration around a closed loop L, $N = \int_L \nabla\phi(\mathbf{r}) \, d\mathbf{r} = 0$. In the presence of topological defect lines, edge or screw dislocations in the case of smectic A (the equivalent of vortex lines in superfluids/superconductors) N need not be zero, but will assume the integral value $N = B$, if the loop encloses a defect line of Burgers' strength B.

The phenomenological free energy describing the NA transition has the following form:

$$F_{NA}(\Psi, \delta S, \delta \mathbf{n}) = \tfrac{1}{2} \int d\mathbf{r}[\chi^{-1}\psi^2 + b\psi^4/2 + c\psi^6/3 - C\psi^2(\delta S) + \chi_N^{-1}(\delta S)^2 + (\xi_\parallel^2/\chi)|\nabla_\parallel \Psi|^2$$

$$+ (\xi_\perp^2/\chi)|(\nabla_\perp - iQ_0 \delta \mathbf{n})\Psi|^2 + K_s(\nabla'\delta \mathbf{n})^2 + K_t(\mathbf{n} \cdot \nabla x\delta \mathbf{n})^2 + K_b(\mathbf{n}x\nabla x\delta \mathbf{n})^2] \tag{20.1}$$

The smectic susceptibility χ diverges as $T \to T_{NA}$ and changes sign at $T = T_{NA}$. This causes the constant ψ for which $F_{NA}(\Psi)$ is equal to its global minimum to change from zero to non-zero, driving the NA transition. Here K_s, K_t and K_b are the nematic Frank elastic constants and S the nematic orientational order parameter. A subscript $\parallel$ means parallel to the layer normal, which we take to be along $\hat{\mathbf{z}}$, while a subscript $\perp$ means in the plane perpendicular to this direction, i.e. the x–y plane. In the nematic the director is constrained to be of unit length so that small-amplitude director fluctuations $\delta \mathbf{n}$ are in the x–y plane and $\mathbf{n} = \hat{\mathbf{z}} + \delta \mathbf{n}$ ($F_{NA}(\Psi)$ is correct to second order in $\delta \mathbf{n}$). The coefficients $\xi_\parallel^2$ and $\xi_\perp^2$ are distinguished from one another to provide for the possibility of anisotropy in the system. We will see shortly that this is justified by experiment. The gradient terms are typical of those put into Landau–Ginzburg free energies in order to promote spatial homogeneity of the order parameter. The ∥ term accounts for changes in ψ and in layer thickness while the $\perp$ term accounts for changes in ψ and couples the layer rotation angle $\theta(\mathbf{r}) = (\nabla_{\perp}\Psi)/iQ_0\Psi = \nabla_{\perp}\phi(\mathbf{r})/Q_0$ to director rotation $\delta\mathbf{n}(\mathbf{r})$, being minimum when $\theta(\mathbf{r}) = \delta\mathbf{n}(\mathbf{r})$.

$F_{NA}(\Psi, \delta S, \delta\mathbf{n})$ is the basis for the description of a variety of phenomena associated with the NA transition. Of particular interest with respect to the behaviour in the aerogels are the following.

20.2.1.1 *Tricritical point*

The smectic A grows into a nematic of order parameter S and the appearance of the smectic may cause a change δS in the nematic ordering, as increasing S will have the effect of increasing the intermolecular interactions and thus further decreasing the energy of the A phase (and lowering the nematic entropy) (McMillan, 1971). To account for this the δS terms are aded to $F_{NA}(\Psi)$, and upon elimination of δS one finds that the effect of the $\Psi - \delta S$ coupling is to reduce the coefficient b of the ψ^4 term: $b \to b' = b - 2C\chi_N$. The nematic susceptiblity χ_N is generally large near the weakly first-order IN transition. Thus it is large near T_{NA} when the McMillan ratio $R = T_{NA}/T_{IN}$ is near $R = 1$. With increasing χ_N the ψ^4 coefficient decreases and the NA transition approaches a tricritical point (TCR), becoming first order for negative b'. Experimetally, it has been found via precision adiabatic calorimetry that the NA transition develops a measurable latent heat for $R_{TCR} > 0.995$, and that if $R > 0.95$ the critical behaviour determined by the effective temperature dependence of $\chi(T)$ approaches tricritical values.

20.2.1.2 *Fluctuation-induced first-order transition*

An interesting example of the influence of nematic fluctuations on smectic ordering was developed by Halperin *et al.* (1974). The $\xi_{\perp}$ term couples $\delta\mathbf{n}$ with Ψ, increasing the energy of the director fluctuations for non-zero Ψ and altering their fluctuation spectrum. The simplest case is when the $\delta\mathbf{n}$ fluctuations are large (e.g. near the Landau tricritical point) and variation of Ψ can be neglected ($\nabla\Psi \approx 0$). In this case we have for F_{NA}

$$F_{NA}(\Psi, \delta\mathbf{n}) = \tfrac{1}{2}\int d\mathbf{r}\{\chi^{-1}\psi^2 + b'\psi^4/2 + c\psi^6/3 + (\xi_{\perp}^2/\chi)(Q_0\delta\mathbf{n}|\Psi|)^2 + K[(\nabla\cdot\delta\mathbf{n})^2$$
$$+ (\nabla\times\delta\mathbf{n})^2]\} \tag{20.2}$$

where we take $K_s = K_t = K_b = K$. The free energy of the $\delta\mathbf{n}$ fluctuations can be obtained by summation over the director modes $\delta\mathbf{n}(\mathbf{q})$ to obtain $\langle\delta\mathbf{n}(\mathbf{r})^2\rangle = \int d\mathbf{q}\langle\delta\mathbf{n}(\mathbf{q})^2\rangle$,

$$\langle\delta\mathbf{n}(\mathbf{q})^2\rangle = (k_B T)\{[(\xi_{\perp}^2/\chi)Q_0^2|\Psi|]^2 + Kq^2\}^{-1} \tag{20.3}$$

which gives

$$\langle\delta\mathbf{n}(\mathbf{r})^2\rangle = 4\pi\int_0^{\infty} q^2\,dq\,(k_B T)[(\xi_{\perp}^2/\chi)Q_0^2|\Psi|^2 + Kq^2]^{-1})$$

$$\sim 4\pi\int_0^{q_{max}} dq\,(k_B T/K) - 4\pi\int_0^{\infty} dq\,(k_B T/K)(\xi_{\perp}^2/\chi)Q_0^2|\Psi|^2\{[(\xi_{\perp}^2/\chi)Q_0^2|\Psi|]^2 + Kq^2\}^{-1}$$

$$\tag{20.4}$$

The first (divergent) term is cut off at q_{max} and the limit of the finite part extended to infinity. When substituted back into eqn (20.2) the first term renormalizes χ^{-1}, Φ shifting T_{NA}, and the second term introduces into $F_{NA}(\psi)$ a ψ^3 term with a negative coefficient $e = -3\pi k_b T_{NA}\chi^{3/2}\xi_{\perp}^3/K^{3/2}$. The result is that the transition becomes first order. This effect has been observed in the vicinity of the Landau tricritical point, extending the range of R values over which the transition is first order (Anisimov *et al.*, 1990).

20.2.1.3 Critical behaviour

Except for materials having McMillan ratios very near 1 as noted above, the NA transition appears to be second order, motivating efforts at universality classification. The principal quantities characterizing the critical behaviour are the exponents α, γ, $v_\parallel$ and $v_\perp$, giving the asymptotic dependence of the heat capacity, smectic susceptibility, and correlation lengths $\xi_\parallel$ and $\xi_\perp$ respectively ($\Delta C_{pNA} = At^{-\alpha}$, $\chi = \chi_0 t^{-\gamma}$, $\xi_\parallel = \xi_{\parallel 0} t^{-v}$ and $\xi_{\perp 0} t^{-v_\perp}$). Here t stands for $T - T_{NA}/T_{NA}$. The two-component nature of Ψ suggests that the thermodynamics of the NA transition ought to be that of the XY model in three dimensions ($\alpha = -0.007$, $\gamma = 1.316$, $v_\parallel = v_\perp = 0.669$). However, this behaviour is not observed and is approached only in systems having a very broad nematic phase, i.e. very small value of R. Measurements of these four exponents are available for roughly 30 different single-component LCs and mixtures having McMillan ratios in the range $0.66 < R < 0.995$, and show the following general trends, as recently reviewed and analysed by Garland and Nounesis (1995):

1 The ΔC_{pNA} exponent $\alpha \to 0$ for small R ($\alpha_{XY} = -0.007$) and $\alpha \to 0.5$ for large R ($\alpha_{TCR} = 0.5$).
2 The χ exponent $\gamma \to 1.3$ for small R ($\gamma_{XY} = 1.316$) and $\gamma \to 1$ for large R ($\gamma_{TCR} = 1.00$), with a maximum $\gamma \approx 1.5$ for $R \approx 0.92$.
3 The $\xi_\parallel$ and $\xi_\perp$ exponents are unequal ($v_\parallel > v_\perp$) leading to highly anisotropic smectic A correlations as $t \to 0$, with the correlation volumes extended along **n**. For small R, $v_\parallel \to \sim 0.7$, $v_\perp \to \sim 0.65$, $v_\parallel/v_\perp \to \sim 1$ ($v_{\parallel XY} = 0.669$, $v_{\perp XY} = 0.669$, $[v_\parallel/v_\perp]_{XY} = 1$), while for large R $v_\parallel \to \sim 0.55$, $v_\perp \to \sim 0.4$, $v_\parallel/v_\perp \to \sim 1.4$ ($v_{\parallel TCR} = 0.50$, $v_{\perp TCR} = 0.50$, $[v_\parallel/v_\perp]_{TCR} = 1$). This large R behaviour represents the most significant deviation with the standard XY and TCR models.
4 The correlation volume exponent $v_\parallel + 2v_\perp \to 2$ for small R ($[v_\parallel + 2v_\perp]_{XY} = 2.007$) and $v_\parallel + 2v_\perp \to \sim 1.4$ for large R ($[v_\parallel + 2v_\perp]_{TCR} = 1.50$), with a maximum at $v_\parallel + 2v_- \approx 2.1$ for $R = 0.92$.

The McMillan ratio effectively parameterizes the relative magnitude of nematic fluctuations at T_{NA}. The approach to XY-like behaviour at small R, where the nematic order parameter is saturated, the Frank constants are large and the nematic director fluctuations small, thus suggests that the coupling of Ψ to the nematic director fluctuations may be responsible for the complex dependence of the exponents on R and the consequent apparently non-universal critical behaviour.

There has been extensive theoretical analysis of the NA critical behaviour directed towards incorporating nematic and smectic fluctuations. These are reviewed by de Gennes and Prost (1993). The recent one-loop self-consistent calculation by Patton and Andereck (1992, 1994), which deals with the coupling between Ψ and δ**n** and their fluctuations in a fully self-consistent way, reproduces the above complex trends. They show explicitly that the anisotropy in the divergence of $\xi_\parallel$ and $\xi_\perp$ is a consequence of having δ**n** constrained to the x–y plane (**n·n** = 1), finding cross-over from isotropic behaviour at large t to an anisotropic fixed point ($v_\parallel = 2v_\perp$) at small t. By relaxing the constraint on $|$**n**$|$, i.e. by making it elastic with the condition of a term $F_{constraint} = \Gamma($**n·n** $- 1)$ to F_{NA}, they find isotropic behaviour for $\Gamma = 0$ and anisotropy otherwise, with a cross-over t depending on Γ. The role of the nematic fluctuations in generating anisotropy is of interest in the present context in aerogels because we find in our 8CB – aerogel experiments that the anisotropy characteristic of bulk 8CB is absent in the aerogel.

The smectic A–nematic transition can also be viewed as a collective thermal ionization of vortex lines (Nelson and Toner, 1981), which generates anisotropic inverted X–Y behaviour. The description of smectic disorder in terms of topological defects may be particularly appropriate for aerogel confinement since, at temperatures well below the bulk T_{NA}, the smectic layering retains only short-range order even though the local layer structure is well established.

20.2.2 X-ray Structure Factor and Correlation Functions

Since the NA transition is a positional ordering it can be effectively probed via X-ray diffraction, which enables the determination of $\xi_\parallel$ and $\xi_\perp$ for the pretransitional short-range ordered smectic fluctuations in the nematic phase ($t > 0$). The scattering intensity $I(\mathbf{q})$ is the Fourier transform (FT) of $\langle n(0)n(\mathbf{r})\rangle$, the density–density correlation function, which is, in turn, obtained from the pair correlation function of Ψ, $S_\Psi(\mathbf{q}) = FT[G_\Psi(\mathbf{r}) \equiv \langle \Psi(0)\Psi(\mathbf{r})\rangle]$, with the $\mathbf{q}$ origin shifted to $\hat{z}Q_0$ by the average layering

$$I(\mathbf{q}) = FT[\langle n(0)n(\mathbf{r})\rangle] \equiv \int d^3r \, \exp(i\mathbf{q}\cdot\mathbf{r})\langle n(0)n(\mathbf{r})\rangle$$

$$= S_\Psi(\mathbf{q} - \hat{z}Q_0) = \langle |\Psi(\mathbf{q} - \hat{z}Q_0)|\rangle^2$$

where $\Psi(\mathbf{q}) = FT[\Psi(\mathbf{r})]$. In order to calculate $\langle |\Psi(\mathbf{q})|\rangle^2$ the Gaussian approximation is used, only terms to second order in Ψ and its gradients are retained in the $F_{NA}(\Psi)$

$$F_{NA}(\Psi, \delta\mathbf{n}) = 1/16\pi^3 \int d\mathbf{q} \, \{\psi(\mathbf{q})^2/\chi + \xi_\parallel^2/\chi \, |q_\parallel \Psi|^2 + \xi_\perp^2 \chi |(q_\perp - iQ_0\delta\mathbf{n})\Psi|^2\} \tag{20.5}$$

and equipartition is used to give $I(\mathbf{q})$ and its approximate (Ornstein–Zernicke) correlation function

$$\langle |\Psi(\mathbf{q})|\rangle^2 = S_{\text{bulk nem}}(\mathbf{q}) = (k_B T/2)\chi[1 + (q_\parallel\xi_\parallel)^2 + (q_\perp\xi_\perp)^2 + c(q_\perp\xi_\perp)^4]^{-1} \tag{20.6a}$$

$$G(\mathbf{r}) \propto [(z/\xi_\parallel)^2 + (r_\perp/\xi_\perp)^2]^{-1/2} \exp\{-[(z/\xi_\parallel)^2 + (r_\perp/\xi_\perp)^2]^{1/2}\} \tag{20.6b}$$

which is successfully used to fit nematic phase X-ray structure factor data for many materials, including 8CB (Davidov *et al.*, 1979). Of relevance to the present experiments is the powder average of $I(\mathbf{q})$ since ther smectic layering in the aerogel is globally isotropic, to be discussed in section 20.7.1.

In the smectic phase $F_{NA}(\Psi, \delta\mathbf{n})$ can be simplified. The density modulation amplitude is constant and the director is always along the local layer normal ($\nabla_\perp - iQ_0\delta\mathbf{n} \to 0$), F_{NA} depending only on the phase fluctuations $\phi(\mathbf{r})$, or equivalently the layer displacement $u(\mathbf{r})$. Twist and bend of $\mathbf{n}$ are effectively forbidden. Only the $q_\parallel^2$ and $q_\perp^4$ terms survive the divergence and $F_{NA}(u) \to \Sigma_\mathbf{q}(1/2) \, [Bq_\parallel^2 + Kq_\perp^4]|u(\mathbf{q})|^2$, where B is the layer compression elastic constant (de Gennes and Prost, 1993). The mean square layer fluctuation obtained from this $F_{NA}(\mathbf{q})$ for a system of dimension L is given by $\langle |u(\mathbf{r})|^2\rangle = k_B T/[2\pi(BK)^{1/2}] \, \ln(L/a)$, which diverges logarithmically with L, indicating that the smectic layering is at its lower marginal dimensionality. That is, with this free energy the smectic order is unstable against layer displacement fluctuations and develops only quasi-long-range order (QLRO). This point is important to us because we may expect any disordering perturbation to destroy the QLRO of the unperturbed smectic A phase since it is poised on the brink of disorder even in the absence of such perturbations. Confinement in the aerogels turns out indeed to destroy the QLRO of the SmA, generating a state of short-ranged smectic A order distinctly different from that of the pretranslational smectic A ordering in the nematic. $F_{NA}(u)$ may be used to obtain the structure factor in the smectic A phase $I(\mathbf{q})$, and its powder average $I(q)$ (Caille, 1972; de Gennes and Prost, 1993)

$$I(\mathbf{q}) = S(\mathbf{q} - \hat{z}Q_0) \sim [(q_z - Q_0)Q_0]^{-(2-\eta)} \qquad \text{for } q_\perp = 0 \tag{20.7a}$$

$$I(\mathbf{q}) = S(\mathbf{q} - \hat{z}Q_0) \sim [q_\perp]^{-(4-2\eta)} \qquad \text{for } q_z = 0 \tag{20.7b}$$

$$I(\mathbf{q}) \sim (q - Q_0)^{-(1-\eta)} \tag{20.7c}$$

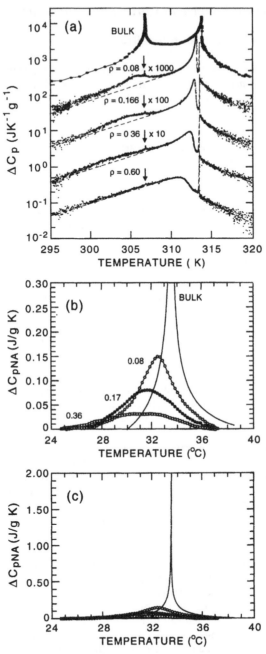

Figure 20.1 (a) Excess heat capacity ΔC_p of the 8CB–aerogel samples (Wu et al., 1995) and of bulk 8CB (Kasting et al., 1980) in the vicinity of the IN and NA transitions. The semi-log plots are shifted vertically by factors of 10 to prevent overlap. The low-temperature IN wings expected in the absence of the NA transition and used as background for the NA transition are indicated by the dashed lines. The sharp bulk IN and NA peaks from the small amount of bulk 8CB in cracks and on surfaces are evident. (b,c) Linear plots of Δ_{pNA}, the heat capacity due to smectic A ordering (IN transition contribution subtracted out) in the 8CB–aerogel samples and bulk 8CB (Wu et al., 1995). The bulk data peaks at $C_{pNA} \approx 2.1$ J/g K (Kasting et al., 1980). The aerogel confinement significantly rounds ΔC_{pNA}.

The power-law character of these scattering peaks is a direct consequence of the thermal layering fluctuations and measurement of the exponent η enables determination of the layering elastic constants. In 8CB, however, this effect is weak, and not measurable without an exceptionally narrow and high-contrast resolution function (Als-Nielset *et al.*, 1980). In the current experiments the smectic A scattering produces resolution essentially resolution-limited Bragg reflections.

20.2.3 *The Nematic–Smectic A Transition in 8CB*

In the experiments presented here we study the phase behaviour of the single-component liquid crystal octyloxycyanobiphenyl (8CB), incorporated in the silica aerogels. The nematic–smectic A transition of 8CB has probably received more attention than that of any other compound. Bulk 8CB exhibits isotropic (I), nematic (N), smectic A (A) and crystal (X) phases as follows (Karat and Madshusudana, 1976; Davidov *et al.*, 1979; Kasting *et al.*, 1980):

$$I \quad (T_{IN} = 40.5\,°C) \quad N \quad (T_{NA} = 33.5\,°C) \quad A \quad (T_{AX} = 21.5\,°C) \quad X$$

for a McMillan ratio $R = 0.978$. The NA transition is second order (Chan *et al.*, 1985; Fisch *et al.*, 1984; Ricard and Prost, 1981; Thoen *et al.* 1982), or possibly very weakly first order (Anisimov *et al.*, 1990; Benzekri *et al.*, 1992), and is accompanied by a λ-like heat capacity anomaly (Kasting *et al.*, 1980; Thoen *et al.*, 1982). High-resolution adiabatic calorimetry established an upper limit of 0.0014 J/g for the NA latent heat, if any, of 8CB, and a net NA enthalpy of 0.76 J/g (Thoen *et al.*, 1982). The bulk excess heat capacity above background associated with the NA transition is shown in Figure 20.1 (Kasting *et al.*, 1980). This anomaly is well fitted (as is the NA anomaly in other materials) by a power-law divergence ($\propto A_\pm$) with a correction to scaling term ($\propto D_\pm$) and a critical contribution to the non-singular term ($\propto B_\pm$) for $T > T_{NA}$ (+) and $T < T_{NA}$ (−):

$$\Delta C_{pNA} = [A_\pm t^{-\alpha}][1 + D_\pm t^{-0.5}] - B_\pm \text{ J/g K} \tag{20.8}$$

The asymptotic divergence as T approaches T_{NA} is power law with exponent $\alpha = 0.30 \pm 0.05$. The 8CB anomaly is rather symmetric about $t = 0$, with $A_+ = 2.9$ J/g K, $A_- = 3.2$ J/g K, $D_+ = 5.31$, $D_- = 3.20$, and $B_\pm = -0.448$ J/g K (Kasting *et al.*, 1980).

The divergence of the correlation lengths for pretranslational short-range smectic ordering in the nematic phase in the direction parallel to the nematic director $\hat{n}$ ($\xi_\parallel$) and normal to $\hat{n}$ ($\xi_\perp$) (Davidov *et al.*, 1979) is shown in Figure 20.2. The asymptotic behaviour of these lengths is also power law with exponents $v_\parallel = 0.67$ and $v_\perp = 0.51$, making the correlation volumes increasingly anisotropic as T approaches T_{NA}, extended like prolate ellipsoids along $\hat{n}$. The susceptibility for smectic fluctuations χ diverges with exponent $\gamma = 1.32$.

The exponents α, $v_\parallel$, $v_\perp$ and γ of 8CB are similar to those of other materials having comparable R and place 8CB in a regime of cross-over to tricritical behaviour. The heat capacity exponent has increased significantly relative to the 3D XY value and the correlation volume exponent $v_\parallel + 2v_\perp = 1.68$ is closer to the TCR value than the XY one. It also agrees well with the hyperscaling relation $2 - \alpha = v_\parallel + 2v_\perp$. The $v_\parallel > v_\perp$ anisotropy is shared by essentially every other NA system.

20.3 Silica Aerogel Structure and Properties

The aerogels in which we confined 8CB were prepared by base-catalysed polymerization of tetramethylorthosilicate in methanol. The result is a condensation of silica particles, typically 30–50 Å in diameter, which in turn aggregate into clusters which are fractal over some range of length scales. These clusters contact to form a gel network from which the solvent is removed

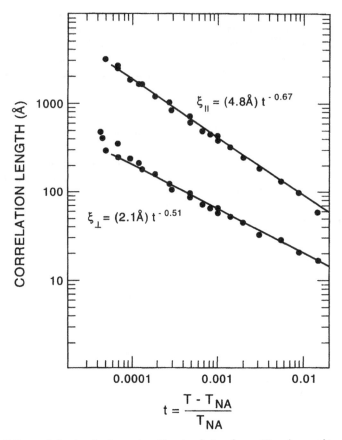

Figure 20.2 Bulk correlation lengths for pretransitional ordering along **n**(∥) and normal to **n**(⊥) in the nematic phase of 8CB obtained from high-resolution X-ray measurements of $I(\mathbf{q})$ of monodomain smectic A samples (Davidov et al., 1979). The ∥ exponent $v_{\parallel} = 0.67$ is larger than $v_{\perp} = 0.51$, generating an increasingly anisotropic correlation volume as $t \to 0$.

by supercritical drying. The solution concentration of the precursor and the drying conditions can be varied to produce samples with different densities: 0.08, 0.17, 0.36 and 0.60 g/cm³ for the present experiments. *These densities (in units of g/cm³) will be used to identify the gels throughout this chapter.* The 0.08 and 0.17 aerogels were purchased from Airglass AB in Sweden. The 0.36 and 0.60 aerogels were made at Lawrence Livermore National Laboratory by T. Tillotson and L. Hrubesh. The 0.60 sample was partially air dried before supercritical extraction.

Some relevant characteristics of these aerogels are shown in Table 20.1. As can be seen, the 'empty' space within these gels which will be filled by liquid crystal, given by the pore volume fraction ϕ, ranges from $\phi = 0.7$ to 0.95, while the average pore chord $\langle p \rangle$ ranges from 100 to 700 Å. The aerogel mass density ρ is obtained via direct density measurement. The solid mass density $\rho_s \equiv \rho/(1 - \phi)$ (which is in general less than the bulk fused silica density of 2.2 g/cm³), average solid chord $\langle s \rangle$, average pore chord $\langle p \rangle$, surface-to-volume ratio of the pore region S/V_p, and ϕ are all obtained from small-angle X-ray scattering experiments (Schaefer and Keefer, 1986; Schaefer 1994; Schaefer et al., 1994; Emmerling and Fricke, 1992) and are reported for these aerogels by Wu et al. (1995).

Table 20.1 Parameters relevant to the aerogels and the 8CB–aerogel system. The aerogel parameters are mass density ρ; pore volume fraction ϕ; solid mss density ρ_s; X-ray scattering length $\langle \ell \rangle$; average solid chord $\langle s \rangle$; average pore chord $\langle p \rangle$, where ρ, ϕ, ρ_s, $\langle s \rangle$ and $\langle p \rangle$ are from Wu et al. (1995). V_p/S is the inverse surface to volume ratio of the pore region of the aerogel. ξ_{peak} is the smectic correlation length at the peak in ΔC_{pNA} and $\xi(t_R)$ is the effective finite size length by extrapolation from $\Delta C_{pNApeak}$. ξ_{sat} and ξ_X are respectively the low-temperature (maximum) correlation length for smectic and crystal ordering. $\xi_{dyn, sat}$ is the mean pore size extracted from the fast intrapore relaxation observed in the DLS correlation functions (see the previous chapter)

ρ (g/cm^3)	0.08	0.17	0.36	0.60
Aerogel parameters:				
ϕ	0.95	0.90	0.79	0.73
ρ_s (g/cm^3)	1.48	1.59	1.73	2.20
$\langle \ell \rangle$ (Å)	40	45	37	33
$\langle s \rangle$ (Å)	42	50	47	45
$\langle p \rangle$ (Å)	700 ± 100	430 ± 65	180 ± 45	120 ± 25
V_p/S (Å)	159	137	51	30
Liquid crystal structural lengths:				
ξ_{peak} (Å)	200	125	60	
$\xi_{\parallel}(t_R)$ (Å)	220	150	90	
$\xi_{\Delta Hsat}$ (Å)	940	400	210	
ξ_{sat} (Å)	1400	520	180	95
ξ_X (Å)	1040	515	255	
Liquid crystal dynamical lengths:				
$\xi_{dyn, sat}$ (Å)	410	250	210	130

20.3.1 Probing Silica Aerogel Structure via X-ray Diffraction

Analysis of aerogel scattering can be understood by considering a chord, a straight line drawn through the aerogel intersecting alternately solid and pore regions. The pore and solid chords, $\langle p \rangle$ and $\langle s \rangle$ respectively, are the average contiguous pore and solid lengths along such a line. Following Debye *et al.* (1957) (DAB), we assume the dispersion of solid material to be random, the electron density–density correlation function along the line being a *simple* expontial:

$$\langle n(0)n(x) \rangle \sim \exp(-x/\langle \ell \rangle) \tag{20.9}$$

analogous to the time correlation function of the random telegrapher's wave. These correlations are short ranged and isotropic ($\langle n(0)n(r) \rangle = \exp(-r/\langle \ell \rangle)$) in which case the X-ray scattered intensity $I(q)$ (static structure factor) is proportional to the FT of $\langle n(0)n(r) \rangle$:

$$I(q) \sim FT[\langle n(0)n(r) \rangle] = \int d^3r \, \exp(iq \cdot r) \langle n(0)n(r) \rangle \tag{20.10a}$$

$$I_{DAB}(q) = K\langle \ell \rangle^4 [1 + (q\langle \ell \rangle)^2]^{-2} \tag{20.10b}$$

and $I_{DAB}(q)$ is the square of a Lorentzian in q, centred at $q = 0$. Here $K = k\rho_s^2 S/V$ is a scattering cross-section depending on n_{ag}^2, the square of the electron density of the solid regions, and therefore ρ_s^2, and on the surface-to-volume ratio S/V. The constant K can be determined via absolute intensity measurements and the geometrical factor k is known, so the product $\sigma = \rho_s^2 S/V = \rho^2(1 - \phi)^2 S/V$ can be measured. The second moment of $I(q)$ can be obtained from the measured $I(q)$

$$\tilde{Q} \equiv \int I(q)q^2 \; dq = \pi K \langle \ell \rangle / 4 \tag{20.11}$$

to enable the determination of $\langle \ell \rangle$ via the known $\tilde{Q}$ and K. Porod's relation, $S/V = (1 - \phi)\phi(4/\langle \ell \rangle)$, a result of viewing the porous solid as a collection of randomly oriented planar interfaces (Porod, 1982), in which case $I(q)$ varies as $I(q) = Kq^{-4}$, can then be used, along with σ and the known density ρ to solve for S/V and ϕ. Finally the lengths $\langle \ell \rangle$, $\langle p \rangle$ and $\langle s \rangle$ are related by

$$1/\langle \ell \rangle = 1/\langle p \rangle + 1/\langle s \rangle \quad \text{with} \quad (\langle p \rangle = \langle \ell \rangle/(1 - \phi) \quad \text{and} \quad \langle s \rangle = \langle \ell \rangle/\phi \tag{20.12}$$

The quantities $\langle \ell \rangle$, $\langle p \rangle$, $\langle s \rangle$, ϕ and ρ_s for our aerogels are given in Table 20.1.

Equations (20.10) and (20.11) indicate that if $\langle p \rangle$ and $\langle s \rangle$ are very different, the shorter of the two, for our aerogels $\langle s \rangle$ because $[3\langle s \rangle] < [\langle p \rangle] < [17\langle s \rangle]$, will largely control the *shape* of the aerogel $I(q)$. The shape in q of $I(q)$ (half width at quarter height), particularly for the lower-density aerogels, therefore provides *no direct information* on the aerogel pore size $\langle p \rangle$, but is nearly the inverse solid chord $1/\langle s \rangle$. $\langle p \rangle$ must be extracted from the absolute intensity data as described above. This rather non-intuitive feature of the aerogel scattering is a result of the identity of the amplitude of the scattered field in adjacent pores and therefore coherence of the scattering from adjacent pores, i.e. both 'self' and 'other' terms contribute to $\langle n(0)n(x) \rangle$.

This situation is fundamentally different for scattering from layer structure within the pores. Although we will analyse the scattering from the smectic in detail in section 20.7, it is useful briefly to compare it here with the aerogel scattering. If we assume that the layer structure loses its order (translational and/or orientational) upon proceeding from pore to pore, then the Bragg scattering amplitude ψ from adjacent pores will be of random relative phase, and therefore incoherent. In this case it is easy to show that only the 'self' term contributes to $G(x)$ and the effective smectic correlation function becomes

$$G_s(x) = \langle \psi(0)\psi(x) \rangle \sim \exp(-x/\langle p \rangle) \tag{20.13}$$

directly containing the pore chord. Because the lineshape of the diffuse Bragg peak due to the short-range layering is related to the Fourier transform of $G_s(x)$, the study of layer ordering in aerogels or other porous media, under conditions that the layer order fills the pores, provides a means of direct measure of pore dimensions, the widths of the layering peak being inversely proportonal to the pore size $\langle p \rangle$.

The aerogel structure can also be characterized in terms of an effective size distribution $P_p(L)$ of pores of dimension L. The condition that the pores fill the empty space within the aerogel leads to the normalization requirement $\int P_p(L)L^3 \; dL = \phi$. A pore–pore self-correlation function of the form of eqn (20.13) requires (see the previous chapter)

$$P_p(L) = (\langle p \rangle L)^{-2} \exp(-L/\langle p \rangle) \tag{20.14}$$

the functional form for $P_p(L)$ to be used when a distribution is required to model the void structure within the aerogel. The volume-averaged mean pore size of this distribution is

$$\langle L \rangle = \int P_p(L)L^4 \; dL / \int P_p(L)L^3 \; dL = 2\langle p \rangle$$

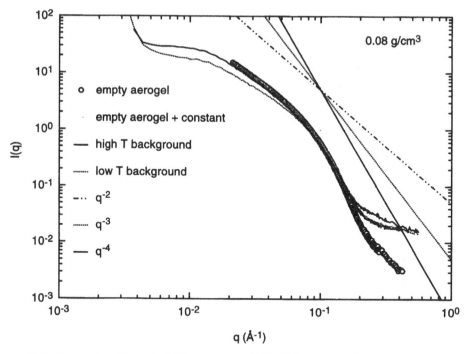

Figure 20.3 X-ray scattered intensity *I(q)* from empty and 8CB-filled 0.08 aerogel against scattering vector *q*, and simple power laws for comparison. Filled aerogel scans are obtained at high temperature (*T* ≈ 40 °C), near the isotropic–nematic phase transition where the smectic layering is absent, and at low temperature (*T* ≈ 0 °C), where the 8CB is frozen. The crystal peak, shown in Figure 20.6, is not plotted here. The fractal [*I(q)* ~ q^{-2}] and Porod [*I(q)* ~ q^{-4}] regimes are evident. Filling adds a *q*-independent constant scattering evident for *q* > 0.2 Å^{-1}, but otherwise does not substantially change *I(q)*. Freezing lowers the slope (mass fractal dimension) in the fractal regime. *I(q)* of the filled aerogels appears to change reversibly through freeze–thaw and temperature cycling.

The expression 'mean pore size' can be applied to many different averaging procedures, depending on the context. We will use it to indicate the pore chord $\langle p \rangle$.

20.3.2 X-ray Structure Factor of the Aerogels Used in this Study

Figures 20.3–20.6 show plots of measured X-ray scattering intensity *I(q)* against scattering wavevector *q* typical of the empty and 8CB-filled aerogels used in these experiments. These scans are scaled in intensity to overlap for comparison of shapes. Experimental details will be provided in section 20.5. Figures 20.3 and 20.5 also indicate the simple power laws *I(q)* ~ q^{-2}, q^{-3}, q^{-4} for comparison purposes. The figures show that the aerogel *I(q)* approch the Porod q^{-4} dependence for *q* > 0.1 Å^{-1}. At smaller *q* (0.01 < *q* < 0.1 Å^{-1}) is a range of lower slope, *I(q)* ~ q^{-D}, corresponding to length scales over which the aggregation clusters are fractal, of mass fractal dimension *D* ≈ 2 (Schaefer and Keefer, 1986; Schaefer, 1994; Schaefer *et al.*, 1994; Emmerling and Fricke, 1992). At low *q* is a shoulder marking the length scale beyond which r.m.s. fraction fluctuations in the volume-averaged density decrease with increasing *V* as $V^{-1/2}$, i.e. beyond which the aerogel is of uniform density. At the smallest *q* the detector begins to encounter stray scattering from the incident beam.

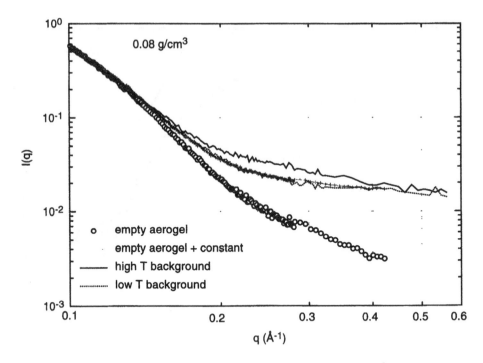

Figure 20.4 Blowup of aerogel background scans in the q region near $q = 0.199$ Å^{-1}, where the smectic A layering peak will appear. The filled low-T background (crystal 8CB) can be obtained from that of the empty aerogel by addition of a constant. The high-$T = 40\,°$C scan has an additional broad component, shown in Figure 20.10, due to remnant pair correlations in the high-temperature nematic. This peak was included in the background for the analysis of the growth of smectic order in the aerogel, i.e. the high-T scan was used as the background.

Figure 20.7 is reproduced from the work of Vashishta and coworkers (Nakano *et al.*, 1993) to give a geometrical picture of the aerogel structure. It shows a visualization of aerogel structures having densities similar to those used here obtained by molecular dynamic computer-simulated condensation of SiO_2. The cube edge dimension is about 50 Å, near the lower end of the fractal range. The connected pore structure is evident and should be structurally self-similar over approximately a decade in increasing scale length.

Figures 20.5 and 20.6 also present typical scans in the smectic A region, $T \approx 17\,°$C, featuring the smectic A Bragg peak near $q = 0.2$ Å^{-1}. Also shown in Figures 20.3, 20.4 and 20.5 are the $I(q)$ for 8CB-filled aerogels: obtained at low temperature, $T \approx 0\,°$C, where the 8CB has frozen into a three-dimensional crystal phase; and at high temperature, $T \approx 40\,°$C, where the bulk would be in the nematic phase, near the IN transition. The crystal has a first-order Bragg reflection at $q = 0.476$ Å^{-1}, evidenced by the sharp peaks in Figure 20.6. These crystal peaks are not plotted in Figures 20.3 and 20.4, in order to emphasize the behaviour of the aerogel background scattering. Figure 20.4 shows empty aerogel, low-T and high-T scans in the region of q where the smectic A peak will appear. Also plotted is the empty aerogel $I(q)$ with a constant added, where the constant is adjusted to match best the sum to the low-T scan. This shows that at low T, in the crystal phase, the observed $I(q)$ in this q region is proportional to that of the empty aerogel plus a q-independent contribution from the 8CB. At high T, however, there is clearly an additional contribution to the scattering, peaking at $q \approx 0.25$ Å^{-1}, present even

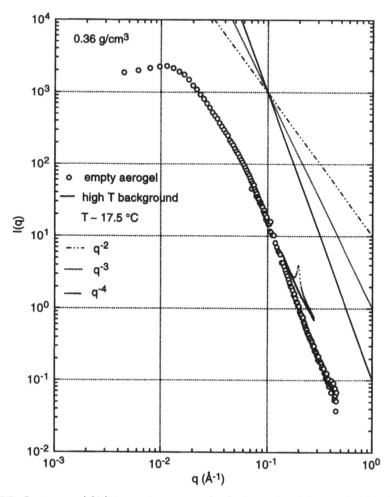

Figure 20.5 Empty aerogel, high-temperature scan used as background, and the smectic A layering peak at $T \approx 17\,°\text{C}$ for the 0.36 sample, along with simple power laws for comparison. Filling both lowers the slope in the Porod region and adds a constant term to the background.

in the isotropic phase and nearly independent of T. This contribution will be discussed in section 20.6.1. Apart from this, comparison of the empty and filled low-T and high-T $I(q)$ over the entire range of q shows that both filling and temperature change alter the aerogel structure to some extent, with freezing tending to lower D. These changes do not significantly alter the aerogel properties given in Table 20.1. It appears, therefore, that the silica structures are not modified significantly on length scales of about 1000 Å or shorter by the insertion of the liquid crystal. Filling does produce cracks in the lowest-density aerogel used here. The aerogel scattering of the filled aerogels appears to change reversibly through freeze–thaw and temperature cycling.

Equations (20.10) and (20.11) indicate that if $\langle p \rangle$ and $\langle s \rangle$ are very different, as is the case for our aerogels, the shorter of two will largely control the *shape* of the aerogel $I(q)$. This is illustrated in Figure 20.6, which plots the 0.08 aerogel $I(q)$ along with the DAB form $I(q) = I_{\text{DAB}}(q) + C$, with $\langle \ell \rangle = 40$ Å, as expected from Table 20.1.

423

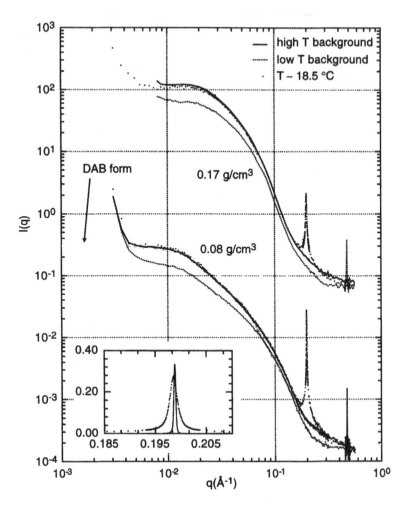

Figure 20.6 Example scans showing typical smectic A layering peaks at $T \approx 18.5\,°\mathrm{C}$ in the 0.08 and 0.17 samples, along with the high-T = 40 °C and low-T = 0 °C backgrounds. The 8CB crystal peak in the low-T scans at $q = 0.476\ \text{Å}^{-1}$ are shown here. This overplot illustrates that in the smectic A regime the scattering away from the SmA peak matches the high-T background much better than the low-T background. Also shown is a fit of eqn (20.2), the Lorentzian squared Debye–Anderson–Brumberger form (Debye et al., 1957) to the background, the width of which gives the characteristic length $\langle \ell \rangle = 40\ \text{Å}$, essentially the solid chord $\langle s \rangle = 42\ \text{Å}$. The aerogel scattering gives no direct information on the pore size. On the other hand, the width of the layering peaks is directly related to pore dimension. The inset shows the 0.08 smectic peak and resolution function on a linear scale.

20.4 The 8CB–Aerogel System

The behaviour of the 8CB–aerogel system for the aerogels of Table 20.1 has also been studied by calorimetry (Bellini et al., 1992; Wu et al., 1995), static light scattering (Bellini et al., 1992), and dynamic light scattering (Bellini et al., 1995). The results of relevance to the X-ray scattering experiments may be summarized as follows.

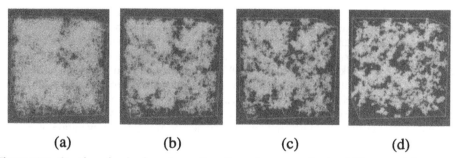

Figure 20.7 Sample molecular dynamics configurations of computer-simulated SiO_2 aerogels having densities comparable with those used in these experiments: (a) 0.8 g/cm³; (b) 0.4 g/cm³; (c) 0.2 g/cm³; (d) 0.1 g/cm³. The cube edge is about 50 Å. These simulated aerogels give $I(q)$ very similar to those of the aerogels used in our experiments. The structure is fractal over roughly the next decade in scale.

The optical properties of liquid crystal–aerogel systems provide direct information on molecular orientation in the aerogels and thus are relevant to the understanding of the smectic translational ordering. Dynamic light scattering shows that in the aerogel, orientational confinement appears gradually with decreasing temperature via a glass-like orientational freeze-out to non-ergodicity, in contrast to bulk liquid crystal systems where confinement appears abruptly at the isotropic–nematic/smectic phase transitions. The results from the light scattering experiments are detailed in the previous chapter.

The 8CB–aerogel samples have the appearance of glass when the liquid crystal is isotropic, but the appearance of chalk when the liquid crystal is nematic. This indicates that the nematic phase in the aerogel is strongly light scattering and that this scattering must arise from nematic orientation fluctuations. Models of scattering from randomly oriented locally ordered nematic domains (Bellini *et al.*, 1992; and the previous chapter) show that the optical mean free path ℓ is strongly related to nematic domain size ξ by

$$\ell^{-1} = \phi(8/9\pi)(\Delta\varepsilon^2/\varepsilon^2)k^4\xi^3 \tag{20.15}$$

so that a measure of ξ can be obtained from turbidity data. These data indicate that locally the nematic is ordered within pores but disordered on length scales larger than $\langle p \rangle$.

In contrast to what is found in the bulk, where light scattered by the nematic is entirely dynamic, for nematic confined in the aerogel, the scattered light is almost entirely static. This is probably a result of physisorption of the liquid crystal molecules onto the surface of the aerogel and the resulting rigid confinement of the liquid crystal in the smallest pores. In both the nemtic and the smectic phase, the residual fluctuating contribution to the scattered intensity yields a correlation function showing distinct relaxation processes, including nematic-like orientation fluctuations within larger pores (intrapore relaxation) and very slow glass-like non-exponential dynamics for pore–pore relaxation.

As the temperature is decreased these nematic ordering processes saturate at $T = 36\,°C$, leaving only a temperature-invariant intrapore nematic relaxation for $36\,°C > T > 33\,°C$. For $T < 33\,°C$, where smectic ordering commences, the intrapore relaxation exhibits a drastic slowing down, over many orders of magnitude in time, in contrast to the bulk where the suppression of bend and twist by the layering accelerates orientational relaxation (de Gennes and Prost, 1993). This 'smectic slowing down' indicates enhanced barriers for nematic relaxation, apparently as a result of smectic ordering within the pores.

The 8CB–aerogel heat capacity data, obtained by Garland and coworkers, show that aerogel confinement significantly alters the thermodynamic characteristics of the liquid crystal ordering (Bellini *et al.*, 1992; Wu *et al.*, 1995). The bulk and the aerogel-confined 8CB excess heat capacity ΔC_p accompanying nematic and smectic A ordering are shown semi-logarithmically in Figure 20.1 (Wu *et al.*, 1995). As can be seen the confinement rounds the anomalies and shifts them to lower T, affecting the NA transition more strongly than the IN, i.e. all but eliminating the NA anomaly in the most dense 0.60 aerogel. In the bulk the IN transition is first order and accompanied by a cusp-like singularity in ΔC_p. In the bulk and aerogel systems the NA anomaly occurs on top of a broad, sloping background IN tail. This tail has been determined for the bulk (Kasting *et al.*, 1980) and is essentially that remaining in the 0.60 sample. ΔC_p also exhibits small bulk-like IN and NA features from residual bulk 8CB on and in cracks in the aerogel (figure 20.1(a), arrows). This and the background can be subtracted out to yield ΔC_{pNA}, the excess heat capacity due to the NA ordering (Wu *et al.*, 1995), shown on a linear scale in Figures 20.1(b) and (c). The relation of ΔC_{pNA} to the evolution of smectic order in the aerogels will be discussed in section 20.8.2.

20.5 Experimental Methods

Our experimental goal was to characterize via X-ray scattering the structural changes due to the appearance of SmA layering in the aerogel-confined 8CB as a function of temperature, and aerogel density. Here we describe the methods and apparatus used in the experiments.

Each of the aerogels, machined to a thickness of 1 millimetre, was filled in air with 8CB from the isotropic phase by capillary action, which eliminated any gas bubbles from the samples within several hours of filling, such that they became transparent in the isotropic phase. The 8CB was used as purchased from BDH. After filling, the aerogels were held above the IN transition temperature while their surface was wiped clean of excess bulk 8CB. It was never possible to remove all of this excess, leading to a small, bulk smectic A peak in our data at some temperatures. However, wiping the samples reduced the size of this peak sufficiently that it was never more than a minor contribution to the scattering and it usefully provided a direct measure of the smectic layer spacing of the bulk material in the aerogel.

The X-ray experiments described here were performed at the X10 beamlines of the X-ray-emitting electron storage ring at the National Synchrotron Light Source. Samples were

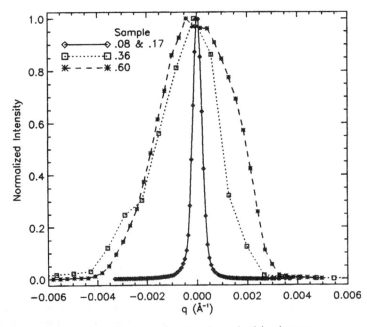

Figure 20.8   Instrumental resolution (arm-zero) functions for each of the data sets.

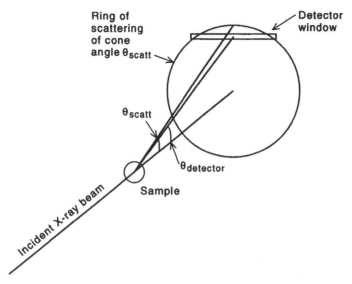

Figure 20.9 Effect of the horizontal dimension of the acceptance window of the detector on the resolution of the scattering from a powdered crystal which scatters into a perfect delta function peak at θ_{scatt}. Photons scattered at θ_{scatt} are detected when $\theta_{detector} < \theta_{scatt}$.

mounted in temperature-controlled ovens on the Huber four-circle diffractometer at either X10A or X10B. The goal of the setups was to optimize both the intensity and the angular resolution of the scattering for the particular aerogel under study. As we will show the Bragg reflection intensity $I(q)$ from the smectic layering is a peak of finite width in wavevector q, indicating that smectic ordering in the aerogels is only short ranged. As the aerogel density increases $I(q)$ broadens in q. Thus the data for the higher-density aerogels (0.36 and 0.60) were acquired at lower angular resolution on X10B, where the resolution was determined principally by two sets of horizontal and vertical slits on the diffractometer detector arm. The data for the lower-density aerogels (0.08 and 0.17) were taken at X10A, where the angular resolution was determined by a Ge(111) analyser crystal on the detector arm. The appropriate resolution functions for each data set are shown in Figure 20.8. These were obtained from 'arm-zero' scans in which we attenuated the incident beam and then scanned the detector through it.

The resolution that is measured by an arm-zero scan is not the only factor determining the resolution that is actually obtained from a powder sample such as an aerogel-confined liquid crystal. Consider a powder sample that scatters X-rays only onto the surface of a cone defined by its half angle θ_{scatt}, as shown in Figure 20.9. The final slits form an acceptance window through which the detector accepts scattered photons. If the horizontal dimension of this window is large then the corners of the window intersect the rim of the scattering cone when the detector is set at angle $\theta_{detector}$ such that $\theta_{detector} < \theta_{scatt}$. This asymmetry will lead to an apparently asymmetric Bragg peak which has larger slope on the high-θ side than on the low-θ side for a given $\theta - \theta_{peak}$. However, if the horizontal dimension of the acceptance window of the detector is set to be small enough then this assymmetry is negligible. In our data the effects of the finite horizontal acceptance are found only in the asymmetry of the peak in the 0.36 data, evident in Figure 20.13 from the small systematic deviation of the peaks from the symmetric Lorentzian fit. We have ignored this problem in our data analysis.

The basic experimental method consisted of changing the temperature in small steps (0.1 to 1 °C, depending on T), usually monotonically, waiting for equilibration, and scanning θ (typically for about 45 minutes) to collect $I(q)$. It was critical that the experiment run as quickly as possible because of the limited amount of time for which we had access to the synchrotron, so that the q spacing between adjacent data points, the T spacing between scans, and the equilibration time between scans were all optimized. For example, the q spacing between adjacent points, different q ranges and at different temperatures was set so that the features of the lineshape were well resolved but not overresolved.

Finally, we verified that our results were not significantly dependent upon the position that we probed on the sample, and that they did not contain any thermal hysteresis effects. In order to check for positional dependencies in our data we scanned two different positions on each of the 0.08 and 0.17 samples. The lineshapes appeared to be independent of position except for a consistent trend in which the lineshapes from the more frequently scanned positions appeared to be very slightly wider than the lineshapes at the same temperatures from the less frequently scanned positions. In other words, the temperatures at which a given lineshape narrowness (i.e. correlation length) was achieved were slightly lower at the more frequently scanned positions. We believe that this is due to higher concentrations of beam-induced impurities at the positions, which should lower all temperatures associated with the NA phase transition. On several occasions we acquired data as the temperature was increasing as well as while it was decreasing. We did not see any hysteresis effects. While the lineshapes from different positions in the samples were similar, the intensity of the scattering was not. When the 8CB–aerogel sample surfaces are wiped to remove bulk 8CB in the isotropic phase and the temperature lowered, the thermal contraction of the 8CB, which is larger than that of the aerogel, causes it generally to retract into the aerogel, leaving some aerogel pores near the surface unfilled. This process is highly irregular (Feder, 1988) producing a variation in the net amount of 8CB encountered by the X-ray beam at different sample positions. However, the lineshape was independent of the extent of filling.

20.6 Data, Fitting and Parameters

In this section we present typical data, discuss the data fitting functions and procedures, and present a summary of the temperature behaviour of the quantities extracted.

20.6.1 *Background Scattering*

Figures 20.5 and 20.6 indicate clearly that in order to characterize quantitatively our 8CB smectic Bragg peak data it was necessary to account for the q dependence of the background scattering due to the structure of the aerogel itself. As discussed in section 20.3 we therefore obtained background scans at temperatures where the smectic A layering peak was absent, at low $T \approx 0\,°C$ where the 8CB was crystalized in the aerogel matrix, and at high $T \approx 40\,°C$ where the system was out of the NA pretransition region in the nematic or isotropic phases. From Figure 20.4 we see that the high-T background contains an additional scattering component, a broad peak at $q \approx 0.25$ Å^{-1}, shown directly in Figure 20.10 by subtraction of the low-T scan from the high-T scan. Fitting a Lorentzian to this peak yields a correlation length of $\xi = 10$ Å and peak position $q = 0.24$ Å^{-1} corresponding to a spatial period $\lambda = 26.2$ Å. This spacing matches the molecular length of 8CB but is significantly shorter than its smectic A layer spacing $d = 31.6$ Å, the latter indicative of the slightly interdigitated, smectic A$_d$-like (de Gennes and Prost, 1993) nature of the 8CB smectic A layering. Because λ matches the molecular length, we interpret this broad peak as arising from remnant pair correlation as is typically found in nematic liquid crystals.

We compared the low-T and high-T scans with those where there is a smectic A Bragg reflection peak ($10\,°C < T\ 36\,°C$). At smaller $q < 0.1$ Å^{-1}, and in the wing of the Bragg peak, the high-T scan matches much better than the low-T crystalline 8CB. Typical comparisons are available in Figures 20.5 and 20.6. This indicates respectively that: the aerogel structure

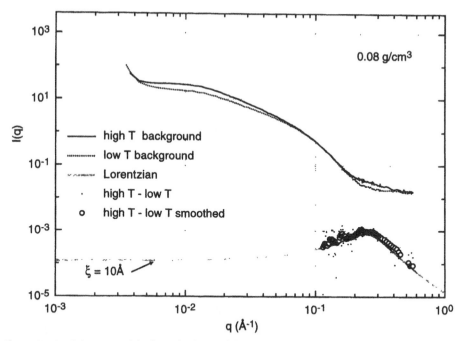

Figure 20.10 Subtraction of the low-T background from the high-T background to give the high-T background correlation peak, fitted to a Lorentzian with parameters $\xi = 10$ Å, $q_o = 0.24$ Å^{-1}. This peak position indicates a correlation spacing of 26 Å, the molecular length. This correlation peak appeared to be present at all T where 8CB was non-crystalline and was therefore included in the background for the analysis of the SmA Bragg scattering.

at high T matches that in the smectic range more closely than if the 8CB is frozen; and that the high-temperature pair correlation peak is present at all T in the smectic A range. We therefore used the high-temperature scan $I_{\mathrm{hi}T}(q)$ to give the background *shape* in q and with this choice the data were fitted to the general form:

$$I(q) = I_{\mathrm{smA}}(q) + I_{\mathrm{bk}} = I_{\mathrm{smA}}(q) + B_{\mathrm{bk}}I_{\mathrm{hi}T}(q) + C_{\mathrm{bk}} \tag{20.16}$$

where $I_{\mathrm{smA}}(q)$ is the smectic A peak fitting function, yet to be discussed, and B_{bk} and C_{bk} are fitting constants, necessary because, respectively; the T dependence of the average 8CB density changes the X-ray scattering contrast with the aerogel and therefore the intensity of the aerogel scattering (about 10% over the experimental T range), and liquid crystal 8CB generates a weak, nearly q-independent scattering with some dependence on T.

20.6.2 *Typical Data and Fitting*

We will first qualitatively compare the data from the 8CB–aerogel samples with what is seen in the bulk. Figures 20.5 and 20.6 show comparative log–log plots of scans at $T \approx 17\,^\circ$C in the three lowest-density aerogels. Figure 20.11 shows linear plots of scans taken below the bulk NA transition temperature $T_{\mathrm{NA}} = 33.7\,^\circ$C from a bulk powder held in a glass capillary and from the most porous (0.08 g/cm^3) and least porous (0.60 g/cm^3) aerogel samples. Figure 20.12 shows linear plots of scans from the 0.36–8CB sample. The peaks in Figures 20.5, 20.6, 20.11 an 20.12 are due to scattering from the smectic A layers, centred at $q \approx 0.198$ Å in all

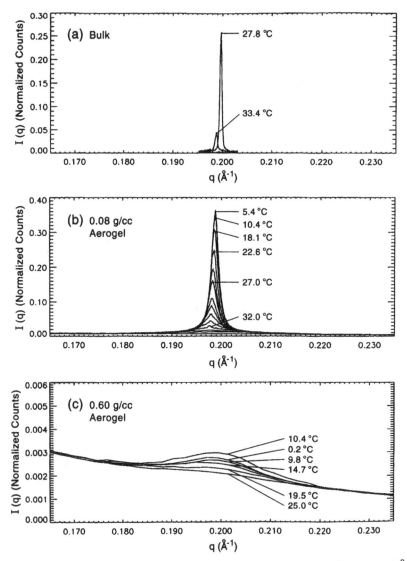

Figure 20.11 Typical data from samples of 8CB in (a) the bulk, (b) the 0.08 g/cm³, $\langle p \rangle = 700$ Å aerogel, and (c) the 0.60 g/cm³, $\langle p \rangle = 120$ Å aerogel. The smectic layering peaks are increasingly widened and grow increasingly slowly with decreasing temperature as the aerogel porosity decreases. The widths of the bulk peaks in (a) are resolution limited at all temperatures $T < T_{NA} = 33.7$ °C. No attempt is made to show the pretransitional narrowing of the bulk peaks as $T \to T_{NA}$ from above. The background scattering from the aerogel itself is slightly visible in (b) and very visible in (c).

three data sets, indicating that there is indeed smectic layering in the aerogels. The background scattering from the aerogel is quite evident in the 0.60 data and less so in the 0.08 data in Figure 20.11.

The width of the capillary bulk peak is resolution limited below T_{NA}. Its height rises quickly over the first few degrees below this temperature to nearly its saturated value as the SmA order parameter ψ grows. The width of the peak from a bulk powder becomes resolution limited in the critical region above the transition in the nematic as $T \to T_{NA}$, as in this region

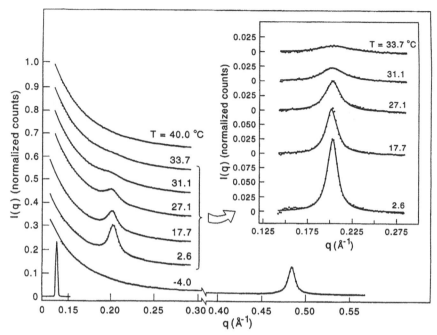

Figure 20.12 X-ray scattering $I(q)$ from the 8CB 0.36 g/cm³ aerogel sample against temperature. For clarity, each $I(q)$ is shifted up relative to that of the next lowest temperature by 0.1 vertical units. The scan at $T = 40\,°C$ is the high-T background. The smectic A layering peak at about 0.198 Å^{-1} narrows gradually in q with decreasing temperature, abruptly disappearing at $T \approx 0\,°C$ to be replaced by crystalline peaks for q in the range 0.476 Å^{-1} < 1.40 Å^{-1}, with the peak at lowest q ($q_x = 0.476$ Å^{-1}) having the largest intensity. The full lines are fits to a sum of a Lorentzian (8CB) peak, the high-T background times a scale factor, and a constant. The $T = -4\,°C$ curve is the low-T background.

the two correlation lengths characterizing the pretransitional layering fluctuations, $\xi_\parallel$ and $\xi_\perp$, diverge (Davidov *et al.*, 1979) (see section 20.2.2). The aerogel peaks, on the other hand, show both narrowing widths and increasing heights starting a few degrees below the bulk transition temperature and continuing slowly over one or more tens of degrees. The width of the aerogel peaks is always much larger than the width of the instrumental resolution function, indicating that the correlation length or domain size in the gels remains finite. The correlation lengths and peak heights are smaller and the temperature range over which they grow is larger in the least porous 0.60 aerogel than in the most porous 0.08 aerogel. This trend is in accord with the intuitive notion that the effect of confinement on the transition should increase with decreasing pore size.

The data from the 0.36 gel were the first to be acquired and fitted (Clark *et al.*, 1993). We discovered empirically that this was fitted rather well by taking $I_{smA}(q)$ in eqn (20.16) to be a single Lorentzian. Explicitly the net fitting function is

$$I_1(q) = A_0\{1 + [\xi(q - q_0)^2]\} + B_{bk}I_{hiT}(q) + C_{bk} \tag{20.17}$$

Here the subscript 1 stands for 'single Lorentzian' and the fit is parameterized by the peak height A_0, position q_0 and effective (single-Lorentzian) correlation length ξ, as well as the background scaling factor B_{bk} and constant C_{bk}. $I_{hiT}(q)$ is an interpolation of the smoothed high-temperature background. A few samples of these fits to the data from the 0.36 aerogel including the aerogel background are shown in Figure 20.12 and the same fits are shown with the aerogel backgrounds subtracted away in Figure 20.13. All of the parameters were fitted at the same time. The Marquart fitting program that we used, the CURVEFIT non-linear routine

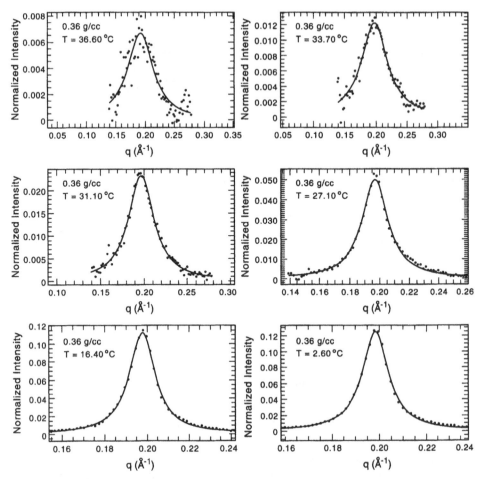

Figure 20.13 Single-Lorentzian fits to typical SmA peak data from the 0.36 sample. The fitted background has been subtracted from the data. The temperatures of the scans are as indicated in the plots.

in the IDL user's library, required fairly accurate initial estimates of the fitting parameters. In order to supply these, we first obtained by hand a reasonably accurate fit of a seed scan. We used the parameters of this hand fit as initial estimates for the fitting routine. The output best-fit parameters were then used as initial estimates for the parameters of the scans immediately above and below the seed in temperature. The rest of the scans were fitted by iterating this process.

We were able to fit the 0.60 data to $I_1(q)$ as well. Examples of these fits are shown in Figure 20.14.

We now turn to the fitting of the 0.08 and 0.17 data sets, discussing specifically only the 0.08 data since the two sets are qualitatively quite similar. The utility of the single-Lorentzian lineshape for describing the 0.36 and 0.60 data sets suggested its further use on the 0.08 and 0.17 data. This effort had only limited success, as the 0.08 and 0.17 data sets were considerable more complex than those of the 0.36 and 0.60, as we now discuss.

The single-Lorentzian fits of the 0.08 and 0.17 data sets were not successful for $T < 35.3\,°C$, as demonstrated in Figure 20.15. Here the background has been fitted and subtracted from the layering peak, which is plotted as $\log[I(q)]$ against $q - q_{peak}$ for two temperatures, $T = 12.9$

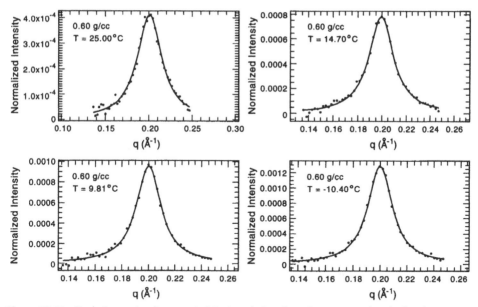

Figure 20.14 Single-Lorentzian fits to typical SmA peak data from the 0.60 sample. The fitted background has been subtracted from the data. The temperatures of the scans are as indicated in the plots.

and 35.6 °C. The full line is a single-Lorentzian fit to the $T = 12.9$ °C scan, which, as the inset shows, is quite good when $q - q_{peak}$ is sufficiently small, but which falls below the data rather badly in the wings. The functional form of the $T = 12.9$ °C data can be usefully visualized in

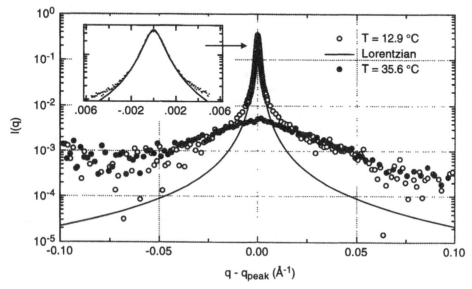

Figure 20.15 Plot of scans at $T = 12.9$ and 35.2 °C in the 0.08 8CB sample. The background has been fitted and subtracted and the resulting $I(q)$ plotted against $q - q_{peak}$. The full line is a single-Lorentzian fit to the $T = 12.9$ °C scan which is excellent at small $q - q_{squeak}$ (inset) but falls below in the wings..

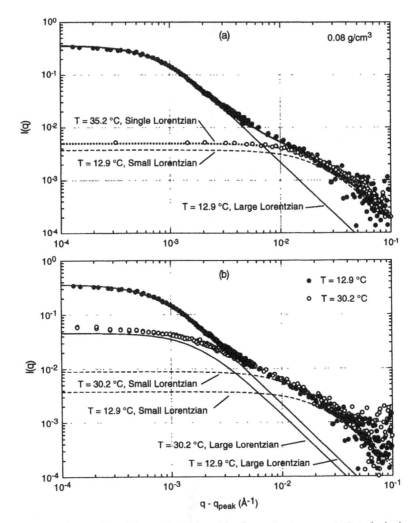

Figure 20.16 (a) The $I(q)$ data of Figure 20.15 plotted log–log against $|q - q_{peak}|$. Data for both $q - q_{peak} > 0$ and $q - q_{peak} < 0$ are plotted and overlap, indicating that the peaks are symmetric. The light black line is a single-Lorentzian fit to the $T = 12.9\,°C$ data. The bimodal lineshape can be fitted to a sum of large and small Lorentzians (heavy grey line), with the dashed line giving the second broad Lorentzian. The data at $T = 35.2\,°C$ can be fitted to a single Lorentzian. Note that the wings of the high- and low-T peaks overplot. (b) Comparison of large and small Lorentzians at $T = 12.9$: and $30.2\,°C$.

a log–log plot of $I(q)$ against $|q - q_{peak}|$, as is shown in Figure 20.16. Here both the high-q and low-q sides of the peak are plotted and overlap, indicating that it is symmetric about q_{peak}. The full line is again the single-Lorentzian fit, with its characteristic level central plateau region and straight wings of limiting slope -2. The data, however, are characterized by an 'S'-shaped decay with the wings levelling off after an initial Lorentzian-like decrease and then sloping over into a noisy $|q - q_{peak}|^2$ final decay. Thus it appears that the portion of the scattering that is missed in the single-Lorentzian fit might be Lorentzian itself, albeit with a larger width and smaller amplitude.

This behaviour is suggestive of a superposition of distinct broad and narrow peaks, so that fits to a sum of two Lorentzians centred near q_{peak} were attempted. These were in fact quite good. Figure 20.16 shows a typical two-Lorentzian fit, indicating the narrow Lorentzian, broad

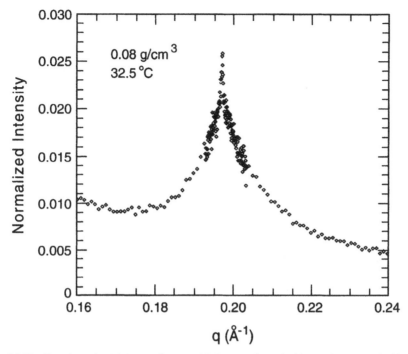

Figure 20.17 Data from the 0.08 aerogel at 32.4 °C showing the spike-like peak from residual bulk 8CB on top of the much broader peak from 8CB in the pores of the aerogel.

Lorentzian, and their sum. Data for $T < 35.3$ °C were then fitted using the two-Lorentzian form:

$$I_2(q) = A_L\{1 + [\xi_L(q - q_L)]^2\}^{-1} + A_S\{1 + [\xi_S(q - q_S)]^2\}^{-1} + B_{bk}I_{hiT}(q) + C_{bk} \qquad (20.18)$$

The subscript 2 indicates a two-Lorentzian sum, while the subscripts L and S distinguish the parameters of the narrow (Large ξ and amplitude) and wide (Small ξ and amplitude) Lorentzians respectively. Fits to $I_2(q)$ vary all weight parameters.

Figure 20.16 also shows the $T = 35.6$ °C data, which appear more single-Lorentzian like, without the sigmoidal wings of the lower-T scan. The heavy dotted line is a single-Lorentzian fit to the $T = 35.6$ °C scan, which is quite good. Thus, at low T, the scans clearly have two components, a narrow large-amplitude and a wide small-amplitude Lorentzian. *The narrow Lorentzian has a strong dependence on T, broadening and decreasing in amplitude as T increased. The wide-Lorentzian parameters are only weakly dependent on T.* At sufficiently large T then, the widths of the two Lorentzians become comparable and the distinction between them is lost. This occurs at $T \approx 35.3$ °C and $T \approx 32.6$ °C in the 0.08 and 0.17 samples respectively. Above these temperatures there is no evidence from the log–log plots, cf. Figure 20.16, that two Lorentzians were necessary. As might be expected from this observation the parameters of the double-Lorentzian fitting behaved erratically at high T; so single Lorentzians were used to fit the data and exhibited none of these problems.

We have carefully inspected the results of fits of the 0.36 and 0.60 data to both $I_1(q)$ and $I_2(q)$ using different background smoothings and temperatures. In general the fitting parameters obtained from single-Lorentzian fits to these data sets are much better behaved as functions of T than the fitting parameters from double-Lorentzian fits without much difference in the closeness of fits to the data. Thus if the 0.36 and 0.60 data are two component, the distinction is lost in the noise and background subtraction process. For these reasons we will report only the results of the single-Lorentzian fits to the 0.36 and 0.60 data sets. For conciseness, we

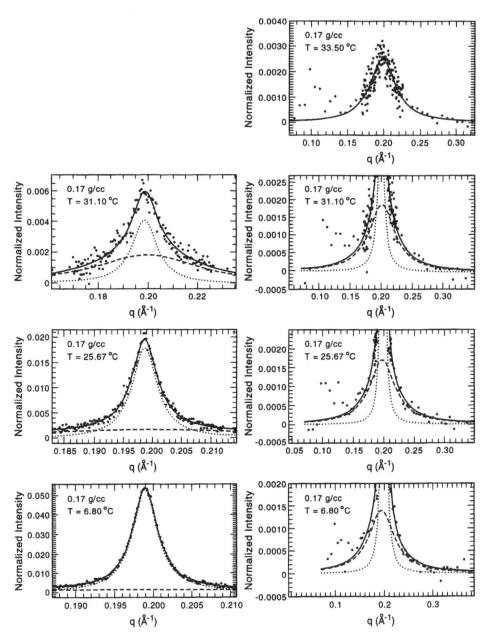

Figure 20.18 Fits to typical SmA peak data from the 0.08 sample. Single and double Lorentzians are used in their respective regimes (see text); . . . , large Lorentzian; – – – –, small Lorentzian; ———, total.

define the *single-Lorentzian regime* to be the regime of T and $\langle p \rangle$ where we report the results of single-Lorentzian fits: $T > 35.6\,°C$ in the 0.08 and $T > 32.5\,°C$ in the 0.17 samples and all T in the 0.36 and 0.60 samples. Similarly, we define the *double-Lorentzian regime* to be where we report the results of double-Lorentzian fits: $T < 35.6\,°C$ in the 0.08 and $T < 32.5\,°C$ in the 0.17 samples.

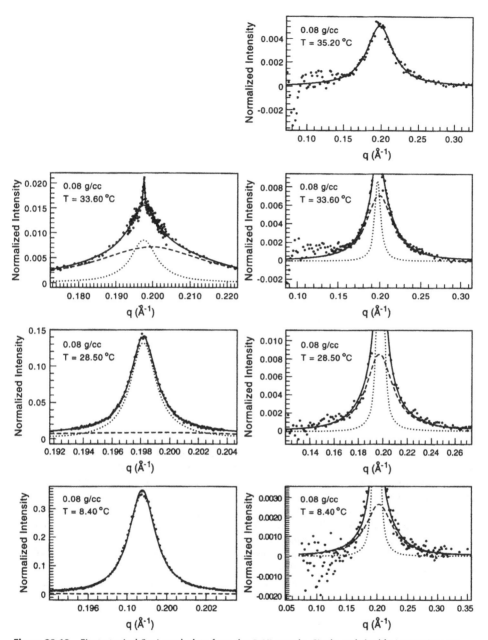

Figure 20.19 Fits to typical SmA peak data from the 0.17 sample. Single and double Lorentzians are used in their respective regimes (see text): ...., large Lorentzian, – – – –, small Lorentzian; ———, total. The bulk spike can be seen at $T = 33.60$ °C.

At temperatures in the range 24 °C $< T < 32$ °C, there appeared in $I(q)$, in addition to the broadened Bragg scattering from the 8CB, a sharp spike, having a slightly larger q_{peak} than that of the narrow Lorentzian, shown in Figures 20.17 and 20.18. This spike is found clearly in the 0.08 and 0.17 data sets, appearing with decreasing T at $T \approx 32$ °C, growing to a maximum amplitude at $T \approx 29$ °C, and then slowly disappearing by $T \approx 24$ °C. It is also found over a narrow T range in the 0.36 data. The shape of this spike turns out to be well approximated

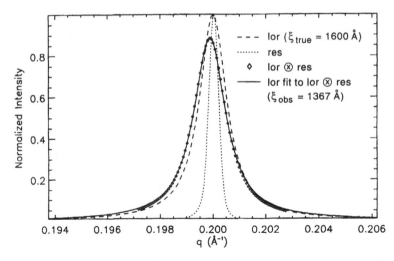

Figure 20.20 Simulation of the resolution broadening of a Lorentzian lineshape. The dashed line is a computer-generalized Lorentzian having $\xi_{true} = 1600$ Å. This simulates the 'true data' lineshape that would be observed with perfect instrumental resolution. The dotted line is the measured instrumental resolution function res(q) for the 0.08 and 0.17 data sets. The diamonds are the simulated 'observed data' lineshape, obtained by convolution of the 'true data' with res(q). The full line is the best-fit Lorentzian to the simulated observed lineshape. It fits well, giving a simulated 'observed' correlation length $\xi_{obs} = 1367$ Å. These lengths are typical of the situation in the 0.08 data at low T and represent the largest resolution broadening among all of the data sets.

by the resolution function, suggesting that it comes from residual bulk material. The presence of such a bulk peaklet is not unexpected, as the calorimetry measurements on this system (Bellini *et al.*, 1992; Wu *et al.*, 1995) reveal small heat capacity peaks at the bulk IN and NA transition temperatures that arise from residual bulk 8CB on the aerogel surfaces or in cracks (see Figure 20.1). Because the spike has a bulk-like dependence of peak intensity on T (see Figure 20.11(a)) it does not become visible until a degree or so below the bulk $T_{NA} = 33.5°$ and disappears with decreasing T because 8CB is contracting relative to the aerogel and would tend to exit the cracks and surfaces first. We accounted for the spike when fitting scans where it was obviously present by adding to $I_2(q)$ the scaled resolution function, shifted to be centred on the spike position. Thus we fitted to $I(q) = I_2(q) + D \, \text{res}(q - q_{spike})$ where res(q) is the instrumental resolution and D and q_{spike} are two more fitting parameters, for a total of 10. The $T = 33.6°C$ scan of Figure 20.19 shows a typical fit to the bulk spike.

Representative fits of the data to single or double Lorentzians using the strategies outlined above are shown in Figures 20.18 and 20.19.

20.6.3 *Resolution Effects*

We have not yet accounted for the effect of X-ray diffractometer resolution on the parameters that we obtained from our fits. The physically important parameters are the integrated intensities and correlation lengths of each of the two Lorentzians. The integrated intensities are independent of the resolution function as long as it is normalized to fixed area but the correlation lengths do depend on the resolution. The observed linewidths are at least three times the resolution width or more over the entire data set. The correlation length corresponding to the half width of the resolution function itself is about 4000 Å for the 0.08 and 0.17 data sets, about 740 Å for the 0.36 set, and about 530 Å for the 0.60 set. The maximum correlation

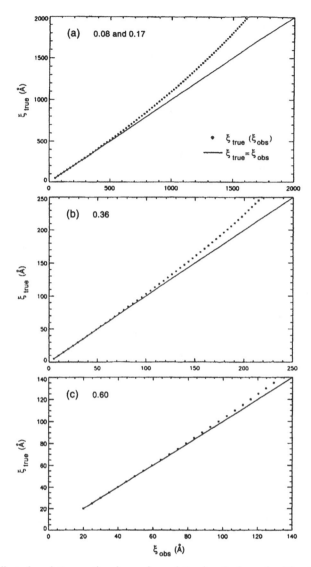

Figure 20.21 Effect of resolution on the observed correlation lengths for each of the data sets as determined via the procedure of Figure 20.20. The dotted curves give the 'true' correlation length ξ_{true} in terms of ξ_{obs}, the 'observed' correlation length obtained from a fit of a Lorentzian to the convolution of a Lorentzian and the resolution function.

lengths in each data set as obtained from our fits were $\xi \approx 1350$ Å in the 0.08, $\xi \approx 500$ Å in the 0.17, $\xi \approx 150$ Å in the 0.36, and $\xi \approx 90$ Å in the 0.60 data sets.

We can understand the effect of resolution on the interpretation of our data via simulation as is shown in Figure 20.20. Here, for the 0.08 data we generate a Lorentzian corresponding to the 'true data' peak that would be observed with perfect resolution for the lowest-T scan (the worst-case scenario), convolve it with the 0.08 resolution function res(q) in Figure 20.8, giving the simulated 'observed data' that would be observed with our res(q), and then fit these 'observed data' to a Lorentzian, simulating our fitting process. The result shows that the convolution with res(q) does not significantly alter the functional form of the peak. It is still Lorentzian but somewhat broader. Since res(q) falls off rapidly, with no tails, the wings of the data peak are not significantly altered. Figure 20.20 shows the results of this process for a 'true data' correlation length $\xi_{true} = 1600$ Å is convolved with res(q) for the 0.08 data set to produce a simulated 'observed data' peak which is well fitted by a Lorentzian of correlation length $\xi_{obs} = 1352$ Å. This value of ξ is in the neighbourhood of the values that we obtain from the 0.08 sample at the lowest temperatures.

We have corrected for the effect of resolution on our fits by following the above procedure to generate tables of ξ_{true} against ξ_{obs} for each of the res(q) in Figure 20.8. The results are shown in Figure 20.21. These tables were used to convert fitted 'observed' Lorentzian widths ξ_{obs} to 'true' Lorentzian widths. The largest such corrections were 17% in the 0.08 data at low T.

20.6.4 Bulk Liquid Crystal 8CB

The data analysis to follow is facilitated by comparison of the 8CB–aerogel behaviour with that of bulk 8CB. Figures 20.1 and 20.2 showed respectively the bulk excess heat capacity and bulk correlation lengths for layer order in the direction parallel to the nematic director. In addition we measured the evolution with T of the position and integrated intensity of the smectic A layering peak in bulk powder 8CB samples (cf. Figure 20.11(a). The 8CB was contained in quartz capillaries for these experiments and the scans carried out with the high-resolution setup (0.08 and 0.17 arm zero in Figure 20.8). These results are included in Figures 20.22 and 20.23. The NA transition in the bulk was found to be at $T = 33.5 \pm 0.1\,°C$, as expected.

20.6.5 Liquid Crystal 8CB–aerogel: Parameters against Temperature

Here we summarize the thermal behaviour of bulk 8CB and the 8CB–aerogel samples, including the principal parameters which can be extracted from the measured $I(q)$.

We begin with the peak position data in Figure 20.22, which plots the bulk q_{peak} obtained from the capillary powder data and, for the 8CB–aerogel, the large Lorentzian peak position in the double-Lorentzian regime ($\bullet$), the peak position in the single-Lorentzian regime ($\odot$), and the position of the bulk peaklet ($\bigcirc$). The bulk data show a remarkable cusp-like minimum at the NA transition, with q_{peak} decreasing as T approaches T_{NA} from either direction. The 8CB–aerogel systems show the same general trend, but with q_{peak} reduced and the minimum in q_{peak} shifted to lower T. The presence of the bulk peaklet (cf section 20.6.2) in the aerogel data can be used as follows:

1　Comparison of the bulk peaklet position and aerogel Lorentzian peak positions for each aerogel shows the 8CB in the pores has a slightly smaller q_{peak} than the bulk 8CB present.
2　Comparisons of the aerogel bulk peaklet positions with that of bulk 8CB shows that overall the bulk 8CB in the aerogels has a smaller peak position than the capillary 8CB sample, probably a result of contamination of the 8CB in the aerogel, which would not be surprising given the large surface-to-volume ratio of the pores. This shift is not monotonic with pore size, but is more severe for the 0.08 sample than the 0.17, suggesting again that the 8CB in the 0.08 aerogel has a higher impurity concentration than that of the 0.17.

In order to compare more directly the capillary 8CB and aerogel Lorentzian q_{peak} behaviour we scale the 0.08, 0.17 and 0.36 aerogel q_{peak} data so that their bulk peaklet positions match that of bulk capillary 8CB. There was no bulk peaklet observed in the 0.60 scans. The required scaling is small, less than 1%. Once this scaling was completed we found that the overall trends with pore size were also not monotonic, but that the minimum in the 0.08 sample was at a lower temperature than that of the 0.17 sample. This we believe to be an additional

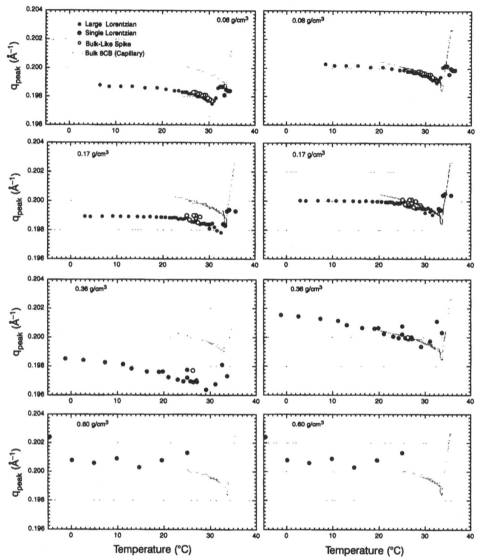

Figure 20.22 *Left column*: Peak position against temperature for bulk 8CB (full curve) and the four 8CB–aerogel samples, showing q_{peak} position in the single-Lorentzian regime (⊙), of the large Lorentzian in the double-Lorentzian regime (◉), and of the bulk peaklet (○). *Right column*: Left column data after scaling q_{peak} of the bulk peaklet to match bulk 8CB and shifting the 0.08 data to have their minimum q_{peak} such that its minimum q_{peak} falls on the curve of (minimum q_{peak}) against $1/\langle p \rangle$ for the bulk and other aerogels.

consequence of its higher 8CB impurity concentration. This observation gave us a means of correcting the temperature scale of the 0.08 data to remove the principal effects of 8CB impurity, based on the assumption that, as is generally the case for liquid crystal phase transitions, low concentrations of impurities simply shift transition temperatures. We shifted the T scale of the 0.08 data up by 1.8 °C, such that its minimum q_{peak} fell on the curve of minimum q_{peak} against $1/\langle p \rangle$ for the bulk and other aerogels. The before and after scaling and this 0.08 shift are shown in Figure 20.22.

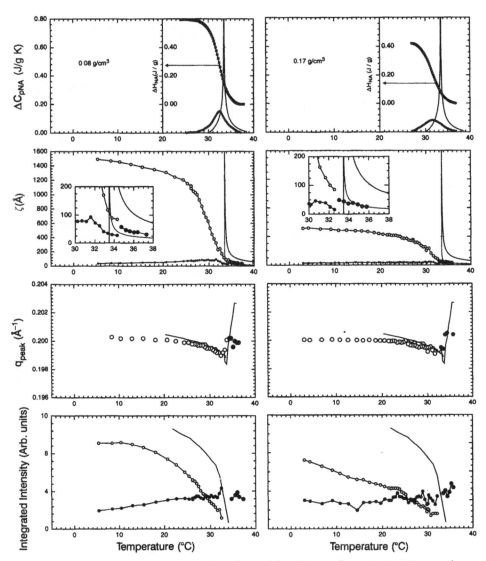

Figure 20.23 (a) Composite temperature dependence of the NA excess heat capacity ΔC_{pNA}, net heat released ΔH_{NA}, correlation lengths ξ, peak positions q_{peak} and integrated intensities for bulk 8CB and for the 8CB 0.08, 0.17 and 0.36 aerogel samples. Correlation lengths are bulk $\xi_\parallel$ (full curve) and $\xi_\perp$ (dotted curve), large Lorentzian (○), small Lorentzian (●) and single Lorentzian (⊙). Integrated intensities are for bulk (full curve), large Lorentzian (○), small Lorentzian (●), total (dotted curve) and single Lorentzian (⊙). Note that the vertical scale of the bulk integrated intensity is arbitrary relative to the aerogel data.

Figures 20.23 and 20.24 present compilations of the final heat capacity, correlation length, peak position, and integrated intensity data as follows:

1 *First (top) panels* give the bulk and aerogel $\Delta C_{pNA}(T)$ from Figure 20.1 and the net heat $\Delta H_{NA}(T) = \int_T^{40} \Delta C_{pNA}(T')\mathrm{d}T'$.
2 *Second panels* give the bulk correlation lengths for layer order the direction parallel to the nematic director ($\xi_{\parallel\mathrm{bulk}}$, full curve) and normal to it ($\xi_{\perp\mathrm{bulk}}$, dotted curve) from Figure 20.2.

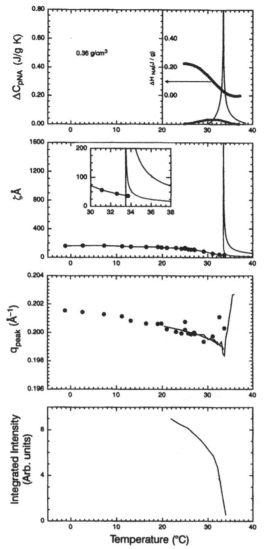

Figure 20.23 (b) Composite temperature dependence of the NA excess heat capacity ΔC_{pNA}, net heat released ΔH_{NA}, correlation lengths ξ, peak positions q_{peak} and integrated intensities for bulk 8CB and for the 8CB 0.08, 0.17 and 0.36 aerogel samples. Correlation lengths are bulk $\xi_{\parallel}$ (full curve) and $\xi_{\perp}$ (dotted curve), large Lorentzian (O), small Lorentzian (●) and single Lorentzian (⊙). Integrated intensities are for bulk (full curve), large Lorentzian (O), small Lorentzian (●), total (dotted curve) and single Lorentzian (⊙). *Note that the vertical scale of the bulk integrated intensity is arbitrary relative to the aerogel data.*

Figure 20.23 gives the 8CB–aerogel correlation lengths of the large (O) and small (●) Lorentzians in the double-Lorentzian regime, and the correlation length in the single-Lorentzian regime (⊙). Figure 20.4 presents bulk and either the large Lorentzian or single-Lorentzian correlation length for each of the aerogels.

3   *Third panels* in Figure 20.23 give the location of q_{peak} for bulk 8CB, and for the 8CB–aerogel the large Lorentzian peak position in the double-Lorentzian regime (●), the peak position in the single-Lorentzian regime (⊙), and the position of the bulk peaklet (O). Figure 20.24

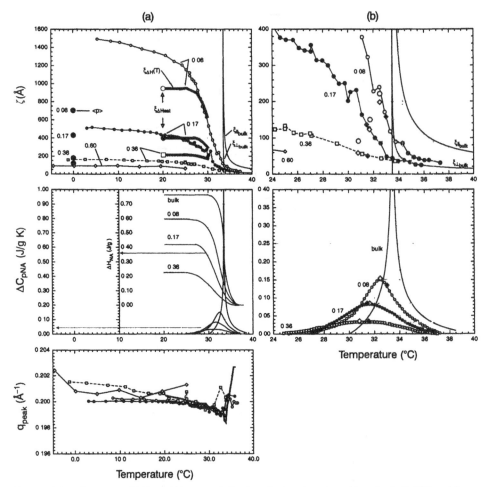

Figure 20.24 Summary of ΔC_{pNA}, ΔH_{NA}, correlation lengths ξ and peak positions for bulk 8CB and for the 8CB 0.08 (○), 0.17 (●), 0.36 (□) and 0.60 (◇) aerogel samples. (a) The heavy grey curves are correlation lengths $\xi_{\Delta H}(T)$, obtained from the temperature dependence of the released heat (see section 20.8.2.2). (b) This figure shows that the measured correlation lengths are comparable with the pore volume-to-surface ratio V_p/S (◆) at the temperatures of the heat capacity peaks (◇). The grey lines indicate schematically the determination of the effective finite size from the ΔC_{pNA} maximum, which produces lengths $\xi_{\parallel}(t_R)$ (○) comparable with the measured ξ at the ΔC_{pNA} peak and to V_p/S.

presents bulk and either the large Lorentzian or single-Lorentzian peak position for each of the aerogels.

4   *Fourth panels* in Figure 20.23 give the relative integrated intensity of the Lorentzians in the double-Lorentzian (large (○) and small (●)) and single-Lorentzian (⊙) regimes on an arbitrary vertical scale. It also shows total integrated intensity and that of the bulk capillary data. *The vertical scale of the bulk integrated intensity is arbitrary relative to the aerogel data.*

20.6.6   *Crystal 8CB–Aerogel*

Bulk 8CB crystallizes near $T = 20\,°C$. In the 8CB–aerogel samples this crystallization is

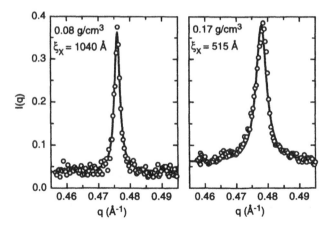

Figure 20.25 Fits of single Lorentzians to the crystal 8CB peaks of Figure 20.6. The crystal correlation lengths ξ_x obtained are slightly larger than the pore chord $\langle p \rangle$ (see Table 20.1). There is no sign of a second Lorentzian in the crystal phase in the aerogel.

suppressed, occurring at successively lower temperatures for increasing aerogel density: at $T \approx 0$ to $5\,°C$ for the 0.08 and 0.17 samples; $T \approx -2$ to $+2\,°C$ for the 0.36 sample; and $T \approx -5$ to $-10\,°C$ for the 0.60 sample. The crystallization is first order characterized by the coexistence of crystal and smectic A peaks, a gradual exchange of intensity between them, and significant hysteresis. Figure 20.25 shows the crystal peaks in the 0.08 and 0.17 aerogel–8CB samples, visible at $q \approx 0.48\ Å^{-1}$ in the low-T scans of Figure 20.6. Attempts were made to fit these and the lowest-q crystal peak of the 0.36 data (Figure 20.12) and all crystal peaks were well fitted by single Lorentzians, with no need for a second Lorentzian. The lowest-q crystal peak position varied slightly among the different aerogels: $q_x = 0.476\ Å^{-1}$ in the 0.08; $q_x = 0.478\ Å^{-1}$ in the 0.17; $q_x = 0.484\ Å^{-1}$ in the 0.36. The single-Lorentzian correlation lengths obtained, corrected for resolution effects as discussed in the last section, were: $\xi_x = 1040\ Å$ for the 0.08; $\xi_x = 515\ Å$ for the 0.17; $\xi_x = 255\ Å$ for the 0.36. These values are comparable with $\langle p \rangle$ (see Table 20.1). Several higher-order crystal peaks were studied for the 0.36 sample and reliable fits could be obtained for the $q_x = 1.148\ Å^{-1}$ peak, for which $\xi_x = 254\ Å$ Clark et al., 1993). The lack of q dependence of ξ_x indicates single crystallites with a mean dimension ξ_x, essentially filling the aerogel pores (see section 20.7.3). This crystal structure with a lowest-order peak at $q \approx 0.48\ Å^{-1}$ is unique to the aerogels. Crystallization of the bulk produces a completely different structure with Bragg reflections at $q_x = 0.25, 1.28$ and $1.43\ Å^{-1}$. Thus aerogel confinement can either change the relative stability of different crystal morphologies, or provide nucleation conditions that favour the formation of particular structures that would not be the first to nucleate in the bulk.

20.7 Results and Analysis

This section will provide an analysis of smectic A ordering in the aerogel pores based on the results presented in Figures 20.23 and 20.24. The data fitting process described in section 20.6 provides information on the lineshape of the smectic A layering peak which can be summarized as follows:

1 The smectic A layering peak is observably broadened in wavevector by aerogel confinement;
2 The smectic A layering peak lineshape can be determined without significant instrumental distortion;

3 The smectic A layering peak is nearly symmetric about its peak wavevector;
4 The smectic A layering peak is well fitted by single or double Lorentzians.

We first use these facts to gain an understanding of the correlation functions for the smectic layering order parameter. This enables the development of a physical picture of smectic ordering in the aerogel pores and guides the interpretation of the fitting parameters in section 20.8. We will first discuss the implications of the single-Lorentzian lineshape which fits the data in the single-Lorentzian regime. Generalization to the double-Lorentzian case follows by superposition.

20.7.1 *Description of Smectic A Correlations in the Aerogel*

The Bragg peaks from the smectic layering are diffuse, indicating that the smectic ordering in the aerogel is short ranged. The X-ray intensity for this kind of isotropic scatterer with short-range order is given by the FT of the electron density correlation function (eqn (20.10a). The sample volume can be partitioned into the aerogel solid region, of uniform density n_{ag}, and the liquid crystal region, of density $\langle n_{lc} \rangle + \delta n_{lc}(\mathbf{r})$, where $\langle n_{loc} \rangle$ is the average density in the liquid crystal region. We define a function $f(\mathbf{r})$, such that $f(\mathbf{r}) = 1$ in the aerogel region and $f(\mathbf{r}) = 0$ in the liquid crystal region. In this case the r-dependent parts of $\langle n(0)n(r) \rangle$ are given by $\langle n(0)n(r) \rangle = (n_{ag} - \langle n_{lc} \rangle)^2 \langle f(0)f(r) \rangle + \langle \delta n_{lc}(0)\delta n_{lc}(r) \rangle$, a sum of liquid crystal and aerogel contributions. In order to evaluate $\langle \delta n_{lc}(0)\delta n_{lc}(r) \rangle$ we note that even at the higher temperatures the width in q of the diffuse Bragg peaks from the smectic layering is typically small compared with their position, q_{peak}, indicating that the smectic layering is locally well defined. If we choose point 0 at random in the liquid crystal we will find at 0 some local layer normal, which we call $\hat{\mathbf{z}}$. Also, by sampling local regions we can obtain the average layer spacing $\langle d \rangle$. Following the McMillan–de Gennes analogy (de Gennes and Prost, 1993), we can then write $\delta n_{lc}(\mathbf{r}) = \psi(\mathbf{r}) \cos[\mathbf{Q}_{loc} \cdot \mathbf{r} + \phi(\mathbf{r})]$, where $\mathbf{Q}_{loc} = \hat{\mathbf{z}}(2\pi/\langle d \rangle) \equiv \hat{\mathbf{z}}\langle Q \rangle$, $\phi(\mathbf{r}) = \langle Q \rangle u(\mathbf{r})$, and $u(\mathbf{r})$ is the local layer displacement parallel to $\hat{\mathbf{z}}$ relative to a perfect smectic. Letting $\delta\phi(\mathbf{r}) = \phi(\mathbf{r}) - \phi(0)$, the resulting correlation function, conditional on the orientation of $\mathbf{Q}_{loc}$, is

$$\langle \delta n_{lc}(0)\delta n_{lc}(\mathbf{r}) \rangle |_{\hat{\mathbf{z}}} = \langle \psi(0)\psi(\mathbf{r})\{\exp[i\mathbf{Q}_{loc} \cdot \mathbf{r} + i\phi(\mathbf{r})] + \text{c.c.}\}\{\exp[-i\phi(0)] + \text{c.c.}\}\rangle/4$$

$$\approx \langle \psi(0)\psi(\mathbf{r}) \rangle (\{\exp[i\mathbf{Q}_{loc} \cdot \mathbf{r}][1 + i\langle \delta\phi(\mathbf{r}) \rangle - \langle \delta\phi(\mathbf{r})^2 \rangle/2 + \ldots]\} + \text{c.c.})/4$$

$$\approx \langle \psi(0)\psi(\mathbf{r}) \rangle\{\exp[-\langle \delta\phi(\mathbf{r})^2 \rangle/2]\} (\cos[\mathbf{Q}_{loc} \cdot \mathbf{r}])/2$$

$$\approx \langle \psi(0)^2 \rangle\langle f(0)f(\mathbf{r}) \rangle\{\exp[-\langle \delta\phi(\mathbf{r})^2 \rangle/2]\} (\cos[\mathbf{Q}_{loc} \cdot \mathbf{r}])/2 \tag{20.19}$$

Here we have assumed that fluctuations in the layering amplitude $\psi(\mathbf{r})$ are uncorrelated with the layer positional fluctuations $\phi(\mathbf{r})$, and we have used $\langle \phi(\mathbf{r}) \rangle = 0$, a result of the random choice of 0, $\langle \delta\phi(\mathbf{r}) \rangle = 0$, a necessary result of C_∞ rotational symmetry about $\hat{\mathbf{z}}$, and our choice of $\mathbf{Q}_{loc}$ to be the mean wavevector. We define $f(\mathbf{r}) = \psi(\mathbf{r})/\langle \psi(0)^2 \rangle^{1/2}$, the fractional fluctuation of $\psi(\mathbf{r})$ about its r.m.s. value. Fourier transformation of $\langle \delta n_{lc}(0)\delta n_{lc}(\mathbf{r}) \rangle |_{\hat{\mathbf{z}}}$ yields the corresponding contribution to $I(\mathbf{q})$, conditional on $\hat{\mathbf{z}}$:

$$I(\mathbf{q})|_{\hat{\mathbf{z}}} = \text{FT}[\langle \delta n_{lc}(0)\delta n_{loc}(\mathbf{r}) \rangle | \hat{\mathbf{z}}] = S_\psi(\mathbf{q}) \otimes S_\phi(\mathbf{q}) \otimes \delta(\mathbf{q} - \hat{\mathbf{z}}\langle Q \rangle) \tag{20.20}$$

where $S_\psi(\mathbf{q}) = FT[G_\psi(\mathbf{r}) \equiv \langle \psi(0)\psi(\mathbf{r}) \rangle]$, $S_\phi(\mathbf{q}) = \text{FT}[G_H(\mathbf{r}) \equiv \exp\{[\langle \delta\phi(\mathbf{r})^2 \rangle/2]\}]$, and $\otimes$ denotes convolution. $I(\mathbf{q})|_{\hat{\mathbf{z}}}$ can be viewed as coming from the average local domain oriented with its wavevector along $\hat{\mathbf{z}}$. The amplitude and phase correlation functions and their structure factors are respectively peaked at $\mathbf{r} = 0$ and $\mathbf{q} = 0$, and are in general anisotropic but must exhibit C_∞ about $\hat{\mathbf{z}}$ and reflection symmetry under $\hat{\mathbf{z}} \rightarrow -\hat{\mathbf{z}}$. The convolution of eqn (20.20) reproduces the effective domain shape structure factor, $S_{\psi\phi}(\mathbf{q}) \equiv S_\psi(\hat{\mathbf{z}}) \otimes S_\phi(\mathbf{q})$, about $\hat{\mathbf{z}}\langle Q \rangle$. Figure 20.26 shows a schematic contour drawing of $I(\mathbf{q})$. To obtain $I(q)$ we must complete the ensemble by averaging over orientations of $\mathbf{Q}_{loc}$. This operation is equivalent to the orientational

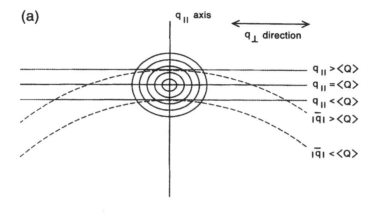

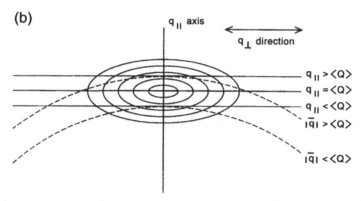

Figure 20.26 Schematic of the geometry of the replacement of the powder average of eqn (20.22) with the planar average of eqn (20.23) for an anisotropic Lorentzian squared $S(\mathbf{q}) = \mathrm{LOR}^2(\mathbf{q} - \mathbf{Q}_{loc})$. Shown are contours of constant $S(\mathbf{q})$ for two values of $\xi_\perp$, the lines of constant q summed over in the exact powder average (eqn (20.22)), and lines of constant q_z averaged over in the approximate planar average (eqn (20.23)). The planar average is always symmetric about Q_{loc} and approximates the powder average more closely in (a) than in (b). The powder average is asymmetric about Q_{loc}, with increasing values of $\xi_\perp$ tending to make it relatively higher for $q > Q_{loc}$ and the $1/q^2$ factor in eqn (20.22) tending to make it relatively higher for $q < Q_{loc}$.

averaging occurring in scattering from a crystalline powder and may be equivalently carried out over $\mathbf{q}$, by summing $I(\mathbf{q})|_{\hat{z}}$ over surfaces of constant q, i.e.

$$I(q) = \int d\Omega\ I(\mathbf{q})|_{\hat{z}} = \int (dA_q/q^2)I(\mathbf{q})|_{\hat{z}} \tag{20.21}$$

as indicated by the dashed lines in Figure 20.26.

We now consider some example powder averages to illustrate the sensitivity of $I(q)$ to features of $S_{\psi\phi}(\mathbf{q})$. If the region (of dimension $\delta q_\perp$) in the $\mathbf{q}_\perp$ direction (perpendicular to $\hat{z}$) where $S_{*H}(\mathbf{q})$ is substantial is sufficiently well localized (generally $\delta q_\perp \ll \langle Q \rangle$) then the constant-$q$ contours in Figure 20.26 approximate planes perpendicular to $\hat{z}$ ($\mathbf{q}_\perp$ direction). With $d\Omega \approx d^2 q_\perp q^{-2}$ the powder average can then be written

$$I(q + \langle Q \rangle) \approx \int d^2q_\perp q^{-2}[S_{\psi\phi}(\mathbf{q})] \tag{20.22}$$

If additionally $S_{\psi\phi}\mathbf{q})$ is localized along $\hat{z}$ ($\delta q_\parallel \ll \langle Q \rangle$), then the powder average reduces to a simple integration over the transverse wavevector $\mathbf{q}_\perp$:

$$I(q + \langle Q \rangle) \approx \langle Q \rangle^{-2} \int d^2q_\perp[S_{\psi\phi}(\mathbf{q})] \tag{20.23}$$

With this result we can now ask what $S_{\psi\phi}(\mathbf{q})$ will generate a Lorentzian $I(q)$ in eqn. (20.23). The general answer is that $S_{\psi\phi}(\mathbf{q})$ must be of the form $S_{\psi\phi}(\mathbf{q}) = \mathrm{LOR}(q_\parallel)S_\perp(q_\perp)$, with $\mathrm{LOR}(q_\parallel) = [1 + (\xi_\parallel q_\parallel)^2]^{-1}$ or be an anisotropic Lorentzian squared $[\mathrm{LOR}^2(\mathbf{q})] = [1 + (\xi_\parallel q_\parallel)^2 + (\xi_\perp q_\perp)^2]^{-2}$, specifically if

$$S_{\psi\phi}(\mathbf{q}) = S(0)\,\mathrm{LOR}(q_\parallel)\,S_\perp(q_\perp) = \langle \psi^2 \rangle 8\pi\xi_\parallel[1 + (\xi_\parallel q_\parallel)^2]^{-1}\xi_\perp^2 S_\perp(q_\perp \xi_\perp) \tag{20.24a}$$

where $\int d^2q_\perp \xi_\perp^2 s_\perp(q_\perp \xi_\perp) = \pi$, or

$$S_{\psi\phi}(\mathbf{q}) = S(0)\,\mathrm{LOR}^2(\mathbf{q}) = \langle \psi^2 \rangle (8\pi\xi_\parallel \xi_\perp^2)[1 + (\xi_\parallel q_\parallel)^2 + (\xi_\perp q_\perp)^2]^{-2} \tag{20.24b}$$

Then eqn (20.23) gives

$$I(q + \langle Q \rangle) = \langle \psi^2 \rangle[32\pi^2\langle Q \rangle^{-2}]\xi_\parallel[1 + (\xi_\parallel q)^2]^{-1} \tag{20.25a}$$

or

$$I(q + \langle Q \rangle) = \langle \psi^2 \rangle[8\pi^2\langle Q \rangle^{-2}]\xi_\parallel[1 + (\xi_\parallel q)^2]^{-1} \tag{20.25b}$$

respectively. Note that in this approximation $I(q)$ depends only on $\xi_\parallel$, i.e. is independent of $\xi_\perp$. The amplitude factor $S(0) = \langle \psi^2 \rangle(8\pi\xi_\parallel \xi_\perp^2)$ is chosen so that $(2\pi)^{-2} \int d^3q\,[S_{\psi\phi}(\mathbf{q})] \equiv G(0) = \langle \psi^2 \rangle$.

To summarize, with $S_{\psi\phi}(\mathbf{q}) \sim \mathrm{LOR}^2(\mathbf{q})$ or $\mathrm{LOR}(q_\parallel)S_\perp(q_\perp)$ having $\xi_\parallel$ and $\xi_\perp$ different but sufficiently large to satisfy the approximations giving eqns (20.22) and (20.23) generates a symmetric Lorentzian $I(q + \langle Q \rangle)$, which the convolution of eqn (20.20) centred at $q_{\mathrm{peak}} = \langle Q \rangle$. Since the latter describes the data we take the anisotropic $\mathrm{LOR}^2(\mathbf{q})$ or $\mathrm{LOR}(q_\parallel)S_\perp(q_\perp)$ forms for $S_{\psi\phi}(\mathbf{q})$ to describe the smectic layer correlations in the aerogel.

Figure 20.26 shows that when $\xi_\parallel\langle Q \rangle$ is small, asymmetry of $I(q)$ about $\langle Q \rangle$, with $I(q)$ becoming relatively larger on the high-q side, develops when $\xi_\perp$ becomes small relative to $\xi_\parallel$. In this case any asymmetry in the $I(q)$ data can be used to evaluate $\xi_\perp$, or at least set limits on it. Examples of the implementation of this idea to typical scans in the 0.08 aerogel at the high and low ends of the smectic A temperature range are shown in Figures 20.27 and 20.28. At $T = 10.4\,°\mathrm{C}$, in the double-Lorentzian regime, the fitting parameters are $\xi_L = 1252$ Å and $\xi_s = 41$ Å. Then, keeping the amplitudes and small Lorentzian parameters fixed, we replaced the large Lorentzian by the powder average of the anisotropic $S_{\psi\phi}(\mathbf{q}) \sim \mathrm{LOR}^2(\mathbf{q})$ from eqn (20.24b), numerically integrating $\mathrm{LOR}^2(\mathbf{q})$ via eqn (20.23) for a range of $\xi_\perp$. As expected for $\xi_\perp \ll \xi_\parallel$ the calculated $I(q)$ is asymmetric and becomes less so as $\xi_\perp$ increases. We can match the data well for any $\xi_\perp > 500$ Å $\sim \xi_\parallel/2$, so that this process enables a lower limit to be determined for $\xi_\perp/\xi_\parallel$. For all of the low-temperature 0.08 and 0.17 data a conservative limit is $\xi_\perp/\xi_\parallel \geqslant 1/3$. At low T we carry out this procedure only for the large Lorentzian because the small amplitude and uncertainties arising from subtraction of the background preclude its application to the small Lorentzian.

Figure 20.28 shows a similar process at the high-temperature end of the range, in the single-Lorentzian region, where, because $\xi_\parallel$ is much smaller, asymmetry of $I(q)$ comes both from the anisotropy of $S_{\psi\phi}(\mathbf{q}) \sim \mathrm{LOR}^2(\mathbf{q})$ and from the q dependence of the angular integration in eqn (20.22), which generates asymmetry of the opposite sign. Here we can establish more rigorous limits on $\xi_\perp/\xi_\parallel$. Figure 20.28 shows fits to the 0.08 data at $T = 35.2\,°\mathrm{C}$, with $\xi_\parallel = 47$ Å for a series of $\xi_\perp$. The best fit is obtained for $\xi_\perp = 30$ Å, indicating that at higher temperatures $\xi_\perp \approx \xi_\parallel$, i.e. that $S_{\psi\phi}(\mathbf{q})$ is nearly isotropic.

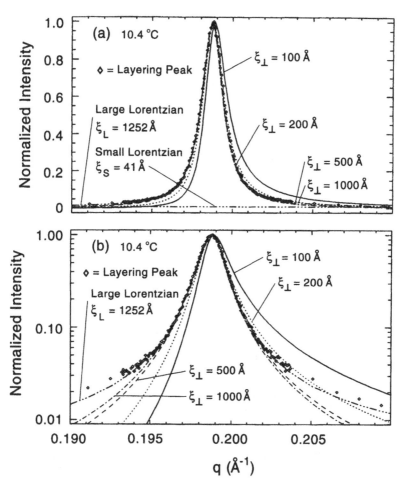

Figure 20.27 Linear (a) and semi-log (b) plots of trial fits of the background-subtracted layering peak *I*(*q*) of the 0.08 sample at *T* = 10.4 °C by the analytically calculated powder average of $S_{\psi\phi}(\mathbf{q}) = \text{LOR}^2(\mathbf{q})$ plus the small Lorentzian. Both the large and small Lorentzians obtained from our standard double-Lorentzian fits are also shown. The value of $\xi_\parallel = 1252$ Å is the same in each trial fit as the best double-Lorentzian fitted value of $\xi_L = 1252$ Å and the double fitted small Lorentzian parameters are also kept fixed. Trial values of $\xi_\perp$ range from 100 to 1000 Å. Reasonble fits are obtained for $\xi_\perp > 500$ Å.

We can now return to the direct space correlation function, defining

$$G(\mathbf{r}) \equiv G_\psi(\mathbf{r})G_\phi(\mathbf{r}) = \langle\psi^2\rangle G_f(\mathbf{r})G_\phi(\mathbf{r}) = \langle\psi^2\rangle\langle f(0)f(\mathbf{r})\rangle\{\exp[-\langle\delta\phi(\mathbf{r})^2\rangle/2]\} \qquad (20.26)$$

so that

$$G(\mathbf{r}) = G_\psi(\mathbf{r})G_\phi(\mathbf{r}) = \text{FT}^{-1}[S_{\psi\phi}(\mathbf{q})]$$
$$= S(0)\,\text{FT}^{-1}[\text{LOR}(q_\parallel)f_\perp(q_\perp)] = \langle\psi^2\rangle\exp(-z/\xi_\parallel)\,\text{FT}^{-1}[f_\perp(q_\perp)] \qquad (20.27a)$$

or

$$G(\mathbf{r}) = S(0)\,\text{FT}^{-1}[\text{LOR}^2(\mathbf{q})] = \langle\psi^2\rangle\exp\{-[(z/\xi_\parallel)^2 + (r_\perp/\xi_\perp)^2]^{1/2}\} \qquad (20.27b)$$

449

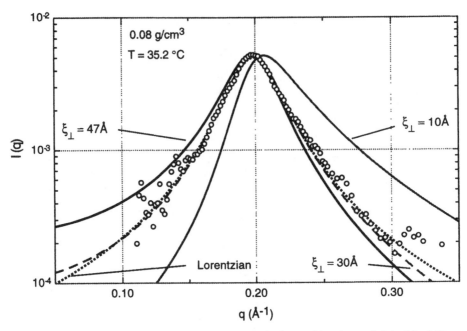

Figure 20.28 Semi-log plots of trial fits of the background-subtracted layering peak $I(q)$ of the 0.08 sample at 35.2 °C by the analytically calculated powder average of $S_{\psi\phi}(\mathbf{q}) = LOR^2(\mathbf{q})$. The Lorentzian obtained from our standard single-Lorentzian fit to $I(q)$ is also shown. The value of $\xi_{\parallel}$ is the same in each trial fit as the best single-Lorentzian fitted value of $\xi = 47$ Å. Trial values of $\xi_{\perp}$ are as indicated. The trial $I(q)$ are asymmetric for small $\xi_{\perp}$ because of the deviation of the powder averaging contours from the $\xi_{\perp}$ plane (Figure 20.26). The asymmetry of opposite sign at larger $\xi_{\perp}$ is due to the q^{-2} factor in the powder average (eqn (20.22)). A reasonable fit is obtained for $\xi_{\perp} = 30$ Å.

The lineshape data indicate that this $G(r)$, with $\xi_{\perp} \approx \xi_{\parallel}$ in the single-Lorentzian regime and $\xi_{\perp} > \xi_{\parallel}/3$ for the large Lorentzian in the double-Lorentzian regime, describes the layering correlations.

If we take $\xi_{\perp} = \xi_{\parallel} = \xi$ then $G(r)$ and $I(q + \langle Q \rangle)$ reduce to the *simple* exponential and its FT:

$$G(r) = \langle \psi^2 \rangle \exp(-r/\xi) \qquad I(q + \langle Q \rangle) \propto \langle \psi^2 \rangle \xi [1 + (\xi q)^2]^{-1} \tag{20.28}$$

These results may be compared with the Ornstein–Zernicke $G(r) \sim (1/r) \exp(-r/\xi)$ characteristic of the bulk nematic scattering (eqn (20.6a)) and the powder average of $S_{bulk}(\mathbf{q})$ using eqn (20.6). Substituting $S_{bulk}(\mathbf{q})$ into eqn (20.23) yields

$$I_{bulk}(q + \langle Q \rangle) \sim -\ln[1 + (\xi q)^2] \tag{20.29}$$

Finally we consider the integrated scattering intensity, I, which may be obtained under the approxmation of eqn (20.23) from either eqn (20.24) or (20.25) as

$$I = \int dq I(q + \langle Q \rangle) = \langle Q \rangle^{-2} \int d^3 q [S_{\psi\phi}(\mathbf{q})] = C \langle \psi^2 \rangle \tag{20.30}$$

where C is a constant. Thus the integrated intensity does not directly depend on correlation length but is determined only by $\langle \psi^2 \rangle$. If $\langle \psi^2 \rangle$ is not spatially uniform then evaluation of I requires an additional volume average $I = C/V \int_V d^3 r \langle \psi^2 \rangle_r$.

20.7.2 *Domain Model of Smectic Layering*

The observed Lorentzian powder-averaged lineshapes found for all values of ζ can also be interpreted in terms of scattering from a distribution of domains, $P_\zeta(L)$, of varying size L. Each domain is assumed to have perfect internal smectic ordering (constant ψ and no phase fluctuations) which is uncorrelated with that of the other domains. The condition that the domains fill the smectic volume leads to the normalization $\int P_\zeta(L)L^3 \, dL = \phi$. For this picture $G(r) = G_\psi(r) = \langle \psi^2 \rangle P_{same}(r)$, where $P_{same}(r)$ is the probability that two points separated by r are in the same domain, i.e. the fractional overlap volume of a domain and its replica displaced by r and averaged over the distribution. Assuming the pores to be shapes with small eccentricity, the $P_\zeta(L)$ which gives the simple exponential $f(r)$ is (see the previous chapter)

$$P_\zeta(L) = (\zeta L)^{-2} \exp(-L/\zeta) \tag{20.31}$$

which also gives eqn (20.14) if $\zeta = \langle p \rangle$. If this $P_\zeta(L)$ gives the simple exponential $G(r)$ of eqn (20.28) then it must also produce an oriented domain structure factor $S_{\psi\phi}(\mathbf{q})$ which is LOR$^2(\mathbf{q})$. For a distribution of uncorrelated internally ordered domains $S_{\psi\phi}(\mathbf{q})$ can be obtained by summing the scattering of single-sized domains, properly weighted by $P_\zeta(L)$. If the domains are spherical, of radius L, their form factor is $F(qL) = 4\pi q^{-3}[\sin(qL) - qL \cos(qL)]$, so we can write the scattering structure factor

$$S_{\psi\phi}(\mathbf{q}) = \langle \psi^2 \rangle \int_0^\infty F^2(qL)P(L)dL = \langle \psi^2 \rangle C\zeta^3[1 + (q\zeta)^2]^{-2} \tag{20.32}$$

and powder-averaged intensity

$$I(q + \langle Q \rangle) = 4\pi C \langle \psi^2 \rangle \zeta[1 + (\zeta q)^2]^{-1} \tag{20.33}$$

where C is a numerical factor and the second equality in eqn (20.32) can be shown via numerical integration. This form is equivalent to that of eqn (20.24) for $\zeta_\parallel = \zeta_\perp$, indicating that the above statistical distribution of spherical domains can account for the LOR^2 lineshape.

The domain model can be generalized to give anisotropic domains. Specifically the anisotropic exponential $G(r)$ of eqn (20.27) can be obtained in the domain model by generalizing the arguments which give the exponential density–density correlation function for the empty aerogel starting from the random telegrapher's wave (eqn (20.9)). If we pick a place at random in the smectic and orient the coordinate system with the $\hat{z}$ axis along the layer normal of the domain in which we fall, then the probability for escaping the domain for a small displacement dr will be

$$P_{esc}(d\mathbf{r} \,|\, \hat{z}) = [(dz/\xi_\parallel)^2 + (dr_\perp/\xi_\perp)^2]^{1/2}$$

and $G(\mathbf{r})$ will be that of eqn (20.27b),

$$G(\mathbf{r}) = \langle \psi^2 \rangle P_{same}(\mathbf{r} \,|\, \hat{z}) = \langle \psi^2 \rangle \exp\{-[(z/\xi_\parallel)^2 + (r_\perp/\xi_\perp)^2]^{1/2}\} \tag{20.34}$$

where $P_{same}(\hat{r} \,|\, \hat{z})$, the probability that points separated by r are in the same domain, is obtained by integration of $P_{esc}(d\mathbf{r} \,|\, \hat{z})$.

20.7.3 *Distinguishing Phase from Amplitude Fluctuations*

Equation (20.26) illustrates that pair correlations in the smectic layering can be limited by either amplitude or phase fluctuations. Because 8CB layering exhibits only the lowest-order Bragg reflection the present experiments cannot distinguish between the two in the absence of a specific model for the disorder. The calculation leading to eqn (20.19), based on the assumption of a *sinusoidal* density modulation, describes the 8CB situation. However, a variety of other liquid crystals have observable second or higher harmonic reflections in the bulk, which requires a more general calculation of the correlations. One approach is to replace the sinusoidal density modulation by a stacking of scattering sheets

$$\delta n_{1c}(\mathbf{r}) = \{\psi_j(\mathbf{r})\delta[\mathbf{r} - \hat{\mathbf{z}}\,(jd + u_j(\mathbf{r})]\}$$

where $\psi_j(\mathbf{r})$ and $u_j(\mathbf{r})$ are the amplitude and displacement of the *j*th sheet. A development parallel to that leading to eqn (20.19) shows that for small displacements the contribution of the *n*th harmonic to $\langle \delta n_{1c}(0)\delta n_{1c}(\mathbf{r})\rangle|_{\hat{z}}$ is obtained by replacing $\langle Q\rangle$ by $n\langle Q\rangle$ in eqn (20.19), in particular increasing the argument in $G_\phi(\mathbf{r})$ by a factor n^2. Thus, if present, higher harmonics can be used to distinguish amplitude and phase fluctuation effects since in the presence of phase fluctuations the higher harmonics will develop more rapid decay of $G_\phi(\mathbf{r})$ and broader lineshapes, as is found in bulk smectics (Safinya *et al.*, 1986).

While 8CB smectic A layering gives only the Bragg fundamental, the 8CB crystal phase the aerogels exhibits a number of different reflections and orders ranging over a factor of four in $\langle Q\rangle$. For these peaks which can be fitted, the lineshape is a symmetric single Lorentzian, appears to be independent of $\langle Q\rangle$, and yields correlation lengths $\xi_X \sim \langle p\rangle$ (cf. section 20.6.4 and Clark *et al.* (1993) indicating the absence of phase fluctuation effects in the confined crystal and an exponential $G_\psi(r)$. In this case the picture of well-ordered finite size domains applies and, to the extent that the crystal domains fill the pores, the observed exponential correlation function can be interpreted as an independent scattering measurement of the pore 'self' correlation function. Since $\xi_X \sim \langle p\rangle$ the crystal domains appear on one hand to fill the pores and on the other hand to be limited by the pore dimensions. Of course other materials may crystallize with a different mean domain size. The exponential $G_\psi(r)$ indicates that the random porous medium DAB model of ordered domains can be used to describe the crystallization in the aerogel (cf. eqn (20.5) and section 20.7.2). The measured ξ_X are shown in Table 20.1.

20.8 Discussion

20.8.1 *Structure: Temperature Dependence*

20.8.1.1 *Temperature regimes*

Figures 20.23 and 20.24 exhibit three distinct structural regimes with decreasing T:

1   *The $\langle p\rangle$ independent high-temperature regime:* For $T > 35\,^\circ$C, $I(q)$ is single Lorentzian in all of the aerogels and ξ is only weakly dependent on $\langle p\rangle$ and T, approaching $\xi_{\perp\text{bulk}}$ in the 0.08, 0.17 and 0.36 samples for $T > 34\,^\circ$C. Even for the shortest-range correlations observed ($\xi \approx 30$ Å at $T \approx 37\,^\circ$C) ξ is significantly smaller than $\xi_\parallel$, and is in fact only one-half of the minimum bulk value of $\xi_\parallel \approx 60$ Å, found at the highest temperatures of the bulk nematic phase. *Evidently, smectic correlations within the aerogels are perturbed over the entire nematic temperature range.* The Lorentzian form for $I(q)$ and the consequent LOR² or LOR$(q_\parallel)S_\perp(q_\perp)$ nature of $S(q)$ in the nematic temperature range is further evidence for this. As noted in eqn (20.29), the bulk smectic correlations in the nematic phase are Ornstein–Zernicke like, generating a log(Lorentzian) lineshape after powder averaging, which does not fit the data.

   The symmetry of the aerogel layering peaks shows that the SmA domains are nearly isotropic for $T > 30\,^\circ C$ (cf Figure 20.28) and, since our measured ξ is essentially $\xi_{\parallel\text{aerogel}}$, as is indicated by our trial fits in Figures 20.27 and 20.28, we arrive at the conclusion that *smectic fluctuations in the nematic phase in the aerogels have a severely suppressed correlation range in the (longitudinal) direction along the local nematic director/layer normal.* Additionally, the equality at high temperatures of the values of ξ in three lowest-density aerogel samples is suprising. The nematic in the 0.08 aerogel ($\langle p\rangle = 700$ Å) ought to be less deformed than it is in the 0.17 ($\langle p\rangle = 430$ Å) and 0.36 gels ($\langle p\rangle = 180$ Å). This is shown by $\xi_{\text{dyn, sat}}$, the nematic correlation lengths determined by the DLS measurements (Table 20.1 and the previous chapter), which, although less than $\langle p\rangle$, increase significantly with increasing $\langle p\rangle$.

Based on this one might expect that the 0.08 nematic should thus allow larger correlated smectic A fluctuations or domains than the 0.17 or 0.36 nematics, but this is not the case. There is apparently a volume disordering operative in even the large aerogel pores which prevents the rod-shaped correlation volumes of the bulk nematic from developing.

2 *Intermediate-temperature steep growth regime:* For 33 °C the double-Lorentzian lineshape develops in the 0.08 and 0.17 samples and the narrow or single Lorentzian ξ grows rapidly with decreasing T and increasing $\langle p \rangle$. At the high-T end of this regime the integrated intensity is dominated by the small Lorentzian. The increase in integrated intensity is associated entirely with the large Lorentzian, while the small Lorentzian decreases in integrated intensity.

3 *Low-T saturation regime:* For $T < 25\,°C$, ξ continues to grow as T decreases, approximately as a linear function of T, but much less steeply than at intermediate temperatures. The final slope increases with increasing pore size, from near zero for the 0.36 and 0.60 samples. *The data show that there is no divergence of ξ but rather saturation-like behaviour as T is lowered, enabling us to conclude that the bulk quasi-long-range smectic ordering is absent in these aerogel hosts.* At the lowest T, the ratio of the saturation correlation length to the pore chord, $\xi_{sat}/\langle p \rangle$, increases with increasing $\langle p \rangle$, from $\xi_{sat}/\langle p \rangle \cong 0.79$ for $\langle p \rangle = 95$ Å to $\xi_{sat}/\langle p \rangle \cong 2.0$ for $\langle p \rangle = 700$ Å. In the 0.08 aerogel at low temperature ξ become larger than either $\langle p \rangle$ or the crystal correlation lengths ($\xi \sim 2\langle p \rangle$, $\xi \sim 1.35\xi_x$) indicating tendency for the nematic and smectic A domains to align with the director along the long axis of the pore or in the direction of connecting passages between pores. The resulting anisotropy in the domain shape at low T would be $\xi_{\parallel}/\xi_{\perp} \sim \xi/\langle p \rangle \sim 2$, which by the discussion of section 20.7.1 would not be detectable. Another possibility is that $\langle p \rangle$, which was determined prior to filling, grew during filling because of internal breakage of the silica aerogel, a process most likely to occur for the lower-density aerogels. However, the near constant ratio of the crystal correlation length ξ_x to $\langle p \rangle$ ($\xi_x/\langle p \rangle \sim 1.4$, see Table 20.1) indicates that this is not happening.

20.8.1.2 *Correlation lengths – dependence on temperature and porosity*

Our correlation length data set bears a resemblance to that of the correlation lengths for magnetic ordering found in the diluted Ising antiferromagnet formed by $RB_2Co_{0.7}Mg_{0.3}F_4$, which has been studied by Birgeneau et al. (1983). In this system the site dilution and a uniform applied magnetic field H produce an effective staggered random field on the antiferromagnetic spin system. This system is similar to ours in that the correlation lengths saturate at low T to a value dependent on the degree of disorder (which increases with increasing H), never developing long-range order, and in that the correlation lengths at high T converge to values independent of disorder. However, there is an important difference between this magnetic and the aerogel systems in that *while the high temperature ξ in the magnetic system are asymptotic to the $H = 0$ (zero random field) critical values, the aerogel correlation lengths are not asymptotic.* Birgeneau et al. were able to fit their correlation length data reasonably well to the phenomenological equation

$$\Gamma = \xi^{-1} = \Gamma_{thermal} + \Gamma_{rf} = A \exp[-C/T\ (K)] + \xi_{rf}^{-1} \tag{20.35}$$

based on the idea that $\Gamma \equiv \xi^{-1}$, the rate of decorrelation, has a thermal contribution, $\Gamma_{thermal} = \xi_{thermal}^{-1}$, and a contribution from the random field, which determines $\Gamma_{sat} \equiv \xi_{sat}^{-1}$, the low-$T$ limit of ξ^{-1}. The thermal decorrelation is assumed to be an activated process. To explore the applicability of such a model we plot $\log(\xi^{-1})$ against $T(K)^{-1}$ in Figure 20.29(a). The steep growth regime of ξ in the 0.08 and 0.17 samples appears roughly as a straight line on such a plot, indicating that eqn (20.35) may be applicable to our system. Fits of eqn (20.35), shown in Figure 20.29, varying A, C and ξ_{sat}^{-1}, were made to the ξ data. The fitted decorrelation barrier for the 0.08 to 060 aerogels are respectively $C \approx 48\,000$, $46\,000$, $28\,000$ and $20\,000$ K,

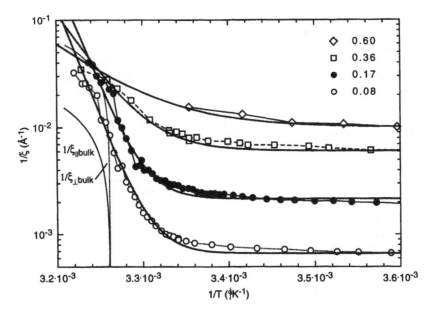

Figure 20.29 (a) Semi-log plot of ξ^{-1} against $T(K)^{-1}$ for all four aerogels. (b) Fit of ξ data to the thermally activated decorrelation model of eqn (20.35). The decorrelation barrier varies from 48 000 K (0.08) to 20 000 K (0.60).

and for each aerogel we find ξ_{sat}^{-1} as expected. Although the fits are reasonably good the large barrier energies which we obtain are unphysical, the result of the need to generate a relatively large change in ξ over a narrow range in $1/T$. We conclude that this activation barrier mechanism for ξ_{thermal} is not operative in our case.

However, the view that the inverse correlation length is a sum of saturation and thermal terms is useful. We plot in Figure 20.30 ζ against $\langle p \rangle$ at a representative sample of temperatures using lin–lin, log–log, log–lin and lin–log axes scalings to check for the functional dependence of ξ on $\langle p \rangle$. Of particular interest in the log–log plot, which indicates that the $\xi(T, \langle p \rangle)$ curves are nearly proportional to one another for $T < 29\,°C$, and that they have a power-law dependence on $\langle p \rangle$ in this range, $\xi \sim \langle p \rangle^{1.56}$. Figure 20.31 illustrates the proportionality of $\xi(T, \langle p \rangle)$ by plotting ξ/ξ_{sat} and $\Gamma/\Gamma_{\text{sat}} = (\xi/\xi_{\text{sat}})^{-1}$, which overplot very well for $T < 30\,°C$, in the steep growth and saturation regimes. The necessary scaling of ξ is $\xi \sim \langle p \rangle^{-1.56}$. Figure 20.31(b) shows that the $\xi(T, \langle p \rangle)$ data satisfy the following scaling relation:

$$1/\xi(T, \langle p \rangle) = \Gamma(T, \langle p \rangle) = \Gamma_{\text{sat}}(\langle p \rangle) + \Gamma_{\text{thermal}}(T, \langle p \rangle) = \Gamma_{\text{sat}}(\langle p \rangle) [1 + g(T)]$$

$$= \lambda(\langle p \rangle/\lambda)^{-1.56}[1 + g(T)] \tag{20.36}$$

where $\lambda = 240$ Å. Here $g(T)$, characterizing the growth of thermal decorrelation as T increases, is a universal function for the four aerogels.

20.8.1.3 Integrated intensities

Figure 20.23 exhibits the background subtracted integrated peak intensities (total, large Lorentzian, small Lorentzian, respectively, σ_{tot}, σ_{L}, σ_{S}) for the three lowest-density 8CB–aerogels, giving the relative values of σ_{tot}, σ_{L}, σ_{S} on an arbitrary vertical scale. Also shown is

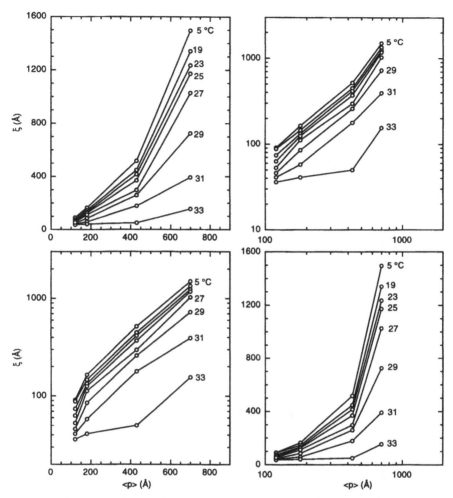

Figure 20.30 Dependence of correlation length on the average pore chord $\langle p \rangle$ and temperature T. We have used lin–lin (a), log–log (b), log–lin (c) and lin–log (d) axes scalings to check functional form.

the 8CB bulk integrated intensity in the smectic A phase. *Note that the vertical scale of the bulk integrated intensity is arbitrary relative to the aerogel data.* The bulk data are very well fitted by the power law $\sigma_{\text{bulk}} \sim (33.5\,^{\circ}\text{C} - T)^{0.42}$. Recalling eqn (20.30) and the discussion following, the integrated intensity measures the local mean square layering order parameter summed over the illuminated volume $\sigma \sim \int d^3r\,[\langle \psi^2(r) \rangle]$. This result leads to difficulties in the direct interpretation of integrated intensity from inhogeneous systems such as the aerogel-confined LCs. For example, a low value of σ can mean homogeneous weak ordering or the presence of occasional strong ordering in an otherwise disordered system. Thus integrated intensity must be interpreted in the framework of the other data. To address this we will consider shortly two limiting behaviours of the integrated intensity. The data show that the overall growth of σ_{tot} expressed as a ratio of limiting low-T to high-T values is larger for the bulk than for the aerogels, even though the net T range of the bulk data is smaller. This indicates that the average order parameter in the aerogels at low T must be substantially smaller than that of the bulk.

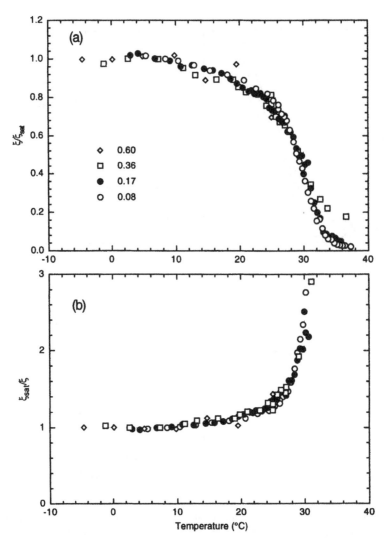

Figure 20.31 (a) Correlation lengths $\xi(T, \langle p \rangle)$ and (b) inverse correlation lengths $\Gamma(T, \langle p \rangle)$ scaled by their saturation values at low T, showing that the scaled $\xi(t)$ overplot well for $T < 30\,°C$. The required scaling of ξ is $\xi \sim \langle p \rangle^{-1.56}$, showing that the data are of the form $\xi^{-1}(T, \langle p \rangle = \Gamma(T, \langle p \rangle) = \lambda(\langle p \rangle/\lambda)^{-1.56}[1 + g(T)]$.

The basic applicable expression for the scattered intensity $I(q + \langle Q \rangle) \sim A[1 + (\xi q)^2]^{-1}$, with $A = \langle \psi^2 \rangle \xi$, was obtained in eqn (20.28) from the smectic correlation function and in (20.33) as scattering from a domain distribution. We consider two distinct processes within the context of this equation, domain coarsening or annealing, in which smaller domains join up to make larger ones, the net domain volume remaining constant, and ordering by the growth of critical correlations. Domain coarsening at constant $\langle \psi^2 \rangle$ is characterized by: $A \sim \xi$; large q behaviour $I(q + \langle Q \rangle) \sim Bq^{-2}$, with a q^{-2} coefficient $B = \langle \psi^2 \rangle/\xi \sim 1/\xi$; and ξ-independent integrated intensity $\sigma \sim \langle \psi^2 \rangle$. By contrast, for ordering by growth of critical fluctuations ($A \sim t^{-\gamma}$, $\xi \sim t^{-\nu}$, with $\gamma \approx 2\nu$) we have $\langle \psi^2 \rangle \sim \xi$ which gives: $A \sim \xi^2$; large-q

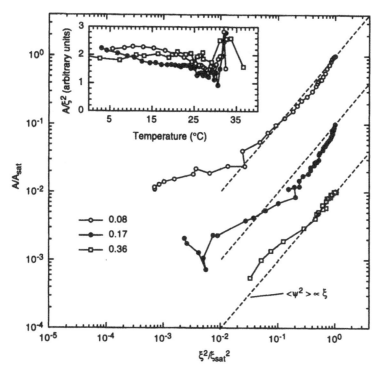

Figure 20.32 Large Lorentzian amplitude against (large Lorentzian correlation length, ξ)^2 for the various aerogels showing that the large Lorentzian grows in with $\langle\psi^2\rangle \sim \xi$, similarly to the bulk pretranslational fluctuations. The inset shows that A/ξ^2 is approximately independent of T.

behaviour with B independent of ξ; and $\sigma \sim \langle\psi^2\rangle \sim \xi$ – relationships which lead to the nearly temperature-independent large-q scattering characteristic of critical ordering.

Characteristics of growth by critical fluctuations are evident in the development of smectic order in the aerogels. Figure 20.15 shows 0.08 scans at low and high T for which the amplitude of the q^{-2} tails are independent of the degree of ordering. This is a feature common to the data in all of the aerogels and a signature of ordering by critical fluctuations as noted above. In Figure 20.32 we plot for the 0.08, 0.17 and 0.36 samples the amplitude A of the large Lorentzian as a function of ξ^2 (the square of the large Lorentzian correlation length), as well as A/ξ^2 against T in the inset. In the main plot A and ξ are normalized by their lowest T values in the respective aerogels and then shifted on the log scale to eliminate overlap. The plots show that $A \sim \xi^2$ and therefore $\langle\psi^2\rangle \sim \xi$ over most of the T ranges where the large Lorentzian is substantial. This behaviour again is characteristic of growth by critical fluctuations.

On the other hand we have evidence from the temperature dependence of the peak position (see Figures 20.23 and 20.24) that the local layering is bulk like during the growth of the large Lorentzian. It might be reasonable then to assume that for the large Lorentzian $\sigma_L \sim \langle\psi^2\rangle \sim (T - T_L)^{0.42}$, the T dependence of $\langle\psi^2\rangle_{bulk}$ with growth starting at $T \approx 33\,°C$. However, as can be seen from Figure 20.23, the plots of σ_L against $T - T_L$ show a much more gradual increase of σ_L than $\langle\psi^2\rangle$ in the bulk. This might be a result of the regions contributing to σ_L filling only part of the available volume for T close to T_L. Additionally the 0.17 σ_L against T plot is qualitatively rather different from that of the 0.08 sample, a difference which was repeatable over several different temperature runs. This difference may be a consequence of

457

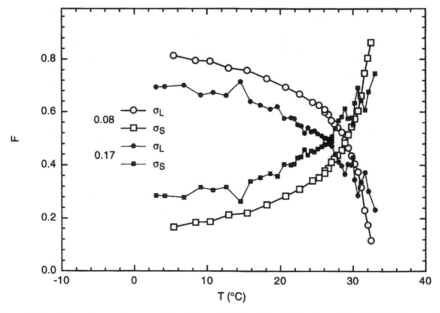

Figure 20.33 Comparison of measured fractions of total scattered intensity scattered into the large and small Lorentzians for the 0.08 and 0.17 samples.

the non-uniformity of the change of interfacial profile of filled pore as the 8CB contracts with lowering T (Feder, 1988), an effect which could strongly influence the σ data. However, the ratios $R_L = \sigma_L/\sigma_{tot}$ and $R_S = \sigma_S\sigma_{tot}$ should be free of such sample-filling artefacts and do indeed behave similarly in the 0.08 and 0.17 samples, as shown in Figure 20.33. These ratios indicate that the small Lorentzian is larger in the 0.17 sample relative to the 0.08 one, supporting the notion that it is associated with the disordering mechanism. The ratios show that the scattered intensity is dominated by the small Lorentzian at large T and that $R_L = R_S$ occurs in the ζ steep growth regime.

20.8.1.4 Small Lorentzian

The small Lorentzian of the 0.08 and 0.17 data sets is a broad feature which qualitatively appears to be a remnant of the single Lorentzian observed at the highest temperatures. With the exception of a narrow T range where the fitting is switched from one Lorentzian to two, the small Lorentzian integrated intensity and correlation length continue smoothly from the high-T single-Lorentzian values. This suggests that the very short-range smectic ordering condition in the aerogels at high temperature ($\xi_S \approx 40$ Å) is maintained in some form throughout the entire smectic T range. In the 0.08 and 0.17 aerogels σ_S decreases in both relative and absolute terms with decreasing T. Since we expect $\langle \psi^2 \rangle$ either to increase or at worst not to change with decreasing T, the only reasonable way to account for decreasing σ_S is via a decrease in the volume or number of entities producing the scattering. Note that although σ_S decreases the fluctuation intensity at large wavevector does not decrease, i.e. the coefficient of the q^{-2} tail is still independent of T. Note also that, if volume is being transferred from small Lorentzian scattering to large Lorentzian scattering, then the deviation of σ_L from the bulk T dependence is even larger. The correlation lengths associated with the 0.17 small Lorentzian range from 30 to 60 Å and are only slightly larger in the 0.08 sample. These dimensions are comparable with the aerogel solid chord $\langle s \rangle$ (silica particle size) and may be a result of a specific layering arrangement at the solid surface (e.g. a homeotropic smectic layer coating a sphere of dimension $\langle s \rangle$). The pore volume-to-surface ratio V_p/S (Table 20.1) is the mean distance to a surface within the pore volume and is 159 and 137 Å in the 0.08 and 0.17 aerogels respectively. V_p/S is much smaller than $\langle p \rangle$, typically $V_p/S \sim \langle p \rangle/4$, indicating that the pore structure is rough, as the picture of Figure 20.7 suggests. In this case disordered layer orientation extending one or two layers from the surface would fill most of the sample volume, as indeed it must if there is be no bulk-like pretransitional ordering for $T > 33.5\,°C$.

20.8.1.5 *Peak position*

The remarkable temperature dependence of the bulk peak position shown in Figure 20.23 has not been reported before. In the nematic phase q_{peak} increases strongly with T, heading in a monotonic fashion to $q_{peak} \approx 0.24$ Å^{-1}, the position of the high-T pair correlation peak discussed in section 20.6.1. There is a cusp-like minimum in q_{peak} at T_{NA}, i.e. an abrupt discontinuity in dq_{peak}/dT to a positive layer expansion coefficient in the smectic A phase. The increase of layer expansivity as T_{NA} is approached suggests a coupling of layer spacing to ψ. These features should be understandable in terms of the Landau–Ginzburg description of the NA transition, with the addition of a $\Psi^*\mathbf{V}_{\|}\Psi - \Psi\mathbf{V}_{\|}\Psi^*$ term to $F_{NA}(\Psi, \delta S, \delta \mathbf{n})$, which will be the topic of a future publication.

The dramatic bulk T dependence of q_{peak} provides a useful 'internal' probe the status of the layer ordering in the aerogel. Figure 20.23 shows that the temperature behaviour of q_{peak} in the aerogels is similar that of the bulk, but with a rounding of the bulk cusp-like q_{peak} minimum. An important common feature of the q_{peak} data is that q_{peak} has essentially the bulk behaviour once ξ reaches about 200 Å, i.e. this extent of short-ranged layering is sufficient to establish bulk-like values and T dependence of q_{peak}. This is consistent with the peaking of the heat capacity anomaly for $\xi < 200$ Å, as will be discussed in section 20.8.2.

20.8.1.6 *Modelling peak shape and correlation length*

The basic feature of $\xi(T, \langle p \rangle)$ is the absence of long-range smectic ordering in the aerogels, which is to say that the smectic layering system in the aerogel is below its lower marginal dimensionality (LMD), i.e. that $d_{LMDag} > 3$. This is not too surprising because, as noted in section 20.2, the bulk is *at* its LMD ($d_{LMDbulk} = 3$), able to develop only quasi-long-range layering order in the presence of thermal fluctuations, and because disordering fields are known to *raise* the LMD for systems with continuous symmetry. There is currently no theoretical model for the temperature evolution of correlation length for $d < d_{LMD}$.

Disorder in a smectic with well-established local layering, apparently the situation in the aerogels for $T < T_{NA}$, can be described in terms of vortex lines (edge and screw dislocations). The smectic A disordering may be direct, a consequence of the induction of vortex lines by the pinning of the layering at the silica surfaces, in which case a vortex glass model, having the phase of the layering randomly pinned on a density of points, is appropriate, or it may be indirect, imposed by disorder in the nematic director field. As pointed out in section 20.2, the coupling of the nematic director fluctuations is a key aspect of the physics of bulk smectic ordering, and so it may be in the aerogel, especially in view of the fact that the nematic itself is only short-range ordered (see the previous chapter).

In order to explore the effect of nematic disorder on the smectic ordering we consider the growth of Ψ in a nematic which is subject to a random field in the form of random point torques τ_i, applied on a lattice of number density n, in the one Frank constant, K, approximation. In a sample of dimension L, the local mean square fluctuation in the director $\mathbf{n}$ will be $\langle \delta \mathbf{n}^2 \rangle \sim (\langle \tau_i^2 \rangle n/K^2)L$, divergent as $L \to \infty$, indicative of short-range order. The effective temperature for a director mode of wavevector q will be $k_B T_{eff}(q) \sim \langle \tau_i^2 \rangle n/Kq^2 \sim q^{-2}$, indicative of an increase of 2 in the nematic LMD (from 2 to 4). We can now examine the effect of such director fluctuations on the smectic ordering by following the Halperin *et al.* (1974) development in eqns (20.3) and (20.4), but with $k_B T$ replaced by $k_B T_{eff}(q)$. The result is the appearance in $F_{NA}(\psi)$ of a term *linear* in Ψ with a positive coefficient $f \sim \langle \tau_i^2 \rangle n \chi^{-1/2} q_0 \xi_\perp / K^{3/2}$. This term suppresses the formation of smectic order.

A consistent feature of the aerogel layering peaks is their Lorentzian shape, implying that $S_{\psi\phi}(\mathbf{q})$ is LOR^2 or $LOR(q_\parallel)S_\perp(q_\perp)$. We discuss these in turn.

$S_{\psi\phi}(\mathbf{q}) = LOR^2$: in this case $G(r)$ is a simple, exponential. At low temperatures, when $\xi \sim \langle p \rangle$, a simple exponential $G(r)$ can be rationalized using the domain model (section 20.7.2) and the DAB random porous solid arguments leading to eqn (20.34) for the mean domain size of $\langle p \rangle$. In the simplest picture, the exponential is just the self-term of the solid-pore random telegrapher's wave characteristic of aerogel. However, Figures 20.23 and 20.24 indicate that in each of the aerogels there is a temperature range of many degrees centigrade below the bulk T_{NA} where $\xi < \langle p \rangle$ and the correlated volumes are much smaller than the pore volumes, wherein the ξ is determined by a combination of aerogel and thermal factors. It is important to understand the origin of the Lorentzian peak shape in this $\xi < \langle p \rangle$ regime. From eqn (20.6) we see that an exponential $G(r)$ could be achieved via exponential amplitude correlations, $\langle f(0)f(\mathbf{r}) \rangle \sim \exp(-\alpha r)$, or phase diffusion, $\langle \delta\phi(\mathbf{r}^2) \rangle \sim r$, or some combination.

One simple scenario yielding exponential $G(r)$ is suggested by random field models (Imry, 1984). We assume a q-dependent random field $h(q)$ conjugate to $\psi(q)$ affected by adding an $(h(q)\psi(q))$ term to $F_{NA}(\Psi, \delta\mathbf{n})$ in eqn (20.5). The response, $\psi_h(q)$, ignoring fluctuations, will be given by

$$\psi_h(q) = h(q)\chi / [1 + (\xi q)^2] \tag{20.37}$$

and the structure factor given by

$$S_h(q) = |\Psi_h(q)|^2 \propto \chi^2 / [1 + (\xi q)^2]^2 \tag{20.38}$$

a Lorentzian squared. Here we have assumed that $h(q)$ is independent of q for $q < \xi^{-1}$, i.e. that the correlation length for the random disordering field is smaller than ξ. Although this idea generated the correct peak shape from eqn (20.5), in the mean field approximation the ξ appearing here is the bulk correlation length, and it is clear from the data that the measured correlation length is radically different from that of the bulk. Further pursuit of this idea requires incorporation of the effects of the disordering field fluctuations on the temperature dependence of ξ.

$S_{\psi\phi}(\mathbf{q}) = LOR(q_\parallel)S_\perp(q_\perp)$: in this case $G(r)$ is of the form $G(z)G(r_\perp)$, and $G(z)$ is simple exponential. It is not clear how a $G(r)$ factored into z- and $r_\perp$-dependent parts could arise. One possibility is that this reflects essentially independent highly anisotropic phase and amplitude correlations, e.g. $\langle f(0)f(\mathbf{r}) \rangle \sim G(r_\perp)$ while $\langle \delta\phi(\mathbf{r}^2) \rangle \sim z$ and $\exp(-\langle \delta\phi(\mathbf{r}^2) \rangle / 2) \sim \exp(-\alpha z)$ (or with z and $r_\perp$ interchanged). If $G(r)$ were longer ranged than α^{-1} this would respond to an ordering process in which complete layers form before developing significant intralayer correlations.

20.8.2 Relationship of the Structural and Thermal Behaviour

20.8.2.1 Finite size effects

Figure 20.24(b) enables us directly to compare the temperature evolution of the correlation lengths with the NA heat capacity anomaly and evolved heat in the 0.08, 0.17 and 0.36 aerogels and in the bulk. We will first describe the bulk case and then contrast this with what is seen in the aerogels. In the bulk nematic, $\xi_\parallel$ and $\xi_\perp$ measure the range of the correlations of thermal fluctuations into the SmA phase, and diverge as $t \to 0$. This divergence is accompanied by a divergence in C_p, because $C_p \propto \langle H^2 \rangle - \langle H \rangle^2$, and this noise in the enthalpy fluctuations will

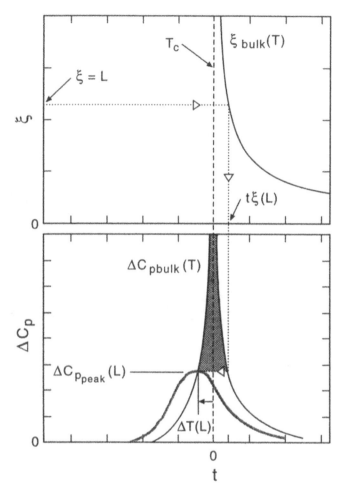

Figure 20.34 Schematic construction showing the relation of the rounding and shifting in ΔC_p as a result of finite sample size L.

increase as the number of fluctuating entities is decreased by the enhanced correlation. As T is lowered in the bulk SmA phase, the fluctuations are increasingly quenched so that ΔC_{pNA} decreases with increasing $T_{NA} - T$. As T is lowered there is a general tendency for enhanced order and therefore lower entropy, the net released heat ΔH_{NA} having a sigmoidal dependence on T, within infinite slope at the inflection point which marks T_{NA}.

In the aerogel the anomaly in ΔC_{pNA} is rounded and shifted to lower T, with increased rounding shifting as $\langle p \rangle$ is reduced. Rounding and shifting of ΔC_p is suggestive of finite size behaviour (Fisher and Ferdinand, 1967; Fisher and Barber, 1972). For a system of finite size L, having a bulk second-order transition, the divergences of ξ and ΔC_p as $t \to 0$ proceed nearly as in the bulk until the correlation length ξ becomes comparable with L, at $t = t\xi(L)$, limiting further growth of correlations. This has the effect of limiting the growth of ΔC_p to the maximum of $\Delta C_{ppeak}(L) \sim \Delta C_{pbulk}(t\xi(L))$, as shown schematically in the graphical construction in Figure 20.34, and it rounds off, also shifting down in T by $\Delta T(L) \sim t\xi(L)T_{NA}$, a shift which scales as the width in t of the rounding. If the finite size L is sufficiently large then ξ and ΔC_p will have their asymptotic power-law behaviour, $\xi \sim t^{-\nu}$ and $\Delta C_p \sim t^{-\alpha}$, and the finite size

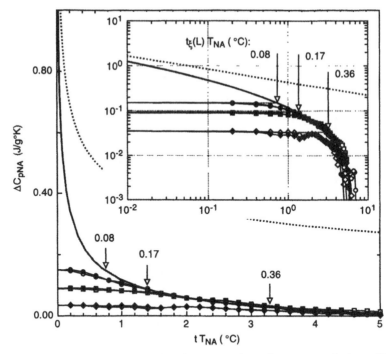

Figure 20.35 ΔC_{pNA} data of Figure 20.1 plotted relative to the peak temperture, lin–lin and log–log. The high-T (open) and low-T (solid) data overlap reasonably well. The dotted curves are the asymptotic $t^{-\alpha}$ component of the fits to the bulk data, showing that the rounding in the aerogels is over a temperature range beyond which the asymptotic scaling will apply. The rounding temperatures $t_R T_{NA}$ are indicated by the arrows.

scaling relations $t\xi(L) \sim \Delta T(L) \sim L^{-1/\nu}$ and $\Delta C_{ppeak}(L) \sim L^{\alpha/\nu}$ emerge from the construction of Figure 20.34.

The rounding and shifting in the 8CB–aerogel data can be characterized by plotting the high- and low-T sides of the peaks of Figures 20.1(b) and (c) against T, measured relative to the peak position T_{peak}, which, as Figure 20.35 shows, can be chosen to make the two sides overlap rather well. The resulting peak shift is $\Delta T = 33.5\,°C - T_{peak}$. The effective rounding width, T_R, can then be obtained by setting $\Delta C_{pbulk}(t_R) = \Delta C_{ppeak}$, the resulting values of $T_R = t_R T_{NA}$ are indicated by the arrows, and are comparable with the shifts $\Delta t(L)$, as Figure 20.36 indicates. However, Figure 20.35 also shows that in the aerogel the rounding is so severe that it occurs well *outside* of the T range where ΔC_{pNA} has its asymptotic form. This can be seen from the dotted line in Figures 20.35 which is just the $t \to 0$ asymptotic part of bulk $\Delta C_{pNA} \sim (2.9\,J/g)\,t^{-\alpha}$, obtained using eqn (20.8) with $D = B = 0$.

We thus do not expect the asymptotic finite size scaling behaviour to relate ΔC_{ppeak}, t_ξ and ξ, but we can try to obtain some insight into the ΔC_p behaviour by assuming that the construction of Figure 20.34, and its inverse, still applies. The result is in Figure 20.24(b) where for each aerogel, starting from ΔC_{ppeak}, we obtain t_R by setting $\Delta C_{pbulk}(r_R) = \Delta C_{ppeak}$ and then get the corresponding ξ by setting $\xi = \xi_\parallel(t_R)$. We use $\xi_\parallel(T)$ because this is the largest of the bulk correlation lengths and is thus the first to encounter the disorder. This construction shows that the shifted peak values of $\Delta C_{pNA}(T = T_{peak})$ in the three aerogels lie nearly on the low wing of $\Delta C_{pNAbulk}(T = T_{peak})$, i.e. $\Delta C_{ppeak} \sim \Delta C_{pbulk}(T_R)$ and $\Delta T \sim T_R$. This is consistent with the shifting and rounding arising from a finite size effect when the ΔC_{pbulk} peak is symmetric about $t = 0$, which is nearly the case with 8CB. Additionally the resulting $\xi_\parallel(r_R)$ values ($\bigcirc$)

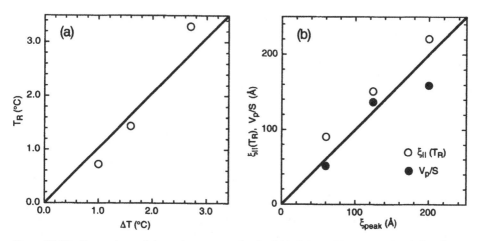

Figure 20.36 Comparisons of thermal parameters for the 0.08, 0.17 and 0.36 aerogels which indicate that a finite size effect with size comparable with the surface-to-volume ratio is responsible for the shifting and rounding of ΔC_{pNA}. (a) Rounding temperature T_R against peak shift ΔT. (b) Measured X-ray correlation length at the peak in ΔC_{pNA} against pore surface-to-volume ratio V_p/S and $\xi_\parallel(T)$, the effective finite size extrapolated from the peak value of ΔC_{pNA} against pore surface-to-volume ratio V_p/S and $\xi_\parallel(T_R)$, the effective finite size extrapolated from the peak value of ΔC_{pNA} via $\xi_\parallel(T)$.

are comparable with ξ_{peak}, the *measured* correlation lengths in the aerogel at the shifted heat capacity peaks ($\diamond$), as shown in Figure 20.24(b).

These features indicate that a finite size effect is operative in the rounding and shifting of the heat capacity, with finite size length $L \sim \xi_\parallel(t_R) \sim \xi_{peak}$. Comparing these values with the various lengths characteristic of the aerogel in Table 20.1 shows that while they are much less than pore size $\langle p \rangle$, they are quite comparable with V_p/S, the pore volume-to-surface ratio, indicated by ($\blacklozenge$) in Figure 20.24(b). V_p/S is a measure of the mean nearest distance to solid material within a pore, and the small values of V_p/S are an indication of the roughness of the pore surface. We conclude that the heat capacity implies a small length scale volume disordering mechanism, which is consistent with the suppression of all but very short-range pretransitional structural ordering.

20.8.2.2 *Net Heat of Transition*

Figures 20.23 and 20.24 show that the net heat released in the NA anomaly decreases from the bulk value of $\Delta H_{NAbulk} = 0.76$ J/g to $\Delta H_{NAag} \sim 0.23$ J/g in the 0.36 aerogel. It is of interest to try to relate the reduction $\delta \Delta H_{NA}(\xi_{sat}) = \Delta H_{NAbulk} - \Delta H_{NAag}$ to the elimination of smectic A long-range order and low-T saturated correlation length ξ_{sat}. Returning to Figure 20.34, if a critical heat capacity anomaly ΔC_p is rounded by finite size L, then ΔH_{NA} is reduced by

$$\Delta H_{NA}(L)_{red} = \int_{\Delta T_-(L)}^{\Delta T_+(L)} \Delta C_{pNA}(T')dT' = \int_{L_-}^{L_+} \Delta C_{pNA}(T')[d\xi(T')/dT']^{-1} \, d\xi \qquad (20.39)$$

the shaded region in Figure 20.34, which in the scaling regime varies as $\Delta H_{NA} \sim L^{(\alpha-1)/\nu}$. We suppose that the net reduction of ΔH_{NA} in the aerogels is this kind of finite size effect, i.e. that $\delta \Delta H_{NA}(\xi_{sat}) = \Delta H_{NA}(L)_{red}$, and determine the resulting L, which we call $\xi_{\Delta Hsat}$, via eqn (20.39). As above, we assume that the relevant bulk correlation length is $\xi_\parallel(T)$. We evaluate L by setting $\Delta H_{NA}(L)_{red} = \delta \Delta H_{NA}$ for each aerogel, using eqn (20.39) with $\xi = \xi_\parallel(T)$ and numerically evaluate the integral because the rounding of ΔC_{pNA} is so severe. The resulting L values are given as $\xi_{\Delta Hsat}$ in Table 20.1 and as the grey data points at $T = 20\,°C$ in Figure 20.24(b), and are comparable with ξ_{sat} for the 0.17 and 0.36 aerogels, but smaller for the 0.08 aerogel for which there is some continued growth of correlations and scattered intensity at temperatures below where ΔH_{NAag}^{lim} saturates.

This procedure can be generalized to develop further the relationship between heat release and ordering by noting that the bulk ΔH_{NA} saturated for $T < 30\,°C$, whereas the aerogels continue to evolve heat. We assume that since the bulk anomaly is completed the local order is saturated, and that the heat released in the aerogels for $T < 30\,°C$ is due to the growth of domain size. For example, at $T = 30\,°C$ the difference between the net heat released from the 0.08 aerogel and the bulk is larger that its saturated value of 0.16 J/g, corresponding to a smaller $\xi_{\Delta H}$. Thus we apply eqn (20.39) as above, but now at all T for $T < 30\,°C$, considering $\xi_{\Delta H}$ to be temperature dependent, decreasing as the difference between the saturated bulk ΔH_{NA} and ΔH_{NAag} increases. The resulting $\xi_{\Delta H}(T)$, shown as the grey points in Figure 20.24(b), approach $\xi_{\Delta Hsat}$ at low T, and compare reasonably well with the measured $\xi(T)$. The basic assumption in this procedure is that the relationship between heat evolved and ordering is the same for the finite size reduction of ΔH as for the heat released by domain coarsening in the aerogel, i.e. heat not released by divergent critical fluctuations can be equivalently released as the established local ordering in the aerogel extends to larger length scale. Given the range of assumptions involved, the favourable comparison of $\xi_{\Delta H}(T)$ with $\xi(T)$ indicates that this idea is qualitatively correct.

20.9 Summary

The following are the highlights of our results:

1 The smectic ordering of 8CB in the aerogels is short ranged, with correlation lengths that smoothly increase with either increasing pore size or decreasing temperature.
2 In the limit of low temperature the smectic A correlations acquire saturation values comparable with the pore size of the densest aerogels and with about twice the pore size in the least-dense aerogel. The approach to saturation in all of the aerogels is well described by a common temperature function, scaled by a power-law dependence on pore size.
3 As the temperature is raised the correlation range decreases smoothly to values well *below* the nematic bulk pretransitional values. The large anisotropy in the shape of the correlation volume for layering fluctuations, characteristic of the bulk nematic, is eliminated in the aerogels.
4 The heat capacity anomalies show characteristic finite size behaviour when the layering correlation length becomes comparable with the aerogel pore volume-to-surface ratio, which is much smaller than the saturation correlation length or pore size. The pore volume-to-surface ratio is a measure of the mean distance to the nearest solid surface and is smaller than the pore dimension because of the rough nature of the pore surfaces.
5 The layering peak, intrinsically a powder average because of the lack of long-range order in the aerogels, is well fitted by a single- or double-Lorentzian lineshape. The Lorentzian shape implies a simple exponential correlation function for the smectic order parameter, which can be rationalized at low temperatures using the domain model description of random porous media. At higher temperatures there is thermal decorrelation, probably in the form of phase diffusion of the smectic order parameter.

These results indicate that a length scales comparable with the pore size the smectic ordering encounters at low temperatures a 'hard' disordering which limits the growth of layering correlation lengths also comparable with the pore size, so it may be the nematic order which ultimately limits the growth of smectic correlations. At temperatures near the bulk NA transition, at length scales comparable with the aerogel pore volume-to-surface ratio, the smectic ordering encounters a volume-filling 'soft' disordering, which limits the growth of layer correlations at temperatures in the bulk nematic and produces the finite size effects in the heat capacity. The volume nature of this disordering suggests that it too may be a result of the disordering of the nematic phase.

Acknowledgements

The authors acknowledge many useful conversations with C. Garland, L. Wu, B. Zhou, D. W. Schaefer and C. D. Muzny. This work was supported by NSF Solid State Chemistry Grant DMR 92-23729.

References

ALIEV, F. M. (1994) Liquid crystals-porous glasses heterogeneous systems as materials for investigation of interfacial properties and finite size effects, *Mol. Cryst. Liq. Cryst.* **243**, 91–105.

ALIEV, F M , MESHKOVSKII, I. K. and KUTNETSOV, V. I. (1984) Change in the phase transition temperatures of liquid crystalline substances in pores of different sizes, *Sov. Phys. Dokl.* **29**, 1009.

ALS-NIELSEN, J., LISTER, J. D., BIRGENAU, R. J., KAPLAN, M., SAFINYA, C. R., LINDEGAARD-ANDERSEN, A. and MATHIESEN, S. (1980) Observation of algebraic decay of positional order in a smectic liquid crystal, *Phys. Rev. B* **22**, 312.

ANISIMOV, M. A., CLADIS, P. E., GORODETSKII, E. E., HUSE, D. A., PODNEKS, V. E., TARATUTA, V. G., VAN SARLOOS, W. and VORONOV, V. P. (1900) Experimental test of a fluctuation-induced first order phase transition: The nematic-smectic A transition, *Phys. Rev. A* **41**, 6749.

ARMITAGE, D. and PRICE, F. P. (1976) Size and surface effects on phase transitions, *Chem. Phys. Lett.* **44**, 305–308.

BELLINI, T., CLARK, N. A., MUZNY, C. D., WU, L., GARLAND, C. W., SCHAEFER, D. W. and OLIVIER, B. J. (1992) Phase behavior of the liquid crystal 8C in a silica aerogel, *Phys. Rev. Lett.* **69**, 788–791.

BELLINI, T., CLARK, N. A. and SCHAEFER, D. W. (1995) Dynamic light scattering study of nematic amectic A liquid crystal ordering in silica aerogel, *Phys. Rev. Lett.* **74**, 2740–2743.

BENZEKRI, M., CLAVERIE, T., MARCEROU, J. P. and ROUILLON, J. C. (1992) Nonvanishing of the layer compression elastic constant at the smectic A to nematic phase transition: a consequence of Landau-Peierls instability?, *Phys. Rev. Lett.* **68**, 2480.

BIRGENAU, R. J., YOSHIZAWA, H., COWLEY, R. A., SHIRANE, G. and IKEDA, H. (1983) Random field effects in the two dimensional Ising antiferromagnet $Rb_2Co_{0.7}Mg_{0.3}F_4$, *Phys. Rev. B* **28**, 1438.

CALILLE, A. (1972) X-ray scattering by smectic A liquid crystals, *CR Acad. Sci., Ser. B* **274**, 891.

CHAN, K. K., DEUTSCH, M., OCKO, B. M., PERSHAN, P. S. and SORENSON, L. B. (1985) Integrated x-ray scattering intensity measurement of the order parameter and the nematic to smectic A transition, *Phys. Rev. Lett.* **54**, 920.

CHOW, L. C. and MARTIRE, D. E. (1969) Thermodynamics of solutions with liquid crystal solvents II. Surface effects with nematogenic compounds, *J. Phys. Chem.* **73**, 1127.

CLARK, N. A., BELLINI, T., MALZBENDER, R. M., THOMAS, B. N., RAPPAPORT, A. G., MUY, C. D., SCHAEFER, D. W. and HRUBESH, L. (1993) X-ray scattering study of smectic ordering in a silica aerogel, *Phys. Rev. Lett.* **71**, 3505–3508.

DAVIDOV, D., SAFINYA, C. R., KAPLAN, M., SCHAETZING, R., BIRGENAU, R. J. and LISTER, J. D. (1979) High resolution x-ray and light scattering study of critical behavior associated with the nematic-smectic A transition in 4-cyano-4'-octylbiphenyl, *Phys. Rev. B* **19**, 1657.

DEBYE, P., ANDERSON, H. R. and BRUMBERGER, H. (1957) Scattering by an inhomogeneous solid. II The correlation function and its application, *J. Appl. Phys.* **28**, 679.

EMMERLING, A. and FRICKE, J. (1992) Small angle scattering and the structure of aerogels, *J. Non-Cryst. Solids* **145**, 113.

FEDER, J. (1988) *Fractals*, New York: Plenum Press.

FISCH, M. R., PERSHAN, P. S. and SORENSEN, L. B. (1984) Absolute measurement of the critical behavior of the smectic elastic constant of bilayer and monolayer smectic A liquid crystals on approaching the transition to the nematic phase, *Phys. Rev. A* **29**, 2741.

FISHER, M. E. and BARBER, M. N. (1972) Scaling theory for finite size effects in the critical region, *Phys. Rev. Lett.* **28**, 1516.

FISHER, M. E. and FERDINAND, A. E. (1967) Interfacial, boundary and size effects at critical points, *Phys. Rev. Lett.* **19**, 169.

GARLAND, C. W. and NOUNESIS, G. (1995) Critical behaviour at nematic smectic-A transitions, *Phys. Rev. E* **49**, 2964.

DE GENNES, P. G. (1972) An analogy between a smectic A and a superconductor, *Solid St. Commun.* **10**, 753.

DE GENNES, P. G. and PROST, J. (1993) *The Physics of Liquid Crystals*, Oxford: Clarendon Press.

HALPERIN, B. I., LUBENSKY, T. C. and MA, S. K. (1974) First order transitions in superconductors and smectic A liquid crystals, *Phys. Rev. Lett.* **32**, 292.

IMRY, Y. (1984) Random external fields, *J. Stat. Phys.* **34**, 849.

KARAT, P. P. and MADSHUSUDANA, N. V. (1980) Elastic and optical properties of some 4'-n-alkyl-4-cyanobiphenyls, *Mol. Cryst. Liq. Cryst.* **36**, 51.

KASTING, G. B., GARLAND, C. W. and LUSHINGTON, K. J. (1980) Critical heat capacity of octylcyobiphenyl (8CB) near the nematic–smectic A transition, *J. Phys.* **41**, 879.

McMILLAN, (1971) Simple molecular model for the smectic A phase of liquid crystals *Phys. Rev. A* **4**, 1238.

NAKANO, A., BI, L., KALIA, R. K. and VASHISHTA, P. (1993) Structural correlations in porous silica: Molecular dynamics simulation on a parallel computer, *Phys. Rev. Lett.* **71**, 8.

NELSON, D. R. and TONER, J. (1981) Bond-orientational order, dislocation loops, and melting of solids and smectic-A liquid crystals, *Phys. Rev. B* **24**, 363.

PATTON, B. R. and ANDERECK, B. S. (1992) Extended isotropic-to-anisotropic crossover above the nematic–smectic A phase transition, *Phys. Rev. Lett.* **69**, 1556.

PATTON, B. R. and ANDERECK, B. S. (1994) Anisotropic renormalization of anisotropic quantities above the nematic–smectic A phase transition, *Phys. Rev. E.* **49**, 1393.

PERSHAN, P. S. (1988) *Structure of Liquid Crystals*, Singapore: World Scientific.

POROD, G. (1982) General theory, *Small Angle X-Ray Scattering*, ed. O. Glatter and O. Krathy, New York: Academic Press.

RAPPAPORT, A. G. (1995) PhD Thesis, University of Colorado.

RAPPAPORT, A. G., CLARK, N. A., THOMAS, B. N. and BELLINI, T. (1995) Smectic ordering in silica aerogels (to be published).

RICARD, L. and PROST, J. (1981) Critical behavior of second sound near the smectic A nematic phase transition, *J. Phys.* **42**, 861.

SAFINYA, C. R., ROUX, D. SMITH, G. S., SINHA, S. K., DIMON, P., CLARK, N. A. and BELLOCQ, A. M. (1986) Steric interactions in a model membrane system: A synchrotron x-ray study, *Phys. Rev. Lett.* **57**, 2718.

SCHAEFER, D. W. (1994) Structure of mesoporous aerogels, *MRS Bull.* **19-4**, 49–53.

SCHAEFER, D. W. and KEEFER, K. D. (1986) Structure of random porous materials: Silica aerogel, *Phys. Rev. Lett.* **56**, 2199.

SCHAEFER, D. W., WILCOXSON, J. P., KEEFER, K. D., BUNKER, B. C., PEARSON, R. K., THOMAS, I. M. and MILLER, D. E. (1987) Origin of porosity in synthetic materials, *Physics and Chemistry of Porous Media II*, ed. J. R. Bonayar et al., *AIP Conf. Proc.* **154**, p. 63.

SCHAEFER, D. W., BROW, R. K., OLIVER, B. J., RIEKER, T., BEAUCAGE, G., HRUBESH, L. and LIN, J. S. (1994) Characterization of porosity in ceramic materials by small-angle scattering: Vycor glass and silica aerogel, *Modern Aspects of Small-Angle Scattering*, ed. H. Brumberger, Dordrecht: Kluwer.

THOEN, J., MARYNISSEN, H. and VAN DAEL, W. (1982) Temperature dependence of the enthalpy and heat capacity of the liquid-crystal octylcyanobiphenyl (8CB), *Phys. Rev. A.* **26**, 2886.

WU, L., ZHOU, B., GARLAND, C. W., BELLINI, T. and SCHAEFER, D. W. (1995) Heat capacity study of nematic–isotropic and nematic–smectic-A transitions for octylcyanobiphenyl in silica aerogels, *Phys. Rev. E* **51**, 2157–2164.

YOKOYAMA, H., KOBAYASHI, S. and KAMEI, H. (1987) Temperature dependence of the anchoring strength at a nematic liquid crystal–evaporated SiO interface, *J. Appl. Phys.* **61**, 4501.

466

The Random Anisotropy Nematic Spin Model

D. J. CLEAVER, S. KRALJ, T. J. SLUCKIN and M. P. ALLEN

21.1 Introduction

The search for devices based on diffraction, rather than polarization, effects has led to increasing interest in liquid crystal systems confined in randomly interconnected networks of pores. In the first experimental study of this type of system, quasi-elastic light scattering was used to probe the dynamics of the liquid crystal 4'-octyl-4-cyanobiphenyl (8CB) in silica gel (Wu et al., 1992). This showed that randomized confinement shifts the nematic–isotropic (N–I) transition to a lower temperature, and that the low-temperature phase has rather different properties to those of a bulk nematic liquid crystal. Specifically, order parameter fluctuations were found to relax much more slowly (by up to a factor of 10) in the confined system. Calorimetric studies using various of the nCBs in aerogel (Bellini et al., 1992; Bellini and Clark, this volume) and Vycor (Iannacchione et al., 1993; Finotello et al., this volume) confirmed this shift in the N–I transition temperature, and also found the onset of nematic ordering to be continuous; by contrast the N–I transition is always first order in a three-dimensional bulk system. The continuous orientational ordering was found to coincide with a rapid increase in the turbidity of the experimental cell. Finally Tripathi et al. (1994) and Tripathi and Rosenblatt (this volume) have monitored the Cotton-Mouton coefficient of a liquid-crystal-forming material in a Vycor-like porous glass, and find sigmoidal behaviour close to the bulk N–I phase transition.

The effect of confinement on liquid crystal systems has long been a subject of considerable interest. Theoretical work in this area has its roots in the Landau-de Gennes theory of Sheng (1982), suitably modified by insights from the theory of wetting in simple fluids and binary mixtures (Sluckin and Poniewierski, 1986). Sheng showed that when a liquid crystal is confined between two parallel ordering walls (i.e. in a slab geometry), the N-I transition is shifted to higher temperatures, weakened, and eventually suppressed for sufficiently thin slabs or strong ordering. Later extensions of this work (Poniewierski and Sluckin, 1987), which included surface disordering terms in the wall–fluid interactions, found that shifts of the opposite sign are also possible. Experimental measurements performed on thin liquid crystal films have successfully detected these shifts, and have found evidence for a critical thickness (Yokoyama, 1988). Indeed, for 5CB confined between rubbed poly(vinyl alcohol) walls, no discontinuous N-I transition is observable at a wall separation of 83 nm. A lattice model simulation study using a system with a slab geometry has also identified a film thickness at which the N–I transition is no longer first order (Cleaver and Allen, 1993).

Experiments have also been carried out on confined liquid crystal systems in different geometries. Golemme *et al.* (1988) have reported that spherical droplets of 5CB in a polymer matrix show a continuous increase in nematic order if the droplet radius is less than 35 nm. Calorimetric studies (Iannacchione and Finotello, 1992, 1993) of the *n*CB series microconfined in the cylindrical pores of an Anopore membrane show substantial shifting and broadening of the N–I specific heat peak. The behaviour of these systems is very rich, and is sensitive both to the pore radius and to the type of surface alignment. Nevertheless, a reasonably complete phenomenological description appears to be possible (Kralj *et al.*, 1991).

The complexity of random interconnected systems creates considerable difficulties both for analytical and computational modelling. In some closed-pore geometries the dominant effect on the thermodynamics will come from the effect of individual pores, and the extra effects of the pore topology may be relatively unimportant. In other, more open, geometries, the pore structure may be thought of more as a random perturbation on a uniform bulk liquid crystal. Possible candidates for such systems are, for example polymer networks (Žumer and Crawford, this volume).

Indeed a decade ago, de Gennes and coworkers (Brochard and de Gennes, 1983; de Gennes, 1984) suggested that experiments on the phase separation of binary fluids in porous media could be interpreted in terms of a random field Ising model. If we recall that the Ising model provides a good paradigm for phase separation in binary fluids, we see that the idea here is that the porous nature of the matrix containing the fluid is accounted for by a random field, equivalent to a local chemical potential coupling randomly to one or other of the fluids. These ideas were used to interpret experiments by Maher *et al.* (1984). How applicable these ideas are in this context is not entirely clear; Ronis (1986), for example, has proposed an alternative model which emphasizes the narrowness of capillaries in a porous medium.

Stimulated by these ideas, a number of authors have suggested (Gingras, 1993) that similar random field ideas may provide a useful context within which to understand the onset of nematic order in similar porous geometries. The crucial difference between these systems and the binary fluids is the existence of orientational order. Fortunately there are a plethora of studies of magnetic systems with either random fields (Binder and Young, 1986) or random anisotropy (Harris *et al.*, 1973).

There has been some discussion (Wu *et al.*, 1992; Bellini *et al.*, 1992; Iannacchione *et al.*, 1993; Tripathi *et al.*, 1994) as to whether the random field models in fact describe experimental results on liquid crystal properties in porous systems. The purpose of this chapter is to try and follow the theoretical consequences of random field models in this context, and to discuss the circumstances in which these simplified models may help to give physical insight. At this stage we remain strictly agnostic as to whether the experimental results actually do support these ideas. We remark simply that in order to substantiate such claims, well-defined theoretical predictions must be available. In this chapter we present both a mean field study, using a rather bastardized set of ideas drawn from different fields, and a limited computer simulation study which gives rise to some rather surprising consequences. In order to help put our results in context we shall also draw from some unpublished studies of random magnets in which one of the authors has been involved (Denholm *et al.*, 1996). We note that another model of nematic behaviour in porous media which uses ideas from random magnetism, but which differs substantially in detail from the present model, has been presented by Maritan *et al.* (1994; and this volume).

The plan of this chapter is as follows. In section 21.2 we present the basic model, and discuss it in a broader theoretical context. In section 21.3 we present a generalized Landau theory of a nematic in a random anisotropy field, together with its consequences on the liquid crystal phase diagram. The Landau model is not entirely adequate, and we discuss also the expected problems which this solution raises. In section 21.4 we present the results of a limited Monte Carlo study of a random anisotropy nematic, and compare them with the Landau theory. Finally, in section 21.5 we draw some conclusions from the study.

21.2 Model

Harris *et al.* (1973) proposed a lattice model of random magnetism in order to provide a framework within which to understand anomalous magnetic behaviour in amorphous rare earth alloys. This model is known as the random anisotropy magnet (RAM) or HPZ model. Unit magnetic spins s_i are placed at sites i of a (usually cubic) lattice. These spins interact ferromagnetically with their nearest neighbours, and there is an anisotropy term of equal magnitude but random direction placed at each site:

$$H = - J \sum_{\langle , \rangle} s_i \cdot s_j - D \sum_i (s_i \cdot n_i)^2 \qquad (21.1)$$

where the magnetic coupling constant is J, and the first sum is taken over nearest-neighbour pairs. The quantity n_i is a unit vector in a random direction marking the local easy axis, and the degree of disorder is given by the parameter D.

The crucial idea is that the principal way in which the amorphous lattice in which the magnetic ions sit affects their magnetic properties is through the randomness introduced in the easy axis direction. The topological disorder is unimportant. The spirit of this model is similar, for example, to the Anderson model for electronic disorder in amorphous and liquid metals (Anderson, 1958). The model has been much studied in the literature on magnetism (Denholm *et al.*, 1992). Correlated anisotropy, coming from, for example, microcrystalline grains, can also be understood within such a model, though some quantities in the theory need to be reinterpreted.

By analogy with the RAM we can define the random anisotropy nematic (RAN) (Gingras, 1993) model in the following way:

$$H = - J \sum_{\langle , \rangle} P_2(s_i \cdot s_j) - D \sum_i P_2(s_i \cdot n_i) \qquad (21.2)$$

where P_2 is the second Legendre polynomial. The $D = 0$ case is the familiar Lebwohl–Lasher (L–L) lattice model of a liquid crystal (Lebwohl and Lasher, 1972).

In the next section we present some rather primitive mean field ideas about the phase behaviour of this model.

21.3 Landau Theory

We first develop an effective Landau–de Gennes free-energy functional appropriate to the Hamiltonian of eqn (21.2). This can be written in the following way (de Gennes and Prost, 1993):

$$f = f_{nr} + f_{rf} + f_{gr} \qquad (21.3)$$

The normal (non-random) free energy f_{nr} takes the usual form

$$f_{nr} = a't Q^2 - c Q^3 + b Q^4 \qquad (21.4)$$

where Q is the (scalar) liquid crystal order parameter, a', b, c are temperature-independent constants, $t = T - T_{NI}$, where T is the temperature and T_{NI} is the nematic–isotropic transition temperature.

The quantity f_{rf} comes from the random anisotropy term in eqn (21.2). Imry and Ma (1975) long ago developed a picture which has been widely used in the study of the statistical mechanics of random systems. (The main application of their ideas to the random anistropy magnet was made by Chudnovsky *et al.* (1982).) Let us recap this picture.

The effect of a random field is averaged over a length scale ξ, over which the orientation is correlated, and which is to be determined subsequently, in a variational calculation. In general, if we take a large number of sites, the mean random field on each site is different and the random fields are in uncorrelated directions. The sum of the random fields over the region of dimension ξ will therefore in general cancel, and so the mean random field per site will be more or less zero. In fact, however, the cancellation will not be exact. The mean magnitude of the sum of the random fields is given by the the sum of the squares of the random fields. By the central limit theorem this is a factor $N^{1/2}$ larger than any individual random field, where N is the number of molecules in a region of dimension ξ. The *effective* random field which couples to the local order parameter is then, roughly speaking, reduced by a factor of $N^{1/2}$. The effective random field per site is thus approximately $D\,(a_0/\xi)^{3/2}$, where a_0 is a molecular length scale.

In our expression we modify this form slightly, by noting that this reduction in random field is only true for correlation lengths much larger than a_0. We redefine ξ slightly so that it is zero, rather than unity, if molecular and physical correlation lengths are the same. An expression which interpolates between D and $D\,(a_0/\xi)^{3/2}$ is

$$f_{rf} = -\frac{Da_0^{3/2}}{(\xi^2 + a_0^2)^{3/4}}\,Q \tag{21.5}$$

The final term f_{gr} comes from the elastic energy $L(\nabla Q)^2$, which now must be included because the random anisotropy is causing the local orientation to wander in space on a length scale ξ; L is an averaged liquid crystalline elastic constant. Now because Q is changing on a length scale ξ, $|\nabla Q| \simeq Q/\xi$. If we again include the short length cut-off a_0, we obtain

$$f_{gr} = \frac{L}{(\xi^2 + a_0^2)}\,Q^2 \tag{21.6}$$

The formula (21.3), which now combines the ideas of Landau and those of Imry and Ma, must be minimized with respect to *both* the orientational order parameter Q *and* the correlation length scale ξ. The short-range cut-off in f_{gr} and f_{rf} is designed so that when the orientations are correlated on the minimum length scale, $\xi = 0$.

In order to carry out calculations on this model it will be useful to scale the variables in the following way:

$$f^* = \left(\frac{16b^3}{c^4}\right)f \qquad Q^* = \left(\frac{2b}{c}\right)Q \qquad t^* = \left(\frac{4ba'}{c^2}\right)t \qquad \xi_c^2 = \frac{4bL}{c^2} \qquad D^* = \left(\frac{8b^2}{c^3}\right)D$$

Now, dropping the asterisks, we obtain a scaled free energy:

$$f = tQ^2 - 2Q^3 + Q^4 + \frac{\xi_c^2}{\xi^2 + a_0^2}\,Q^2 - \frac{Da_0^{3/2}}{(\xi^2 + a_0^2)^{3/4}}\,Q \tag{21.7}$$

The quantity ξ_c is the relaxation length over which the order parameter relaxes and from now on we shall take $\xi_c = a_0$. In these scaled units the isotropic–nematic transition ($T = T_{NI}$) takes place at $t = 1$ to a phase in which $Q = 1$. The limit of stability of the isotropic phase ($T = T^*$) occurs at $t = 0$.

21.3.1 Nematic in a Non-random Field

As a point of reference, it will be useful to recall what behaviour is expected for a nematic placed in a non-random field, i.e. a field in a uniform direction. This problem was first studied by Fan and Stephen (1970), and there is now an extensive literature, including recent experimental verification (Lelidis and Durand, 1993). The free-energy functional is

$$f = tQ^2 - 2Q^3 + Q^4 - DQ \tag{21.8}$$

This functional now favours the formation of a nematic phase; even the isotropic phase acquires some order and is transformed into a *paranematic* phase. The paranematic–nematic phase transition occurs at $t = D$. In the scaled units the entropy change at the transition is $(1 - 2D)^{1/2}$. It thus reduces to zero at $D = \frac{1}{2}$, at which point there is a nematic–paranematic critical point. For $D > \frac{1}{2}$ the order parameter Q will increase smoothly as temperature is decreased. A further external potential of the form $\frac{1}{2}UQ^2$, by contrast, reduces the phase transition temperature by $\Delta t = \frac{1}{2}U$.

21.3.2 *Nematic in a random field*

Before we start, it will be useful to sum up our expectations as to how the system in a random anisotropy field should behave. One expects that, by contrast with the uniform field case, the isotropic–nematic transition temperature should be *lowered*. This is because even though the local field will induce local order, thus lowering the free energy locally, the free-energy cost of changing director orientation will outweigh this. Thus the isotropic phase, or more strictly, a random paranematic phase, is favoured.

It is convenient to think of the free-energy functional as

$$f = tQ^2 - 2Q^3 + Q^4 + v(Q) \tag{21.9}$$

where all the effect of the random anisotropy has been incorporated in the potential $v(Q)$. The potential $v(Q)$ is what remains after $f_{rf} + f_{gr}$ has been minimized with respect to ξ at constant Q. We find the following result for $\xi(Q)$:

$$\xi = 0 \tag{21.10a}$$

when $Q < \frac{3}{4} D$; and

$$\xi = [(4Q/3D)^4 - 1]^{1/2}a_0 \tag{21.10b}$$

when $Q > \frac{3}{4}D$. The value of $\xi(Q)$ is continuous at $Q = \frac{3}{4}D$.

The effective potential $v(Q)$ takes the following functional form:

$$v(Q) = -DQ + Q^2 \tag{21.11a}$$

when $Q < \frac{3}{4} D$; and

$$v(Q) = -\tfrac{27}{256} D^4 Q^{-2} \tag{21.11b}$$

when $Q > \frac{3}{4} D$. The potential $v(Q)$ changes its form at $Q = \frac{3}{4} D$. Nevertheless $v(Q)$ is continuous at $Q = \frac{3}{4} D$, with continuous derivatives, but not second derivatives; it is shown schematically in Figure 21.1.

For low order parameters Q the correlation length ξ is zero, and director orientations are correlated only on molecular length scales. The free-energy advantage is as though there were a fixed nematic field on the molecules, and thus negatively linear in Q. There is a free-energy cost in changing molecular orientation from point to point, but this is negligible because it is proportional to Q^2. Thus for the isotropic phase the effect is roughly the same, whether the imposed nematic field is random or fixed.

At higher values of Q, whether these are imposed directly through a strong field, or indirectly through the presence of a nematic phase, the cost of changing the orientation from point to point is increased. Now the system responds by increasing the length scale ξ over which the director orientations are correlated, but at the same time the effective mean orientational field (actually the root mean square field) over that length scale is decreased. Clearly in the high order parameter and low disorder limit, the true nematic phase with infinite ξ is recovered. From eqn (21.10b) for high order parameters $\xi \sim Q^4$, and in this regime the low field behaviour is given by $\xi \sim D^4$. We now also have a rather different effective nematic field, which is now

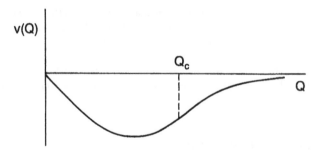

Figure 21.1 Schematic form of the effective potential $v(Q)$ discussed in eqns (21.9) and (21.11).

inversely proportional to Q^2 as the root mean nematic field is averaged over larger and larger regions. Thus the free-energy gain due to the imposed field is low, by contrast with the isotropic field case, and the consequence is to depress the nematic–isotropic phase transition.

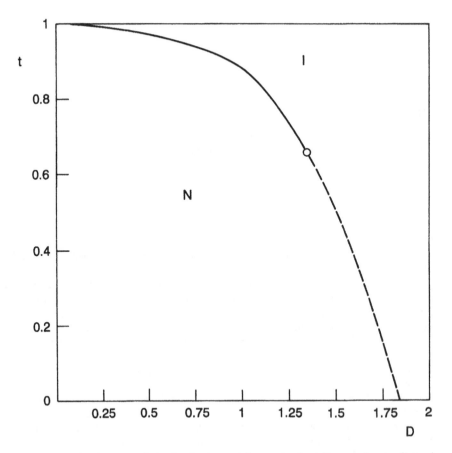

Figure 21.2 Landau theory prediction for the 'isotropic'–nematic phase diagram, showing first-order phase transition (full line), tricritical point $\bigcirc$, and continuous phase transition (dashed line).

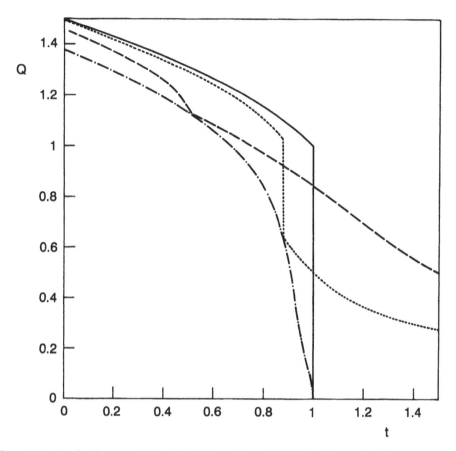

Figure 21.3 Landau theory predictions for $Q(T)$ for different D. Full line: $D = 0$; dotted line: $D = 1$; dashed line: $D = 1.5$. The chain line corresponds to the line of phase transitions.

21.3.3 Specific Predictions of the Landau Theory

At non-zero D there is a paranematic–nematic phase transition. At low D, $\Delta f_\mathrm{I} \sim - D^2$, whereas $\Delta f_\mathrm{N} \sim - D^4$, which is negligible by comparison with Δf_I. We thus expect that the shift Δt in the phase transition temperature $\sim - D^2$. In Figure 21.2 we show the calculated phase diagram as a function of D and t. It confirms this prediction. The first-order transition temperature is reduced from $t = 1$ at $D = 0$ to $t = 0.65$ at $D = 1.35$. At $D_\mathrm{c} \approx 1.35$ there is a tricritical point, beyond which the transition becomes continuous.

This continuous transition is the temperature where the correlation length ξ first becomes non-zero. Just below the transition the model predicts that $\xi \sim \Delta t^{1/2}$. However, the continuous transition is an unphysical consequence of the procedure, which considers ξ as a variational parameter. Other theories which posit structure on length scales set by the problem itself, and which have a lower cut-off length, crop up when discussing the structure of lyotropic phases of one sort or another (de Gennes and Taupin, 1986; Widom, 1984) and produce the same unphysical tricritical points and continuous transitions. An important thermodynamic signature, however, of the porous systems which RAN hopes to model is the rounding of the specific heat peak associated with the first-order isotropic–nematic phase transition. By contrast, the existence of the continuous phase transition in the mean field solution to this model leads to the specific heat discontinuity characteristic of a second-order phase transition. We do not,

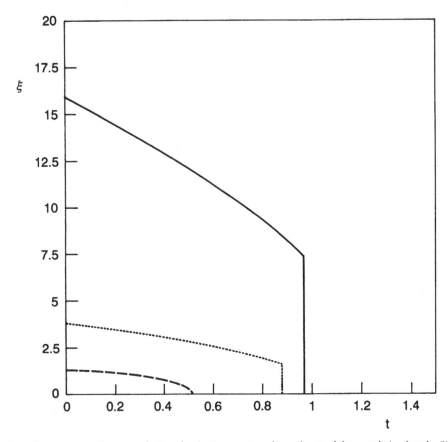

Figure 21.4 Landau theory predictions for the temperature dependence of the correlation length: $\xi(t)$, with ξ measured in units of a_0. Full line: $D = 0.5$; dotted line: $D = 1$; dashed line: $D = 1.5$.

however, take this particular prediction of the model seriously; it is simply an artefact of the artificial procedure of considering ξ as a variational parameter in the theory. It does have some physical significance, but we must not push this idea too far.

In Figure 21.3 we show the order parameter profiles as a function of temperature for a few typical values of D. For $D < D_c$ there remains a discontinuity in $Q(T)$ at the first-order phase transition, whereas for $D > D_c$, there is merely a discontinuity in dQ/dT. We also show (chain curve) the locus of points for different D, for which the correlation length $\xi = 0$.

An interesting point to note is that in the isotropic phase – more generally for low Q – the disorder *increases* the order parameter, and larger D increases it more. This order is, however, local and not orientationally correlated over longer than molecular length scales. By contrast, in the nematic phase – at high Q – disorder *decreases* the order parameter. We can see this in eqn (21.11b); in the high-Q regime $v(Q) \sim D^4 Q^{-2}$. Now the price of the disorder is to introduce a heterogeneous texture causing the change in the order parameter $\sigma Q \sim -D^4$. Thus the disorder once again is seen to discourage the nematic phase.

In Figure 21.4 we show the behaviour of the correlation length ξ as a function of temperature for a number of different values of D. As expected, ξ increases as T decreases, indeed as Q increases. For $D > D_c$ we see the unphysical square root singularity as ξ increases from zero. As a function of D at constant T, we expect the usual Imry–Ma behaviour $\xi \sim D^{-2}$.

We postpone further discussion of the results of the Landau theory to the discussion, because preliminary results from the Monte Carlo simulation of the model suggest that there are serious shortcomings in this mean field formulation. It is to this that we turn in the next section.

21.4 Monte Carlo Simulations

In this section we present the results of an initial computational study of the RAN model. In our simulations, we examine the case $D^* = D/J = 1$, and compare the resulting statistical mechanics with those of the $D^* = 0$ (Lebwohl and Lasher (L–L), 1972) non-random model. We work with a cubic simulation cell, subject to periodic boundary conditions and with linear dimension L, with $16 \leqslant L \leqslant 64$. We define a reduced temperature $T^* = k_B T/J$ where k_B is the Boltzmann constant. All energies and heat capacities are implicitly expressed in units of J.

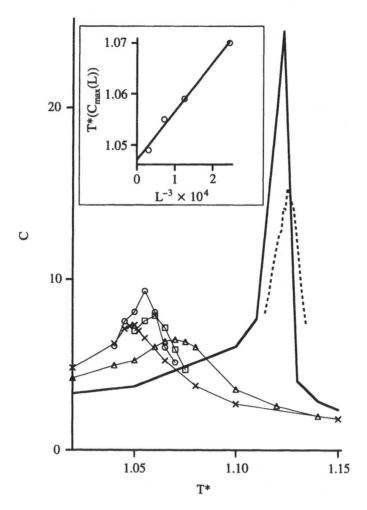

Figure 21.5 Monte Carlo results. Specific heat per spin of the RAN model for $L = 16$ (triangles); 20 (squares); 24 (circles); and 32 (crosses) with thin lines to guide the eye. Also shown are the equivalent $L = 30$ (thick full line) and 16 (thick dashed line) data for the L–L model. Inset: plot of $T^*(C_{max}(L))$ against L^{-3} with line of best fit.

Our simulations have been carried out using standard Metropolis Monte Carlo (MC) techniques on both a massively parallel DAP 510-8C and an IBM RS/6000 workstation. In obtaining the results reported here, we have run with a single, fixed anisotropy configuration for each L. Our correlation function and specific heat data have all been obtained from production runs of at least 10^5 MC sweeps (where one MC sweep corresponds to one attempted move per spin). At each new data point, equilibration runs of 5×10^4 MC sweeps have been performed prior to production.

We have measured the energy, U, and specific heat per spin, C, of the random anisotropy system at a range of T^* for $L = 16$, 20, 24 and 32. We show the specific heat data in Figure 21.5. Also shown in the figure are the equivalent data taken from previous simulations of the $L = 16$ (Cleaver and Allen, 1993) and $L = 30$ (Fabbri and Zannoni, 1986) (i.e. $D^* = 0$) systems. The sharp peaks in the L–L model data correspond to the first order N–I transition. At a first-order phase transition, finite size effect predictions state that the peak heights, $C_{max}(L)$. of these curves should scale as L^3 (Landau, 1990); the L–L model has been shown to obey this scaling rule for $L \geqslant 20$ (Zhang *et al.*, 1992, 1993). The specific heat peaks for the RAN model systems are all shifted to lower temperatures and are significantly broader than the L–L peaks. As L is increased from 16 to 24, the corresponding $C_{max}(L)$ values also increase slightly. A decrease is seen on increasing L to 32, however. This inconsistency is probably a result of our having data for only one anisotropy configuration for each L: to determine any general trend will require an average over several such configurations for each L. We note that the temperatures at which the peak specific heats occur, $T^*(C_{max}(L))$, show a stronger L dependence than do those of the L–L systems. This time the trend is monotonic in L, and our four data points are consistent with the form $T^*(C_{max}(L)) = T^*(C_{max}(\infty)) + \text{constant} \times L^{-3}$. We plot our $T^*(C_{max}(L))$ values and the line of best fit to this form in the inset to Figure 21.5. This fit gives a value of 1.047 for $T^*(C_{max}(\infty))$.

In order to clarify the significance of this specific heat peak, we have used the histogram technique (Cleaver and Allen, 1993; Zhang *et al.*, 1992, 1993; McDonald and Singer, 1967; Ferrenberg and Swendson, 1988) to investigate the form of the energy distribution function at $T^*(C_{max}(L))$ for $L = 24$ and 32. To do this, we have performed a run of 10^6 MC sweeps at $T^* \simeq T^*(C_{max}(L))$ for each of the two L and so measured the energy distribution functions $\mathscr{P}_L(U)$. We have then rescaled these distribution functions using Boltzmann reweighting factors to obtain the equivalent distributions, $\mathscr{P}'_L(U)$ at the peak height temperatures. Lee and Kosterlitz (1990, 1991) have shown that at a first-order transition, $\mathscr{P}'_L(U)$ has a double-peaked structure which becomes increasingly pronounced with increase in L. Our $\mathscr{P}'_L(U)$ are both virtually indistinguishable in shape from a single Gaussian, however. This technique has been shown to be more sensitive than scaling laws in the determination of transition order (Lee and Kosterlitz, 1990, 1991) so we conclude that for the system sizes accessible to this study, there is no indication of a first-order transition in the $D^* = 1$ RAN model. We note that this is consistent with the results of the experimental studies performed on randomly confined liquid crystals.

We have also calculated the orientational pair correlation function

$$g_2(r) = \left\langle \frac{\sum_{i,j} P_2(\hat{s}_i \cdot \hat{s}_j) \delta(r_{i,j} - r)}{\sum_{i,j} \delta(r_{i,j} - r)} \right\rangle \tag{21.12}$$

where $r_{i,j}$ is the minimum imaged distance between spins i and j and we use large angle brackets to indicate an ensemble average. When this function was measured by Fabbri and Zannoni (1986) for the $L = 30$ L–L model below the N–I transition, it was found to give excellent fits to the form

$$g_2(r) = \frac{G}{r} \exp(-kr) + \bar{P}_2^2 \tag{21.13}$$

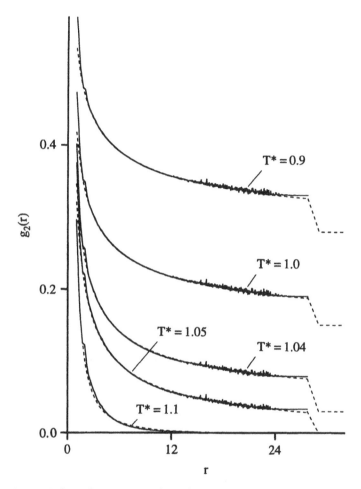

Figure 21.6 Monte Carlo results. Orientational correlation function, $g_2(r)$, data (full lines), and fits to eqn (21.16) (dashed lines) for the RAN model with $L = 32$ at a range of temperatures T^*. Also shown are the limiting values A_L given by the fits (horizontal dashed lines).

where G and k are fitting parameters and $\bar{P}_2$ is the non-zero nematic order parameter (the $1/r$ Goldstone boson contribution is not, in practice, observable). This order parameter is given by

$$\bar{P}_2 = \frac{1}{L^3} \left\langle \sum_{i=1}^{L^3} P_2(\hat{\mathbf{d}} \cdot \hat{\mathbf{s}}_i) \right\rangle \tag{21.14}$$

where $\hat{\mathbf{d}}$ is a unit vector along the nematic director. The $g_2(r)$ curves calculated by Fabbri and Zannoni (1986) indicate a fairly rapid decay in orientational correlations for r up to about five lattice spacings, but very little further decay beyond this distance (i.e. $\bar{P}_2$ is only weakly L dependent).

In Figure 21.6 we show $g_2(r)$ curves obtained at various temperatures using an $L = 32$ RAN model system. We find that in the high-temperature regime, above the heat capacity maximum $T^*(C_{max}(L))$, $g_2(r)$ decays exponentially in the absence of long-range order. At lower temperatures, however, a qualitative difference is seen from the results of Fabbri and Zannoni in that the decay in the correlation function is on the same length scale as L: we find that in this

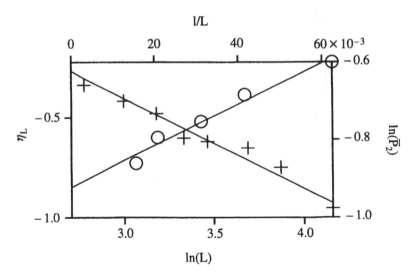

Figure 21.7 Monte Carlo results for the point $T^* = 1.0$ at a range of system sizes. We show a logarithmic plot of $\bar{P}_2$ against L (crosses) and a fit of form $\ln(\bar{P}_2) = $ constant $\times \ln(L)$. Also shown is a plot of η_L against L^{-1} (circles) and the straight line of best fit.

low-temperature regime, the random anisotropy destroys the long-range order and replaces it with an algebraic correlation function

$$g_2(r) = \frac{B}{r^{\eta+1}} \tag{21.15}$$

For $T^* = 1.0$, by using methods described below, we find the exponent $\eta \simeq -0.85$.

There are two independent methods for determining η at fixed T^*. For simulations of given L, we can evaluate $\bar{P}_2$; algebraic correlations then imply that $\bar{P}_2(L) \propto L^{-(\eta+1)/2}$ (Binder and Heermann, 1988). Alternatively, we can examine the observed correlation function for given L; this is always of the form

$$g_2^L(r) = \frac{B_L}{r^{(\eta_L+1)}} + A_L \tag{21.16}$$

and we can extrapolate the exponent η_L to the case when $L = \infty$. In Figure 21.6 we show fits to our $L = 32$ data with this algebraic form, and indicate the limiting values A_L. As L increases, A_L decreases, tending to zero as $L \to \infty$ and thus confirming that the algebraic form for the correlation function is indeed correct (Cleaver, 1996).

In order to obtain the correct value for η, we have performed an extensive investigation of the point $T^* = 1.0$ at a range of different L. For the largest system ($L = 64$), data accumulation at this single point required several weeks of CPU time on an RS/6000 workstation. We show the primary results of this investigation in Figure 21.7. Here we present a logarithmic plot of $\bar{P}_2$ against L. This has the expected linear behaviour, indicating power-law correlations. The gradient of the fit to these data gives $\eta = -0.846$. We also present a plot of η_L against L^{-1}. This shows approximately linear behaviour, leading to unambiguous extrapolation to $L = \infty$. A linear fit to this graph gives $\eta = -0.848$. This is, of course, an empirical behaviour replacing separate extrapolations for each value of r. It is, nevertheless, noteworthy that both methods yield essentially the same value for the exponent η, and interesting that the direct evaluation of η in a finite L simulation gives poor results.

21.5 **Discussion**

Our results are significant both in the context of the physics of liquid crystals in random systems, and in the context of random anisotropy models in general. Experience in the random anisotropy magnet suggests that HPZ models may be applicable even when there is strong but dilute random anisotropy. Aerogel systems with an open structure have similar properties; there is likely to be strong but random anchoring at a number of sites.

We have made mean field calculations of the statistical mechanics of a random anisotropy nematic, and for reasons more fully discussed below there are reasonable grounds for supposing that at least some of the basic physics of aerogel systems is included. The Imry–Ma picture of random systems is widely trusted, and we have found a reduced temperature at the transition, together with a first-order transition whose strength diminishes as the randomness increases. Formally we predict that the first-order transition ends at a tricritical point, beyond which the transition becomes continuous. However, we have compelling reasons for not believing this last feature, so the mean field theory leads us to expect merely a critical point beyond which there is simply a build-up of local order on increasing length scales.

Surprisingly, our simulations cast doubt on this whole picture. We discuss this in greater detail elsewhere (Denholm *et al.*, 1996). They suggest a structure factor $S(k) \sim k^{-(2-\eta)} \sim k^{-2.85}$ in the low-k limit. At the very least, we should expect nematic correlations on length scales considerably longer than the pore size determined in the absence of the nematic fluid. This seems consistent with the results of Wu *et al.* (1992), although more detailed experiments seem desirable.

The destruction of long-range order is expected and occurs in many analogous magnetic models (Chudnovsky *et al.*, 1982). Indeed, the RAN model bears such a family resemblance to its magnetic analogue that we must expect our results to yield some insight into the low-T behaviour of the RAM model (which is as yet incompletely understood). The algebraic decay in correlations has its echo in experiments on rare earth alloys by Pickart *et al.* (1977), who find $S(k) \sim k^{-2.4}$. This result has been reproduced in a simulation by Fisch (1991).

At this stage it is not clear whether the exponent η is universal in a particular model, or is temperature dependent (as is the case in the non-random Berezinski–Kosterlitz–Thouless phase (Kosterlitz and Thouless, 1972)). We would also expect that the algebraic decay would set in at some phase transition temperature, though whether this transition exhibits any thermodynamic singularities and its relationship and location with respect to the specific heat maximum remain unknown.

We might expect that at lower temperatures and larger D^* than those we have investigated, a thermal pinning transition might occur (Denholm *et al.*, 1996) to a nematic glass state (although in experimental systems this might be pre-empted by other liquid crystalline phases (Clark *et al.*, 1993). Also, we can ask whether the first-order transition at $D^* = 0$ is preserved for low D^* and only destroyed at some critical D_c^* (as is the case in the presence of an external field) or whether the first-order transition is destroyed at all D^*. It may also be that we have done, as it were, our *modelling* incorrectly! The mean field theory may be correct, but for a different microscopic Hamiltonian.

These comments concern the behaviour of the model. It is, of course, a sensible question to ask in what kind of systems the model might actually apply. The simplest kind of porous system consists of long individual capillaries, with sufficiently many connections between the capillaries to form a percolating network, a rabbit warren in effect. The statistical mechanics of such a system will surely be dominated by the properties of the individual pores, which has already been the subject of intense study (Poniewierski and Sluckin, 1987). The effect of the joins between the pores, where the nematic director must change direction, will be to reduce slightly the phase transition temperature, which might increase overall because of strongly ordering pore boundaries. This may, for example, be occurring in a recent differential scanning calorimetry study (Dadmun and Muthukumar, 1993) of PAA confined to controlled pore glasses.

As the volume fraction of liquid crystal increases, the confining material presumably becomes sheet like and eventually line like. The rabbit warren is then not the liquid crystal, but rather the polymeric material confining it. The matrix is now like the inside of the Rock of Gibraltar: apparently solid, but actually only a small per cent polymer and really rather empty. Now the properties of the system are dominated by the liquid crystal itself. The confining material provides surface forces which in any given region favour a particular direction, which, however, changes from place to place. There is thus a random field. However, the open nature of the structure means that the bulk of liquid crystal interaction is with other liquid crystal, and the model proposed in this chapter may have some application. In fact, the local random fields only exist close to the confining material, or equivalently at a few, correlated but sparse, sites on the lattice. How this would affect the statistical mechanics is unknown.

At this stage it is an open question whether the detailed model introduced in this chapter, or that of Maritan *et al.* (1994; and this volume) or, indeed, either model, will be more fruitful. In our model the properties depend crucially on the explicit nematic nature of the phase; however, we do not pretend to have a detailed understanding of the statistical mechanics of the model at this stage. The model of Maritan *et al.*, by contrast, uses a Potts model to mimic some of what they believe to be the most important nematic properties. However, once chosen, the statistical mechanics of their random field model has been treated essentially exactly. Further studies, of the model and the modelling process, as well as further experiments, are clearly desirable.

Acknowledgements

We thank W. M. Saslow, D. R. Denholm, C. J. Walden, M. J. P. Gingras, D. Finotello and G. S. Iannacchione for useful discussions, and are particularly grateful to M. J. P. Gingras for sending a preprint prior to its publication.

The computational part of this work was supported by the UK Science and Engineering Research Council (now EPSRC) through funding for an AMT DAP 510-8C provided through the Computer Science Initiative, and by a fellowship to DJC during his time at Southampton (grant GR/G/44193). We gratefully acknowledge IBM for provision of an RS/6000 320 made available through an academic influence agreement, and Professor Dominic Tildesley for making this resource available.

The contribution of SK while at Southampton was supported financially by the Ministry of Education of the Republic of Slovenia. Most of the writing of this chapter was carried out by TJS while a participant at a workshop on topological defects taking place in the Newton Centre for Mathematical Sciences in Cambridge. TJS is grateful to Professor Alan Bray for a number of influential discussions during this period.

Finally we should like to thank Daše Žumer and Greg Crawford for the invitation to participate in this volume. Bill Doane's contribution to the liquid crystal community – as a participant, as an organizer, as a technologist and as a colleague – is widely recognized. We are happy to be able to mark his 60th birthday with the study described in this chapter.

References

ANDERSON, P. W. (1985) *Phys. Rev.* **109**, 1492.

BELLINI, T., CLARK, N. A., WU, L., GARLAND, C. W., SCHAEFER, D. and OLIVIER, B. (1992) *Phys. Rev. Lett.* **69**, 788.

BINDER, K. and HEERMANN, D. W. (1988) *Monte Carlo Simulation in Statistical Physics: An Introduction*, Berlin: Springer-Verlag.

BINDER, K. and YOUNG, A. P. (1986) *Rev. Mod. Phys.* **58**, 801; and references therein.

BROCHARD, F. and DE GENNES, P. G. (1983) *J. Phys. Lett.* **L44**, 85.

CHUDNOVSKY, E. M., SASLOW, W. M. and SEROTA, R. A. (1982) *Phys. Rev. B* **33**, 2697.

CLARK, N. A., BELLINI, T., MALZBENDER, R. M., THOMAS, B. N., RAPPAPORT, A. G., MUZAY, C. D., SCHAEFER, D. W. and HRUBESH, L. (1993) *Phys. Rev. Lett.* **21**, 3505.

CLEAVER, D. J. (in preparation).

CLEAVER, D. J. and ALLEN, M. P. (1993) *Mol. Phys.* **80**, 253.

DADMUN, M. D. and MUTHUKUMAR, M. (1993) *J. Chem. Phys.* **98**, 4850.

DE GENNES, P. G. (1984) *J. Phys. Chem.* **86**, 6469.

DE GENNES, P. G. and PROST, J. (1993) *The Physics of Liquid Crystals*, Oxford: Clarendon Press.

DE GENNES, P. G. and TAUPIN, C. (1986) *J. Phys. Chem.* **86**, 2294.

DENHOLM, D. R., SLUCKIN, T. J. and RAINFORD, B. D. (1992) *Acta Phys. Pol. B* **25**, 219.

DENHOLM, D. R., SLUCKIN, T. J. and SASLOW, W. M. (1995) (in preparation).

FABBRI, U. and ZANNONI, C. (1986) *Mol. Phys.* **58**, 763.

FAN, C. P. and STEPHEN, M. J. (1970) *Phys. Rev. Lett.* **25**, 500.

FERRENBERG, A. M. and SWENDSON, R. H. (1988) *Phys. Rev. Lett.* **61**, 2653; (1989) *Phys. Rev. Lett.* **63**, 1658(E).

FISCH, R. (1991) *Phys. Rev. Lett.* **66**, 204.

GINGRAS, M. J. P. (1993) private communication.

GOLEMME, A., ŽUMER, S., ALLENDER, D. W. and DOANE, J. W. (1988) *Phys. Rev. Lett.* **61**, 2937.

HARRIS, H., PLISCHKE, M. J. and ZUCKERMANN, M. J. (1973) *Phys. Rev. Lett.* **31**, 160.

IANNACCHIONE, G. S. and FINOTELLO, D. (1992) *Phys. Rev. Lett.* **69**, 2094.

IANNACCHIONE, G. S. and FINOTELLO, D. (1993) *Liq. Cryst.* **4**, 1135.

IANNACCHIONE, G. S., CRAWFORD, G. P., ŽUMER, S., DOANE, J. W. and FINOTELLO, D. (193) *Phys. Rev. Lett.* **71**, 2595.

IMRY, Y. and MA, S. K. (1975) *Phys. Rev. Lett.* **35**, 1399.

KOSTERLITZ, J. M. and THOULESS, D. J. (1972) *J. Phys.* **C6**, 1181.

KRALJ, S., ŽUMER, S. and ALLENDER, D. W. (1991) *Phys. Rev. A* **43**, 2943.

LANDAU, D. P. (1990) *Finite Size Scaling and Numerical Simulation of Statistical Systems*, ed. V. Privman, Singapore, World Scientific, p. 225.

LEBWOHL, P. A. and LASHER, G. (1972) *Phys. Rev. A* **6**, 426.

LEE, J. and KOSTERLITZ, J. M. (1990) *Phys. Rev. Lett.* **65**, 137.

LEE, J. and KOSTERLITZ, J. M. (1991) *Phys. Rev. B* **43**, 3265.

LELIDIS, I. and DURAND, G. (1993) *Phys. Rev. E* **48**, 3822.

MAHER, J. V., GOLDBURG, W. I., POHL, D. W. and LANZ, M. (1984) *Phys. Rev. Lett.* **53**, 60.

MARITAN, A., CIEPLAK, M., BELLINI, T. and BANAVAR, J. R. (1994) *Phys. Rev. Lett.* **72**, 4113.

MCDONALD, I. R. and SINGER, K. (1967) *Discuss. Faraday Soc.* **43**, 40.

PICKART, S. J., ALPERIN, H. A. and RHYNE, J. J. (1977) *Phys. Lett. A* **64**, 37.

PONIEWIERSKI, A. and SLUCKIN, T. J. (1987) *Liq. Cryst.* **2**, 281.

RONIS, D. (1986) *Phys. Rev. A* **33**, 4319.

SHENG, P. (1992) *Phys. Rev. A* **26**, 1610.

SLUCKIN, T. J. and PONIEWIERSKI, A. (1986) *Fluid Interfacial Phenomena*, ed. C. A. Croxton, Chichester: John Wiley, p. 215.

TRIPATHI, S., ROSENBLATT, C. and ALIEV, F. M. (1994) *Phys. Rev. Lett.* **72**, 2725.

WIDOM, B. (1984) *J. Chem. Phys.* **81**, 1030.

WU, X.-L., GOLDBURG, W. I., LIU, M.-X. and XUE, J.-Z. (1992) *Phys. Rev. Lett.* **69**, 4870.

YOKOYAMA, J. (1988) *J. Chem. Soc. Faraday Trans. II* **84**, 1023.

ZHANG, Z., MOURITSEN, O. G. and ZUCKERMANN, M. J. (1992) *Phys. Rev. Lett.* **69**, 1803.

ZHANG, Z., ZUCKERMANN, M. J. and MOURITSEN, O. G. (1993) *Mol. Phys.* **80**, 1195.

Nematic–Isotropic Transition in Porous Media

A. MARITAN, M. CIEPLAK and J. R. BANAVAR

22.1 Introduction

The properties of liquids and solids confined in very small pores are interesting both from fundamental and practical standpoints (Klafter and Drake, 1989; Padday, 1978, Israelachvili, 1985; Banavar *et al.*, 1987; Wong, 1988). At the fundamental level, the interactions with surfaces and the effects of finite size and geometrical interconnection can be explored in these systems. At the practical level, these systems are of great importance in areas such as interfacial adhesion, lubrication, rheology, tribology and materials engineering.

The formulation of concepts such as universality, broken symmetry and scaling near continuous transitions has led to a deep and detailed understanding of phase transitions in homogeneous systems. However, the effects of quenched or frozen randomness have only recently begun to receive significant attention. The elucidation of both the equilibrium (i.e. the nature of the equilibrium ordered phases) and the non-equilibrium properties of random systems remains a challenge (Fisher *et al.*, 1988). Unlike continuous phase transitions, the key features of first-order transitions are nucleation phenomena, limits of superheating and supercooling and questions concerning the nature of the singularities in thermodynamic quantities at the transition. The confined geometries of the porous medium have interconnected pore and grain spaces (Banavar *et al.*, 1987; Wong, 1988). Nevertheless, in some porous media such as Vycor glass, a size scale corresponding to that of a characteristic pore is present. Some porous media have fractally rough pore grain interfaces over a range of length scales, while others do not. The geometry of the pore grain interface is complex and is an example of quenched disorder. It follows therefore that these systems provide a fertile playground for exploring many aspects of phase transitions in random restricted geometries. A judicious choice of porous media and pore filling material enables one to control the geometry, the connectivity, the strength of the interactions and the pore size distribution. Porous media are becoming increasingly important in many applications such as nuclear waste storage, desiccation, catalysis and in oilfields. A detailed understanding of the geometry of porous media and the effects of the medium on the material contained within is crucial for a full development of these applications.

Recently, the nematic ordering of liquid crystals contained within a porous medium has been studied by a variety of techniques (Bellini *et al.* 1992, 1995; Iannacchione *et al.* 1993, Dadmun and Muthukumar, 1993, Wu *et al.*, 1992, 1995; Aliev *et al.*, 1991; Tripathi *et al.*, 1994; Kralj *et al.*, 1993). These experiments indicate that the first-order transition in the absence of the confining porous medium is replaced by a smooth evolution to a glassy state in which the correlation length does not exceed the characteristic pore size. Even though the relaxation time was found to become very large compared with experimentally accessible times, no hysteresis was observed. Here, we study this problem theoretically – a range of models is considered and studied using mean field theory enabling the determination of the simplest model that captures the physics of the liquid crystal system. Our mean field studies (Maritan *et al.*, 1994) are complemented by Monte Carlo simulations in three dimensions in order to eliminate certain spurious features of the mean field analysis, to assess the effect of fluctuations and to study the dynamics. Taken together, mean field theory and computer simulations for the model system yield results in accord with experiment.

22.2 Experimental Results

The isotropic to nematic phase transition of thermotropic liquid crystals embedded in various kinds of porous media has been investigated by a range of experimental techniques: calorimetry (Bellini *et al.*, 1992; Iannacchione *et al.*, 1993; Dadmun and Muthukumar, 1993; Wu *et al.*, 1995), dynamic light scattering (Wu *et al.*, 1992; Bellini *et al.*, 1995), static light scattering (Bellini *et al.*, 1992; Aliev *et al.*, 1991), magnetic birefringence (Tripathi *et al.*, 1994) and NMR (Iannacchione *et al.*, 1993; Kralj *et al.*, 1993).

We will focus on those experiments in which the random host material may be modelled as inducing a random-field-like disturbance on the liquid crystal. This does not seem to be a general property of porous materials. Some of them, such as porous membranes, polymeric matrices and sintered porous silica glasses (such as Vycor), have locally smooth surfaces and exert an overall ordering effect on the molecules adjacent to the solid interface (Iannacchione *et al.*, 1992). On the other hand, silica aerogels have a structure of multiply connected strands and a density as low as a few per cent in volume fraction (or conversely a high porosity). These systems have a surface roughness on the length scale of the diameter of the silica strands, which is typically about 30–50 Å (Schaefer, 1994). On a larger length scale, the randomly oriented silica strands have a disordering effect on the nematic and smectic phases of liquid crystals. Evidence for this is provided (Bellini *et al.*, 1992; Wu *et al.*, 1995, Clark *et al.*, 1993) by the depression of the temperatures at which the 'phase transitions' occur with respect to the bulk T_{NI} and T_{NA} (T_{NI} and T_{NA} denote the nematic–isotropic transition temperature and the nematic–smectic A transition temperature respectively). Studies of liquid crystal alignment near random surfaces obtained by silica evaporation lead to a similar conclusion (Faetti *et al.*, 1985).

It has been generally observed that lecithin-treated silica surfaces show weak coupling with the liquid crystal molecules at the interface (Iannacchione *et al.*, 1992) whereas bare silica surfaces strongly induce flat alignment on the surfaces. The weakening effects on the silica–liquid crystal coupling due to the lecithin treatment have also been observed in liquid crystals placed in silica aerogel (Kralj *et al.*, 1993).

We now turn to a brief review of the calorimetric and dynamical light scattering (DLS) experiments. An extensive discussion of the specific heat results is presented by Bellini and Clark in this volume. We outline only the main features connected to the theoretical analysis presented later on in this chapter. Upon increasing the aerogel density ρ (or equivalently lowering the porosity), there are changes in the NI specific heat data:

1   The peak position on the temperature axis decreases, roughly linearly with increasing ρ.
2   The integrated peak, i.e. the NI enthalpy, also decreases.
3   Within the experimental resolution, the peaks associated with the low-density aerogel suggest a first-order NI transition, whereas at higher values of ρ the specific heat peaks appear rounder, which suggests no phase transition at all.

The experimental NI specific heat curve exhibits a single well-defined peak. A simple interpretation of the calorimeteric results is as follows (Bellini *et al.* 1995) on lowering the temperature from the isotropic phase, paranematic pretransitional clusters tend to grow in size – when they reach a size of L_{max} (with the bulk $L_{max} \approx 100$ Å) the first-order transition occurs. Studies of the optical properties of the low-temperature phase indicate that this phase has a multidomain nematic structure, with the average domain size determined by the aerogel density (Bellini *et al.*, 1992). If the average domain size is larger than L_{max}, then the growth of the local nematic domains is discontinuous and a specific heat spike is observed. On the other hand, if the average domain size is smaller than L_{max}, the specific heat peak shows a finite size rounding.

To date, two DLS studies of liquid crystal nematic ordering in random hosts have been performed (Wu *et al.* 1992; Bellini *et al.*, 1995). Their results differ significantly in several respects. This could be because the silica matrices used in the two experiments have quite different structures: the silica volume fraction in the aerogel used by Wu *et al.* (1992) and the experimental sample of Bellini *et al.* (1995) had a volume fraction of about 20%. The main difference in the dynamical behaviour of the nematic phase is that in the low-density aerogel fast relaxation processes are observed, indicating pretransitional-like orientational fluctuations, but the high-density aerogel leads to no rapid relaxation effects. In spite of these differences, there are interesting similarities observed in the long-time dynamics. In both experiments, a drastic slowing down is observed when the temperature is decreased. At sufficiently low temperatures, the correlation functions do not decay to their uncorrelated values within the experimental time range. In the low-density aerogel, the long-time correlation function decays to a non-vanishing plateau, the size of which increases on lowering the temperature (Bellini *et al.*, 1995).

The two DLS studies offer differing explanations for the slowing-down phenomenon:

1 Wu *et al.* observe that, by stretching the logarithm of the time axis, they obtain a fair superposition of the relaxation curves. They suggest that such a stretching may be a consequence of a random-field-like activated dynamics. The stretching constant diverges on lowering the temperature, leading to a complete freeezing of the local nematic ordering.
2 Bellini *et al.*, on the other hand, observe a non-frozen dynamics even within the local smectic phase. They find that the long-time correlation function may be well described by stretched exponentials and that the slowing down is associated with a decrease in the value of the stretching exponent. Since such a behaviour is analogous to the one taking place in spin glasses, they suggest that a slow dynamics may arise from the collective reorientations of the coupled nematic domains. This implies the existence of 'dynamical' correlation length which is larger than the average domain size.

We refer the reader to the chapters by Finotello *et al.*, Aliev, Bellini and Clark, Rappaport *et al.* and Tripathi and Rosenblatt in this volume for further details on experimental results pertaining to liquid crystals confined in porous media.

22.3 Previous Theoretical Developments

Liquid crystals are soft objects – the energy associated with the onset of a long-range order is rather small. Thus disorder of any kind is expected to alter the phase transition in a significant way. Dadmun and Muthukumar (1993) have proposed a simple lattice polymer model which corresponds to the liquid crystal placed within the pores of a porous medium. The properties of this model have been studied by Monte Carlo simulations. Specifically, the liquid crystal polymer chains are modelled by connected trimers which are constrained to lie on the sites of a cubic lattice and the presence of porous medium is implemented by declaring some sites to be inaccessible for the trimers. Thus the only effect of the porous medium sites is to provide excluded volume. The double occupancy of a site is forbidden to mimic steric interactions. The trimers consist of three beads and two bonds and they can exist in two states: either the bonds bend at a right angles or they form a straight segment. In the latter case the intramolecular energy is assigned to be $\varepsilon_b < 0$ and in the former the energy is zero. Thus the trimers prefer to be straight. Furthermore, intermolecular interactions create a preference for parallel alignments of the trimers: two neighbouring segments attract with energy $\varepsilon_s < 0$ if they are parallel to each other. Otherwise the energy of the segment–segment interaction is zero. For simplicity, $\varepsilon_s = \varepsilon_b$. In this model, the intermolecular interactions show some dependence on the instantaneous shapes of the trimers.

The simulation starts with the trimers added in a totally ordered state at a concentration of 81.25% and their subsequent dynamics is defined in terms of the modified reptation model – the end of a trimer is moved to a randomly selected vacant neighbouring site and the whole trimer follows. The new configuration is accepted or rejected according to the Metropolis criterion. Dadmun and Muthukumar (1992) have focused on studies of the specific heat, as determined from the fluctuations in the total energy of the system. They find that in the pure system the transition is first order (the maximum of the specific heat grows linearly with the volume of the system). On the other hand, introduction of the excluded volume impurities, at a concentrations greater than 2.5%, changes the nature of the transition and reduces the transition temperature. In a subsequent experimental paper Dadmun and Muthukumar (1993) have studied p-azoxyanisole in porous glasses using differential scanning calorimetry and find qualitative agreement with their Monte Carlo simulations. They observe also that the pore wall anchoring raises T_{NI} while finite size effects lower it. In the experimental system studied, the typical pore size was small and, therefore, the finite size effects were stronger and led to the broadening of the heat capacity peak.

Another approach to study the effect of disorder on the isotropic–nematic (IN) and nematic–smectic A (NA) transitions has been proposed by Gingras (1995). He considers the lattice version of the Maier–Saupe model (Maier and Saupe, 1959), i.e. the Lebwohl–Lasher model (Lebwohl and Lasher, 1972) described by the Hamiltonian

$$H_{IN} = -\sum_{\langle i,j\rangle} J_{ij}(\mathbf{S}_i \cdot \mathbf{S}_j)^2 \tag{22.1}$$

where $J_{ij} = J$ on the nearest-neighbour sites and $\mathbf{S}_i$ is a unit three-dimensional vector describing the orientation of a rod-like molecule of the liquid crystal. The order parameter for the IN transition is the second-rank tensor $Q_{ab} = \langle 3S_{ia}S_{ib} - \delta_{ab}\rangle$. At high temperatures $Q_{ab} = O$. At a temperature of $T_{IN} = 0.749J/k_B$, it is known that the system undergoes a weak first-order transition to an ordered uniaxial nematic phase.

Gingras (1995) observes that the porous medium acts on the liquid crystal in two ways. First, it provides excluded volume, and, second, chemical affinity effects introduce a preferential molecular orientation at the pore walls. The first effect should be equivalent to the presence of a random J_{ij} and is considered to be of secondary importance. On the other hand, Gingras suggests that the second effect should be represented by the Hamiltonian

$$H_r = -\sum_i \sum_{m=1}^{m=\infty} g_{2m}(\mathbf{h}_i \cdot \mathbf{S}_i)^{2m} \tag{22.2}$$

The field h_i is expected to have short-range correlation of the order of the mean pore size and may be modelled by a Gaussian variable of zero mean and dispersion h. If the disorder is not too large then the combined Hamiltonian $H_{IN} + H_r$ leads to a phase transition between the isotropic and speronematic states. In the latter, the frozen director wanders in space. This appears to be superficially similar to what happens in spin glasses and it has been argued by Wu *et al.* (1992) that the problem may be in the spin glass universality class. Gingras argues, however, that the symmetry properties of H_r are such that a *random field* description of the IN transition is more appropriate here. The reason for this is that Q_{ab} is quadratic in S_i and a term like $(h_i \cdot S_i)^2$ is only linear in Q_{ab} and is conjugate to the order parameter.

Gingras also predicts that weak disorder destroys the nematic to smectic A transition and converts it into a spin-glass-like transition at non-zero temperature and that this transition should be in the same universality class as the vortex–glass transition in disordered superconductors.

Zhang and Chakrabarti (1995) have recently performed Monte Carlo studies of the model

$$H = -\varepsilon \sum_{<ij>} \tfrac{1}{2} [3(\mathbf{S}_i \cdot \mathbf{S}_j)^2 - 1] - \sum_i (\mathbf{S}_i \cdot \mathbf{h}_i)^2 \tag{22.3}$$

where ε is the strength of the interaction between molecules at nearest-neighbour sites. Evidence for activated dynamics has been found for strong random fields. Specifically, the order parameter autocorrelation function has been demonstrated to scale as $\exp(-x^2)$ where $x = (\ln t / \ln \tau)^2$ and τ is the relaxation time. They argue that this does not agree with the results of experiments (Wu *et al.*, 1992). They show that introducing a correlation in the orientation of the random fields leads to the dynamics becoming slower and fitted by a $1/(1 + x^2)$ functional form, which is more consistent with the experimental findings. In a related paper, Bhattacharyat *et al.* (1995) have studied the kinetics of phase ordering of a nematic liquid crystal in a parallelepiped pore. They study the generation and evolution of a variety of defect structures with different anchoring conditions at the walls and show how these defects give rise to a variety of growth behaviours. Finally, we note that the theory of nematic ordering in randomly confining systems is discussed in an excellent companion chapter in this volume by Cleaver *et al.*

22.4 Random Field Model

We associate the effects of the porous medium with that of a random field (Brochard and de Gennes, 1983). Unlike conventional spins, the director of a liquid crystal is a headless vector (Maier and Saupe, 1959; Lebwohl and Lasher, 1972); the spin may be thought of as a rod with its orientation being specified in a half sphere of orientational space.

We consider the Hamiltonian

$$H = H_0 + H_1 = -J \sum_{<i,j>} (\mathbf{S}_i \cdot \mathbf{S}_j)^2 - \sum_i \mathbf{h}_i \cdot \mathbf{S}_i, \tag{22.4}$$

where $\mathbf{S}_i$ is an n-component unit vector (in the liquid crystal context $n = 3$) and the h_i are independently chosen quenched fields distributed according to a probability density $P(|\mathbf{h}|)$ which is rotationally invariant. The Hamiltonian H_0 is the Lebwohl–Lasher model (Lebwohl and Lasher, 1972). In the context of field theory it is called the RP^{n-1} model (Hikami and Maskawa, 1982; for reviews see Ohno *et al.* (1990) and Kunz and Zumbach (1989, 1992)). The free energy of a generalized version of the Hamiltonian in (22.4)

$$H' = -J \sum_{<i,j>} (\mathbf{S}_i \cdot \mathbf{S}_j)^2 - \sum_i \varepsilon_i \mathbf{h}_i \cdot \mathbf{S}_i, \tag{22.5}$$

where the ε_i are site variables that take on values $+1$ or -1, is identical to the free energy of (22.1) for any arbitrary set of $\{\varepsilon_i\}$. This follows from the local gauge invariance of H_0. In other words, redefining $\varepsilon_i \mathbf{S}_i = \mathbf{t}_i$, and noting that the trace over the $\mathbf{S}_i$ variables is equivalent to the trace over the $\mathbf{t}_i$ variables and $\varepsilon_i^2 = 1$, the free energy is the same for each of the 2^N sets of choices of $\{\varepsilon_i\}$ (N, here, denotes the number of sites). Thus our Hamiltonian is essentially equivalent to a random axis anisotropy model. We exploit the equality of the free energies by tracing over the ε_i variables (since each of the 2^N partition functions is exactly equal, the individual partition function is $(1/2^N)$ of this trace – this results in a trivial contribution to the entropy) to show that our coupling H_1 is exactly equivalent to $\Sigma_i \ln \cosh (\beta S_i \cdot h_i)$ which is a special case of the generic form proposed by Gingras (1995) of $\Sigma_m \Sigma_i g_{2m} (h_i \cdot S_i)^{2m}$. Thus, even though our Hamiltonian (22.1) has an external field term coupled linearly to the director, the symmetries of H_0 make it exactly equivalent to biquadratic and higher-order even couplings. (Note that this result obtains for any realization of the disorder in h_i.)

22.5 Mean Field Theory

We will begin with the infinite range version of (22.4),

$$H = -\frac{J}{N} \sum_{i,j} (S_i \cdot S_j)^2 - \sum_i h_i \cdot S_i \tag{22.6}$$

and generalize the method of Schneider and Pytte (1977); see also Sherrington, (1976)). We introduce the matrix

$$\sigma_{\alpha\beta} = \frac{1}{n-1} \left(\frac{n}{N} \sum_i S_i^\alpha S_j^\beta - \delta^{\alpha\beta} \right) \tag{22.7}$$

where S_i^α indicates the αth component of the vector S_i. The Hamiltonian (22.6) can be rewritten as

$$H = -NJ \left(\frac{n-1}{n} \right)^2 \mathrm{Tr}\sigma^2 - \sum_i h_i \cdot S_i - \frac{N}{n} \tag{22.8}$$

The partition function can be evaluated using the Hubbard–Stratonovich transformation

$$Z = \int \mathcal{D}S \exp(-\beta H) = \mathcal{N} \int \sum_{\alpha\beta} dx_{\alpha\beta} \exp\left(-N\beta J \, \mathrm{Tr}(x\tilde{x}) - \sum_i V(x, h_i) \right) \tag{22.9}$$

where, redefined as J, $\tilde{x}$ is the transpose matrix of x and

$$\exp[-V(x, h)] = \int d^{m-1}\hat{S} \, \exp\left[\beta \left(2J \sum_{\alpha\beta} x_{\alpha\beta} \frac{nS^\alpha S^\beta - \delta^{\alpha\beta}}{n-1} + h \cdot S \right) \right] \tag{22.10}$$

where $\hat{S}$ is defined as $S/|S|$. In eqn (22.9) in the $N \to \infty$ limit, we obtain

$$\sum_i V(x, h_i) = N \int d^n h \, P(h) V(x, h) = N \langle V(x, h) \rangle_h \tag{22.11}$$

The integral can be calculated using the saddle point method, i.e. the free energy per site, $-(\beta^{-1}/N) \ln Z$, is given by the minimum of

$$F = \mathrm{Tr}(x\tilde{x}) + \beta^{-1} \langle V(x, h) \rangle_h \tag{22.12}$$

where from now on we will measure β^{-1} and h in units of J. At the stationary points of F one has

$$x_{\alpha\beta} = \left\langle \left\langle \frac{nS_\alpha S_\beta - 1}{n-1} \right\rangle_0 \right\rangle_h \tag{22.13}$$

where $\langle \ldots \rangle_0$ denotes the average with respect to the single spin Boltzmann weight

$$\exp\left[\beta \left(2 \sum_{\alpha\beta} x_{\alpha\beta} \frac{nS^\alpha S^\beta - 1}{n-1} + h \cdot S \right) \right]$$

Thus x is a symmetric and traceless $n \times n$ matrix.

We now specialize to the case where only a fraction p of the sites have field $h = h_0 \hat{h}$, i.e.

$$P(h) = (1-p)\delta^n(h) + p\delta(h - h_0)C \tag{22.14}$$

with $C^{-1} = h_0^{n-1} S_n$, where S_n is the surface area of a unit sphere in n-dimensional space. This choice corresponds to the field being imposed on a fraction p of the sites – the strength of the field is h_0, but its orientation is random. Physically, these sites represent the pore-gain interface. Thus, p is larger when the interface area per unit volume of the porous medium is higher. With this choice (22.5) yields on further specializing to $h_0 \to \infty$

$$V(x, h) = -\frac{2\beta n}{n-1} \sum_{\alpha\beta} x_{\alpha\beta} \hat{h}^\alpha \cdot \hat{h}^\beta + x \quad \text{independent terms} \tag{22.15}$$

(The conclusions of this analysis are independent of whether the interaction with the external field is of the biquadratic form or of the form $S_i \cdot h_i$ as long as $h_0 \to \infty$. The linear coupling is much easier to handle within mean field theory.) We now take the quenched average of (22.15). We make use of the fact that $\langle \hat{h}^\alpha \cdot \hat{h}^\beta \rangle_h = \delta^{\alpha\beta}/n$ due to the random orientation of the unit vector $\hat{h}$. Since $\text{Tr } x = 0$ we get

$$F(\beta, 1-p, x) = \text{Tr } x^2 - \frac{1-p}{\beta} \left\langle \ln \int d^{n-1}\hat{S} \exp\left(2\beta \sum_{\alpha\beta} x_{\alpha\beta} \frac{nS^\alpha S^\beta - \delta^{\alpha\beta}}{n-1} \right) \right\rangle$$

$$\equiv (1-p)^2 F(\beta(1-p), 1, x/(1-p)) \tag{22.16}$$

where $F(\beta, 1, M)$ is the free energy of the pure system ($p = 0$). In this case, since M (or x) is symmetric, it can be diagonalized. If $\hat{1}$ is the axis of preferred orientation, then

$$M_{\alpha\beta} = \left\langle \frac{nS^\alpha S^\beta - \delta^{\alpha\beta}}{n-1} \right\rangle = \delta_{\alpha\beta} m_\alpha \tag{22.17}$$

with

$$m_\alpha = Q \quad \text{for } \alpha = 1 \quad \text{and} \quad m_\alpha = -\frac{Q}{n-1} \quad \text{for } \alpha > 1 \tag{22.18}$$

The choice (22.18) ensures that the matrix $M_{\alpha\beta}$ is traceless. For the pure system, it is well known that a first-order transition occurs at β_c ($p = 0$). This follows from the expansion of the free energy in powers of Q of the form $AQ^2 + BQ^3 + CQ^4 \ldots$. Since symmetry considerations do not dictate that $B \equiv 0$, the Landau criterion indicates a first-order transition. Thus, for our case, with $h_0 \to \infty$, and $p \neq 0$,

$$\beta_c^{-1}(p) = \beta_c^{-1}(0)(1-p) \tag{22.19}$$

and

$$Q(\beta, 1-p) = (1-p)Q((1-p)\beta, 1) \tag{22.20}$$

It is immediately clear, physically, that relaxing the condition $h_0 \to \infty$ will only raise the transition temperature, but still lead to a first-order phase transition at non-zero temperature for all $p \neq 1$. In the limit of infinite range exchange, the depressed transition temperature compared with the pure system may be readily understood in terms of a fraction $(1-p)$ of the spins coupled ferromagnetically to each other with the effect of the spins with the $h_0 \to \infty$ field cancelling out in the thermodynamic limit – one merely has the effect of dilution!

22.6 Random Field Potts Model – Mean Field Theory

We now switch to an n-state Potts model (Wu, 1982) in a random field described by the Hamiltonian

$$H = -J \sum_{\langle ij \rangle} s_i \cdot s_j - h_0 \sum_i s_i \cdot h_i \tag{22.21}$$

where the s_i and h_i are one among the vectors $\sigma_1, \ldots, \sigma_n$ with

$$\sigma_\alpha \cdot \sigma_\beta = \frac{n\delta_{\alpha\beta} - 1}{n - 1} = \begin{cases} 1 & \alpha = \beta \\ -1/(n-1) & \alpha \neq \beta \end{cases} \tag{22.22}$$

The σ point from the center to the vertices of a 'tetrahedron' in $(n - 1)$ dimensions. The random field distribution is

$$P(h_i) = (1 - p)\delta(h_i) + \frac{p}{n} \sum_\alpha \delta(h_i - \sigma_\alpha) \tag{22.23}$$

Since

$$s_i \cdot s_j = \frac{n\delta_{s_i \cdot s_j} - 1}{n - 1} \tag{22.24}$$

the model (22.21) is similar to (22.4), except that the possible directions of the molecular orientation are 'quantized'. We point out for the reader that the symmetry of the ordered phase and the universality class of (22.4) and (22.21) are different. Nevertheless, we will show that, within the mean field approximation, an equation identical in form to eqn (22.16) is obtained for the simpler Potts model. The analogue of the order parameter (22.17) is

$$\langle s_i \cdot \sigma_\alpha \rangle = \frac{\langle n\delta_{s_i \cdot \sigma_\alpha} - 1 \rangle}{n - 1} = m_\alpha \quad \text{with} \quad \sum_\alpha m_\alpha = 0 \tag{22.25}$$

The infinite range version of the Potts model is obtained on substituting the first term in (22.21) with

$$(J/2N) \sum_{ij} s_i \cdot s_j \tag{22.26}$$

Proceeding as before, on making the ansatz (22.18), with $J = 1$, one finds

$$F(Q) = \frac{1}{2}Q^2 - \frac{1}{\beta}V(Q, h_0) \tag{22.27}$$

with

$$V(Q, h_0) = (1 - p)\ln[\exp\{\beta Q\} + (n - 1)\exp[-\beta Q/(n - 1)]] + \frac{p}{n}\sum_{\alpha=1}^{n}\ln\{\exp[\beta(Q + h_0\delta_{\alpha, 1})]$$

$$+ \exp[-\beta Q/(n - 1)][n - 1 + (1 - \delta_{\alpha 1})(\exp(\beta h_0) - 1)]\} \tag{22.28}$$

In the limit $h_0 \rightarrow \infty$

$$\lim_{h_0 \to \infty} V(Q, h_0) = (1 - p)\ln[\exp(\beta Q) + (n - 1)\exp(-\beta Q/n - 1)] + \text{constant} \tag{22.29}$$

and thus

$$\lim_{h_0 \to \infty} F(\beta, 1 - p, Q) = \frac{Q^2}{2} - \frac{1 - p}{\beta}\ln[\exp(\beta Q) + (n - 1)\exp(-\beta Q/n - 1)]$$

$$+ \text{constant} \equiv (1 - p)^2 F(\beta(1 - p), 1, Q/(1 - p)) \tag{22.30}$$

which is identical to (22.16). Thus the same conclusions apply here too.

22.7 Random Field Potts Model – Monte Carlo Simulations

The infinite range model does not incorporate fluctuations, nor does it yield a percolation threshold. It is physically clear that when $p > (1 - p_c)$, where p_c is the percolation threshold, the connectivity between the remaining spins (with no field) is no longer present and thus the system is unable to sustain a non-zero temperature transition. In order to assess the importance of fluctuations, a finite range interaction and a non-trivial percolation threshold, and to monitor the short-time dynamics, we have undertaken Monte Carlo simulations of the three-state Potts model on a cubic lattice with a fraction p of the sites under the influence of an infinitely strong random field favouring one of the three states randomly. Note that our simulations have been carried out for the simple discrete Potts model and *not* for the Hamiltonian (22.1). While our results are qualitatively consistent with experiments, a direct study of model (22.1) or the Lebwohl–Lasher model with a biquadratic coupling to the field has been recently carried out (Zhang and Chakrabarti, 1995). The Hamiltonian of the model is

$$H = - \sum_{\langle ij \rangle} \delta_{\tilde{t}_i, \tilde{t}_j} \tag{22.31}$$

with the constraint that a fraction p of the sites have spins *frozen* randomly into one of the three states. The spins $\tilde{t}_i$ take on one of three values a, b or c. Equation (22.31) is the same as

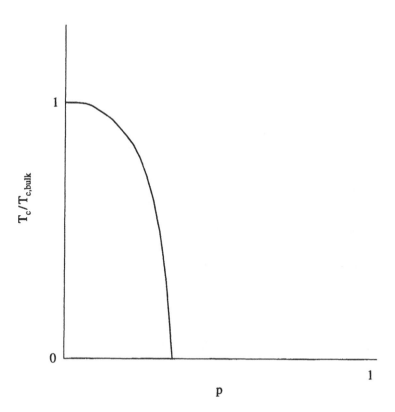

Figure 22.1 Schematic plot of the transition temperature normalized to the bulk transition temperature versus p, the concentration of sites with random, infinite field. In the mean field limit, the transition temperature would be zero at $p = 1$.

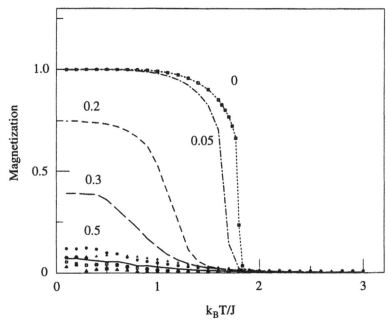

Figure 22.2 The magnetization M plotted as a function of temperature. The values of p are indicated. The bulk case corresponds to $p = 0$. For $p = 0.5$, in addition to the average results (shown by the full line), the results for five individual samples are indicated by the symbols (stars, squares, etc.).

(22.21) with $h_0 \to \infty$ and an exchange interaction equal to $(n - 1)/n$. In (22.31), we have set the exchange equal to 1 and, in the figures, the temperature is measured in units of this exchange. Single spin-flip dynamics with the standard Glauber scheme wefe used. It is important to note that the three-state Potts model in the pure limit undergoes a first-order transition in three dimensions (Wu, 1982).

We define the magnetization, coinciding with the definition (22.25) and the ansatz (22.18), as

$$M = \frac{3}{2N\tau}\left(\text{Max}\left\{\sum_{i,t}\delta_{t_i,a}, \sum_{i,t}\delta_{t_i,b}, \sum_{i,t}\delta_{t_i,c}\right\}\right) - 1/2 \tag{22.32}$$

and the maximum occupation probability (MOP) as

$$\text{MOP} = \frac{3}{2N\tau}\left(\sum_i \text{Max}\left\{\sum_t\delta_{t_i,a}, \sum_t\delta_{t_i,b}, \sum_t\delta_{t_i,c}\right\}\right) - 1/2 \tag{22.33}$$

where now N is the total number of *unfrozen* spins (in the previous sections N was the total number of spins, both frozen and unfrozen), $\tilde{t}_i$ is a function of the time t, the summation over t represents a sum over time intervals to the total time τ and the sum over i is over the unfrozen spins. The equilibrium spin–spin correlation function (SSCF) defined as

$$\left(\frac{3}{2N}\sum_i \delta_{t_i(0),\, t_i(\tau)} - 1/2\right)$$

starts with a value of 1 at $\tau = 0$ and quickly (within about 100 passes) drops to a plateau (SSCFP) around which it fluctuates mildly.

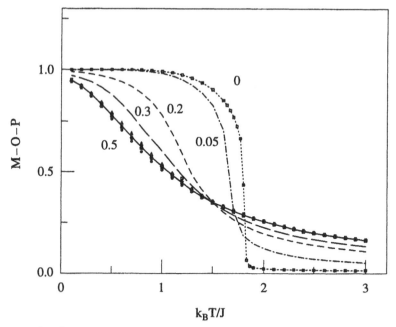

Figure 22.3 Plot of MOP against temperature for the p indicated. For $p = 0.5$, the average results as well as those for individual samples are shown as in Figure 22.2.

Our results are presented in Figures 22.1–22.5. Figure 22.11 shows a schematic phase diagram deduced from the simulations. Unlike the mean field prediction that the transition temperature goes to zero at $p_{th} = 1$, the Monte Carlo results yield a value of $p_{th} < 1 - p_c$ at which the transition temperature becomes zero. (The percolation threshold, p_c, on a three-dimensional simple-cubic lattice, approximately 0.307.) Indeed we find that $p_{th} = 0.5$. Our results for $p_{th} = 0.5$ were obtained with five independent runs on a $16 \times 16 \times 16$ lattice with 20 000 Monte-Carlo passes used for obtaining averages and 5000 passes to reach equilibrium from a nearby temperature. The bulk results employed a $12 \times 12 \times 12$ lattice while systems with other values of p were studied on a $14 \times 14 \times 14$ lattice.

Figures 22.2–22.5 show plots of M, MOP, SSCFP and the specific heat as a function of temperature for various values of p. Our results clearly indicate that at $p = 0.5$, there is no long-range order but sluggish dynamics as evidenced by the large MOP. Nevertheless, there is no evidence of hysteresis or history dependence for any of the quantities – independent runs carried out starting from different configurations led to practically identical results as did cycling the temperature. The behaviour of MOP versus temperature strongly suggests that unlike the random anisotropy Heisenberg ferromagnet (see e.g. Harris *et al.*, 1987) there is no evidence here of a transition to a spin glass phase. The results for the specific heat are in qualitative accord with experimental data (Bellini *et al.*, 1992; Wu *et al.*, 1995). Our simulations indicate that, for small p, the peak position decreases roughly linearly on increasing p in agreement with recent experimental results (Wu *et al.*, 1995) and consistent with an analytic p dependence of T_{NI}.

The lower critical dimensionality of the Lebwohl–Lasher model is probably different from that of the Potts model. In the standard Imry–Ma (Imry and Ma, 1975) argument there are many aspects which are left out (Nattermann and Villain, 1988) (like entropy) which could be important here. If the argument is applied naïvely one would find the lower critical

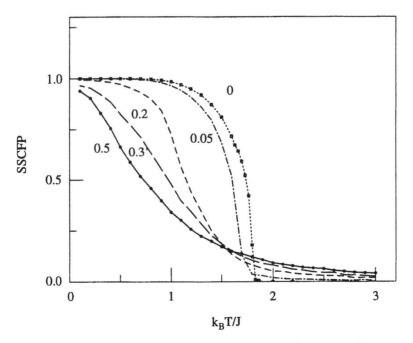

Figure 22.4 Plot of SSCFP versus temperature for various values of *p* indicated. Only the average results are shown.

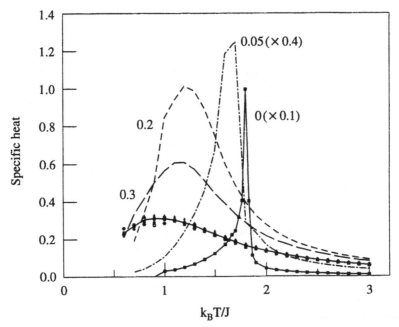

Figure 22.5 Plot of specific heat against temperature for various values of *p*. The data rescaling factors are shown in the brackets For *p* = 0.5, the results for individual samples are indicated by the data points. The label 0.5 has been omitted.

dimensionality of 4 for the random field Lebwohl–Lasher model. Our Monte Carlo studies of the Potts model at $p = 0.05$ are indeed suggestive of a phase transition at non-zero temperatures, but the system sizes studied may not be large enough for the effects of the random pinning fields to be felt.

In summary, we have presented and analysed simple models that are possibly capable of describing the isotropic–nematic transition in porous media or lack thereof. Even though our studies do not take into account the correlated geometry of the aerogel, they yield results in qualitative accord with existing experiments.

Acknowledgements

We are indebted to our coauthor Tommaso Bellini for writing the section on experimental results and for many stimulating discussions. This work was supported by NATO, the Petroleum Research Fund administered by the American Chemical Society, ONR (USA), KBN (Poland), INFN (Italy), and a grant of computer time by the Center for Academic Computing of The Pennsylvania State University. JRB gratefully acknowledges the warm hospitality of Alan Bray at the University of Manchester, of Julia Yeomans at Oxford and the support of the Fulbright Foundation and EPSRC.

References

ALIEV, F. M., VERSHOVSKAYA, G. Yu. and ZUBKOV, L. A. (1991) Optical properties of a liquid crystal isotropic phase in pores, *Sov. Phys. JETP* **72**, 846–850.

AMIT, D. J. (1984) *Field Theory, The Renormalization Group, Critical Phenomena*, Sinapore: World Scientific, section 2.5.2.

BANAVAR, J. R., KOPLIK, J. and WINKLER, K. (eds) *Physics and Chemistry of Porous Media II*, New York: AIP.

BELLINI, T. (1995) Private communication.

BELLINI, T., CLARK, N. A., MUZNY, C. D., WU, L., GARLAND, C. W., SCHAEFER, D. W. and OLIVER, B. J. (1992) Phase behaviour of the liquid crystal 8CB in a silica aerogel, *Phys. Rev. Lett.* **69**, 788–791.

BELLINI, T., CLARK, N. A. and SCHAEFER, D. W. (1995) Dynamic light scattering study of nematic and smectic A liquid crystal ordering in silica aerogel, *Phys. Rev. Lett.* **74**, 2740–2743.

BHATTACHARYA, A., RAO, M. and CHAKRABARTI, A. (1995) Preprint.

BROCHARD, F. and DE GENNES, P. G. (1983) Phase transition of binary mixtures in randon media, *J. Phys. Lett.* **55**, 1681–1684.

CLARK, N. A., BELLINI, T., MALZBENDER, R. M., THOMAS, B. N., RAPPAPORT, A. G., MUNZY, C. D., SCHAEFER, D. W. and HRUBESH, L. (1993) X-ray scattering study of smectic ordering in a silica aerogel, *Phys. R. Lett.* **71**, 788–791.

DADMAN, M. and MUTHUKUMAR, M. (1992) The response of semiflexible liquid crystals to quenched random disorder, *J. Chem. Phys.* **97**, 578–585.

DADMUN, M. D. and MUTHUKUMAR, M. (1993) The nematic to isotropic transition of a liquid crystal in porous media, *J. Chem. Phys* **98**, 4850–4852.

FAETTI, S., GATTI, M., PALLESCHI, V. and SLUCKIN, T. J. (1985) Almost critical behaviour of the anchoring energy at the interface between a liquid crystal and a SiO substrate, *Phys. Rev. Lett.* **55**, 1681–1684.

FISHER, D. S., GRINSTEIN, G. and KHURANA, A. (1988) Theory of random magnets, *Phys. Today* **41**, (12), 56–67.

GINGRAS, M. J. P. (1995) Preprint.

HARRIS, A. B., CAFLISCH, R. and BANAVAR, J. R. (1987) Random anistropy axis magnet with infinite anisotropy, *Phys. Rev. B* **35**, 4929–4934.

HIKAMI, S. and MASKAWA, T., (1982,) Nonlinear sigma models on symmetric spaces large N limit, *Prog. Theor. Phys.* **67**, 1038–1052.

IANNACCHIONE, G. S., CRAWFORD, G. P., DOANE, J. W. and FINOTELLO, D. (1992), Orientational effects on confined 5CB, *Mol. Cryst. Liq. Cryst.* **222**, 205–213.

IANNACCHIONE, G. S., CRAWFORD, G. P., ŽUMER, S., DOANE, J. W. and FINOTELLO, D. (1993) Randomly constrained orientation order in porous glass, *Phys. Rev. Lett.* **71**, 2595–2598.

IMRY, Y. and MA, S. K. (1975) Random-field instability of the ordered state of continuous symmetry, *Phys. Rev. Lett* **35**, 1399–1401.

ISRAELACHVILI, J. N. (1985) *Intermolecular and Surface Forces*, London: Academic Press.

KLAFTER, J. and DRAKE, J. M. (eds) (1989) *Molecular Dynamics in Restricted Geometries*, New York: John Wiley.

KRALJ, S., LAHAJNAR, G., ZIDANŠEK, A., VRBANČIČ-KOPAČ, N., VILFAN, M., BLINC, R. and KOSEČ, M., (1993) Deuterium NMR of a pentylcyanobiphenyl liquid crystal confined in a silica aerogel matrix, *Phys. Rev. E* **48**, 340–349.

KUNZ, H. and ZUMBACH, G. (1989) First order phase transition in the two-dimensional and three-dimensional RP^{N+1} and CP^{N-1} models, in the large N limit, *J. Phys. A: Math. Gen.* **22**, L1043–L1048.

KUNZ, H. and ZUMBACH, G. (1992) Phase transition in a nematic N-vector model – the large N limit, *J. Phys. A: Math. Gen.* **25**, 6155–6162.

LEBWOHL, P. A. and LASHER, G. (1972) Nematic-liquid-crystal order – A Monte Carlo calculation, *Phys. Rev. A* **6**, 426–429.

MAIER, W. and SAUPE, A. (1959) Eine einfache molekular-statistische Theorie der nematischen kristallinfluessigen Phase, Teil I, *Z. Naturforch.* **14A**, 882–889; (1960) Teil II, **15A**, 287–292.

MARITAN, A., CIEPLAK, M., BELLINI, T. and BANAVAR, J. R. (1994) Nematic isotropic transition in porous media, *Phys. Rev. Lett.* **72**, 4113–4116.

NATTERMANN, T. and VILLAIN, J. (1988) Random field Ising systems – A summary of current theoretical views, *Phase Transitions* **11** 5–51.

OHNO, K., CARMESIN, H. O., KAWAMURA, H. and OKABE, Y. (1990) Exact theories of M-component quadripolar systems showing a first order transition, *Phys. Rev. B* **42**, 10360–10380.

PADDAY, J. F. (ed.) *Wetting, Spreading and Adhesion*, London: Academic Press.

SCHAEFER, D. W. (1994) Structure of mesoporous aerogels, *MRS Bull.* **19**, 49–53.

SCHNEIDER, T. and PYTTE, E. (1977) Random-field instability of the ferromagnetic state, *Phys. Rev. B* **15**, 1519–1522.

SHERRINGTON D. (1976) A class of exactly soluble random spin systems, *Phys. Lett* **58A**, 36–38.

TRIPATHI, S., ROSENBLATT, C. and ALIEV, F. M. (1994) Orientational susceptibility in porous glass near a bulk nematic–isotropic phase transition, *Phys. Rev. Lett* **72**, 2725–2728.

WONG, P. Z. (1988) The statistical physics of sedimentary rock, *Phys. Today* **41** (12), 24–32; and references therein.

WU, F. Y. (1982) The Potts model, *Rev. Mod. Phys.* **54**, 235–268.

WU, L., ZHOU, B., GARLAND, C. W., BELLINI, T. and SCHAEFER, D. W. (1995) Heat capacity study of nematic–isotropic and nematic–smectic A transition for octylcyanobiphenyl in silica aerogels, *Phys. Rev. E* **51**, 2157–2164.

WU, X., GOLDBURG, W. I., LIU, M. X. and XUE, J. Z. (1992) Slow dynamics of isotropic–nematic phase transition in silica gels, *Phys. Rev. Lett.* **69**, 470–473.

ZHANG, Z. and CHAKRABARTI, A. (1995) Preprint.

496

Index

Printed and bound by CPI Group (UK) Ltd, Croydon, CR0 4YY

24/10/2024

01778287-0004